High-Speed Wireless Communications

The escalating demand for high-speed wireless services in the face of the technical and physical limitations of wireless channels presents a significant technological challenge. To analyse and design ultra-wideband (UWB), 3G, 4G, and other emerging wireless applications successfully, you need to master the underlying principles involved and the engineering complexities of their implementation.

This book describes the theory and major applications of high-speed multimedia wireless communications, covering recent developments and identifying directions for future research. Jiangzhou Wang covers UWB wireless systems, 3G, and 4G mobile networks. He discusses the overlay (interference) problem in UWB and how to deal with it; shows different forms of wireless broadband access and demonstrates that OFDM is not the best wireless access technology for high-speed transmission. Wang also presents a new space-time frequency MIMO architecture for future wireless systems. Other key topics, such as hybrid ARQ, advanced channel coding, and modulation and transmit diversity, are also discussed in detail.

Providing both the fundamental theory that underlies the technology, and the latest developments in key applications, this book is an invaluable resource for graduate students of electronic, communications, and computer engineering, as well as practitioners in the wireless communications industry.

Jiangzhou Wang is Professor and Chair of Electronics, University of Kent; Editor for *IEEE Transactions on Communications*; and a Guest Editor for *IEEE Journal on Selected Areas in Communications*. He is the author of more than 200 technical papers and three books in the area of wireless communications.

High-Speed Wireless Communications

Ultra-wideband, 3G Long-Term Evolution, and 4G Mobile Systems

JIANGZHOU WANG

CAMBRIDGE
UNIVERSITY PRESS

University Printing House, Cambridge CB2 8BS, United Kingdom

Cambridge University Press is part of the University of Cambridge.

It furthers the University's mission by disseminating knowledge in the pursuit of education, learning and research at the highest international levels of excellence.

www.cambridge.org
Information on this title: www.cambridge.org/9780521881531

First published 2008

A catalogue record for this publication is available from the British Library

Library of Congress Cataloguing in Publication data
Wang, Jiangzhou.
High-speed wireless communications : ultra-wideband, 3G long-term evolution, and 4G broadband mobile systems / Jiangzhou Wang.
p. cm.
Includes bibliographical references and index.
ISBN 978-0-521-88153-1 (hardback)
1. Wireless communication systems. 2. Ultra-wideband devices. 3. Broadband communication systems. I. Title.
TK5103.2.W367 2008
004.6'6–dc22 2008019823

ISBN 978-0-521-88153-1 Hardback

To
Leiping, April, Angela, and Larry

Contents

Preface

Wireless communications services are penetrating into our society at an explosive growth rate, and demands for a variety of high-speed wireless multimedia services continue to increase. It is everyone's wish that wireless could act like a wired connection with the same quality as fixed networks. To realize true high-speed wireless systems, sustained technical innovation on many fronts will be required. The physical limitations on and problems with wireless channels (bandwidth and power constraints, multipath fading, noise and interference) present a fundamental technical challenge to reliable high-speed wireless communications. This book is an ideal reference for graduate students and practitioners in the wireless industry.

The text of this book has been developed through years of research by the author and his graduate students. The aim of this book is to provide an R&D perspective on the field of high-speed wireless multimedia communications by describing the recent research developments in this area and also by identifying key areas in which further research will be needed.

The book is organized into four parts: introduction, ultra-wideband (UWB) communications, evolved 3G mobile communications and 4G mobile communications, with twelve chapters.

Acknowledgements

First, I would like to thank my former graduate students and researchers for their hardwork and dedication to research. Without them, this book would have never appeared. In particular, many thanks to Dr Xiangyang Wang, Dr Bin Xia, Dr Yiqing Zhou, Dr Yang Liu, Dr Yuanliang Huang, Mr Tat-Tung Wong and Dr Huiling Zhu.

I am deeply indebted to Professor Laurence Milstein at the University of California, San Diego, USA, for his knowledge, mentoring and friendship. He ignited my initial interest in CDMA overlay through his deep insight and research experience. His selfless supervision style has influenced my entire career both directly and indirectly.

I am also grateful to Professor Mamoru Sawahashi and Dr Kenichi Higuchi, who have enriched my knowledge in wireless communications over the years through collaboration with NTT DoCoMo. I would like to thank NTT DoCoMo Inc., Japan, whose continual support of my research led to this book.

Finally, I wish to thank my wife, Leiping, and three children, April, Angela and Larry, for their love and support.

Abbreviations

3-D	three dimensional
3G	third generation
3GPP	third generation partner project
3GPP2	third generation partner project 2
4G	fourth generation
ACF	autocorrelation function
ACK	acknowledgement
A/D	analog to digital
AM	amplitude modulation
ARQ	automatic repeat request
AWGN	additive white Gaussian noise
BER	bit error rate
BPF	bandpass filter
BPSK	binary phase shift keying
BS	base station
BW	bandwidth
CDF	cumulative distribution function
CDMA	code division multiple access
CGRA	complex-value Gaussian random variable
CIR	carrier to interference ratio
CRC	cyclic redundancy check
CSI	channel state information
D/A	digital to analog
DAB	digital audio broadcast
DS	direct sequence
DS-CDMA	direct sequence code division multiple access
DSP	digital signal processing
DS-SS	direct sequence spread spectrum
DVB	digital video broadcast
EGC	equal gain combining
FCC	federal communications commission
FDD	frequency division duplex
FDMA	frequency division multiple access

FEC	forward error correction
FFT	fast Fourier transform
FIR	finite impulse response
FH	frequency hopping
FM	frequency modulation
GPS	global position systems
GSC	generalized selective combining
GSM	global systems for mobile
HSDPA	high-speed downlink packet access
I	in phase
IC	interference cancellation
IEEEE	Institute of Electrical and Electronic Engineering
IFFT	inverse fast Fourier transform
IR	impulse radio
IR-UWB	impulse radio ultra-wideband
ISI	inter-symbol interference
ISM	industrial, scientific and medical
I/Q	in-phase/quadrature
ITU	international telecommunications union
J/S	jammer to signal power ratio
LAN	local area network
LDPC	low density parity check
LMS	least mean square
LPF	lowpass filter
LST	layered space-time
LSTF	layered space-time-frequency
LTE	long term evolution
MA	multiple access
MAC	medium access control
MAI	multiple access interference
MAP	maximum a posteriori
Mbps	mega bits per second
MC	multicarrier
MC-CDMA	multicarrier code division multiple access
MCI	multicode interference
MF	matched filter
MIMO	multiple input multiple output
ML	maximum likelihood
MLD	maximum likelihood detection
MIP	multipath intensity profile
MMSE	minimum mean squared error
MPI	multipath interference
MQAM	multilevel quadrature amplitude modulation
MRC	maximum ratio combining/combiner

MRRC	maximum ratio receiving combining
MS	mobile station
MTJ	multitone jamming
MTLL	mean time to lose lock
MUI	multiuser interference
MUX	multiplexing
NBI	narrowband interference
NCC	notch filter in cascade with code correlator
NCK	negative acknowledgement
ND	normalized deviation
OFCDM	orthogonal frequency and code division multiplexing
OFDM	orthogonal frequency division multiplexing
OOK	on-off keying
OVSF	orthogonal variable spreading factor
PAM	pulse amplitude modulation
PAR	peak-to-average ratio
PBJ	partial-band jamming
pcMMSE	per carrier minimum mean square error
PDA	personal digital assistant
PDF	probability density function
PG	processing gain
PIC	parellel interference cancellation
PN	pseudo noise
PPM	pulse position modulation
P/S	parallel to serial
PSAM	pilot symbol assisted modulation
PSD	power spectral density
Q	quadrature
QAM	quadrature amplitude modulation
QoS	quality of service
QPSK	quadrature phase shift keying
RF	radio frequency
SC	selection combining
SF	spreading factor
SIC	soft interference cancellation
SINR	signal-to-interference-plus-noise ratio
S/P	serial to parallel
SIR	signal-to-interference ratio
SISO	single input single output
SMC	selection maximal combining
SNR	signal-to-noise ratio
SS	spread-spectrum
STBC	space-time block code
STF	space-time-frequency

STTC	space-time trellis code
STTD	space-time based transmit diversity
TD	transmit diversity
TD-STBC	transmit diversity space-time block code
TDMA	time division multiple access
TH	time hopping
TH-IR	time-hopping impulse radio
UWB	ultra-wideband
VSF	variable spreading factor
WCDMA	wideband code division multiple access
WiFi	wireless fidelity
WMSA	weighted multi-slot averaging
WSSUS	wide-sense-stationary uncorrelated-scattering
WWRF	wireless world research forum
ZF	zero forcing

Part I

Introduction

1 Introduction to high-speed wireless communications

Wireless communications and internet services have been penetrating into our society and affecting our everyday life profoundly during the last decade far beyond any earlier expectations. In addition, the demand for wireless communications is still growing rapidly and wireless systems that support voice communications have already been deployed with great success. Further wireless mobile and personal communication systems are expected to support a variety of high-speed multimedia services, such as high-speed internet access, high-quality video transmission and so on. To meet the demand for high data rate services in broadband wireless systems, various systems and/or technologies have been proposed, such as the ultra-wideband (UWB) system, and evolved third generation (3G) and fourth generation (4G) mobile communications systems.

1.1 UWB communications

In the foreseeable future, the development of low-power, short-range and high-speed transmission systems is going to play a significant role in the area of wireless communication, due to a blooming growth in demand for information sharing and data distribution tools to be used in hot-spot layer and personal network layer communications. At the same time, the radio frequency (RF) spectrum suitable for wireless links is limited, so efficient spectrum utilization is a challenging problem in physical-layer communication engineering [1]. All these have motivated the exploration of the UWB transmission system.

Recently, there has been a growing interest in the research and development of novel technologies aimed at allowing new services to use the radio spectrum already allocated to established services, but without causing noticeable interference to existing users. UWB systems [1]–[3], using bandwidths in excess of 500 MHz with very low power spectral density, are currently attracting much interest as a means of obtaining additional capacity by overlaying the narrowband signals that currently occupy various portions of the spectrum.

If the emissions from UWB devices are regulated to avoid causing significant interference to licensed narrowband services, then it becomes possible to allow UWB systems to operate on an unlicensed basis, enabling UWB technology to support a diverse range of short distance applications, such as wideband multimedia services for home, radar, automotive and medical imaging systems. Currently, mobile phones and unlicensed wireless

LANs are not allowed to be used in hospitals, because of the fear that they will interfere with medical equipment. However, future UWB devices might be used in hospitals since they have very low power spectral density, i.e., their emissions have very low interference potential.

The term "ultra-wideband" originated from the Department of Defense of the United States of America in 1989 [4], although the research and development of the related radio technology had been taking place for decades. A common accepted definition of UWB is that the system signal occupies a bandwidth greater than 500 MHz or 25% of the center frequency. Numerically, this is given by

$$\text{fractional bandwidth} = \frac{f_U - f_L}{f_C} \geq 0.25 \tag{1.1}$$

where f_U and f_L are the upper and lower frequencies, respectively, of the -10 dB emission point. The center frequency f_C is defined as

$$f_C = \frac{f_U + f_L}{2} \tag{1.2}$$

There are existing technologies capable of offering short-range wireless services. Bluetooth technology, using the 2.4 GHz spectrum of the industrial, scientific and medical (ISM) band, supports the data rate of 700 kilo bits per second (kbps). The IEEE's standards for wireless local area networks (LAN), IEEE 802.11a and IEEE 802.11b (so-called wireless fidelity (WiFi)), using the 5 GHz and 2.4 GHz spectra of the ISM band, support data rates of up to 54 mega bits per second (Mbps) and 11 Mbps, respectively. In comparison to these short-range communication devices, future UWB devices will offer much higher data rates of up to around 500 Mbps, which offers opportunities for expansion of wireless communications in many areas. For example, UWB technology can be used for cable displacement in the construction of high-speed home and business networking; establishing high throughput links for hand-held devices and various consumer electronic appliances such as notebook computers, digital cameras, portable music players and personal digital assistants; supplying short-range voice, simultaneous video, audio and Internet services in campus areas, libraries, and even medical and elderly care facilities; as well as providing high-speed data distribution within airport terminals or railway stations.

The relative advantage of UWB technology in comparison to the conventional narrowband system is not limited to the increase in transmission rate. As the UWB system bandwidth would be much larger than the actual data rate, recall Shannon's equation for channel capacity:

$$\text{capacity} = \text{bandwidth} \cdot \log\left(1 + \frac{\text{signal power}}{\text{noise power}}\right) \tag{1.3}$$

The expansion of operating bandwidth allows a lower system signal-to-noise ratio for the same capacity. As a result, UWB devices can have a much lower operating power than conventional narrowband communication systems, as illustrated in Figure 1.1. Low emission power is not only beneficial in power-constrained scenarios such as battery-operated devices, but also allows cheap RF components.

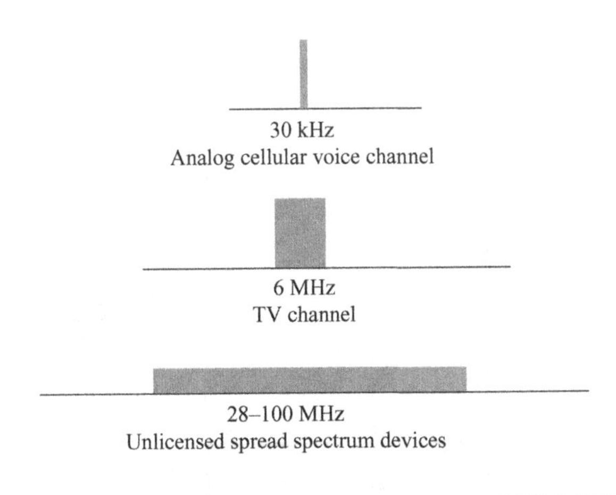

Figure 1.1 Bandwidth of typical wireless technologies.

Multipath fading is another main concern for the use of indoor wireless systems, since the transmitted signal usually reaches the receiver by more than one path due to reflection, refraction and scattering of radio waves by obstacles. This problem is much relieved for UWB systems due to fine time resolution, since the bandwidth of the UWB signal would be much greater than the coherence bandwidth of the channel response. The greater multiple propagation path resolvability allows better control of fading as multipath components can be combined constructively, which is less satisfactory in narrowband communication systems.

In the USA, the standardization and regulation processes are undergoing change and the Federal Communications Commission (FCC) has introduced spectral masks for the operation of different types of UWB device. UWB devices for communication must operate with their -10 dB (fractional) bandwidth of at least 500 MHz within the frequency band between 3.1 and 10.6 gigahertz (GHz), as shown in Figure 1.2. The region of the spectrum below approximately 2 GHz should be well attenuated because it is the most heavily occupied region of the spectrum, containing services for public safety, aeronautical and maritime navigation and communications, AM, FM and TV broadcasting, private and commercial mobile communications, medical telemetry, amateur communications, and GPS operations. Under such arrangement, these GHz-bandwidth devices should have very low emission power so as to prevent noticeable interference to the numerous established narrowband systems scattered over the allocated spectrum for UWB devices. Overlaying can meet the goal of efficient use of scarce spectrum resources, but the deployment of UWB technology has aroused a lot of controversy since systems sharing the spectra are under the threat of mutual interference. Lack of coordination among the systems involved could further worsen the situation since UWB devices would operate on an unlicensed basis due to the large expected market size.

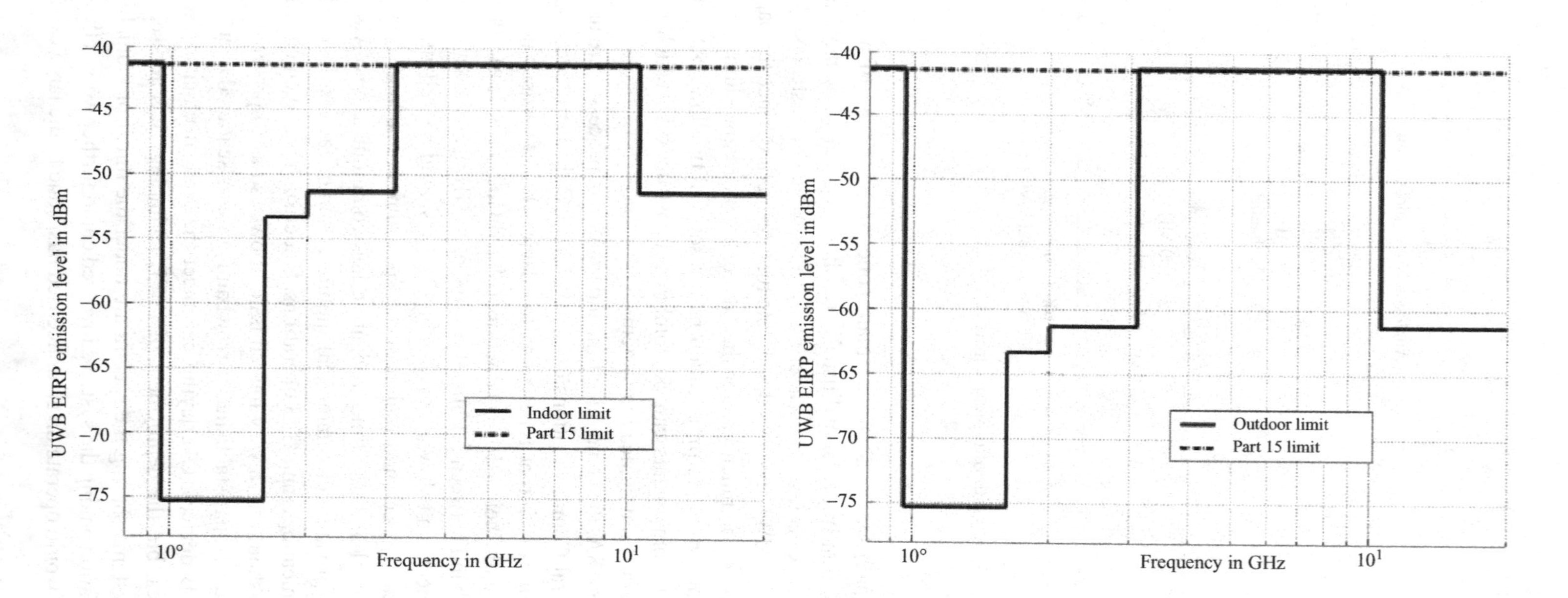

Figure 1.2 FCC spectral mask

In accordance with the characteristics described, the UWB wireless communication system should be very effective and robust in gathering multipath energy, suppressing narrowband interference, and combating multiple access interference of other UWB systems. The overlaid UWB system should cause negligible harm to existing systems, have a frequency spectrum suited to the operational requirements, and achieve acceptable performance in low transmission energy. Therefore, the deployment of UWB technology is a controversial issue and there are fierce discussions over the methods of its implementation, for example in the standardization of IEEE 802.15.3a, which considers the ultra-wide bandwidth physical channel. It should be pointed out that UWB technology is just an innovative way of using a valuable resource, but should not be confined to any specific scheme. There is no restriction on the modulation techniques for UWB technology as long as the operating requirements are met.

1.1.1 Multicarrier-CDMA-based UWB

When considering overlay, interference reduction and interference suppression are the key issues for sharing the spectrum in harmony between the established narrowband systems and the overlaid UWB system. The way the allocated resources are shared among the UWB devices and the rich number of multipaths are used are also important issues.

The multiple-access scheme is key for UWB wireless communication since the system is to be used at short range, supporting numerous users, devices or services. Code division multiplexing is suggested to be a suitable scheme for UWB systems over frequency division multiplexing (the fractional bandwidth requirement cannot be met), time division multiplexing (synchronization among all system users is needed) or carrier-sense multiple access (inefficient channel sensing and back-off due to collision). Direct sequence code division multiple access (DS-CDMA) is a spread-spectrum modulation technique. The origin of the technology can be dated back to pre-World-War II years, and was initially used in military communications. CDMA communication systems for commercial use have appeared in the past decade.

Figure 1.3 illustrates the fundamental components of a DS-CDMA system. Under this modulation scheme, at the transmitter, the data signal is spread by a higher rate random code to give the spread signal, which is then passed to the RF modulator where the signal spectrum is shifted to the assigned frequency location. The resultant signal is amplified and transmitted. At the receiver, the signal is RF demodulated by reference carrier and despread by the locally generated code to recover the data. The spreading and despreading processes result in the attractive features of DS-CDMA system: multiple access and multipath resolution capabilities, as well as the abilities of anti-jamming and anti-interference, while decades of effort and experience have proved that CDMA is an efficient way for wireless communication as well as a suitable candidate for overlaying purposes. All of these are necessary for the UWB communication environment.

Based on these features, DS-CDMA is a probable multiple access scheme, but it is less feasible when applied directly to UWB communications because the ultra-wide bandwidth demands circuitry with an ultra-fast sampling rate and analog-to-digital conversion

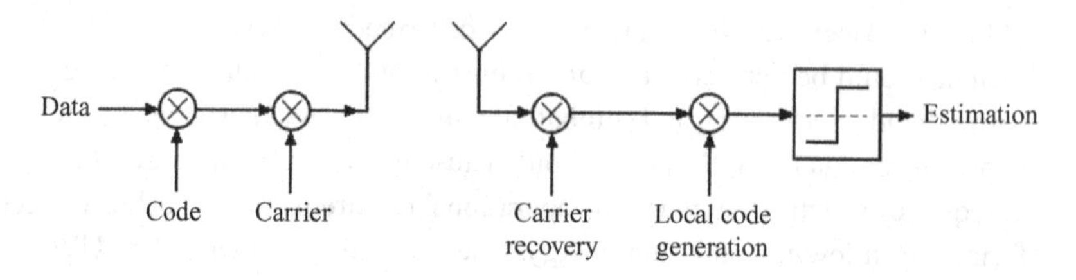

Figure 1.3 Block diagram of casual DS-CDMA system.

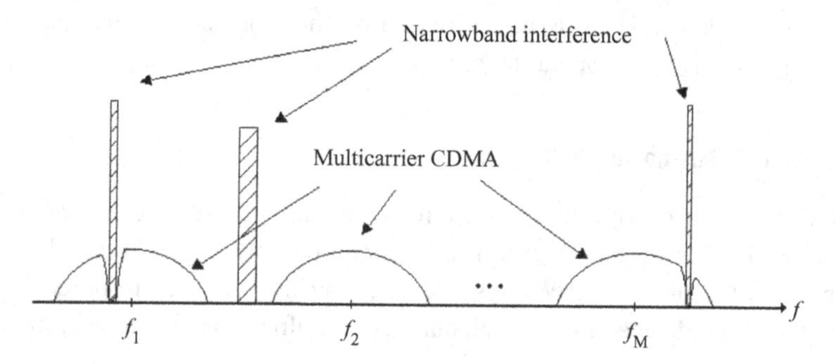

Figure 1.4 Spectrum of multicarrier CDMA UWB system.

components with a very large dynamic range are required due to the fact that the received signal exhibits large variation. Furthermore, if a casual DS-CDMA scheme was adopted, the UWB system would have to cope with the negative effects of all the established narrowband systems scattering along the allocated bandwidth. As a result, the hybrid of casual DS-CDMA – multicarrier CDMA, is a promising choice.

Under multicarrier CDMA, after the data signal is spread by a higher rate random code, the resultant signal is then modulated on multiple carriers. Multicarrier CDMA offers the following advantages. First, the rate of signal processing is no longer directly related to the entire occupied spectra since the ultra-wide bandwidth is sliced into a number of wide bandwidth portions, so lower speed units can be used. Second, the signal spectra of different carriers can be disjoint from each other, and they can be shifted to an appropriate frequency region, with the reinforcement of band-limiting techniques so that important spectra such as ISM bands or those reserved for emergency use can be easily avoided. Resource allocation can be flexible where trade-off between overall transmission rate and quality of service from frequency diversity is allowable. Third, various techniques developed for DS-CDMA signal processing can be applied.

UWB systems are overlaid on the established services, scattered over the operating bandwidth. In order to share the resource harmoniously, the idea of filtering at the transmitter has been suggested, in which notches are placed on appropriate frequency locations so that the emission power of the UWB system is suppressed over the spectra occupied by the established systems (Figure 1.4), in order to fulfill the interference avoidance requirement.

Notch filtering is initially employed for jamming-signal suppression at the receiver and the related technology has been well explored over the past few decades. It was proposed that notch filtering at the transmitter can be achieved by passing the spreading code through a transversal filter before signal spreading takes place. Limited complexity is an attractive feature of the transversal filter, and the tap weights adaptation of the filter can be based on either the available information in advance for fixed spectra assignment in narrowband systems, or feedback of statistics from the receiving end of the UWB device based on an observation of the captured signal for a totally unknown situation. Since the overlaid UWB system should not disrupt the established services, but there is no co-ordination with those systems, this kind of passive avoidance approach is suitable for a UWB operating scenario.

Alternatively, a convenient method to introduce notch filtering at the transmitter can be done by chip shape modification. The chip shape is used for band-limiting the transmitted signal, and its modification for other interference suppression purposes has been discussed [5], [6]. In this book, modification of the spectrum of the transmitted chip signal is employed for interference avoidance with established services, and the details are discussed further in Chapter 2.

On the receiving end, the UWB signal can be corrupted by very strong narrowband interference from existing narrowband users, and can also be distorted by the channel response which results in a large number of multiple propagation paths. In a conventional Rake system for exploring multipath diversity, the tapped delay line receiver with tap weights forms the spreading code, while the time separation between the taps is the chip interval of the code. The receiver collects the resolvable paths and combines the statistics coherently to give the data estimation.

However, the processing gain of the DS-CDMA system can only tolerate a certain level of jamming signal. As the emission power of a UWB system is extremely low in comparison to those of the established services, the UWB system would malfunction under an ultra-strong jamming signal. Further increasing the processing gain is not an appropriate solution as the data transmission rate is sacrificed, while simply discarding a sub-carrier that is experiencing narrowband interference is also too inefficient. The use of an adaptive Rake can be a better choice where the determination of the tap weights takes the channel condition into account and eventually the taps can be adjusted to despread the UWB signal as well as suppress the narrowband interference. In this work, a pre-combining minimum-mean-squared-error (MMSE) adaptive Rake is considered, such that the receiver only depends on the spreading code cross correlations and the average power profile of the channels so as to relieve the possible severe tracking problem. The resultant structure is thus capable of jointly gathering the multipath energy and suppressing narrowband interference. To illustrate the effectiveness of the adaptive Rake receiver when used in UWB communication, a performance comparison is also made between the adaptive Rake receiver as well as the conventional Rake receiver and the one with a notch filter mounted in the front end.

A characteristic feature of indoor wireless communications is the large number of multiple propagation paths. Combining all available paths is too inefficient as this makes the receiver much too complex. Selective maximal combination (SMC) is thus employed

Figure 1.5 Impulse radio.

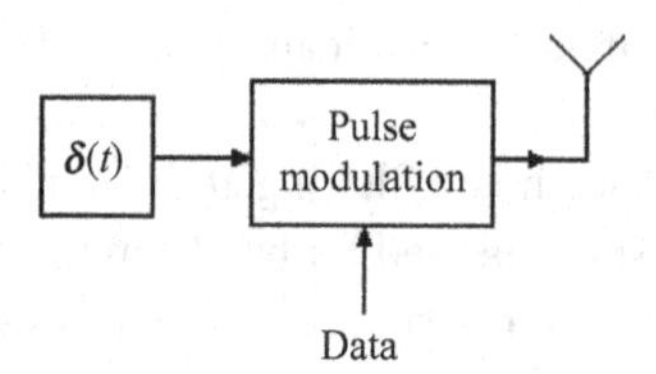

Figure 1.6 Transmitter of direct impulse excitation UWB.

for the design in which only the contributions from significant paths are considered, which are usually the signals from the line-of-sight path and several strong non-line-of-sight paths.

1.1.2 Impulse-radio-based UWB

From the earlier work on electromagnetic signals for radio transmission and radar, UWB has been related to the carrier-free baseband signal, called impulse radio (IR). Direct impulse excitation has been regarded as the conventional approach for UWB signal generation. Figure 1.5 shows the signal from the direct impulse excitation approach UWB. The extremely short duration pulse (a duty cycle of nanoseconds) provides the ultra-wide bandwidth and it is characterized with extremely low power spectral densities. Information is conveyed by changing the characteristics of the pulse, such as pulse amplitude modulation (PAM), pulse position modulation (PPM), on-off keying (OOK), etc.

A typical functional block diagram of the IR-UWB system is shown in Figure 1.6. The main advantage of IR is that it does not need RF components in transceivers. Therefore, implementation is relatively simple and the cost is low. Simplicity and low cost are the attractive features of direct impulse approach UWB. Low power consumption can be achieved and costly analog components can be avoided as higher frequency translation is bypassed. The antenna provides band-pass filtering and pulse shaping of the radiated signal. A high degree of digital implementation is possible. However, direct synthesis of the very wide bandwidth, from near zero to several gigahertz, means that very high speed signal processing units are required, but the low spectral region has to be dropped according to the regulations. PAM, PPM and OOK allow simplicity in transceiver design but yield signals with spectral lines, which is undesirable. Antenna design would be challenging since minimizing the signal distortion over a bandwidth of several gigahertz is difficult. Thus obvious constraints are placed on management and utilization of the allocated spectrum, which limits the overall efficiency of the system.

Since the receiver of time hopping IR (TH-IR) is operated by time-gating matched to the pulse duration, this time-gating reduces the power of continuous-time interference to the duty cycle of IR. Therefore, TH-IR inherently has the capability of narrowband interference suppression. When narrowband interference is very strong, however, we may need to use a notch filter to help reject it. The IR overlay is investigated in Chapter 3.

Accurate synchronization plays a cardinal role in the efficient utilization of any spread-spectrum system, and the process of synchronization is performed in two steps: (1) code acquisition, and (2) tracking via code-tracking loop. Much work on the study of acquisition in CDMA had been done about 40 years ago, and the work can be summarized as follows:

Parameters of an acquisition subsystem in direct spread-spectrum systems include detection threshold, correlation interval, number of tests per code chip, and system complexity determined by searching strategy and verification scheme. There are also some implicative parameters, such as the input signal-to-noise ratio (SNR), chip rate, period of spread code, length of the uncertainty region and false alarm penalty time [7].

Detection is the core unit of any acquisition subsystem, and it may be coherent or non-coherent, depending on whether acquisition is behind carrier demodulation or not. Different selections of detection threshold, such as fixed or adaptive, are determined by the adopted detection criteria, such as the Bayes rule, the Neyman–Pearson rule, and others. Furthermore, the detection dwell time, which may be fixed or variable, is the time interval allotted to each decision. In addition, the first detection level may or may not be followed by a verification logic that is used to ensure the initial synchronization indication and to prevent eventual false alarm, alternatively known as "multiple-dwell-time detector" or "single-dwell-time detector." The single-dwell detector can employ either full-period or partial-period correlation. In contrast, the multiple-dwell detector almost invariably employs a short-time partial-period correlation to expedite the acquisition process since the negative statistical effects of partial-period correlation can be removed at higher stages of the verification mode. Such a verification structure can employ immediate-rejection logic or nonimmediate-rejection logic. Passive integrations, such as matched-filter and active correlator, are two alternatives for correlation detection.

Different methods of code acquisition have been proposed. The maximum likelihood (ML) technique is similar to an optimal multiuser detection problem and gives excellent synchronization performance at prohibitive complexity. The sequential estimation method gathers information from past cell-decisions to improve the current cell-decision with a sequential search, but its merits decrease with decreasing SNR. The sliding correlator technique is considered exclusively for low-SNR environments and moderate system complexity.

Regarding sliding correlator techniques, serial search, parallel search, hybrid search, and matched-filter are included. For fast acquisition, a parallel search scheme or matched-filter is considered. As a trade-off between system complexity and acquisition speed, a hybrid search scheme combines serial and parallel searches. Every cell decision is made at the chip rate in the matched-filter, enabling fast acquisition. In the serial search scheme, more sophisticated serial-search strategies, such as Z-search, expanding-window, straight-line, etc., are considered.

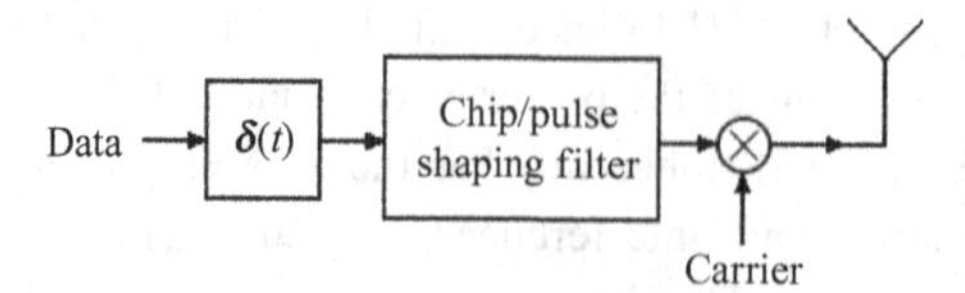

Figure 1.7 Transmitter of spectrally filtered approach UWB.

In frequency-selective fading channel environments, acquisition gives only the information that one of those paths has been found, and it ends when the threshold of the comparator is exceeded. An additional sweep should then be performed over a short timing window to find the existence and location of different multipaths and their complex coefficients. In [8], the optimal decision rule was proposed for frequency-selective fading channels based on the ML criterion.

One of the design challenges provided by the wide bandwidth property of IR-UWB signals is timing acquisition. Because of the high resolution in time required to locate the narrow pulses employed in IR systems, a considerable number of possible pulse positions in the uncertainty region must be searched in order to acquire the phase of the received signal. Therefore, a rapid acquisition algorithm is very important in IR-UWB communication systems. A novel rapid acquisition is proposed and analyzed in Chapter 4.

Although IR-UWB played a leading position during the early stages in the exploration of UWB technology, the limitation of IR-UWB motivated researchers to seek alternative methods of UWB signal generation. The carrier-based or spectrally filtered method, which is a conventional approach to narrowband communication, has been considered as a potential candidate for UWB application. Figure 1.7 is the functional block of the spectrally filtered UWB system, which involves translation of the baseband signal to a higher frequency, in contrast with IR.

The data sequence is chip or pulse shaped and passes to the frequency conversion component, which can be implemented by analog RF mixer, or the digital pseudo carrier. The resultant signal is then bandpass filtered and transmitted to the antenna. The spectrally filtered approach may suffer from an increase in production cost and system complexity in comparison to the IR approach, but this approach allows a better and more active control of the transmitting spectrum under bandpass filtering and frequency conversion, which is important to fulfil the spectrum mask requirement imposed by the FCC. Lower-speed processing units can be used since the ultra-wide bandwidth can be achieved indirectly. Information transmission can make use of phase, frequency and amplitude modulation, but modifications and adaptations are required to apply the spectrally filtered approach to UWB wireless communications.

1.1.3 MAC layer in UWB networks

From a network topological point of view, two types of UWB network are generally considered in practical scenarios. One is the infrastructure network and the other is the *ad hoc* network [9].

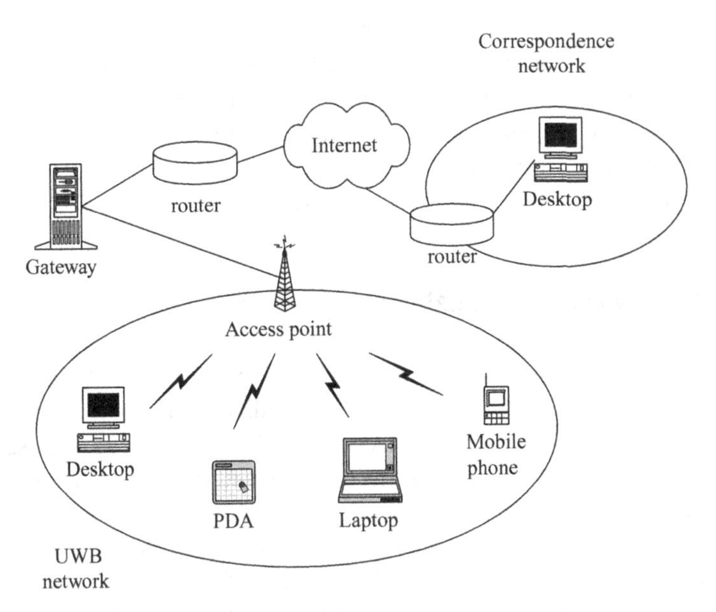

Figure 1.8 An example of an infrastructure UWB network.

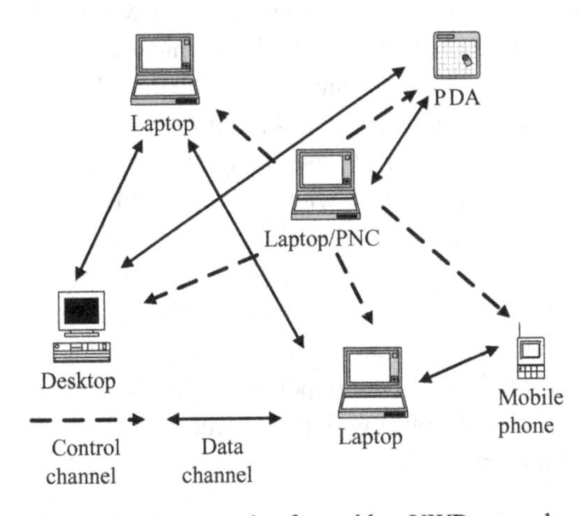

Figure 1.9 An example of an *ad hoc* UWB network.

In Figure 1.8 an example of an infrastructure UWB network is shown. Via an access point, mobile nodes (desktops, laptops, personal digital assistants (PDAs) and mobile phones) embedded with UWB transceivers can be connected to the Internet in order to communicate with other remote users. Moreover, the access point can also deliver packets for the nodes in the same UWB network.

In Figure 1.9 an example of an *ad hoc* UWB network is shown. Since an *ad hoc* network could work without setting any base station or access point in advance, people are able to share large files or have high-quality video conferences easily in a small area.

Medium access control (MAC) provides a fundamental method to coordinate the channel access among competing devices. Key research challenges in the area of MAC are in resource allocation and quality of service (QoS) provisioning. In [9], a centralized MAC scheme for the infrastructure network and a distributed scheme for the *ad hoc* network are investigated. Since this book is mainly devoted to the physical layer, please refer to [9] for detailed discussions of the MAC layer.

1.2 Evolved 3G mobile communications

The wireless channel resources of bandwidth and time need to be shared by all users. Frequency division multiple access (FDMA), time division multiple access (TDMA) and code division multiple access (CDMA) are three major solutions proposed for this multiple access problem. TDMA and FDMA try to separate the users by dividing time and bandwidth, respectively, while CDMA allows all users to access the whole bandwidth all the time and differentiates between users by using pre-assigned spreading codes. CDMA beats other digital and analog technologies on every front, including signal quality, security, power consumption and reliability. CDMA allows universal reuse of spectrum, flexible asynchronous access (e.g., voice activation), soft capacity, soft handoff, and large time-bandwidth signaling that can combat multipath channel fading with Rake combining [10]. Due to many overwhelming advantages, DS-CDMA has been chosen as the multiple access scheme by the International Telecommunications Union (ITU) for third generation (3G) wireless cellular systems. The wideband CDMA (WCDMA) from 3rd Generation Partnership Project (3GPP) [11] and CDMA2000 from 3GPP2 [12] are two major standards developed for 3G mobile systems. A wideband CDMA system is defined as one where the spread bandwidth of the underlying waveforms in the system typically exceeds the coherence bandwidth of the channel over which the waveforms are transmitted. Moreover, in order to realize completely the advantages of this time-bandwidth sharing and to enhance the inherent multiple access capability of CDMA, many recently developed efficient signal processing technologies have been designed and adopted in both standards. Some important technologies involved in the 3G system standards include: multicode transmission, transmit diversity, packet switched transmission, adaptive modulation and coding.

1.2.1 Transmit diversity

For the wireless channel, there usually are multiple propagation paths from the transmitter to the receiver and these paths have randomly time-varying amplitudes and delays depending on the geographical location and mobility of the receiver relative to the transmitter. This time variation in signal characteristics is usually referred to as fading, and can lead to significant performance degradation compared to the performance in a traditional deterministic channel with only additive white Gaussian noise (AWGN). Theoretically, transmitter power control is the most effective technique to mitigate multipath fading; however, the diversity technologies are more applicable and have been widely adopted in practical communication systems.

Receiving diversity

Until the current decade, nearly all diversity technologies actually refer to receiving diversity schemes. Diversity signaling is a powerful technique to combat short-term fading and has been thoroughly studied. This technique provides several independently faded replicas of the same information signal at the receiver and significantly reduces the probability that all the signal replicas suffer from amplitude fading at the same time. Depending on the propagation mechanism, some commonly used forms of receiving diversity techniques include: space diversity, frequency diversity, time diversity and multipath diversity (as used in the Rake receiver).

In space diversity systems, multiple copies of the signal are obtained by using multiple spaced receive antennas. In frequency and time diversities, the same transmission is repeated at multiple frequencies or at multiple time instants. Similarly, in multipath diversity, multiple copies of the signal are obtained from different resolvable paths. It is observed that under fairly general conditions, a channel affected by fading can be turned into an AWGN channel by increasing the number of diversity branches [13]. After obtaining the necessary signal at each diversity branch, these signals need to be combined to obtain the best result. For most communications, there are three basic linear combining methods: selection combining (SC), maximal ratio combining (MRC) and equal gain combining (EGC).

SC is the simplest of all the schemes. An ideal selection combiner chooses the branch signal with the highest instantaneous signal-to-noise-ratio (SNR); thus the output SNR is equal to that of the signal from the best branch.

MRC is the optimal linear combining scheme, where signals from all branches are weighted proportionally to their individual signal voltage-to-noise power ratios and summed together. The MRC combiner adaptively adjusts both the magnitude and phase of weights for each branch in order to maximize the total SNR at the output of the combiner.

EGC is a suboptimal but simple combining method. It is similar to MRC, but there is no attempt to weight the signals before addition. In other words, each weighting factor has phase opposite to that of the signal in the respective branch, but all weighting factors have equal and unity amplitude.

It is readily seen that the complexity of MRC and EGC receivers depends on the varying number of diversity branches. In addition, MRC and EGC are sensitive to weighting factor errors, and these errors tend to be more important when the instantaneous SNR is low. On the other hand, SC takes advantage of only one branch out of all available branches and hence does not fully exploit the amount of diversity offered by the channel. Therefore, there has been an interest in bridging the gap between these two extremes, i.e. MRC/EGC and SC, by proposing generalized selection combining (GSC), which adaptively selects and combines (following the rules of MRC or EGC) the several strongest branches among all available ones [13]. In the context of coherent DS-CDMA systems, the GSC Rake schemes generally offer less complexity than the conventional EGC and MRC methods. In addition, the GSC scheme is expected to be more robust toward weighting factor errors and less sensitive to the so-called "combining loss" of the very weak branches.

Multi-element antenna arrays are widely deployed at base stations to improve the uplink performance through traditional receiving diversity and to increase the potential

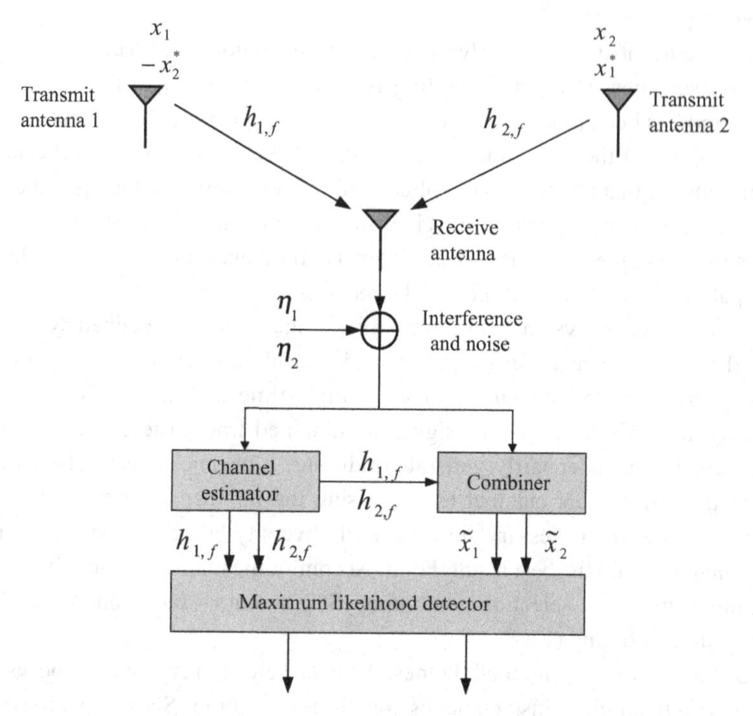

Figure 1.10 Alamouti space-time block code based transmit diversity scheme.

system capacity through some space-time processing techniques. However, employing multiple antennas at the mobile terminal contradicts the requirements of the future pocket communicators with respect to the size, limitation of the power consumption and complexity of the receiver.

Transmit diversity

Transmit diversity (TD) refers to the use of multiple transmit antennas in order to provide additional copies of the signal over independent channels. TD has been studied extensively as an efficient technique to achieve spatial diversity for downlink with multiple transmit antennas at the base station rather than multiple receive antennas at the mobile stations. Since future mobile multimedia services will be involved with high data rate transmission, it is desirable for the TD technique to provide improvement of both link-level performance and system capacity on the high-speed downlink without needing to increase the total transmit power or expand the bandwidth [13].

Although a number of TD approaches had been proposed, Alamouti presented the first space-time block code (STBC) to achieve full diversity gain with a simple linear processing decoding algorithm [14]. The Alamouti TD, which is based on STBC, exploits two transmit antennas and one receive antenna, as shown by the baseband representation in Figure 1.10. It provides the full diversity order of two and the full rate one, as it transmits two symbols in two time epochs.

This TD scheme takes one block, i.e., two consecutive symbol periods, to convey one pair of finite-alphabet modulated symbols x_1 and x_2. During the first symbol period of

Table 1.1 The encoding and transmission sequence for the Alamouti TD scheme

	Antenna 1	Antenna 2
time t	x_1	x_2
time $t+T$	$-x_2^*$	x_1^*

one given STBC coding block, two signals x_1 and x_2 are simultaneously transmitted from antenna 1 and antenna 2, respectively. During the second symbol period, signal $(-x_2^*)$ is transmitted from antenna 1 while x_1^* is transmitted from antenna 2, where the superscript * stands for the complex conjugate operation. The encoding sequence is shown in Table 1.1.

It is clear that the encoding is done in both the space and time domains. Let us denote the transmit sequence from antennas 1 and 2 by $\mathbf{x}^1$ and $\mathbf{x}^2$, respectively,

$$\mathbf{x}^1 = [x_1 \quad -x_2^*] \tag{1.4}$$

$$\mathbf{x}^2 = [x_2 \quad x_1^*] \tag{1.5}$$

The key feature of the Alamouti TD-STBC scheme is that the two transmit sequences from the two transmit antennas are orthogonal, since the inner product of the sequences $\mathbf{x}^1$ and $\mathbf{x}^2$ is zero, i.e., $\mathbf{x}^1 \bullet \mathbf{x}^2 = 0$. It is assumed that the channel is flat fading. At time t the fading channel from transmit antennas 1 and 2 to the receive antenna is modeled by complex multiplicative coefficients $h_1(t)$ and $h_2(t)$, respectively. Assuming that fading of each channel is constant across two consecutive symbol periods, they can be expressed as follows:

$$h_1(t) = h_1(t+T_s) = h_1 \tag{1.6}$$

$$h_2(t) = h_2(t+T_s) = h_2 \tag{1.7}$$

where T_s is the symbol period. The received signal during the first and second symbol intervals, respectively, can be expressed by

$$y_1 = y(t) = x_1 \cdot h_1 + x_2 \cdot h_2 + \eta_1 \tag{1.8}$$

$$y_2 = y(t+T_s) = (-x_2^*) \cdot h_1 + x_1^* \cdot h_2 + \eta_2 \tag{1.9}$$

where η_1 and η_2 represent AWGN and interference at time t and $t+T_s$. Moreover, they are assumed to be independent from each other.

The combiner to decode the Alamouti STBC is shown in Figure 1.10. Assuming coherent combining and ideal channel state information available at the receiver side, the two decision variables are constructed as

$$\begin{aligned}\tilde{x}_1 &= y_1 \cdot h_1^* + y_2^* \cdot h_2 \\ &= x_1 \cdot (|h_1|^2 + |h_2|^2) + \eta_1 \cdot h_1^* + \eta_2^* \cdot h_2 \end{aligned} \tag{1.9}$$

$$\begin{aligned}\tilde{x}_2 &= y_1 \cdot h_2^* - y_2^* \cdot h_1 \\ &= x_2 \cdot (|h_1|^2 + |h_2|^2) - \eta_2^* \cdot h_1 + \eta_1 \cdot h_2^* \end{aligned} \tag{1.10}$$

where $|\cdot|$ stands for the magnitude or the square root of the norm. The combined decision variables are then sent to the maximum likelihood detector.

It has been proven that the Alamouti TD-STBC scheme is equivalent to the conventional maximum ratio receiving combining (MRRC) scheme [14]. However, the equivalence is only valid when the pair of channels is independent non-frequency selective and when the total transmission power is actually twice that of the MRRC scheme. Furthermore, the Alamouti scheme can be readily extended to the system configuration with two transmit and M receive antennas, through which the same diversity order as a $2M$-branch MRRC is obtained with simply additional linear processing at the receiver. In other words, using two antennas at the transmitter, the scheme doubles the diversity order of systems employing only one transmit antenna and multiple receive antennas.

Other TD schemes

The concept of space-time code for open-loop TD was first introduced by Tarokh *et al.* [15]. This approach, now called space-time trellis code (STTC), is a joint design of error control coding, trellis-coded modulation and space-time diversity appropriate to multiple transmit antennas. STTC provides the best possible trade-off between constellation size, data rate, diversity advantage and trellis complexity. It offers a substantial diversity improvement, coding gain and spectral efficiency over flat fading channels. STTC performs well in a slowly fading environment, although the decoding complexity increases exponentially with the number of antenna elements and expected diversity order. Therefore, space-time trellis coding schemes are widely exploited to combat the effects of channel fading in multiple input multiple output (MIMO) systems.

In addressing the issue of decoding complexity in the Alamouti TD-STBC scheme [14], the concept of STBC first emerged, which is optimal for the complex modulation alphabet and two transmit antennas. The theory of STBC was further developed by Tarokh *et al.* [15]. They defined STBCs in terms of orthogonal code matrices. The properties of these matrices ensure full diversity equal to the number of transmit antennas, and a linear maximal likelihood detection. STBC actually generalizes the Alamouti scheme to an arbitrary number of transmit antennas and is able to achieve the full diversity promised by transmit and receive antennas. It is proven in [15], however, that the complex-value orthogonal design exists only for two transmit antennas. In other words, the Alamouti scheme is unique because it is the only STBC with a square complex transmission matrix to achieve the full rate one. For complex modulation signal constellations and more than two transmit antennas, the rate of orthogonal STBC is less than one.

Therefore, some specific orthogonal STBC schemes for three and four transmit antennas were also proposed in [15] and consequent performance results were given in [15], which can provide full diversity and half or three-quarter code rate. Similarly, STBC approaches were presented specifically for a system with four transmit antennas. All the above attempts concentrate in generalizing the complex orthogonal designs that support STBC with full diversity and a high transmission rate.

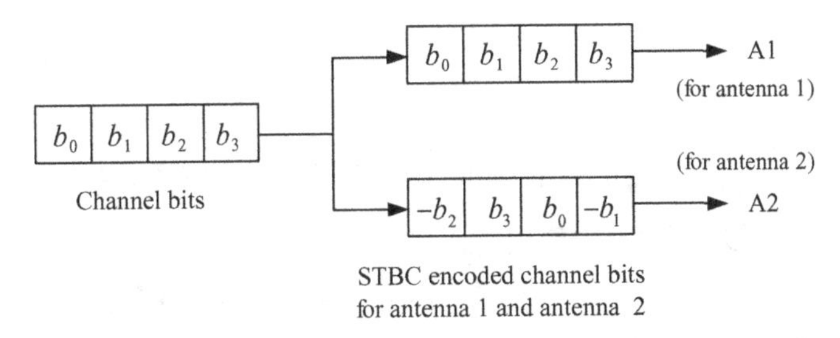

Figure 1.11 Generic block diagram of the STTD encoder.

Transmit diversity in 3G systems

As a method to combat impairments in wireless fading channels, TD is particularly appealing because of its relative simplicity of implementation and the feasibility of multiple antennas at the base station. Moreover, the cost of multiple transmit chains at the base station can be amortized over numerous mobile stations in service.

When the number of transmit antennas is fixed, the decoding complexity of STTC, which is measured by the number of trellis states at the decoder, increases exponentially as a function of both the diversity level and the transmission rate. In comparison to decoding complexity, the TD-STBC is therefore attractive since only linear processing is required at the receiver. These codes are able to achieve the full diversity order possible for a given number of antennas, without bandwidth expansion, although they do not offer the coding gain possible with STTC. Indeed, due to its significant advantages of simplicity and performance, the two-antenna complex modulation TD-STBC scheme originally proposed by Alamouti has already been adopted as an open-loop TD scheme in the 3G standards [11], [12].

In order to provide high data rate services for a large number of users in the system, both open- and closed-loop downlink TD techniques are incorporated into the 3G standards. The open-loop TD specifications are divided into two categories: space-time block coding based transmit antenna diversity (STTD) and time switched transmit diversity [11]. The STTD is the TD-STBC scheme proposed by Alamouti in essence. The STTD encoder operates on blocks of every four consecutive channel bits for QPSK (quadrature phase shift keying) modulation. The block diagram of the generic STTD encoder for channel bits b_0, b_1, b_2 and b_3 is shown in Figure 1.11. The channel bit b_j $(j = 0, 1, 2, 3)$ has been channel coded, rate matched and interleaved with a real value of ± 1.

Besides the above STTD encoding diagram, the detailed block diagram of post-processing specified by the 3G standards is also illustrated for transmit antenna 1 and antenna 2 in Figures 1.12 and 1.13, respectively. It is shown that, after the STTD encoding, two streams of coded bits will be allocated individually to the two processing branches corresponding to the two transmit antennas. The STTC coded bits will be first serial-to-parallel converted, spread by the complex spreading secondly, then split into real and imaginary parts, and finally QPSK modulated onto the carrier frequency.

It is worth noting that the same complex-valued spreading code is employed for both transmit antenna 1 and antenna 2; in other words, no expense of extra spreading codes is paid for the system with TD-STBC compared to the system without TD-STBC. In

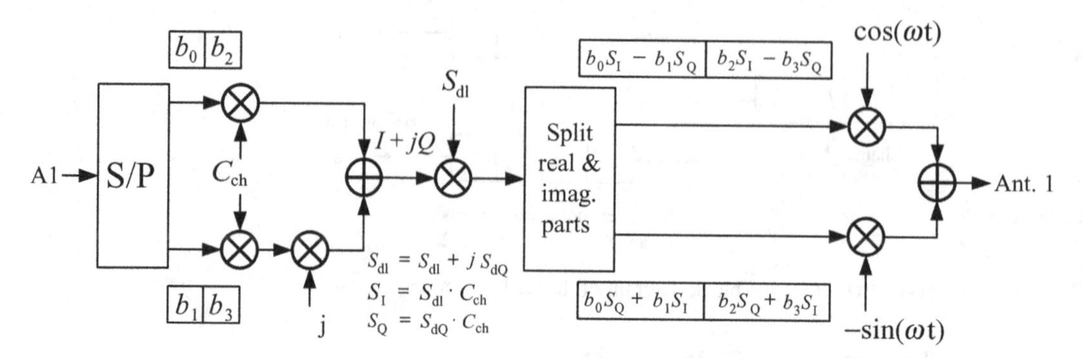

Figure 1.12 Post-STTD processing for transmit antenna one.

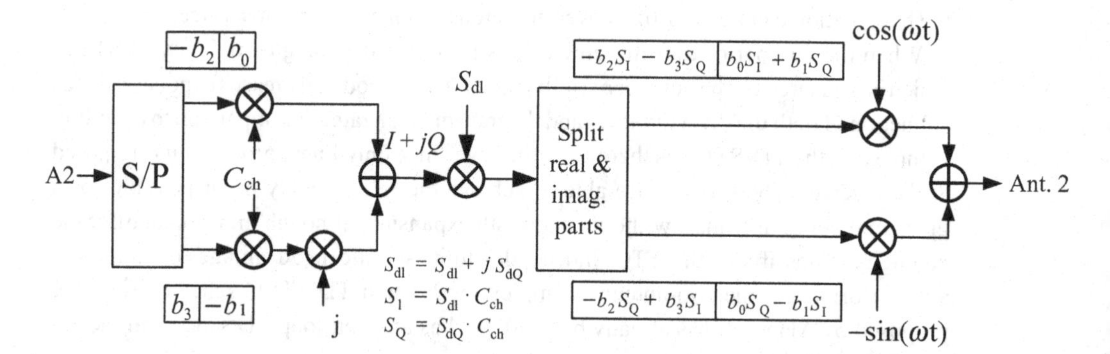

Figure 1.13 Post-STTD processing for transmit antenna two.

both Figures 1.12 and 1.13 the complex-valued spreading codes ($S_{\mathrm{I}} + \mathrm{j}S_{\mathrm{Q}}$) are actually constructed by multiplying the downlink scrambling codes ($S_{\mathrm{dI}} + \mathrm{j}S_{\mathrm{dQ}}$) by the orthogonal variable spreading factor (OVSF) channelization codes C_{ch} [10].

In Chapters 5 and 6, the performance of TD applied to 3G systems is presented in perfect and imperfect channel estimations.

1.2.2 Adaptive modulation

In order to provide a wireless high data rate service, efficient multi-level quadrature amplitude modulation (MQAM) techniques have to be adopted. However, the broadband wireless channel is a frequency selective fading channel. Therefore, adaptive modulation is necessary. When SINR is large, MQAM (e.g., 256QAM, 64QAM or 16QAM) should be used, whereas when SINR is small, a simple modulation scheme (QPSK or 8PSK) can be selected. The introduction of QAM to wireless CDMA results in several problems. One of them is that QAM is very sensitive to channel estimation quality, especially when the Doppler frequency shift is large. This is because the QAM demodulator must scale the received signal to normalize channel fading so that its decision regions correspond to the transmitted signal constellation. If the channel estimate contains any error, the demodulator will improperly scale the received signal, leading to an incorrect decision even in the absence of channel noise.

MQAM systems with antenna diversity

MQAM was first proposed half a century ago, although developments in its use for wireless communications were initially slow. Studies were mainly on constellation design, and implementation issues for fixed channels or satellite links. The first major paper considering MQAM for mobile radio applications was published in 1987 [16], which employed MQAM for voice transmission in Rayleigh fading channels. After that, MQAM for wireless mobile radio communications became a very hot topic due to its high spectral efficiency. Extensive studies for MQAM systems have been developed, including some on the potential for adaptive modulation, which can significantly improve system throughput by changing the modulation scheme adaptively according to different channel conditions [17]. Other research covers the performance evaluation of MQAM systems [18–20], which is the basis for the study of adaptive modulation systems. Therefore, it is necessary to provide a general bit error rate (BER) expression for the performance of MQAM systems in wireless environments.

In wireless communications, the transmitted symbol is distorted by amplitude and phase fluctuations due to the channel fading. To recover the transmitted MQAM symbol from the distorted signal at the receiver there are two traditional approaches: differential detection and coherent detection [18], [19]. It is found that, compared with coherent detection, differential detection suffers a 3 dB penalty [20]. Thus, coherent detection is more attractive due to its better performance, although it requires a knowledge of channel information to compensate for amplitude and phase fluctuations.

Coherent detection for MQAM systems has been studied in [18] under the assumption of perfect channel estimation. In practice, since MQAM is very sensitive to amplitude and phase fluctuations [19], channel estimation becomes a key technique for MQAM systems. In pilot symbol assisted modulation (PSAM) MQAM systems in wireless fading channels, pilot symbols are periodically inserted in the data stream. Then channel information can be extracted from the received signal with the help of pilot data [18]. Later studies have been developed to investigate the effect of imperfect channel estimation on PSAM MQAM systems in terms of symbol error rate (SER). In general the transformation from SER to BER is not straightforward. To clearly indicate the transmission capability of a system, a direct BER measurement/calculation is needed. Some studies on BER performance were mainly based on computer simulations, though an analytical result is provided for a special case (e.g. perfect channel estimation and two-branch space diversity). Hanzo *et al.* presented the first analytical BER expression for 16-QAM in Rayleigh fading channels in 1990 [21]. But perfect channel estimation was assumed. In [19], Tang *et al.* analytically investigated the effect of imperfect pilot symbol assisted channel estimation on the BER performance of 16/64-QAM systems. The integration operation for calculating the BER in [19] requires a heavy computational load, however, and is thus impractical to be extended to diversity cases. Therefore, analytical investigation of the effect of channel estimation error on MQAM systems over fading channels is necessary.

Moreover, considering the poor performance of high-order modulations (e.g. 64/256-QAM) in fading channels, an enhanced technique, antenna diversity, is employed to improve the system performance. Antenna diversity is achieved by using multiple antennas at the receiver to combat fading without expanding the bandwidth. Using the characteristic function method in [20], a general BER for PSAM MQAM systems will be provided in Chapter 7.

MQAM modulated multicode CDMA systems

In the 3G mobile standard, QPSK modulation was adopted in the original CDMA air interface. To support different high-quality data services with various bit rates, multicode transmission has been considered to satisfy the requirements and has been adopted in the 3G standards [11].

In multicode WCDMA systems, the data stream is first split into a number of parallel substreams, and then each substream is spread on a code-multiplexed channel. Since the channelization codes are generated from orthogonal variable spreading factor (OVSF) codes [10], their orthogonal properties prevent the different channels from interfering with each other in Gaussian or flat fading channels. In frequency selective fading channels, however, multipath interference occurs due to nonorthogonality of different code channels in the presence of multipath propagation delays. The performance of multicode WCDMA systems in frequency selective fading channels has been addressed in [22], [23]. It is found that the performance is seriously degraded by the multipath interference.

So far, several algorithms have been proposed to suppress the multiuser interference (MUI) in the uplink of CDMA or WCDMA systems [22]. Although the multiuser detector for CDMA systems provides the best BER performance, its complexity increases exponentially with an increase in the number of code/user channels. By using an interference regeneration and subtraction method, the interference cancellation technique is found to efficiently combat multipath interference. Because of its low complexity and performance improvement, interference cancellation has been extensively studied. A performance evaluation of systems employing interference cancellation technique was made in [22], [23] by means of simulation and analytical approaches. These algorithms have been applied to suppress the MUI or multipath interference.

On the demand for much higher speed package transmission in downlink, in addition to the multicode transmission, MQAM has been applied to high-speed downlink packet access (HSDPA) due to its high spectral efficiency [11], [12], although few studies have been investigated on CDMA systems with high-order modulations. Simulation results for MQAM modulated downlink multicode WCDMA systems have been presented in [23] with a Rake receiver and interference cancellation.

So far, no explicit BER analytical results have been given for QAM modulated multicode CDMA systems with imperfect interference regeneration and cancellation resulting from inaccurate channel estimation and tentative data decisions. In multicode WCDMA systems, channel estimation suffers a lot from multipath interference. Since it is used not only for coherent detection, but also for multipath interference regeneration, it affects the system performance much more than that for traditional MQAM systems. Therefore, it is important to investigate the effect of estimation error on these issues. In Chapter 8, an analytical BER expression for MQAM modulated multicode WCDMA systems with interference cancellation is derived.

1.3 4G mobile communications

3G long term evolution (LTE), also referred to as Super-3G, has been standardized to provide higher data rate services [11]. Its target peak data rates are 100 Mbps and 50 Mbps for downlink and uplink, respectively. Although LTE can provide a wireless multimedia service, its data rate is still limited. In order to provide richer multimedia services via wireless, 4G mobile communications, also referred to as LTE-Advanced, need to use a much broader bandwidth; for example, 100 MHz bandwidth is required to achieve a data rate of up to 1 Gbps in conjunction with advanced wireless access technologies, MIMO technologies, hybrid ARQ and so on.

1.3.1 OFCDM – a promising wireless access technique

A popular modulation technique known as orthogonal frequency division multiplexing (OFDM) has emerged as an efficient method to transfer high-speed data over frequency selective channels. The Wireless World Research Forum (WWRF) considers OFDM the most important technology for a future public cellular radio access system. Several wireless networking systems (e.g., IEEE 802.11 and 802.16) and wireless broadcasting systems (e.g., DVB-T, DAB) have already been developed using OFDM technology and are now available in mature commercial products. In the OFDM system, data is multiplexed on many narrowband sub-carriers, which can be easily generated by using highly efficient digital signal processing based on fast Fourier transform (FFT). By inserting a cyclic prefix (CP) between adjacent OFDM symbols, intersymbol interference (ISI) is virtually eliminated if the maximum channel delay spread is less than the time interval of the CP.

With a wide bandwidth over a wireless channel, a transmit signal experiences a broadband channel with many multipaths, which may cause severe multipath interference. Although OFDM has been widely accepted, a new multicarrier access scheme, i.e., orthogonal frequency and code division multiplexing (OFCDM), has been proposed by NTT DoCoMo for future wireless communication systems [24], [25]. It has been shown that OFCDM, or orthogonal multicarrier code division multiple access (MC-CDMA), exhibits better performance than the conventional DS-CDMA scheme in the broadband channel. Based on conventional OFDM, OFCDM systems employ a large number of orthogonal sub-carriers to transmit symbols in parallel, so that the symbol duration is increased substantially and the system can combat the multipath interference.

In order to have frequency diversity, the same data in OFCDM modulates a number of interleaved sub-carriers in terms of frequency domain spreading. Since time domain spreading is also introduced, multiple codes in both domains can be used to transmit a high data rate with high efficiency. The spreading scheme with both time and frequency domain spreading is called *two-dimensional spreading*. The total spreading factor (N) is the product of the time spreading factor (N_T) and the frequency spreading factor (N_F), i.e., $N = N_T \times N_F$.

In order to work in different cell environments, a variable spreading factor (VSF) is adopted for two-dimensional spreading, and the resultant OFCDM systems are called *VSF-OFCDM* systems, where N, N_T and N_F can be changed flexibly to provide the desired quality of service (QoS) according to variable conditions, such as cell structures and radio link states.

MMSE detection

Using two-dimensional spreading, $N = N_T \times N_F$, there are N code channels available for information transmission. In downlink, to provide high-speed multimedia services, multicode transmission is employed, i.e., multiple code channels are assigned to one user to achieve a high data rate. Using OVSF, the multicode channels are orthogonal in either time domain or frequency domain in an AWGN or quasi-static flat fading channel. However, in a realistic mobile channel, the orthogonality no longer remains in time domain because of possible fast fading or in frequency domain because of independent fading among sub-carriers. Therefore, severe multicode interference (MCI) may occur. In order to improve the system performance, MCI should be mitigated as much as possible.

In VSF-OFCDM systems, each data symbol is two-dimensionally spread in both frequency and time domains. At the receiver, detection (or combining) techniques are needed to collect the signals from the spread chips. These techniques have been well-studied for the OFCDM system with only frequency domain spreading [24, 25], including EGC, MRC, optimal detection or maximum likelihood detection (MLD), and iterative detection. Among these detection techniques, EGC is a very simple one and has good performance at low SNR because it will not enhance the noise power. However, when the orthogonality between code channels is seriously distorted and MCI is dominant, the performance of EGC is very poor since it cannot reduce the MCI. Therefore, EGC is not suitable for a multicode system when the code orthogonality is totally distorted. When the MCI is not serious, however, EGC can be considered because of its simplicity.

Although MLD provides the optimum performance, it is also the most complex detection technique. The basic idea of MLD is to evaluate the Euclidean distance between the received sequence and all possible transmitted sequences and find the most likely transmitted sequence by minimizing the Euclidean distance. So its complexity increases exponentially with the number of code channels [20].

Traditional MMSE is also optimum in terms of mean square error when both noise and interference are taken into account. However, this detection technique involves the inverse operation of a matrix, which is very complicated when the number of code channels is large. Fortunately, it can be simplified in case of full load and slow fading channels [20]. The simplified MMSE has been well studied in [24], [25]. MLD and MMSE will be studied for OFCDM in Chapter 9.

Hybrid detection

Although MMSE is useful due to its simplicity, its performance is not satisfactory in suppressing MCI. Thus it is desirable to evaluate the performance of the VSF-OFCDM system when the MCI cancellation technique is adopted. An efficient interference cancellation method is linear multistage interference cancellation [26], where the complexity of the canceller grows linearly with the number of code channels. Basically, the multistage MCI cancellation is implemented in an iterative way until a specified number of stages are reached.

Although interference cancellation and MMSE detection are well-developed techniques, so far they have not been investigated for the VSF-OFCDM system. The objective of our research is to analytically investigate the performance of VSF-OFCDM with hybrid MCI cancellation and MMSE detection. Chapter 10 will focus on this topic.

1.3.2 OFDM MIMO multiplexing systems

One of the ambitious design goals of future wireless communications is to provide reliable very high data rate transmission; for example, a target peak rate of 1 Gbps in local areas such as a very small cell with high traffic density, hotspot area or indoor environment. One way to get very high bit rates under a scattering-rich wireless channel is to use the multiple-antenna technique [27], which is capable of realizing spectral efficiency that far exceeds that of single-input single-output (SISO) systems. Much work on the study of MIMO techniques has been done, and several schemes have been proposed to exploit its potential, such as: MIMO multiplexing or layered space-time (LST) architecture [28], orthogonal STBC [14] and STTC [15]. MIMO multiplexing is aimed at the highest bandwidth efficiency and is thus the most suitable for ultra high data rate transmission. However, MIMO techniques are suitable for flat fading channels, while the broadband wireless channel is the frequency selective channel in which frequency diversity becomes available. The technique of OFDM MIMO multiplexing [29], [30], is called layered space-time-frequency (LSTF) architecture.

MIMO multiplexing can be viewed as a synchronous CDMA in which the number of transmit antennas is equal to the number of users. Similarly, the interference between transmit antennas is equal to the multiple access interference (MAI) in CDMA systems, while the complex fading coefficients correspond to the spreading sequences. This analogy can be further extended to receiver strategies so the multiuser receiver structures derived for CDMA can be directly applied to MIMO multiplexing systems. Under this scenario, the optimum receiver for an uncoded MIMO multiplexing system is the ML multiuser detector [31] operating on a trellis and computing ML statistics, as in the Viterbi algorithm, but its complexity is exponential in the number of the transmit antennas. Another approach is the minimum mean-square error (MMSE)- or zero forcing (ZF)-nulling/cancelling detector [32], which uses the linear nulling and successive interference cancellation processes to estimate transmitted symbols. The MMSE- or ZF-nulling/cancelling detector has lower complexity than the ML detector but sub-optimal performance.

For coded MIMO multiplexing systems, the optimum receiver performs joint detection and decoding on an overall trellis obtained by combining the trellises of the layered space-time code and the channel code. However, this optimum receiver is almost infeasible because its complexity is an exponential function of the product of the number of transmit antennas and the code memory order. Thus, one approach is to use the non-iterative receiver structure, which consists of the detector for uncoded MIMO multiplexing and the channel decoder. In [33], the non-iterative receiver consisting of complexity-reduced type ML detector and turbo decoder has been shown to achieve satisfactory performance and feasible complexity. On the other hand, the iterative processing technique for joint multiuser detection and decoding can be applied to coded MIMO multiplexing. The complexity of the multiuser detector constitutes a major part of the overall complexity of iterative receivers. Three multiuser detection algorithms that provide a trade-off between performance and complexity have been proposed in the literature. The first is based on the maximum *a posteriori* (MAP) probability rule [34]; the second is the iterative receiver with parallel interference canceller (PIC) [35]; and the third is the iterative minimum-mean-square-error soft-interference-cancellation (MMSE-SIC) detector, which makes use of both soft interference cancellation and instantaneous linear MMSE filtering [36]. In Chapter 11, an iterative receiver with convolutional coding is studied for LSTF.

1.3.3 Hybrid ARQ

Besides error correcting codes, the automatic repeat request (ARQ) is another technique for controlling transmission errors in high-speed wireless communication. It was introduced in the early days of data communication as a result of the development of parity-check codes [37], [38]. The main feature of this technique is that it can adapt to the channel conditions at low complexity. In an ARQ system, each information block is encoded with a good error detection code. At the receiver, error detection is carried out first. The information block will be accepted if no error is detected, and a positive acknowledgment (ACK) signal will be sent to the transmitter via a feedback channel, indicating a successful transmission. If an erroneous code word is detected, a negative acknowledgment (NAK) signal will be sent to the transmitter to ask for a retransmission of the same data. This process continues until the information block is successfully accepted or the maximum retransmission attempts are reached.

There are three basic ARQ protocols, i.e., the stop-and-wait ARQ, the go-back-N_g ARQ and the selective-repeat ARQ [38]. In a stop-and-wait ARQ system, the transmitter sends a code word to the receiver and waits for the acknowledgement signals. In a go-back-N_g ARQ system, the code words are transmitted continuously. When the transmitter receives an NAK signal, it goes back to the code word that is negatively acknowledged and resends that code word and $N_g - 1$ succeeding code words transmitted during the round-trip delay. A selective-repeat ARQ system also transmits the code words continuously. But only those negatively acknowledged code words will be resent. Among these three protocols, selective-repeat ARQ is the most efficient but also the most complex.

Pure ARQ protocols only employ error detection codes. In 1960, forward-error-correction (FEC) codes were introduced into ARQ protocols [37], and the results are now known as hybrid ARQ protocols. Hybrid systems can achieve throughput similar to FEC systems and provide the good reliability and flexibility of pure ARQ protocols. There are two main types of hybrid ARQ protocol, i.e., type I hybrid ARQ and type II hybrid ARQ [38]. In a type I hybrid ARQ system, the erroneously received code words will be discarded. In a type II hybrid ARQ system, the erroneously received code words will be stored in buffers and optimally combined with the retransmitted code words. Incremental redundancy concepts can also be exploited in the type II hybrid ARQ protocol. In the first transmission, only part of the parity bits is transmitted with information bits. When retransmission is needed, another part of the parity bits is transmitted instead of resending the same data as the previous transmission. A new code word with a lower rate can then be constructed with the increased parity bits, which makes this a more powerful method of correcting any transmission errors. The type II hybrid ARQ is efficient but it needs large buffer size at the receiver.

Many powerful FEC codes have been considered for hybrid ARQ systems, such as Turbo codes [39] and convolutional codes [40], [41]. Turbo codes have been shown to have excellent performance [42]. However, turbo decoding is very complicated. Convolutional codes are still attractive in real applications due to their simple realization. Convolutional codes can be obtained by shift register techniques, and exhibit good error correction capability by use of the maximum likelihood decoding scheme, i.e., the Viterbi decoding scheme [43]. Convolutional codes have been widely employed in many communication systems, such as deep-space and satellite communication systems, second generation (2G) and third generation (3G) mobile communication systems [11], [12], and so on. A typical convolutional encoder is involved with one-bit input, a few-bit output and a constraint length. The performance of convolutional codes is determined by the Viterbi decoding, which selects survival paths in making decisions. Thus, convolutional codes have a property of burst errors.

A conventional complete-packet ARQ scheme is inefficient with some codes, such as convolutional codes, because the whole packet will be retransmitted even though there are only one or two bit errors. This problem can be improved if sub-packet schemes are employed. In sub-packet transmissions, only those sub-packets that include errors need to be retransmitted, so that the system throughput can be improved. Therefore, it is interesting to study the performance of convolutional-coded hybrid ARQ systems with sub-packet transmission. Chapter 12 will investigate this topic in detail.

References

[1] Ultra Wideband Working Group, www.uwb.org/standards.htm.

[2] M. L. Welborn, "System considerations for ultra-wideband wireless networks," in *Proc. IEEE VTC'01-Spring*, Rhodes, Greece.

[3] T. Mitchell, "Broad is the way," *IEE Review*, pp. 35–39, Jan. 2001.

[4] T. T. Wong, "Multicarrier CDMA overlay for ultra-wideband wireless communications," M.Phil. Thesis, University of Hong Kong, 2003.

[5] J. Wang and L. B. Milstein, "Multicarrier CDMA overlay for ultra-wideband communications," *IEEE Trans. Commun.*, vol. COM-52, pp. 1664–1669, Oct. 2004.

[6] T. T. Wong and J. Wang, "MMSE receiver for multicarrier CDMA overlay in ultra-wideband communications," *IEEE Trans. Vehicular Tech.*, vol. VT-54, pp. 603–614, March 2005.

[7] Y. L. Huang, "Study of advanced techniques in high speed wireless transmissions," Ph.D. Thesis, University of Hong Kong, 2006.

[8] R. R. Rick and L. B. Milstein, "Optimal decision strategies for acquisition of spread-spectrum signals in frequency-selective fading channels," *IEEE Trans. Commun.*, vol. 46, pp. 686–694, May 1998.

[9] Y. Liu, "Resource management techniques for high performance ultra wideband wireless networks," Ph.D. Thesis, University of Hong Kong, 2006.

[10] F. Adachi, M. Sawahashi and H. Suda, "Wideband DS-CDMA for next-generation mobile communications systems," *IEEE Comm. Mag.*, vol. 36, no. 9, pp. 56–69, Sept. 1998.

[11] Third generation partnership project, www.3gpp.org.

[12] Third generation partnership project 2, www.3gpp2.org.

[13] X. Y. Wang, "Transmit diversity in CDMA for wireless communications," Ph.D. Thesis, University of Hong Kong, 2003.

[14] S. M. Alamouti, "A simple transmit diversity technique for wireless communications," *IEEE J. Select. Areas Comm.*, vol. 16, no. 8, pp. 1451–1458, Oct. 1998.

[15] V. Tarokh, H. Jafarkhani and A. R. Calderbank, "Space-time block codes from orthogonal designs," *IEEE Trans. Inform. Theory*, vol. 45, no. 5, pp. 1456–1467, July 1999.

[16] C. W. Sundberg, W. C. Wong and R. Steel, "Logarithmic PCM weighted QAM transmission over Gaussian and Rayleigh fading channels," *IEE Proc.*, vol. 134, pp. 557–570, Oct. 1987.

[17] A. J. Goldsmith and S. G. Chua, "Variable-rate variable-power MQAM for fading channels," *IEEE Trans. Commun.*, vol. 45, pp. 1218–1230, Oct. 1997.

[18] L. Yang and L. Hanzo, "A recursive algorithm for the error probability evaluation of M-QAM," *IEEE Commun. Lett.*, vol. 4, pp. 304–306, Oct. 2000.

[19] X. Tang, M.-S. Alouini and A. J. Goldsmith, "Effect of channel estimation error on M-QAM BER performance in Rayleigh fading," *IEEE Trans. Commun.*, vol. 47, pp. 1856–1864, Dec. 1999.

[20] B. Xia, "Enhanced techniques for broadband wireless communications," Ph.D. Thesis, University of Hong Kong, 2004.

[21] L. Hanzo, R. Steel and P. M. Fortune, "A subband coding, BCH coding, and 16-level QAM system for mobile radio communications," *IEEE Trans. Veh. Technol.*, vol. 39, pp. 327–339, Nov. 1990.

[22] J. Chen, J. Wang and M. Sawahashi, "MCI cancellation for multicode wideband CDMA systems," *IEEE J. Select. Areas Commun.*, vol. 20, pp. 450–462, Feb. 2002.

[23] K. Higuchi, A. Fujiwara and M. Sawahashi, "Multipath interference canceller for high-speed packet transmission with adaptive modulation and coding scheme in W-CDMA forward link," *IEEE J. Select. Areas Commun.*, vol. 20, pp. 419–432, Feb. 2002.

[24] N. Maeda, Y. Kishiyama, H. Atarashi and M. Sawahashi, "Variable spreading factor-OFCDM with two dimensional spreading that prioritizes time domain spreading for forward link broadband wireless access," in *Proc. IEEE VTC2003-Spring*, Korea, pp. 127–132, April 2003.

[25] N. Maeda, H. Atarashi, S. Abeta and M. Sawahashi, "Pilot channel assisted MMSE combing in forward link for broadband OFCDM packet wireless access," *IEICE Trans. Fundamentals*, vol. E85-A, no. 7, pp. 1635–1646, July 2002.

[26] Y. Q. Zhou, "Advanced techniques for high speed wireless communications," Ph.D. Thesis, University of Hong Kong, 2003.

[27] G. J. Foschini, "Layered space-time architecture for wireless communication in a fading environment when using multi-element antennas," *Bell Labs. Tech. J.*, vol. 1, no. 2, pp. 41–59, 1996.

[28] G. J. Foschini and M. J. Gans, "On limits of wireless communications in a fading environment when using multiple antennas," *Wireless Pers. Commun.*, vol. 6, pp. 311–335, 1998.

[29] H. Bolcskei, D. Gesbert and A. J. Paulraj, "On the capacity of OFDM-based spatial multiplexing systems," *IEEE Trans. Commun.*, vol. 50, no. 2, pp. 225–234, Feb. 2002.

[30] B. Lu, G. Yue and X. Wang, "Performance analysis and design optimization of LDPC-coded MIMO OFDM systems," *IEEE Trans. Signal. Process.*, vol. 52, no. 2, pp. 348–361, Feb. 2004.

[31] S. Verdu, *Multiuser Detection*. Cambridge: Cambridge University Press, 1998.

[32] G. D. Golden, G. J. Foschini, R. A. Valenzuela and P. W. Wolniansky, "Detection algorithm and initial laboratory results using the V-BLAST space-time communication architecture," *Electronics Lett.*, vol. 35, no. 1, pp. 14–15, Jan. 1999.

[33] K. Higuchi, H. Kawai, N. Maeda *et al.*, "Likelihood function for QRM-MLD suitable for soft-decision turbo decoding and its performance for OFCDM MIMO multiplexing in multipath fading channel," *IEICE Trans. Commun.*, vol. E88-B, no. 1, pp. 47–57, Jan. 2005.

[34] M. Moher, "An iterative multiuser decoder for near-capacity communications," *IEEE Trans. Commun.*, vol. 46, pp. 870–880, July 1998.

[35] S. Marinkovic, B. Vucetic and A. Ushirokawa, "Space-time iterative and multi-stage receiver structures for CDMA mobile communication systems," *IEEE J. Select. Areas Commun.*, vol. 19, no. 8, pp. 1594–1604, Aug. 2001.

[36] X. Wang and H. V. Poor, "Iterative (Turbo) soft interference cancellation and decoding for coded CDMA," *IEEE Trans. Commun.*, vol. 47, pp. 1046–1061, July 1999.

[37] S. B. Wicker, *Error Control Systems for Digital Communication and Storage*. New Jersey: Prentice Hall, 1995.

[38] S. Lin and D. J. Costello Jr., *Error Control Coding: Fundamentals and Applications*. Englewood Cliffs, NJ: Prentice-Hall, 1984.

[39] K. R. Narayanan and G. L. Stuber, "A novel ARQ technique using the Turbo coding principle," *IEEE Commun. Lett.*, vol. 1, pp. 49–51, March 1997.

[40] H. Yamamoto and K. Itoh, "Viterbi decoding algorithm for convolutional codes with repeat request," *IEEE Trans. Inform. Theory*, vol. IT-26, pp. 540–547, Sept. 1980.

[41] B. A. Harvey and S. B. Wicker, "Packet combining systems based on the Viterbi decoder," *IEEE Trans. Commun.*, vol. 42, pp. 1544–1557, April 1994.

[42] C. Berrou and A. Glavieux, "Near optimum error correcting coding and decoding: Turbo-codes (1)," in *Proc. of ICC'93*, pp. 1064–1070, May 1993.

[43] A. J. Viterbi, "Error bounds for convolutional codes and an asymptotically optimum decoding algorithm," *IEEE Trans. Inform. Theory*, vol. IT-13, pp. 260–269, April 1967.

Part II

UWB communications

2 Multicarrier CDMA overlay for UWB communications

This chapter investigates the use of promising measures incorporated in the multicarrier CDMA overlay to meet the goals of UWB communications, as well as on their impacts on the parties involved. At the transmitter, interference reduction on the established narrowband systems is done using notch filters. A multipath Nakagami fading channel is assumed. At the receiver, the interference suppression from those narrowband systems is fulfilled by MMSE detection techniques. Numerical results show that pre-combining MMSE with selective-maximal combining provides the UWB system with much better performance than a receiver made up of a notch filter in cascade with code correlator (NCC).

2.1 Transmitter, channel and narrowband interference

The multicarrier CDMA system is assumed to contain M sub-carriers and K active users. The block diagram of the transmitter is shown in Figure 2.1, where $c_k^{(u)}$ is the uth element of the spreading sequence, which is assumed to be random. $b_k^{\lfloor u/N \rfloor}$ is the $\lfloor u/N \rfloor$th element of the data sequence ($\lfloor x \rfloor$th stands for the integer part of x).

The expectations of the data sequence and spreading sequence elements have the following properties:

$$E\left[b_{k_1}^{u_1} b_{k_2}^{u_2}\right] = \begin{cases} 1, & \text{for} \quad u_1 = u_2; \quad k_1 = k_2 \\ 0, & \text{otherwise} \end{cases} \tag{2.1}$$

$$E\left[c_{k_1}^{u_1} c_{k_2}^{u_2}\right] = \begin{cases} 1/N, & \text{for} \quad u_1 = u_2; \quad k_1 = k_2 \\ 0, & \text{otherwise} \end{cases} \tag{2.2}$$

where N is the processing gain of each sub-channel, given by T_b/T_c where T_b and T_c are bit duration and chip duration, respectively. The product of the two sequences results in the spread sequence of user k, given by

$$d_k^{(u)} = b_k^{\lfloor u/N \rfloor} c_k^{(u)} \tag{2.3}$$

The idea of using notch filter at transmitter has been outlined in [1], [2] such that the frequency bands occupied by narrowband systems are notched out. In this study a low-complexity scheme is suggested whereby the notch action is achieved by the chip-shaping modification via a two-sided transversal-type filter, so the resultant chip

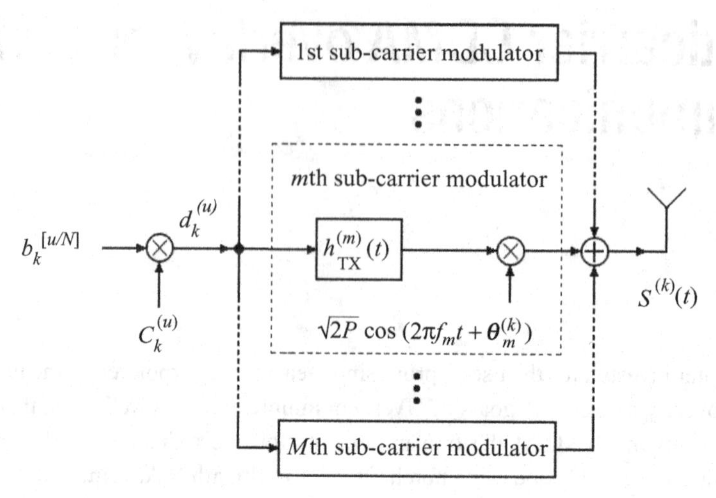

Figure 2.1 Transmitter.

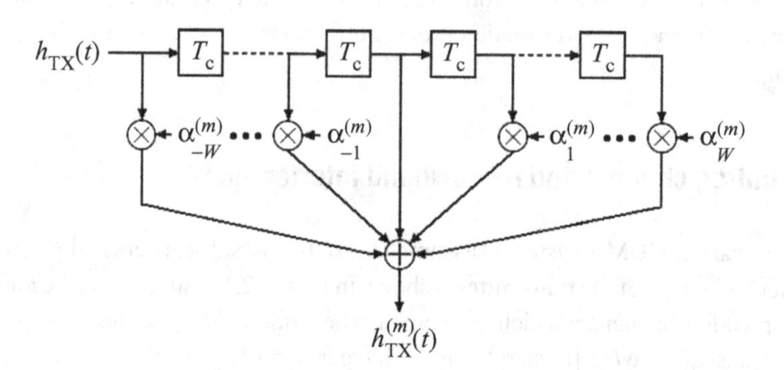

Figure 2.2 Notch filter.

shaping can simultaneously achieve band-limiting and interference avoidance to the established service. In the proposed transmitter, the source signal passes through the chip-shaping filter to give the baseband signal before frequency up conversion, but as a derivative of a conventional multicarrier CDMA transmitter. Independent chip shaping filters are employed in the modulators of the sub-carriers due to the different interference avoidance requirements for each sub-band. The chip-shaping filter is shown in Figure 2.2, and its impulse response is given by

$$h_{TX}^{(m)}(t) = F^{-1}\{H_{TX}^{(m)}(f)\} = F^{-1}\{H_{TX}(f)H_W^{(m)}(f)\} \tag{2.4}$$

where $H_W^{(m)}(f)$ is the frequency response of the two-sided transversal-type notch filter with $2W+1$ taps. For the mth sub-carrier the filter frequency response is

$$H_W^{(m)}(f) = \sum_{w=-W}^{W} \alpha_w^{(m)} \exp(-j2\pi f w T_c) \tag{2.5}$$

The tap coefficients are adapted to suppress the spectra with narrowband systems via the chip-shaping modification, but coordination is impossible among the coexisting systems. This leads to a need for a dynamic scheme where the taps are determined based on the observation at the receiver, driven by the existence of the narrowband system such that any adjustment reflects the actual channel condition. As described in [3], since the present value of the narrowband process can be predicted from past and future values, but is not applicable to the UWB signal and channel noise as they are wideband process, the estimation of the location of the narrowband system can be done by observation of the past and future values of the received signal samples. Here perfect information is assumed by the feedback of parameters from the receiver, and the details of coefficient formulation are provided in a later sub-section (Section 2.3.1). When there is no narrowband interference, no notch is inserted and the filter reduces to an all-pass filter, i.e., $H_W^{(m)}(f) = 1$, then the undistorted raised-cosine chip shape is transmitted, as if the notch filter did not exist. The impact of notch filtering on the chip shape is illustrated and discussed in Section 2.5. In (2.4), $H_{\mathrm{TX}}(f)$ denotes the frequency response of the unfiltered chip-shaped filter at the transmitter, and with the chip-matched filter $h_{\mathrm{RX}}(t)$ at the receiver ($H_{\mathrm{RX}}(f)$ denotes its frequency response) gives the raised-cosine [4] chip shape $H_{\mathrm{RC}}(f)$ with roll-off factor α for the purpose of band-limiting the transmitted signal:

$$\begin{aligned} H_{\mathrm{RC}}(f) &= H_{\mathrm{TX}}(f) H_{\mathrm{RX}}(f) \\ &= \begin{cases} T_c & 0 \le |f| \le \dfrac{1-\alpha}{2T_c} \\ T_c \left\{ 1 + \cos\left[\pi \,{}^{T_c}\!/_{\alpha} \left(f - \dfrac{1-\alpha}{2T_c} \right) \right] \right\} \quad \text{for} & \dfrac{1-\alpha}{2T_c} \le |f| \le \dfrac{1+\alpha}{2T_c} \\ 0 & |f| \ge \dfrac{1+\alpha}{2T_c} \end{cases} \end{aligned} \tag{2.6}$$

Thus, the transmitted signal of the kth user can be written as

$$\begin{aligned} S^{(k)}(t) &= \sqrt{2P} \sum_{u=-\infty}^{\infty} \left(\sum_{m=1}^{M} d_k^u h_{\mathrm{TX}}^{(m)}(t - uT_c) \cos\left(2\pi f_m t + \theta_m^{(k)}\right) \right) \\ &= \sum_{m=1}^{M} \left(s_m^{(k)}(t) \right) \end{aligned} \tag{2.7}$$

where P is the transmit power per sub-carrier, and f_m and $\theta_m^{(k)}$ are the carrier frequency and random phase of the mth sub-carrier, respectively.

The channel response of UWB communications is characterized by a rich number of multiple propagation paths. With reference to the stochastic tapped-delay-line model of UWB indoor application suggested in [5], [6], channel variation is assumed to be slow so that the parameters are time-invariant over a bit period. The channel response of the mth sub-band for the kth user can be expressed as

$$h_m^{(k)}(t) = \mathrm{Re}\left[\sum_{l=1}^{L} \beta_{m,l}^{(k)} \delta\left(t - \tau_{m,l}^{(k)}\right) \exp\left(\mathrm{j}\phi_{m,l}^{(k)}\right) \right] \tag{2.8}$$

where L is the number of resolvable paths for each carrier of the UWB system, with the assumption of slow and frequency selective fading. $\phi_{m,l}^{(k)}$, $\tau_{m,l}^{(k)}$ and $\beta_{m,l}^{(k)}$ denote the phase shift, delay and attenuation factor of the lth path, respectively. All resolvable paths for the same sub-carrier of the same user are assumed to be time-separated by chip duration such that

$$\tau_{m,l}^{(k)} = \tau_m^{(k)} + (l-1)T_c \tag{2.9}$$

where $\tau_m^{(k)}$ is the random delay uniformly distributed over $[0, T_b)$. The phase shifts are independently and uniformly distributed over $[0, \pi)$. The attenuation factors are assumed to be independent, identical and Nakagami distributed and, unless for the same path from the same sub-carrier of the same user, the attenuation factors are uncorrelated to each other such that

$$E\left[\beta_{m,l}^{(k)}\beta_{m',l'}^{(k')}\right] = \begin{cases} \Omega_{l,m}^{(k)}, & \text{for} \quad k = k'; \quad l = l'; \quad m = m' \\ 0, & \text{otherwise} \end{cases} \tag{2.10}$$

where $\Omega_{m,l}^{(k)}$ is the normalized power-decay profile (PDP), which is the second moment of the attenuation, and is related to the PDP of the initial path for the same user $\Omega_{1,m}^{(k)}$ via [5]

$$\Omega_{m,l}^{(k)} = \Omega_{1,m}^{(k)}\kappa \exp\left(-\left(\tau_{m,l}^{(k)} - \tau_{2,m}^{(k)}\right)/\delta_{m,l}^{(k)}\right) \tag{2.11}$$

where $\Omega_{1,m}^{(k)}$ is related to the total transmitted power as

$$\Omega_{1,m}^{(k)} = \frac{1}{1+\kappa\left(1-\exp\left(T_c/\delta_{m,l}^{(k)}\right)\right)} \tag{2.12}$$

and $\delta_{m,l}^{(k)}$ is the decay rate for the lth path. κ is the power ratio of the second PDP to the direct path, which is assumed to be the strongest path and falls into the first observation bin. The PDPs decay exponentially with delay starting from the second bin. The probability density function attenuation factor is given by [7]

$$p\left(\beta_{m,l}^{(k)}\right) = \frac{2\left(\beta_{m,l}^{(k)}\right)^{2\mu_{m,l}^{(k)}-1}}{\Gamma\left(\mu_{m,l}^{(k)}\right)}\left(\frac{\mu_{m,l}^{(k)}}{\Omega_{m,l}^{(k)}}\right)^{\mu_{m,l}^{(k)}} \exp\left(-\frac{\mu_{m,l}^{(k)}}{\Omega_{m,l}^{(k)}}\right)\left(\beta_{m,l}^{(k)}\right)^2 \tag{2.13}$$

where $\mu_{m,l}^{(k)}$ denotes the fading figure, given by

$$\mu_{m,l}^{(k)} = \frac{\left(\Omega_{m,l}^{(k)}\right)^2}{E\left[\left(\beta_{m,l}^{(k)}\right)^2 - \left(\Omega_{m,l}^{(k)}\right)^2\right]} \tag{2.14}$$

The established narrowband systems along the operating spectrum of the UWB system are assumed to be band-limited Gaussian processes with a flat power spectrum. For the sake of simplicity, single-path propagation is assumed for the narrowband system, and at most one narrowband system is assumed in any given sub-band. The collection of established narrowband systems along the operating spectrum of the UWB system can be represented by

$$j(t) = \sum_{m=1}^{M} j_m(t) \tag{2.15}$$

where $j_m(t)$ denotes the narrowband system within the mth sub-band, and the power spectral density of the narrowband system can be expressed as

$$P_j^{(m)}(f) = \begin{cases} P_j^{(m)}, & 0 \leq \left| f - \dfrac{q_m}{T_c} \right| \leq \dfrac{p_m}{2T_c} \\ 0, & \text{otherwise} \end{cases} \tag{2.16}$$

where $P_j^{(m)}$ is the power of the narrowband system within the spectrum region of the mth sub-carrier at the UWB receiver. We define p_m as the ratio of the narrowband system bandwidth to the UWB system sub-band bandwidth, such that $0 \leq p_m \leq 1$. q_m is defined as the offset of the center frequency of the narrowband system signal from the mth sub-carrier's center frequency in the UWB system, which is normalized by the sub-band bandwidth of the UWB system, such that $-0.5(1+\alpha) \leq q_m \leq 0.5(1+\alpha)$.

2.2 Receiver

The received signal can be expressed as

$$r(t) = \sum_{k=1}^{K} \sum_{m=1}^{M} \sqrt{2P} S_m^{(k)}(t) * h_m^{(k)}(t) + j(t) + n(t) \tag{2.17}$$

where $*$ stands for convolution. $n(t)$ is an additive white Gaussian noise (AWGN) with two-sided power spectral density of $\eta_o/2$. The block diagram of the proposed UWB receiver is shown in Figure 2.3.

After frequency down conversion, chip-matching and low-pass filtering, the recovered baseband signal can be written as

$$\begin{aligned} x_m^{(k)}(t) &= \left(r(t)\cos\left(2\pi f_m t + \theta_m^{(k)}\right)\right) * h_{\mathrm{RX}}\left(t - \tau_m^{(k)}\right) \\ &= \sum_{k=1}^{K} \sum_{l=1}^{L} \sqrt{2P} \beta_{m,l}^{(k)} s_m^{(k)}\left(t - \tau_m^{(k)}\right) \cos\left(\Phi_{m,l}^{(k)}\right) * h_{\mathrm{RX}}\left(t - \tau_m^{(k)}\right) \\ &\quad + \left(j(t)\cos\left(2\pi f_m t + \theta_m^{(k)}\right)\right) * h_{\mathrm{RX}}\left(t - \tau_m^{(k)}\right) \\ &\quad + \left(n(t)\cos\left(2\pi f_m t + \theta_m^{(k)}\right)\right) * h_{\mathrm{RX}}\left(t - \tau_m^{(k)}\right) \\ &= s_m^{(k)}(t) + j_m^{(k)}(t) + n_m^{(k)}(t) \end{aligned} \tag{2.18}$$

where $s_m^{(k)}(t)$, $j_m^{(k)}(t)$ and $n_m^{(k)}(t)$ are the contributions from UWB signal, narrowband system and background noise, respectively, and these terms are independent from each other. $\Phi_{m,l}^{(k)}$ is equivalent to the effective phase shift due to the random phase introduced at the transmitting end and the channel distortion, i.e. $\Phi_{m,l}^{(k)} = \theta_m^{(k)} + \phi_{m,l}^{(k)}$. The chip-matched filter $h_{\mathrm{RX}}(t)$ should be synchronized with one of the paths for the kth user,

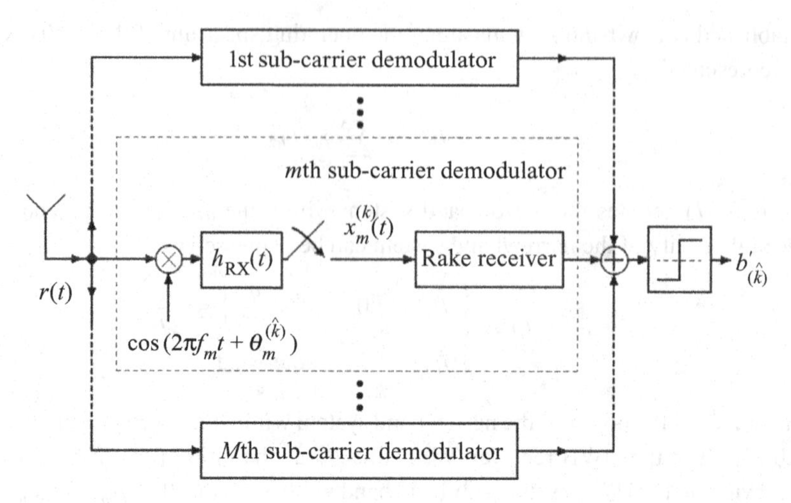

Figure 2.3 Receiver.

i.e. the first arrival path of the desired user. The output of the low-pass filter is sampled at the chip rate, and the discrete-time sample of the recovered baseband signal is then passed to the Rake receiver for data recovery as well as to the estimation circuitry for notch filter coefficient formulation.

2.2.1 Tap weights of the transversal-type notch filter

Referring to the previous discussion of the functional block of the notch filter at the transmitter (illustrated in Figure 2.2), the estimator at the receiver should have a similar structure. If the time reference is set at $t = 0$, $\boldsymbol{X}_m^{(k)}$ is defined as the array of past W received samples and future W received samples from the chip-matched filtered output for the mth sub-carrier such that

$$\boldsymbol{X}_m^{(k)\mathrm{T}} = \left[x_m^{(k)}(-WT_\mathrm{c}) \quad \ldots \quad x_m^{(k)}(-T_\mathrm{c})x_m^{(k)}(T_\mathrm{c}) \quad \ldots \quad x_m^{(k)}(WT_\mathrm{c})\right] \tag{2.19}$$

where $x_m^{(k)}(t)$ is the chip-matched filter output defined in (2.16), and its frequency representation can take the form

$$\begin{aligned} X_m^{(k)}(f) &= F\left\{x_m^{(k)}(t)\right\} \\ &= F\left\{s_m^{(k)}(t) + j_m^{(k)}(t) + n_m^{(k)}(t)\right\} \\ &= S_m^{(k)}(f) + J_m^{(k)}(f) + N_m^{(k)}(f) \end{aligned} \tag{2.20}$$

Then the filter coefficients for the mth sub-carrier can be written as [3]

$$\begin{aligned} \alpha_0^{(m)} &= 1 \\ \boldsymbol{\alpha}^{(m)} &= \left[\alpha_{-W}^{(m)} \ldots \alpha_{-1}^{(m)} \quad \alpha_1^{(m)} \ldots \alpha_W^{(m)}\right] \\ &= \boldsymbol{\rho}_m''^{(k)-1} \boldsymbol{\rho}_m'^{(k)} \end{aligned} \tag{2.21}$$

and are selected to minimize the mean-square output of the estimator at the receiver, so as to provide the desired frequency response to the transmitter filter for suppressing the spectrum of the narrowband system, and the taps are assumed to be sufficiently trained to achieve the optimal values. $\rho'^{(k)}_m$ is a $2W \times 1$ matrix:

$$\rho'^{(k)}_m = E\left\lfloor x^{(k)}_m(0)\boldsymbol{X}^{(k)^{\mathrm{T}}}_m \right\rfloor \tag{2.22}$$

and the uth element of $\boldsymbol{X}_m$ can be written as

$$\begin{aligned}
\rho'^{(k)}_m(u) &= E\left[x^{(k)}_m(0)\boldsymbol{X}^{(k)^{\mathrm{T}}}_m(u)\right] \\
&= E\left[x^{(k)}_m(0)x^{(k)}_m(u)\right] \\
&= E\left[s^{(k)}_m(0)s^{(k)}_m(u)\right] + E\left[j^{(k)}_m(0)j^{(k)}_m(u)\right] + E\left[n^{(k)}_m(0)n^{(k)}_m(u)\right] \\
&= \mathrm{Re}\left(E\left[\int_{-\infty}^{\infty} P^{(m)}_j(f)H_{\mathrm{RC}}(f)\exp(\mathrm{j}2\pi f u T_{\mathrm{c}})\mathrm{d}f\right]\right)
\end{aligned} \tag{2.23}$$

since the UWB signal and the noise signal have zero cross-correlation in time. Similarly $\rho''^{(k)}_m$ is a $2W \times 2W$ matrix:

$$\rho''^{(k)}_m = E\left\lfloor \mathbf{X}^{(k)}_m\mathbf{X}^{(k)^{\mathrm{T}}}_m \right\rfloor \tag{2.24}$$

and the element at the uth row and vth column of $\rho''^{(k)}_m$ can be written as

$$\begin{aligned}
\rho''^{(k)}_m(u,v) &= E\left\lfloor \mathbf{X}^{(k)}_m(u)\mathbf{X}^{(k)^{\mathrm{T}}}_m(v) \right\rfloor \\
&= E\left[x^{(k)}_m(u)x^{(k)}_m(v)\right] \\
&= E\left[s^{(k)}_m(u)s^{(k)}_m(v)\right] + E\left[j^{(k)}_m(u)j^{(k)}_m(v)\right] + E\left[n^{(k)}_m(u)n^{(k)}_m(v)\right]
\end{aligned} \tag{2.25}$$

since the UWB signal, narrowband system signal and background noise are mutually independent. As the UWB signal and the noise signal have non-zero autocorrelation, (2.25) can be expressed as:

(for $u \neq v$)

$$\begin{aligned}
\rho''^{(k)}_m(u,v) &= \mathrm{Re}\left(E\left[\int_{-\infty}^{\infty} P^{(m)}_j(f)\exp(\mathrm{j}2\pi f(u-v)T_{\mathrm{c}})\mathrm{d}f\right]\right) \\
&= \mathrm{Re}\left(E\left[\int_{-\infty}^{\infty} P^{(m)}_j(f)H_{\mathrm{RC}}(f)\exp(\mathrm{j}2\pi f(u-v)T_{\mathrm{c}})\mathrm{d}f\right]\right)
\end{aligned} \tag{2.26}$$

(for $u = v$)

$$\begin{aligned}
\rho''^{(k)}_m(u,v) &= \mathrm{Re}\left(E\left[\int_{-\infty}^{\infty} S^{(k)^2}_m(f)\mathrm{d}f\right]\right) + \mathrm{Re}\left(E\left[\int_{-\infty}^{\infty} P^{(m)}_j(f)\mathrm{d}f\right]\right) \\
&\quad + \mathrm{Re}\left(E\left[\int_{-\infty}^{\infty} N^{(k)^2}_m(f)\mathrm{d}f\right]\right) \\
&= \mathrm{Re}\left(\frac{P}{N}\int_{-\infty}^{\infty}\left(\sum_{k=1}^{K}\sum_{l=1}^{L} E\left[\begin{array}{c}\exp^2\left(\mathrm{j}\left(\Delta\phi^{(k)}_{m,l}\,|^{(\hat{k})}_{m,r} - 2\pi f\Delta\tau^{(k)}_{m,l}\,|^{(\hat{k})}_{m,r}\right)\right) \\ \times\,\Omega^{(k)}_{m,l}H^2_{\mathrm{RC}}(f)\left(H^{(m)}_W(f)\right)^2\end{array}\right]\right)\mathrm{d}f\right)
\end{aligned}$$

$$+\mathrm{Re}\left(\frac{\eta_0}{4}E\left[\int_{-\infty}^{\infty}H_{\mathrm{RC}}(f)\mathrm{d}f\right]\right)$$
$$+\mathrm{Re}\left(E\left[\int_{-\infty}^{\infty}P_j^{(m)}(f)H_{\mathrm{RC}}(f)\mathrm{d}f\right]\right) \tag{2.27}$$

where $\Delta\phi_{m,l}^{(k)}|_{m,r}^{(\hat{k})}$ denotes the phase difference between the paths, which is uniformly distributed over $[0,\pi)$, and $\Delta\tau_{m,l}^{(k)}|_{m,r}^{(\hat{k})}$ denotes the delay difference between the paths, which is uniformly distributed over $[0, T_c)$, given by

$$\Delta\phi_{m,l}^{(k)}|_{m,r}^{(\hat{k})} = \phi_{m,\ell}^{(k)} - \phi_{m,r}^{(\hat{k})} \tag{2.28}$$
$$\Delta\tau_{m,l}^{(k)}|_{m,r}^{(\hat{k})} = \left(\tau_{m,l}^{(k)} - \tau_{m,r}^{(\hat{k})}\right) - \left\lfloor\tau_{m,l}^{(k)} - \tau_{m,r}^{(\hat{k})}\right\rfloor \tag{2.29}$$

2.2.2 Rake structure for data recovery

For each sub-carrier, the R highest $(R \le L)$ power paths out of all resolvable paths are selected for decision making. If we consider the mth sub-carrier of the $\hat{k}$th user, the batch of samples over one bit period to be processed by the $\hat{r}$th correlator can be expressed in array form as

$$\boldsymbol{x}_{m,\hat{r}}^{(\hat{k})} = \beta_{m,\hat{r}}^{(\hat{k})}\sqrt{P}b_{(\hat{k})}F^{-1}\left\{H_W^{(m)}(f)H_{\mathrm{RC}}(f)\right\}\cdot\boldsymbol{c}_{m,\hat{r}}^{(\hat{k})} + \boldsymbol{i}_{m,\hat{r}}^{(\hat{k})} + \boldsymbol{j}_{m,\hat{r}}^{(\hat{k})} + \boldsymbol{n}_{m,\hat{r}}^{(\hat{k})} \tag{2.30}$$

where the emboldened notations are $N \times 1$ arrays:

$$\boldsymbol{c}_{m,\hat{r}}^{(\hat{k})^T} = \left[c_{\hat{k}}^{\left(\tau_{m,\hat{r}}^{(\hat{k})}\right)} \quad \cdots \quad c_{\hat{k}}^{\left(\tau_{m,\hat{r}}^{(\hat{k})}+(N-1)T_c\right)}\right] \tag{2.31}$$
$$\boldsymbol{i}_{m,\hat{r}}^{(\hat{k})^T} = \left[i_{m,\hat{r}}^{(\hat{k})}\left(\tau_{m,\hat{r}}^{(\hat{k})}\right) \quad \cdots \quad i_{m,\hat{r}}^{(\hat{k})}\left(\tau_{m,\hat{r}}^{(\hat{k})}+(N-1)T_c\right)\right] \tag{2.32}$$
$$\boldsymbol{j}_{m,\hat{r}}^{(\hat{k})^T} = \left[j_m^{(\hat{k})}\left(\tau_{m,\hat{r}}^{(\hat{k})}\right) \quad \cdots \quad j_m^{(\hat{k})}\left(\tau_{m,\hat{r}}^{(\hat{k})}+(N-1)T_c\right)\right] \tag{2.33}$$
$$\boldsymbol{n}_{m,\hat{r}}^{(\hat{k})^T} = \left[n_m^{(\hat{k})}\left(\tau_{m,\hat{r}}^{(\hat{k})}\right) \quad \cdots \quad n_m^{(\hat{k})}\left(\tau_{m,\hat{r}}^{(\hat{k})}+(N-1)T_c\right)\right] \tag{2.34}$$

which denote the spreading code of the desired user, multipath interference (MPI) plus multiple access interference (MAI), narrowband interference (NBI) and channel noise (AWGN), respectively. $i_{m,\hat{r}}^{(\hat{k})}(t)$ is the sum of MPI and MAI, expressed as

$$i_{m,\hat{r}}^{(\hat{k})}(t) = s_m^{(\hat{k})}(t) - \sqrt{2P}\beta_{m,\hat{r}}^{(\hat{k})}s_m^{(k)}\left(t-\tau_{m,\hat{r}}^{(\hat{k})}\right)\cos\left(\Phi_{m,\hat{r}}^{(\hat{k})}\right) * h_{\mathrm{RX}}\left(t-\tau_m^{(\hat{k})}\right)$$
$$= \sum_{\substack{l=1\\ l\neq\hat{r}}}^{L}\sqrt{2P}\beta_{m,l}^{(\hat{k})}s_m^{(k)}\left(t-\tau_{m,l}^{(\hat{k})}\right)\cos\left(\Phi_{m,l}^{(\hat{k})}\right) * h_{\mathrm{RX}}\left(t-\tau_m^{(\hat{k})}\right) \tag{2.35}$$
$$+ \sum_{\substack{k=1\\ k\neq\hat{k}}}^{K}\sum_{l=1}^{L}\sqrt{2P}\beta_{m,l}^{(k)}s_m^{(k)}\left(t-\tau_{l,m}^{(k)}\right)\cos\left(\Phi_{m,l}^{(\hat{k})}\right) * h_{\mathrm{RX}}\left(t-\tau_m^{(\hat{k})}\right)$$

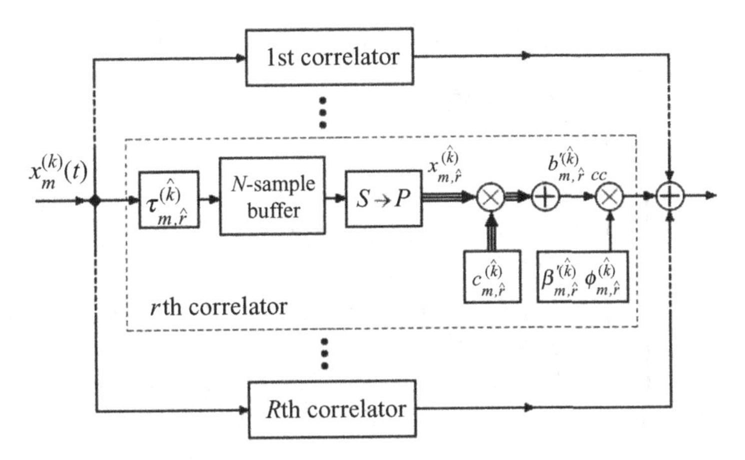

Figure 2.4 CC receiver.

In this section, a comparison of the performance of the receiver of the code correlator, the receiver of the notch filter in cascade with code correlator (NCC) and the receiver of the minimum mean square error (MMSE) adaptive correlator is provided, and these receiving strategies are presented in the following sub-section.

Code correlator (CC)

The conventional method for data recovery in a DS-CDMA system is correlating the received signal with the synchronized spreading code, as illustrated in Figure 2.4. The decision variable of the $\hat{r}$th finger on the mth sub-carrier of the $\hat{k}$th user is given by

$$\hat{b}_{m,\hat{r}\mathrm{CC}}^{(\hat{k})} = \boldsymbol{c}_{m,\hat{r}}^{(\hat{k})\mathrm{T}} \boldsymbol{x}_{m,\hat{r}}^{(\hat{k})} \tag{2.36}$$

and its mean and variance can be expressed as

$$E[b'^{(\hat{k})}_{m,\hat{r}}]_{\mathrm{CC}} = \sqrt{P}\beta_{m,\hat{r}}^{(\hat{k})} E\left(\int_{-\infty}^{\infty} H_W^{(m)}(f) H_{\mathrm{RC}}(f)\mathrm{d}f\right) \tag{2.37}$$

$$\mathrm{Var}[b'^{(\hat{k})}_{m,\hat{r}}]_{\mathrm{CC}} = E[\boldsymbol{c}_{m,\hat{r}}^{(\hat{k})}(\boldsymbol{I}_{m,\hat{r}}^{(\hat{k})} + \boldsymbol{J}_{m,\hat{r}}^{(\hat{k})} + \boldsymbol{N}_{m,\hat{r}}^{(\hat{k})})\boldsymbol{c}_{m,\hat{r}}^{(\hat{k})}] \tag{2.38}$$

where $\boldsymbol{I}_{m,\hat{r}}^{(\hat{k})}$, $\boldsymbol{J}_{m,\hat{r}}^{(\hat{k})}$ and $\boldsymbol{N}_{m,\hat{r}}^{(\hat{k})}$ are independent $N \times N$ matrices due to the autocorrelation function of the signal corresponding to the multipath plus multiple access interference, narrowband interference and channel noise, respectively. If the frequency domain representation of the sum of MPI and MAI is given by $I_{m,\hat{r}}^{(\hat{k})}(f)$, the element at the uth row and vth column of matrix $\boldsymbol{I}_{m,\hat{r}}^{(\hat{k})}$ can be expressed as

$$\begin{aligned}\boldsymbol{I}_{m,\hat{r}}^{(\hat{k})}(u,v) &= E\lfloor \boldsymbol{i}_{m,\hat{r}}^{(\hat{k})}(u)^* \boldsymbol{i}_{m,\hat{r}}^{(\hat{k})}(v)\rfloor \\ &= E[i_{m,\hat{r}}^{(\hat{k})}(\tau_{m,\hat{r}}^{(\hat{k})} + (u-1)T_{\mathrm{c}})^* i_{m,\hat{r}}^{(\hat{k})}(\tau_{m,\hat{r}}^{(\hat{k})} + (v-1)T_{\mathrm{c}})]\end{aligned} \tag{2.39}$$

With random code assumption, for $u \neq v$, $\boldsymbol{I}_{m,\hat{r}}^{(\hat{k})}(u, v) = 0$, and for $u = v$:

$$\boldsymbol{I}_{m,\hat{r}}^{(\hat{k})}(u, v) = \mathrm{Re}\left(E\left[\int_{-\infty}^{\infty} \left(I_{m,\hat{r}}^{(\hat{k})}(f)\right)^2 \mathrm{d}f\right]\right)$$

$$= \mathrm{Re}\left(E\left[\begin{array}{l}\displaystyle\int_{-\infty}^{\infty}\left(\frac{P}{N}\sum_{\substack{l=1\\l\neq r}}^{L}\exp^2\left(\mathrm{j}\Delta\phi_{m,l}^{(\hat{k})}\,|_{m,r}^{(\hat{k})}\right)\Omega_{m,l}^{(\hat{k})}H_{RC}^2(f)\left(H_W^{(m)}(f)\right)^2\right)\mathrm{d}f \\ \displaystyle +\int_{-\infty}^{\infty}\left(\begin{array}{l}\displaystyle\frac{P}{N}\sum_{\substack{k=1\\k\neq\hat{k}}}^{K}\sum_{l=1}^{L}\left[\exp^2\left(j\left(\Delta\phi_{m,l}^{(k)}\,|_{m,r}^{(\hat{k})}-2\pi f\Delta\tau_{m,l}^{(k)}\,|_{m,r}^{(\hat{k})}\right)\right)\right. \\ \qquad\left.\times\,\Omega_{m,l}^{(k)}H_{\mathrm{RC}}^2(f)\left(H_W^{(m)}(f)\right)^2\right]\end{array}\right)\mathrm{d}f\end{array}\right]\right) \tag{2.40}$$

The element at the uth row and vth column of matrices $\boldsymbol{J}_{m,\hat{r}}^{(\hat{k})}$ and $\boldsymbol{N}_{m,\hat{r}}^{(\hat{k})}$ can be expressed as

$$\begin{aligned}\boldsymbol{J}_{m,\hat{r}}^{(\hat{k})}(u, v) &= E\left[\boldsymbol{j}_{m,\hat{r}}^{(\hat{k})}(u)^*\boldsymbol{j}_{m,\hat{r}}^{(\hat{k})}(v)\right] \\ &= E\left[j_m^{(\hat{k})}\left(\tau_{m,\hat{r}}^{(\hat{k})}+(u-1)T_{\mathrm{c}}\right)^* j_m^{(\hat{k})}\left(\tau_{m,\hat{r}}^{(\hat{k})}+(v-1)T_{\mathrm{c}}\right)\right] \\ &= \mathrm{Re}\left(E\left[\int_{-\infty}^{\infty} P_j^{(m)}(f)H_{\mathrm{RC}}(f)\exp(\mathrm{j}2\pi f(u-v)T_{\mathrm{c}})\mathrm{d}f\right]\right)\end{aligned} \tag{2.41}$$

$$\begin{aligned}\boldsymbol{N}_{m,\hat{r}}^{(\hat{k})}(u, v) &= E\left[\boldsymbol{n}_{m,\hat{r}}^{(\hat{k})}(u)^*\boldsymbol{n}_{m,\hat{r}}^{(\hat{k})}(v)\right] \\ &= E\left[n_m^{(\hat{k})}\left(\tau_{m,\hat{r}}^{(\hat{k})}+(u-1)T_{\mathrm{c}}\right) * n_m^{(\hat{k})}\left(\tau_{m,\hat{r}}^{(\hat{k})}+(v-1)T_{\mathrm{c}}\right)\right] \\ &= \begin{cases}\mathrm{Re}\left(E\left[\dfrac{\eta_o}{2}\displaystyle\int_{-\infty}^{\infty} H_{\mathrm{RC}}(f)\mathrm{d}f\right]\right) & \text{for} \quad u = v \\ 0 & \text{for} \quad u \neq v\end{cases}\end{aligned} \tag{2.42}$$

After weighting by the estimated attenuation factor, the signal-to-interference-plus-noise ratio (SINR) of the symbolic decision for the $\hat{r}$th finger on the mth sub-carrier of the $\hat{k}$th user is given by

$$\mathrm{SINR}_{\mathrm{CC}}\left(b'^{(\hat{k})}_{m,\hat{r}}\right) = \frac{E^2\left[b'^{(\hat{k})}_{m,\hat{r}}\right]_{\mathrm{CC}}}{\mathrm{Var}\left[b'^{(\hat{k})}_{m,\hat{r}}\right]_{\mathrm{CC}}} = \left(\beta_{m,\hat{r}}^{(\hat{k})}\right)^2\gamma_{m,\hat{r}\mathrm{CC}}^{(\hat{k})} \tag{2.43}$$

where

$$\begin{aligned}\gamma_{m,\hat{r}\mathrm{CC}}^{(\hat{k})} &= \frac{P\left[E\left(\int_{-\infty}^{\infty}\left(H_W^{(m)}(f)\right)H_{\mathrm{RC}}(f)\mathrm{d}f\right)\right]^2}{E\left[\mathbf{c}_{m,\hat{r}}^{(\hat{k})}\left(\mathbf{I}_{m,\hat{r}}^{(\hat{k})}+\mathbf{J}_{m,\hat{r}}^{(\hat{k})}+\mathbf{N}_{m,\hat{r}}^{(\hat{k})}\right)\mathbf{c}_{m,\hat{r}}^{(\hat{k})\mathrm{T}}\right]} \\ &= \frac{P\left[E\left(\int_{-\infty}^{\infty}\left(H_W^{(m)}(f)\right)H_{\mathrm{RC}}(f)\mathrm{d}f\right)\right]^2}{\sum\mathrm{diag}\left(E\left[\left(\mathbf{I}_{m,\hat{r}}^{(\hat{k})}+\mathbf{J}_{m,\hat{r}}{}^{(\hat{k})}+\mathbf{N}_{m,\hat{r}}{}^{(\hat{k})}\right)\right]\right)}\end{aligned} \tag{2.44}$$

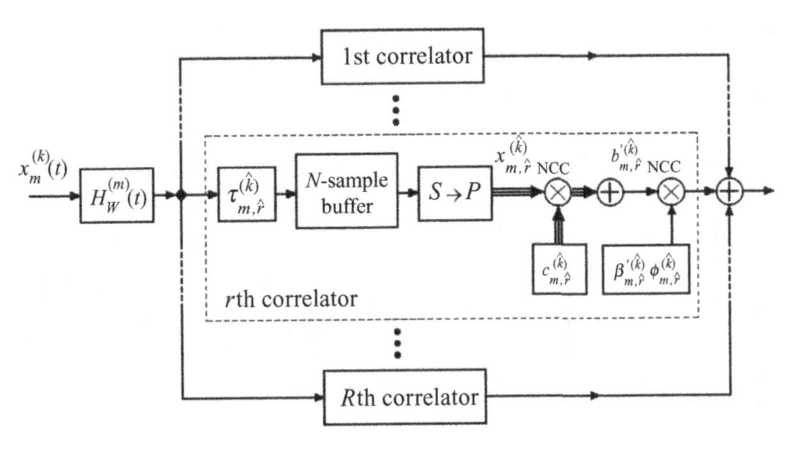

Figure 2.5 NCC receiver.

as under random code assumption $E\left[\boldsymbol{c}_{m,\hat{r}}^{(\hat{k})}(u)\boldsymbol{c}_{m,\hat{r}}^{(\hat{k})}(v)\right] = 1$ for $u = v$, and $E\left[\boldsymbol{c}_{m,\hat{r}}^{(\hat{k})}(u)\boldsymbol{c}_{m,\hat{r}}^{(\hat{k})}(v)\right] = 0$ otherwise. In (2.44) $\sum \text{diag}(\boldsymbol{x})$ stands for the sum of the diagonal elements of matrix $\boldsymbol{x}$.

Notch filter in cascade with code correlator (NCC)

The use of a notch filter at the receiver for narrowband interference suppression in addition to the processing gain of the CDMA system has been described in [3], such that the transversal notch filter is inserted between the chip-matched filter and the Rake structure, as shown in Figure 2.5. The notch filter response at the receiver should be the same as that at the transmitter since their taps are adapted using the same information. The output from the receiver notch filter is

$$x_m^{(\hat{k})}{}_{\text{NCC}}(t) = F^{-1}\{H_W^{(m)}(f)X_m^{(\hat{k})}(f)\} \tag{2.45}$$

where $X_m^{(\hat{k})}(f)$ is the frequency domain representation of (2.18):

$$x_m^{(\hat{k})}(t) = F^{-1}\{X_m^{(\hat{k})}(f)\} \tag{2.46}$$

The signal is then processed by the code-matched Rake receiver, i.e.

$$\boldsymbol{b}'^{(\hat{k})}_{m,\hat{r}_{\text{NCC}}} = \boldsymbol{c}_{m,\hat{r}}^{(\hat{k})^{\text{T}}}\boldsymbol{x}_{m,\hat{r}_{\text{NCC}}}^{(\hat{k})} \tag{2.47}$$

and its mean and variance can be expressed as

$$E\left[b'^{(\hat{k})}_{m,\hat{r}}\right]_{\text{NCC}} = \sqrt{\text{PE}}\left(\int_{-\infty}^{\infty} \left(H_W^{(m)}(f)\right)^2 H_{\text{RC}}(f)\text{d}f\right) \tag{2.48}$$

$$\text{Var}\left[b'^{(\hat{k})}_{m,\hat{r}}\right]_{\text{NCC}} = E\left[\boldsymbol{c}_{m,\hat{r}}^{(\hat{k})}\left(\boldsymbol{I}_{m,\hat{r}_{\text{NCC}}}^{(\hat{k})} + \boldsymbol{J}_{m,\hat{r}_{\text{NCC}}}^{(\hat{k})} + \boldsymbol{N}_{m,\hat{r}_{\text{NCC}}}^{(\hat{k})}\right)\boldsymbol{c}_{m,\hat{r}}^{(\hat{k})}\right] \tag{2.49}$$

where the emboldened terms $\boldsymbol{I}_{m,\hat{r}_{\text{NCC}}}^{(\hat{k})}$, $\boldsymbol{J}_{m,\hat{r}_{\text{NCC}}}^{(\hat{k})}$ and $\boldsymbol{N}_{m,\hat{r}_{\text{NCC}}}^{(\hat{k})}$, which correspond to MPI/MAI, NBI and AWGN (analogous to those of the CC receiver), can be written as

$$
\mathbf{I}^{(\hat{k})}_{m,\hat{r}\mathrm{NCC}}(u,v) \tag{2.50}
$$

$$
= \begin{cases} \mathrm{Re}\left(E\left[\begin{array}{l} \frac{P}{N}\int\limits_{-\infty}^{\infty}\sum\limits_{\substack{l=1\\ l\neq r}}^{L}\exp^2\left(j\Delta\phi^{(\hat{k})}_{m,l}\mathrm{l}^{(\hat{k})}_{m,r}\right)\Omega^{(\hat{k})}_{m,l}H^2_{\mathrm{RC}}(f)\left(H^{(m)}_W(f)\right)^4\mathrm{d}f \\ +\frac{P}{N}\int\limits_{-\infty}^{\infty}\left(\begin{array}{l}\sum\limits_{\substack{k=1\\ k\neq\hat{k}}}^{K}\sum\limits_{l=1}^{L}\left[\exp^2\left(j\left(\Delta\phi^{(k)}_{m,l}\mathrm{l}^{(\hat{k})}_{m,r}-2\pi f\Delta\tau^{(k)}_{m,l}\mathrm{l}^{(\hat{k})}_{m,r}\right)\right)\right. \\ \left.\times\,\Omega^{(k)}_{m,l}H^2_{\mathrm{RC}}(f)\left(H^{(m)}_W(f)\right)^4\right]\end{array}\right)\mathrm{d}f \end{array}\right]\right), & \text{for} \quad u=v \\ 0, & \text{for} \quad u\neq v \end{cases}
$$

$$
\boldsymbol{J}^{(\hat{k})}_{m,\hat{r}\mathrm{NCC}}(u,v)=\mathrm{Re}\left(E\left[\int\limits_{-\infty}^{\infty}\left(H^{(m)}_W(f)\right)^2P^{(m)}_j(f)H_{\mathrm{RC}}(f)\exp(j2\pi f(u-v)T_c)\mathrm{d}f\right]\right) \tag{2.51}
$$

$$
\mathbf{N}^{(\hat{k})}_{m,\hat{r}\mathrm{NCC}}(u,v)=\begin{cases}\mathrm{Re}\left(E\left[\frac{\eta_o}{2}\int\limits_{-\infty}^{\infty}\left(H^{(m)}_W(f)\right)^2H_{\mathrm{RC}}(f)\mathrm{d}f\right]\right), & \text{for} \quad u=v \\ 0, & \text{for} \quad u\neq v\end{cases} \tag{2.52}
$$

Similarly, the SINR of the symbolic decision for the $\hat{r}$th finger on the mth sub-carrier of the $\hat{k}$th user after weighting by the estimated attenuation factor takes the form

$$
\mathrm{SINR}_{\mathrm{NCC}}\left(b'^{(\hat{k})}_{m,\hat{r}}\right)=\frac{E^2\left[b'^{(\hat{k})}_{m,\hat{r}}\right]}{\mathrm{Var}\left[b'^{(\hat{k})}_{m,\hat{r}}\right]}=\left(\beta^{(\hat{k})}_{m,\hat{r}}\right)^2\gamma^{(\hat{k})}_{m,\hat{r}\mathrm{NCC}} \tag{2.53}
$$

where

$$
\gamma^{(\hat{k})}_{m,\hat{r}\mathrm{NCC}}=\frac{P\left[E\left(\int\limits_{-\infty}^{\infty}\left(H^{(m)}_W(f)\right)^2H_{\mathrm{RC}}(f)\mathrm{d}f\right)\right]^2}{E\left[\mathbf{c}^{(\hat{k})}_{m,\hat{r}}\left(\boldsymbol{I}^{(\hat{k})}_{m,\hat{r}\mathrm{NCC}}+\boldsymbol{J}^{(\hat{k})}_{m,\hat{r}\mathrm{NCC}}+\boldsymbol{N}^{(\hat{k})}_{m,\hat{r}\mathrm{NCC}}\right)\boldsymbol{c}^{(\hat{k})\mathrm{T}}_{m,\hat{r}}\right]}
=\frac{P\left[E\left(\int\limits_{-\infty}^{\infty}\left(H^{(m)}_W(f)\right)^2H_{\mathrm{RC}}(f)\mathrm{d}f\right)\right]^2}{\sum\mathrm{diag}\left(E\left[\left(\boldsymbol{I}^{(\hat{k})}_{m,\hat{r}\mathrm{NCC}}+\boldsymbol{J}^{(\hat{k})}_{m,\hat{r}\mathrm{NCC}}+\boldsymbol{N}^{(\hat{k})}_{m,\hat{r}\mathrm{NCC}}\right)\right]\right)} \tag{2.54}
$$

MMSE adaptive correlator

According to [8], [9], [11], adaptive Rake filtering allows joint exploration of the path diversity in casual CDMA systems as well as the suppression of interference by single-user linear MMSE filtering, and the pre-combining approach allows rapid adaptation to the channel condition. The structure of the MMSE adaptive correlator shown in Figure 2.6 is similar to that of the code correlator except the synchronized spreading code is replaced by tap weight. The decision variable of the $\hat{r}$th finger on the mth

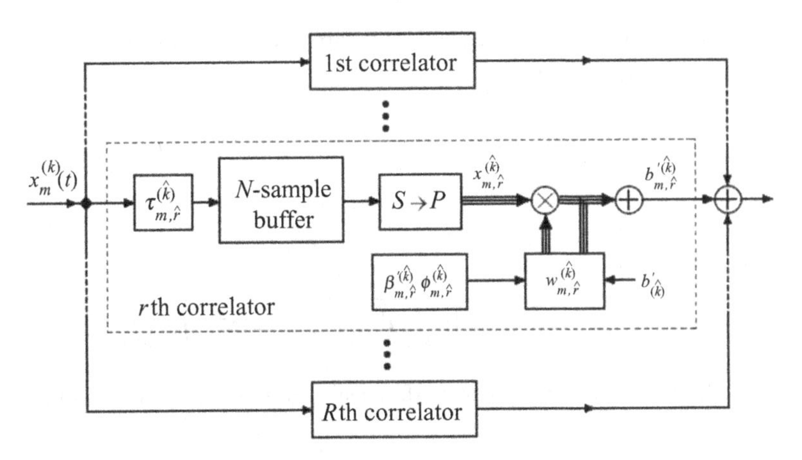

Figure 2.6 MMSE receiver.

sub-carrier of the $\hat{k}$th user is

$$b'^{(\hat{k})}_{m,\hat{r}_{\mathrm{MMSE}}} = \boldsymbol{w}^{(\hat{k})\mathrm{T}}_{m,\hat{r}}\boldsymbol{x}^{(\hat{k})}_{m,\hat{r}} \tag{2.55}$$

where $\boldsymbol{w}^{(\hat{k})}_{m,\hat{r}}$ represents the $N \times 1$ array of tap weights of the MMSE filter given by

$$\boldsymbol{w}^{(\hat{k})}_{m,\hat{r}} = \left[\, w^{(\hat{k})}_{m,\hat{r}}(1) \quad \ldots \quad w^{(\hat{k})}_{m,\hat{r}}(N) \,\right] \tag{2.56}$$

The tap weights are determined by minimizing the mean-squared error between the correlator output with the product of the estimated attenuation factor $\beta'^{(\hat{k})}_{m,\hat{r}}$, the power and the feedback estimated data symbol $b'_{(\hat{k})}$, which is equivalent to [12], [13]

$$\mathrm{MSE} = E\left[\left|\boldsymbol{w}^{(\hat{k})T}_{m,\hat{r}}\boldsymbol{x}^{(\hat{k})}_{m,\hat{r}} - \beta'^{(\hat{k})}_{m,\hat{r}}\sqrt{P}b'_{(\hat{k})}\right|^2\right] \tag{2.57}$$

The feedback of the estimated data symbol, $b'_{(\hat{k})}$, determined by the equal gain combination of the outputs from all R branches of the adaptive Rake from all M sub-carriers, can be written as

$$b'_{(\hat{k})} = \sum_{m=1}^{M}\sum_{r=1}^{R} b'^{(\hat{k})}_{m,r} \tag{2.58}$$

since the outputs of the correlators have been weighted by their own channel attenuation factor as shown in the following. To minimize (2.57), by Wiener's solution, the tap weight array can be expressed as

$$\boldsymbol{w}^{(\hat{k})}_{m,\hat{r}} = \beta'^{(\hat{k})}_{m,\hat{r}}\sqrt{P}\left(\boldsymbol{x}^{(\hat{k})}_{m,\hat{r}}\boldsymbol{x}^{(\hat{k})\mathrm{T}}_{m,\hat{r}}\right)^{-1}\boldsymbol{c}^{(\hat{k})}_{m,\hat{r}} \tag{2.59}$$

where the channel estimation with the feedback estimated data symbol provides the steering force to adapt the spreading code. Perfect channel estimation is assumed, i.e. $\beta'^{(\hat{k})}_{m,\hat{r}} = \beta^{(\hat{k})}_{m,\hat{r}}$. The decision statistic consists of the contribution from the desired signal, MPI, MAI of the UWB system, narrowband interference and noise, while all the terms are assumed to be independent from each other. The mean and variance of the decision variable of the $\hat{r}$th finger on the mth sub-carrier of the $\hat{k}$th user, conditioned on the attenuation factor, are given by [10]

$$E\left[b'^{(\hat{k})}_{m,\hat{r}}\right] = \sqrt{P}\beta^{(\hat{k})}_{m,\hat{r}} E\left(\int_{-\infty}^{\infty} H_W^{(m)}(f) H_{\mathrm{RC}}(f)\mathrm{d}f\right) \tag{2.60}$$

$$\begin{aligned}\mathrm{Var}\left[b'^{(\hat{k})}_{m,\hat{r}}\right] &= E\left[\left|\boldsymbol{w}^{(\hat{k})\mathrm{T}}_{m,\hat{r}}\boldsymbol{x}^{(\hat{k})}_{m,\hat{r}} - \beta'^{(\hat{k})}_{m,\hat{r}}\sqrt{P}b'_{(\hat{k})}\right|^2\right] \\ &= \frac{1}{E\left[\boldsymbol{c}^{(\hat{k})}_{m,\hat{r}}\left(\boldsymbol{I}^{(\hat{k})}_{m,\hat{r}} + \boldsymbol{J}^{(\hat{k})}_{m,\hat{r}} + \boldsymbol{N}^{(\hat{k})}_{m,\hat{r}}\right)^{-1}\boldsymbol{c}^{(\hat{k})\mathrm{T}}_{m,\hat{r}}\right]}\end{aligned} \tag{2.61}$$

As a result, the average signal-to-interference-plus-noise ratio (SINR) of $b'^{(\hat{k})}_{m,\hat{r}}$ is

$$\begin{aligned}\mathrm{SINR}_{\mathrm{MMSE}}\left(b'^{(\hat{k})}_{m,\hat{r}}\right) &= \frac{E^2\left[b'^{(\hat{k})}_{m,\hat{r}}\right]}{\mathrm{Var}\left[b'^{(\hat{k})}_{m,\hat{r}}\right]} \\ &= \left(\beta^{(\hat{k})}_{m,\hat{r}}\right)^2 \gamma^{(\hat{k})}_{m,\hat{r}\mathrm{MMSE}}\end{aligned} \tag{2.62}$$

where

$$\begin{aligned}\gamma^{(\hat{k})}_{m,\hat{r}_{\mathrm{MMSE}}} &= P\left(\int_{-\infty}^{\infty} H_W^{(m)}(f)H_{\mathrm{RC}}(f)\mathrm{d}f\right)^2 E\left[\boldsymbol{c}^{(\hat{k})}_{m,\hat{r}}\left(\boldsymbol{I}^{(\hat{k})}_{m,\hat{r}} + \boldsymbol{J}^{(\hat{k})}_{m,\hat{r}} + \boldsymbol{N}^{(\hat{k})}_{m,\hat{r}}\right)^{-1}\boldsymbol{c}^{(\hat{k})\mathrm{T}}_{m,\hat{r}}\right] \\ &= P\left(\int_{-\infty}^{\infty} H_W^{(m)}(f)H_{\mathrm{RC}}(f)\mathrm{d}f\right)^2 \sum \mathrm{diag}\left(E\left[\left(\boldsymbol{I}^{(\hat{k})}_{m,\hat{r}} + \boldsymbol{J}^{(\hat{k})}_{m,\hat{r}} + \boldsymbol{N}^{(\hat{k})}_{m,\hat{r}}\right)^{-1}\right]\right)\end{aligned} \tag{2.63}$$

2.3 Probability of error

The path-selective approach is adopted so that only the contributions from significant paths are considered. A total of $M \cdot R$ contributions are added for decision making, and they can be added together to provide the overall symbolic decision. Since for the receiver of an MMSE correlator, with the notch filter in cascade with a code-matched correlator and a causal code-matched correlator, the average SINR of the decision statistic is separable in terms of $\beta^{(k)}_{m,r}$ and an independent component $\gamma^{(k)}_{m,r}$, and the conditional average bit error probability can be written as

$$\begin{aligned}P_{\mathrm{e}} &= \int_0^{\infty} Q(\sqrt{\zeta_{\hat{k}}})p(\zeta_{\hat{k}})\mathrm{d}\zeta_{\hat{k}} \\ &= \underbrace{\iint\cdots\int}_{M\cdot R\ \text{folds}} Q\left(\sqrt{\sum_{m=1}^{M}\sum_{r=1}^{R}\left(\beta^{(k)}_{m,r}\right)^2\gamma^{(k)}_{m,r}}\right) p\left(\beta^{(k)}_{1,1},\beta^{(k)}_{1,2},\ldots,\beta^{(k)}_{M,R}\right) \\ &\quad\times \mathrm{d}\beta^{(k)}_{1,1}\mathrm{d}\beta^{(k)}_{1,2}\ldots d\beta^{(k)}_{M,R}\end{aligned} \tag{2.64}$$

where $Q(x)$ is the Q-function and $p(\beta_{1,1}^{(k)}, \beta_{1,2}^{(k)}, \ldots, \beta_{M,R}^{(k)})$ is the joint probability density function of the attenuation factors of the selected multipaths among the sub-carriers. As the path selection is independent among the sub-carriers, $p(\beta_{1,1}^{(k)}, \beta_{1,2}^{(k)}, \ldots, \beta_{M,R}^{(k)})$ can be expressed as

$$p(\beta_{1,1}^{(k)}, \beta_{1,2}^{(k)}, \ldots, \beta_{M,R}^{(k)}) = \prod_{m=1}^{M} p(\beta_{m,1}^{(k)}, \beta_{m,2}^{(k)}, \ldots, \beta_{m,R}^{(k)}) \tag{2.65}$$

which is the product of M joint probability density functions corresponding to each sub-carrier, and the joint probability density function for the mth sub-carrier $p(\beta_{m,1}^{(k)}, \beta_{m,2}^{(k)}, \ldots, \beta_{m,R}^{(k)})$ is [13]

$$p(\beta_{m,1}^{(k)}, \beta_{m,2}^{(k)}, \ldots, \beta_{m,R}^{(k)}) = \frac{L!}{(L-R)!}\left[P(\beta_{m,R}^{(k)})\right]^{L-R} \prod_{l=1}^{R} p(\beta_{m,l}^{(k)}) \tag{2.66}$$

with the assumption that $\beta_{m,1}^{(k)} \geq \beta_{m,2}^{(k)} \geq \ldots \geq \beta_{m,R}^{(k)}$ and $P(\beta_{m,R}^{(k)})$ is the probability distribution function of the Nakagami variable. The derivative of the probability distribution function is given by [14]

$$\begin{aligned} P(\beta_{m,R}^{(k)}) &= \int_0^x p(\beta_{m,R}^{(k)})\mathrm{d}x \\ &= \frac{\gamma\left(\mu_{m,R}^{(k)}, \frac{\mu_{m,R}^{(k)}}{\Omega_{m,R}^{(k)}}(\beta_{m,R}^{(k)})^2\right)}{\Gamma(\mu_{m,R}^{(k)})} \end{aligned} \tag{2.67}$$

where $\gamma(a, b)$ and $\Gamma(a)$ are the lower incomplete gamma function and complete gamma function, respectively:

$$\gamma(a, b) = \int_0^b t^{a-1}\mathrm{e}^{-t}\mathrm{d}t \tag{2.68}$$

$$\Gamma(a) = \int_0^\infty x^{a-1}\mathrm{e}^{-x}\mathrm{d}x \tag{2.69}$$

The $M \cdot R$-fold integration of (2.64) is a complicated computational process. For the sake of simplicity, it is assumed that the decision statistics from all the correlators of the same sub-carrier for the same user share the common $\gamma_m^{(k)}$. Since MPI is at most equivalent to MAI from an additional user for $K \gg 1$, the common $\gamma_m^{(k)}$ is the simplified version of $\gamma_{m,r}^{(k)}$ formulated in the previous sub-section. For the CC receiver and MMSE receiver [1], [15]:

$$\gamma_m^{(k)}{}_{\mathrm{CC}} = \frac{P\left[E\left(\int_{-\infty}^{\infty} (H_W^{(m)}(f))H_{\mathrm{RC}}(f)\mathrm{d}f\right)\right]^2}{\sum \mathrm{diag}\left(E\left[(\boldsymbol{I}_m^{(k)} + \boldsymbol{J}_m^{(k)} + \boldsymbol{N}_m^{(k)})\right]\right)} \tag{2.70}$$

$$\gamma_m^{(k)}{}_{\mathrm{MMSE}} = P\left[E\left(\int_{-\infty}^{\infty} H_W^{(m)}(f)H_{\mathrm{RC}}(f)\mathrm{d}f\right)\right]^2 \sum \mathrm{diag}(E[(\boldsymbol{I}_m^{(k)} + \boldsymbol{J}_m^{(k)} + \boldsymbol{N}_m^{(k)})^{-1}]) \tag{2.71}$$

where $\boldsymbol{J}_m^{(\hat{k})} = \boldsymbol{J}_{m,\hat{r}}^{(\hat{k})}$ and $\boldsymbol{N}_m^{(\hat{k})} = \boldsymbol{N}_{m,\hat{r}}^{(\hat{k})}$, while $\boldsymbol{I}_m^{(\hat{k})}$ is a simplified form of $\boldsymbol{I}_{m,\hat{r}}^{(\hat{k})}$ and its element at the uth row and vth column is given by

$$\boldsymbol{I}_m^{(\hat{k})}(u,v) = \begin{cases} \operatorname{Re}\left(E\left[\int_{-\infty}^{\infty}\left(\frac{P}{N}\sum_{k=1}^{K}\sum_{l=1}^{L}\left[\exp^2\left(\mathrm{j}\left(\Delta\phi_{m,l}^{(k)}\,|_{m,r}^{(\hat{k})} - 2\pi f\Delta\tau_{m,l}^{(k)}\,|_{m,r}^{(\hat{k})}\right)\right) \times \Omega_{m,l}^{(k)} H_{\mathrm{RC}}^2(f)\left(H_W^{(m)}(f)\right)^2\right]\right)\mathrm{d}f\right]\right), & \text{for } u = v \\ 0, & \text{for } u \neq v \end{cases} \tag{2.72}$$

For the NCC receiver

$$\gamma_m^{(k)\mathrm{NCC}} = \frac{P\left[E\left(\int_{-\infty}^{\infty}\left(H_W^{(m)}(f)\right)^2 H_{\mathrm{RC}}(f)\mathrm{d}f\right)\right]^2}{\sum \operatorname{diag}\left(E\left[\left(\boldsymbol{I}_m^{(k)}{}_{\mathrm{NCC}} + \boldsymbol{J}_m^{(k)}{}_{\mathrm{NCC}} + \boldsymbol{N}_m^{(k)}{}_{\mathrm{NCC}}\right)\right]\right)} \tag{2.73}$$

Similarly, $\boldsymbol{J}_{m\mathrm{NCC}}^{(\hat{k})} = \boldsymbol{J}_{m,\hat{r}\mathrm{NCC}}^{(\hat{k})}$ and $\boldsymbol{N}_{m\mathrm{NCC}}^{(\hat{k})} = \boldsymbol{N}_{m,\hat{r}\mathrm{NCC}}^{(\hat{k})}$, and for $\boldsymbol{I}_{\hat{r}\mathrm{NCC}}^{(\hat{k})}$ (the simplified form of $\boldsymbol{I}_{m,\hat{r}\mathrm{NCC}}^{(\hat{k})}$), its element at the uth row and vth column is given by

$$\mathbf{I}_m^{(\hat{k})}{}_{\mathrm{NCC}}(u,v) = \begin{cases} \operatorname{Re}\left(E\left[\frac{P}{N}\int_{-\infty}^{\infty}\left(\sum_{k=1}^{K}\sum_{l=1}^{L}\left[\exp^2\left(\mathrm{j}\left(\Delta\phi_{m,l}^{(k)}\Big|_{m,r}^{(\hat{k})} - 2\pi f\Delta\tau_{m,l}^{(k)}\Big|_{m,r}^{(\hat{k})}\right)\right) \times \Omega_{m,l}^{(k)} H_{\mathrm{RC}}^2(f)\left(H_W^{(m)}(f)\right)^4\right]\right)\mathrm{d}f\right]\right), & \text{for } u = v \\ 0, & \text{for } u \neq v \end{cases} \tag{2.74}$$

Therefore, inserting (2.65) into (2.64) and replacing $\gamma_{m,r}^{(k)}$ by $\gamma_m^{(k)}$, the average bit error probability can be rewritten as

$$P_{\mathrm{e}} = \underbrace{\int_1\int_2\cdots\int_3}_{M\cdot R\text{ folds}} Q\left(\sqrt{\sum_{m=1}^{M}\sum_{r=1}^{R}\left(\beta_{m,r}^{(k)}\right)^2\gamma_m^{(k)}}\right)\prod_{m=1}^{M} \times p\left(\beta_{m,1}^{(k)},\beta_{m,2}^{(k)},\ldots,\beta_{m,R}^{(k)}\right)\mathrm{d}\beta_{1,1}^{(k)}\mathrm{d}\beta_{1,2}^{(k)}\cdots\mathrm{d}\beta_{M,R}^{(k)} \tag{2.75}$$

From [16], the Q-function $Q(x)$ can be expressed as

$$Q(x) = \frac{1}{\pi}\int_0^{\frac{\pi}{2}}\exp\left(-\frac{x^2}{2\sin^2\Theta}\right)\mathrm{d}\Theta \tag{2.76}$$

Substituting $x = \sqrt{\sum_{m=1}^{M}\sum_{l=1}^{R}\left(\beta_{m,r}^{(k)}\right)^2\gamma_m^{(k)}}$ into the above equation gives

$$\begin{aligned} Q\left(\sqrt{\sum_{m=1}^{M}\sum_{l=1}^{R}\left(\beta_{m,r}^{(k)}\right)^2\gamma_m^{(k)}}\right) &= \frac{1}{\pi}\int_0^{\frac{\pi}{2}}\exp\left(-\frac{\sum_{m=1}^{M}\sum_{l=1}^{R}\left(\beta_{m,r}^{(k)}\right)^2\gamma_m^{(k)}}{2\sin^2\Theta}\right)\mathrm{d}\Theta \\ &= \frac{1}{\pi}\int_0^{\frac{\pi}{2}}\prod_{m=1}^{M}\prod_{l=1}^{R}\exp\left(-\frac{\left(\beta_{m,r}^{(k)}\right)^2\gamma_m^{(k)}}{2\sin^2\Theta}\right)\mathrm{d}\Theta \end{aligned} \tag{2.77}$$

The order of integration can be changed and an inner decomposition of the error probability equation can be done, converting to a single integration on the product of the M independent (sub-carrier) R-fold (multipath) integration:

$$P_{\mathrm{e}} = \frac{1}{\pi}\int_0^{\frac{\pi}{2}} \left\{ \prod_{m=1}^{M} \left[\int_0^{\infty} \cdots \int_{\beta_{m,r}^{(k)}}^{\infty} \cdots \int_{\beta_{m,2}^{(k)}}^{\infty} \left(\begin{array}{l} \prod_{r=1}^{R} \exp\left(-\frac{\left(\beta_{m,r}^{(k)}\right)^2 \gamma_m^{(k)}}{2\sin^2\Theta} \right) \\ \times p\left(\beta_{m,1}^{(k)}, \beta_{m,2}^{(k)}, \ldots, \beta_{m,R}^{(k)}\right) \end{array} \right) \times \mathrm{d}\beta_{m,1}^{(k)} \mathrm{d}\beta_{m,2}^{(k)} \cdots \mathrm{d}\beta_{m,R}^{(k)} \right] \right\} \mathrm{d}\Theta$$

$$= \frac{1}{\pi}\int_0^{\frac{\pi}{2}} \prod_{m=1}^{M} \left\{ \frac{L!}{(L-R)!} \int_0^{\infty} \cdots \int_{\beta_{m,r}^{(k)}}^{\infty} \cdots \int_{\beta_{m,2}^{(k)}}^{\infty} \left[\begin{array}{l} \left(\prod_{r=1}^{R} \exp\left(-\frac{\left(\beta_{m,r}^{(k)}\right)^2 \gamma_m^{(k)}}{2\sin^2\Phi} \right) p\left(\beta_{m,r}^{(k)}\right) \right) \\ \times \left(P\left(\beta_{m,R}^{(k)}\right) \right)^{L-R} \end{array} \right] \times \mathrm{d}\beta_{m,1}^{(k)} \mathrm{d}\beta_{m,2}^{(k)} \cdots \mathrm{d}\beta_{m,R}^{(k)} \right\} \mathrm{d}\Theta$$

$$= \frac{1}{\pi}\left(\frac{L!}{(L-R)!} \right)^M \int_0^{\frac{\pi}{2}} \prod_{m=1}^{M} R\left(\beta_{m,1}^{(k)}, \beta_{m,2}^{(k)}, \ldots, \beta_{m,R}^{(k)}\right) \mathrm{d}\Theta \tag{2.78}$$

where $R\left(\beta_{m,1}^{(k)}, \beta_{m,2}^{(k)}, \ldots, \beta_{m,R}^{(k)}\right)$ is the term related to the mth sub-carrier components and can be expressed as

$$\begin{aligned} & R\left(\beta_{m,1}^{(k)}, \beta_{m,2}^{(k)}, \ldots, \beta_{m,R}^{(k)}\right) \\ &= \int_0^{\infty} \left(P\left(\beta_{m,R}^{(k)}\right)\right)^{L-R} \int_{\beta_{m,r}^{(k)}}^{\infty} \cdots \int_{\beta_{m,2}^{(k)}}^{\infty} \\ &\quad \times \left(\prod_{l=1}^{R} \exp\left(-\frac{\left(\beta_{m,r}^{(k)}\right)^2 \gamma_m^{(k)}}{2\sin^2\Theta} \right) p\left(\beta_{m,r}^{(k)}\right) \right) \mathrm{d}\beta_{m,1}^{(k)} \mathrm{d}\beta_{m,2}^{(k)} \ldots \mathrm{d}\beta_{m,R}^{(k)} \\ &= \int_0^{\infty} \left(P\left(\beta_{m,R}^{(k)}\right)\right)^{L-R} R'\left(\beta_{m,R}^{(k)}\right) \\ &\quad \times \int_{\beta_{m,R}^{(k)}}^{\infty} R'\left(\beta_{m,R-1}^{(k)}\right) \cdots \int_{\beta_{m,3}^{(k)}}^{\infty} R'\left(\beta_{m,2}^{(k)}\right) \int_{\beta_{m,2}^{(k)}}^{\infty} R'\left(\beta_{m,1}^{(k)}\right) \mathrm{d}\beta_{m,1}^{(k)} \mathrm{d}\beta_{m,2}^{(k)} \cdots \mathrm{d}\beta_{m,R-1}^{(k)} \mathrm{d}\beta_{m,R}^{(k)} \end{aligned} \tag{2.79}$$

where $R'\left(\beta_{m,r}^{(k)}\right)$ is a function of $\beta_{m,r}^{(k)}$ as $\gamma_m^{(k)}$ and Θ are constant within the R-fold integral $R\left(\beta_{m,1}^{(k)}, \beta_{m,2}^{(k)}, \ldots, \beta_{m,R}^{(k)}\right)$ such that

$$R'\left(\beta_{m,r}^{(k)}\right) = \exp\left(-\frac{\left(\beta_{m,r}^{(k)}\right)^2 \gamma_m^{(k)}}{2\sin^2\Theta} \right) p\left(\beta_{m,r}^{(k)}\right) \tag{2.80}$$

With the assumption of the same fading parameter among the multipaths, one obtains

$$R''(y) = \int_y^{\infty} R'(x)\mathrm{d}x \Rightarrow \frac{\mathrm{d}R''(y)}{\mathrm{d}y} = -R'(y) \tag{2.81}$$

Insert the above relations into (2.79) and iteratively the R-fold integration can be reduced into a single integration with respect to $\beta_{m,R}^{(k)}$ such that [17]

$$R\left(\beta_{m,1}^{(k)}, \beta_{m,2}^{(k)}, \ldots, \beta_{m,R}^{(k)}\right) = \frac{1}{(R-1)!} \int_0^{\infty} \left[P\left(\beta_{m,R}^{(k)}\right)\right]^{L-R} R'\left(\beta_{m,R}^{(k)}\right)\left[R''\left(\beta_{m,R}^{(k)}\right)\right]^{R-1} \mathrm{d}\beta_{m,R}^{(k)} \tag{2.82}$$

Thus the error probability can be written in the form of a triple-level integration:

$$P_{\mathrm{e}} = \frac{1}{\pi}\left(\frac{L!}{(R-1)!\,(L-R)!}\right)^{M} \times \int_0^{\frac{\pi}{2}} \prod_{m=1}^{M} \left[\int_0^{\infty} \left[P\left(\beta_{m,R}^{(k)}\right)\right]^{L-R} R'\left(\beta_{m,R}^{(k)}\right)\left[R''\left(\beta_{m,R}^{(k)}\right)\right]^{R-1} \mathrm{d}\beta_{m,R}^{(k)}\right] \mathrm{d}\Theta \tag{2.83}$$

2.4 Numerical results

The effects of various parameters on the system performance are illustrated in this section. The impact of the use of a notch filter is studied and different receiving strategies (MMSE, NCC and CC) are compared. Without being explicitly specified, the plotting is done based on the results obtained in this section, and the following system parameters are assumed: the number of sub-carriers $M = 8$, the processing gain $N = 32$, the chip period $T_{\mathrm{c}} = 2$ ns, the roll-off factor of the square-root raise-cosine filter $\alpha = 0.25$, the number of taps in the two-sided transversal filter $2W + 1 = 25$, the number of correlators within the Rake receiver per sub-carrier $R = 3$, the number of active users $K = 20$, and the number of paths per sub-band $L = 10$. The decay rate δ, the fading figure μ and the power ratio κ are assumed to be $10^{-1.61}$, 3.5 and $10^{-0.4}$, respectively [5].

2.4.1 Significance of the transmitter filtering

Figure 2.7 shows the spectra of an unmodified chip shape (raised-cosine) and the spectra of a modified chip shape corresponding to narrowband interference at three different frequency locations ($q_m = 0$, 0.2 and 0.3, respectively). Two different bandwidth ratios $p_m = 0.05$ and $p_m = 0.1$ are considered. It is assumed that the power ratio of narrowband interference to the UWB signal $P_j^{(m)}/P = 50$ dB and the signal-to-noise ratio $P/\eta_{\mathrm{o}} = 10$ dB. It can be seen that in addition to raised cosine shaping, narrowband notching is created; however, the most efficient notching appears when the narrowband interference is located at the center frequency of the UWB sub-band ($q_m = 0$). Due to the periodicity of the estimating function (2.5), for $q_m \neq 0$, the notching is created not only in the position of narrowband interference, but also in the symmetric location about the origin, such that the partial clean spectrum has to be notched. In comparison to the undistorted raised-cosine signal spectrum, it can be seen that the larger the bandwidth ratio p_m, the greater the extent of the distortion of the spectrum.

Interference reduction to the established services can be achieved via the use of a notch filter, but this brings side-effects on the chip-matching mechanism due to the distortion, and worsens the inter-chip interference in the despreading process. The correlation function of the original chip shape and modified chip shape are plotted under the

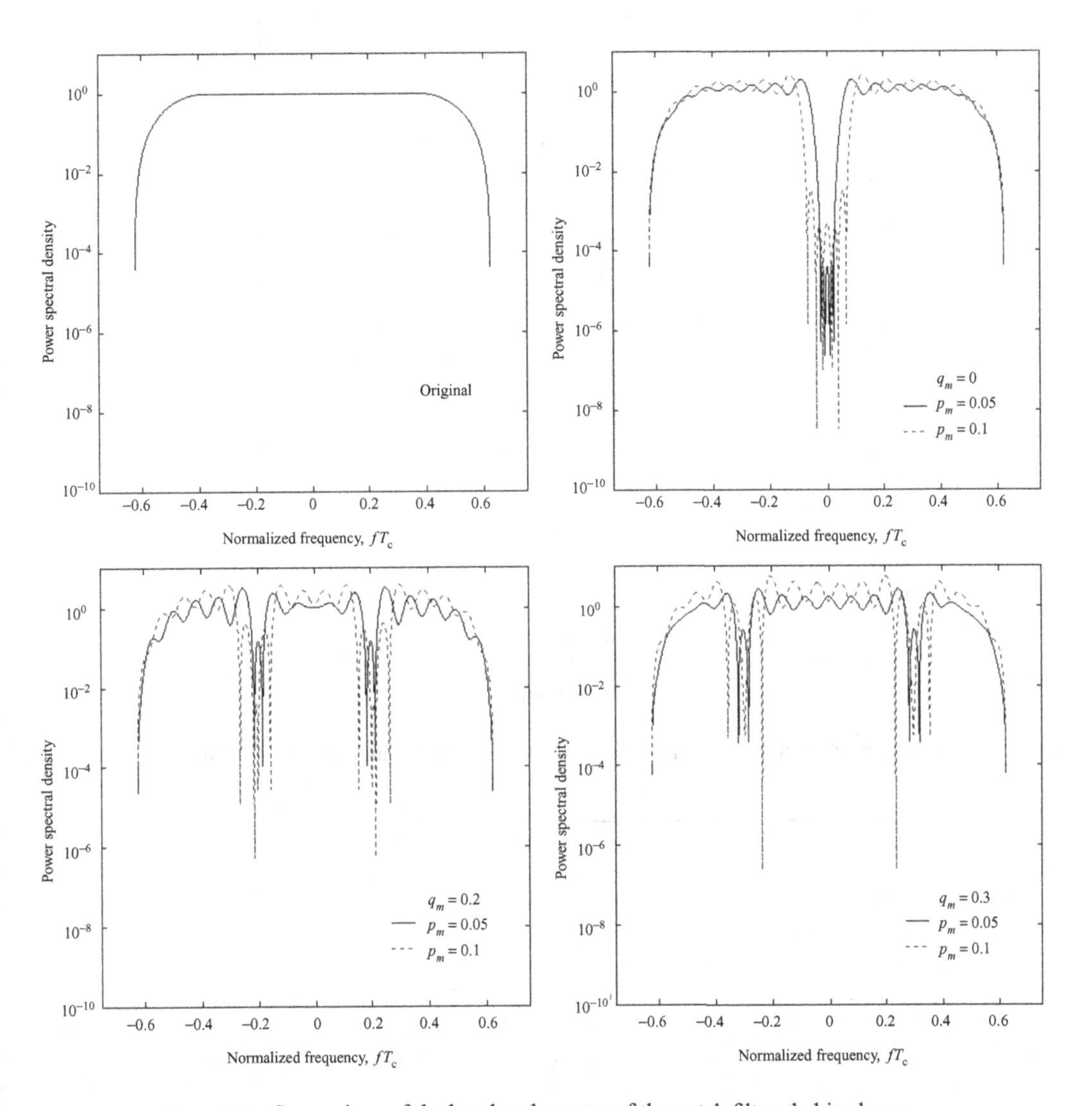

Figure 2.7 Comparison of the baseband spectra of the notch filtered chip shapes.

conditions of the power ratio of narrowband interference to the UWB signal $P_j^{(m)}/P =$ 60 dB and the signal-to-noise ratio, $P/\eta_0 = 10$ dB. With the undistorted chip shape as reference, Figure 2.8 presents the correlation as the function of the normalized time shift with fixed frequency offset $q_m = 0.1$ and Figure 2.9 presents the correlation as the function of the normalized time shift with fixed bandwidth ratio, $p_m = 0.05$.

As shown in the plots, the derivations from the autocorrelation function of the original (fixed line) indicates that the introduction of notch filtering to the pulse shape inevitably causes distortion, but the original correlation property is slightly altered when the time offset is within the single chip interval. As the time offset increases, the original zero-crossing property is changed and the distortion is greater, and these increase the MPI and MAI in the decision statistic, but the processing gain of the random code can suppress this.

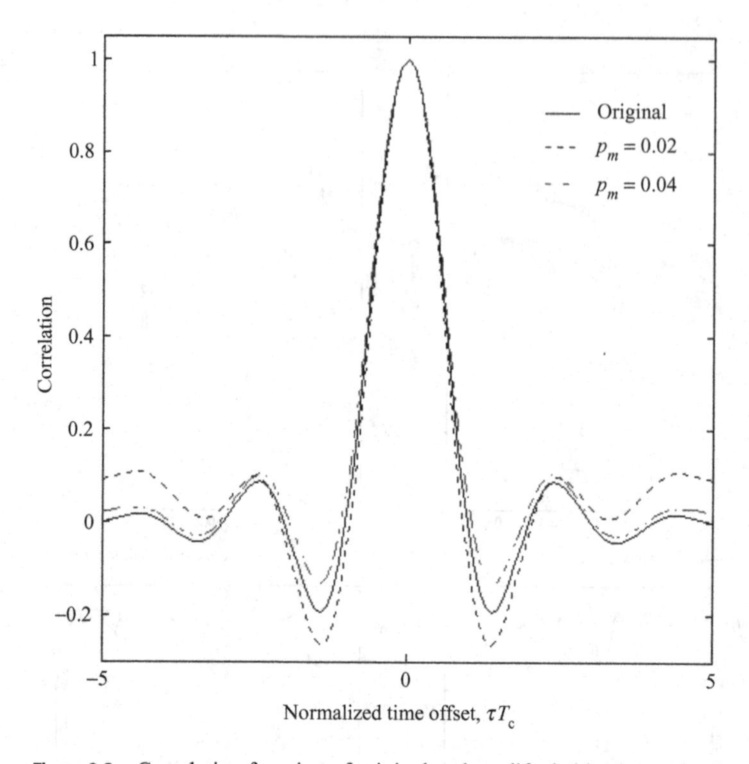

Figure 2.8 Correlation function of original and modified chip shapes for fixed q_m.

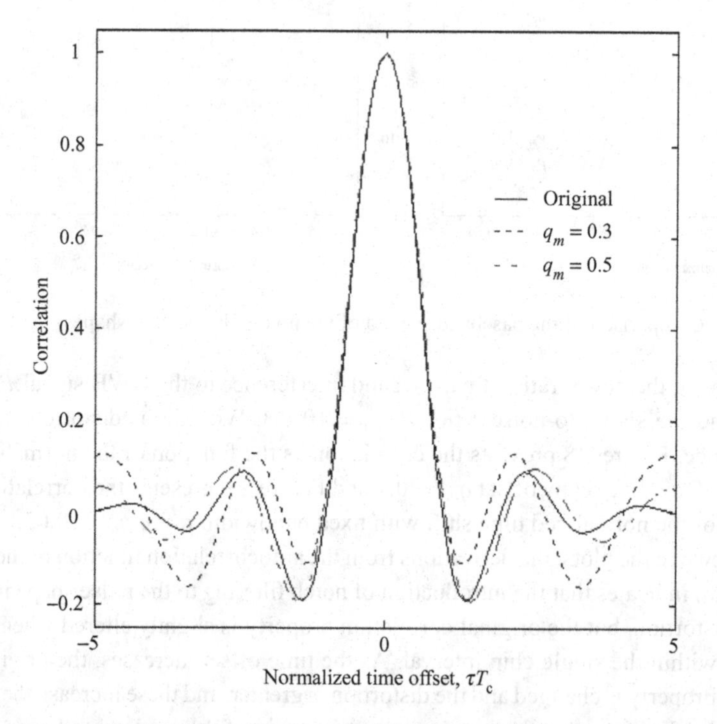

Figure 2.9 Correlation function of original and modified chip shapes for fixed p_m.

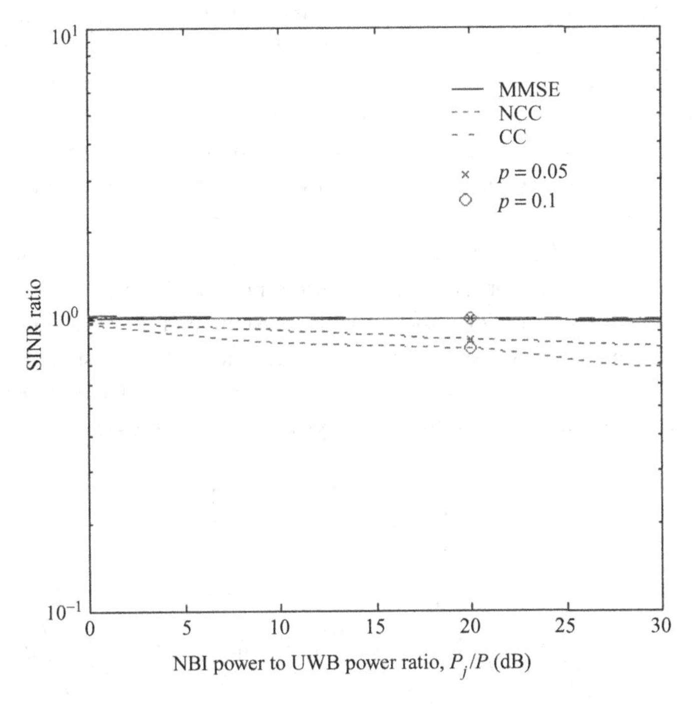

Figure 2.10 Ratio of the average SINR with and without transmitter notch filter.

To investigate the effect of the use of transmitter filtering on UWB system performance, the ratio of the average SINR of the received signal with and without the use of a notch filter at the transmitter is plotted (Figure 2.10) against the power ratio of narrowband interference to the UWB signal $P_j^{(m)}/P$ with $P/\eta_o = 10$ dB, $p_m = 0.05$ and $q_m = 0$. It can be observed that there is a tiny difference in the average SINR for the MMSE and CC receivers before and after the introduction of the notch filter, but the average SINR drop is obvious for the NCC receiver.

For the MMSE and CC receivers, the jamming signal should play a more significant role and its effect will overshadow the effect of using a notch filter at the transmitter. Meanwhile the impact of the transmitter notch filter on the NCC receiver is nevertheless greater than that on the MMSE and CC receivers, but this is just a small attenuation, caused by the double notching (transmitter and receiver), which results in a double distortion of the desired signal waveform. As the change of the average SINR is limited, it can be concluded that the use of a notch filter at the transmitter, which results in the scarification of a certain amount of the transmitted power, does not cause significant degradation of the system performance. This is because even if no notch is inserted in the spectra of the transmitted signal, the power of the frequency region occupied by the narrowband system is corrupted and no longer useful to the decision statistic.

Although under normal operation the emission power of UWB devices would be very low such that its impact on the existing narrowband systems would be insignificant, transmitter notch filtering is still necessary. The motivation for the introduction of notch filtering at a UWB device transmitter is that UWB devices are designed to operate without causing trouble to existing systems, and it is clear that the existing systems do not have any specific measures to deal with the interference from UWB devices. It is the responsibility of UWB devices to avoid causing any possible interference.

In the following, we are going to show that notch filtering at the transmitter of the multicarrier CDMA UWB system can be beneficial to narrowband systems when the power difference with the UWB system is not too great; for example, when they are suffering from the near–far problem. To illustrate this, for simplicity, the narrowband communication system is assumed to be a BPSK system, and the overlaid sub-band of the UWB system is regarded as wideband noise by the narrowband system. By Gaussian approximation, the SINR of the decision statistic for the narrowband system over the spectrum of the mth sub-carrier of the UWB system can be expressed as

$$\xi_j^{(m)} = \frac{\left(\sqrt{2P_j^{(m)}} \int_{\frac{p_m}{2T_c}-\frac{q_m}{T_c}}^{\frac{p_m}{2T_c}+\frac{q_m}{T_c}} J(f)\mathrm{d}f\right)^2}{P\sum_{k=1}^{K}\left(\sum_{l=1}^{L}\Omega_{m,l}^{(k)}\right)\int_{\frac{p_m}{2T_c}-\frac{q_m}{T_c}}^{\frac{p_m}{2T_c}+\frac{q_m}{T_c}} \left(H_W^{(m)}(f)\right)^2 H_{\mathrm{RC}}(f)\mathrm{d}f + {\eta_o}^2 \int_{\frac{p_m}{2T_c}-\frac{q_m}{T_c}}^{\frac{p_m}{2T_c}+\frac{q_m}{T_c}} \mathrm{d}f} \tag{2.84}$$

and the error probability P_e of the narrowband system is given by

$$P_e = Q\left(\sqrt{\xi_j^{(m)}}\right) \tag{2.85}$$

Figure 2.11 shows the plot of probability of the bit error of the narrowband system versus the number of UWB active users K, with $P/\eta_o = 5$ dB, $p_m = 0.05$ and $q_m = 0$. The power ratio $P_{j,m}/P$ is increased from 10 dB to 30 dB, which represents the situation when the power of the narrowband device is relatively weak up to a typical situation.

Two main observations are noted from the plot. First, without the use of the notch filter at the UWB device's transmitter, the performance of the narrowband system is heavily downgraded even when the number of UWB system users is small, but the situation is much improved after the protective measure is introduced. Second, the significant improvement diminishes as K increases due to the fact that the portion of the UWB signal spectrum overlaid on the narrowband system is suppressed but not completely removed, so the impact of the residue increases as K grows, which means that there is a limit to the number of UWB system users that the narrowband system can sustain, but such tolerance is much increased in comparison to the case without notch filtering at the transmitter. It is interesting to point out that while the introduction of the notch filter at the UWB system transmitter provides effective protection to the established systems, there should be capacity for the UWB system to increase its overall transmitted power in the non-overlaying spectra, which allows a smaller power difference between the established system and the overlaid UWB system once the notches are properly inserted.

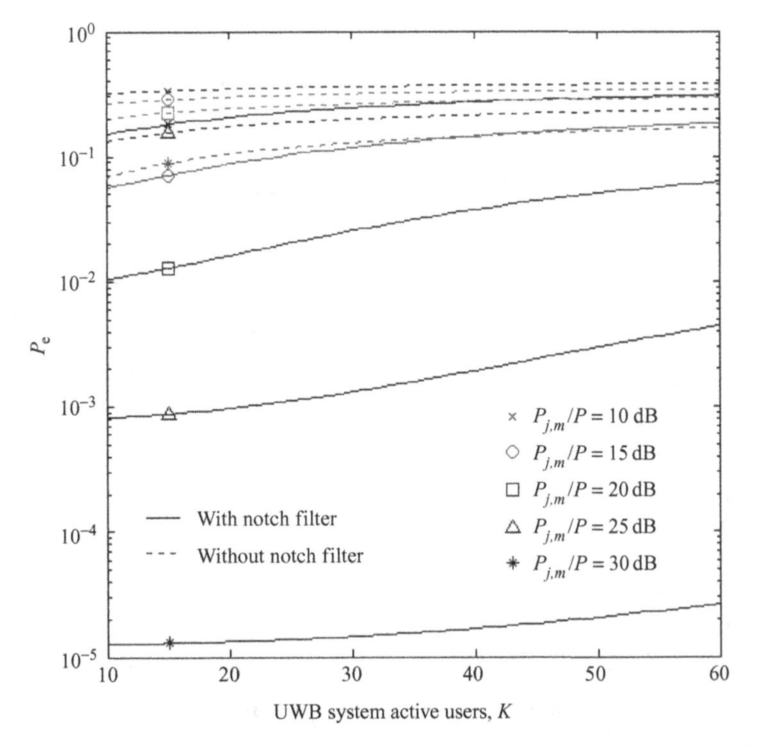

Figure 2.11 Probability of bit error of the narrowband system versus the number of UWB active users.

2.4.2 Comparison of the receiving strategies

Figure 2.12 shows error probabilities as a function of $P_j^{(m)}/P$ for different numbers of narrowband interferers ($M_i = 2$, $M_i = 5$ and $M_i = 7$, respectively). It is assumed that $P/\eta_o = 10$ dB, $p_m = 0.05$ and $q_m = 0$. Satisfactory performance is achieved for small values of M_i irrespective of the type of receiver used, but the situation is different for a large value of M_1.

For the CC receiver, the performance degrades significantly when narrowband interference exists. The curve levels off quickly, which means that the decision statistics from the polluted sub-bands are unreliable even at low $P_j^{(m)}/P$, and the symbol decision can only rely on those sub-bands that remain clean.

For the NCC receiver, the performance curve has a typical S-shape, which represents three different phases: at low $P_j^{(m)}/P$ the probability of error increases very slowly, because the double-sided transversal-type filter at the front end provides sufficient attenuation to the narrowband interference. When $P_j^{(m)}/P$ increases beyond 40 dB, the error rate increases rapidly, and this observation can be explained by the fact that the notch introduced by the filter is too shallow for interference suppression, so the reliability of the decision statistic from the polluted sub-band decreases as $P_j^{(m)}/P$ increases. The curve eventually levels off as $P_j^{(m)}/P$ further increases (> 80dB) and coincides with the curve

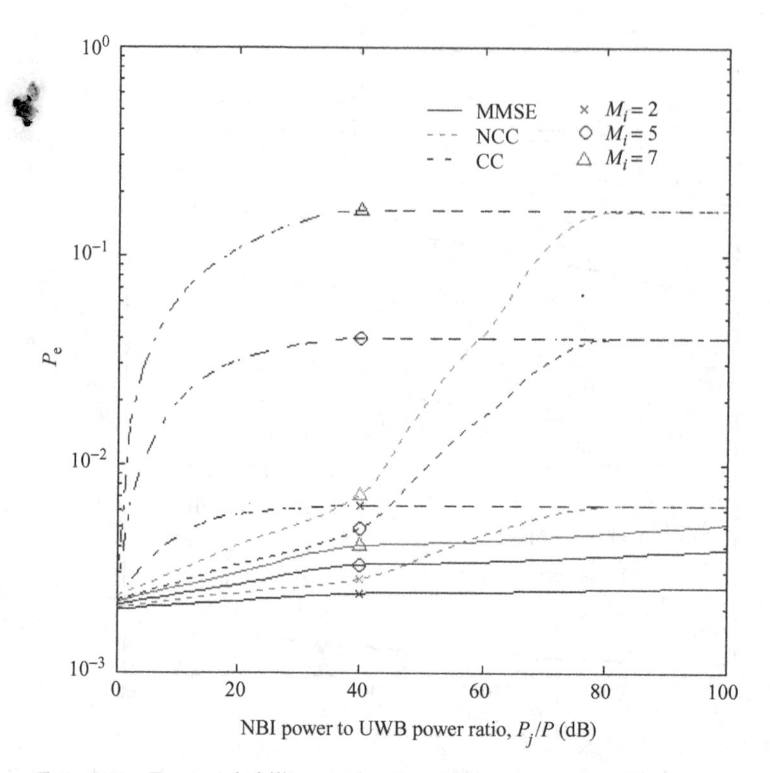

Figure 2.12 Error probability as a function of the power ratio of NBI to UWB signal.

of the CC receiver, which indicates that the contributions from the polluted sub-bands are no longer useful.

In contrast, for the MMSE receiver, the increase of $P_j^{(m)}/P$ only leads to a moderate growth in error probability. This means that the MMSE receiver is still able to provide valid protection under very large $P_j^{(m)}/P$, which is typical for the UWB application scenario. Furthermore, the performance improvement is more significant as the number of sub-bands overlaid with the narrowband system (i.e., M_i) increases.

To summarize, it can be seen from the figure that for a given number of narrowband interferers, the performance of the NCC receiver degrades dramatically when $P_j^{(m)}/P$ increases beyond 40 dB, whereas the performance of the MMSE receiver is almost independent of the size of $P_j^{(m)}/P$. Furthermore, the performance of the NCC receiver and CC receiver are much affected by the number of narrowband interferers. However, the MMSE receiver is robust to the number of interferers.

Multipath diversity can be further explored by adding correlators in the Rake receiver, but this results in an increase in the complexity of the receiver design, so it is desired to seek for a balance point. The effect of the number of correlators in the Rake receiver per sub-carrier R on the performance of the multicarrier CDMA system is illustrated in Figure 2.13, from single-path combing ($R = 1$) to all-path combing ($R = L = 10$). The conditions of $P/\eta_o = 5$ dB, $P/\eta_o = 10$ dB and $P/\eta_o = 15$ dB are picked for comparison, and the other system parameters are $P_j^{(m)}/P = 60$ dB, $p_m = 0.05$, $q_m = 0$ and

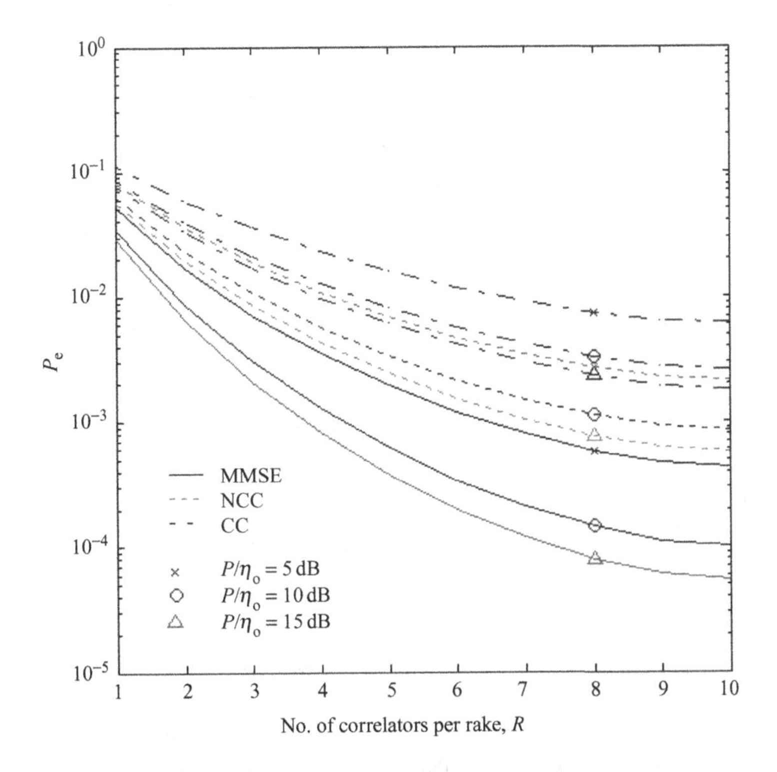

Figure 2.13 Relationship between error probability and R.

$M_i = 4$. It is seen that when the number of correlators R increases, the error probability decreases sharply at the beginning, but the curve becomes flat as R approaches L. In other words, the marginal performance improvement diminishes for a large R; thus a low-complexity Rake receiver can achieve acceptable performance. The MMSE receiver always outperforms the NCC and CC receivers. For example, at $P/\eta_o = 5$ dB, the MMSE receiver already has a better performance than both the NCC and CC receivers with a transmission power of $P/\eta_o = 15$ dB. As P/η_o increases, the performance improvement of the MMSE receiver is much greater than that of the NCC or CC receivers.

Figure 2.14 illustrates the error probability as a function of the number of active users K for $P_j^{(m)}/P = 30$ dB and $P_j^{(m)}/P = 60$ dB, respectively. It is assumed that $p_m = 0.05$, $q_m = 0$, $M_i = 3$ and $P/\eta_o = 5$ dB. The MMSE receiver is significantly superior to both the NCC and CC receivers for the multiple access capability when a strong narrowband system exists. For a probability of error at 10^{-2}, with a narrowband system of $P_j^{(m)}/P = 60$ dB, the multicarrier CDMA system capacity when using the NCC and CC receivers is 25 users and 20 users, respectively, but the capacity can be over 30 users for the MMSE receiver. The number of users supported by the UWB system is much affected by $P_j^{(m)}/P$ for the NCC and CC receivers, but it is less dependent on $P_j^{(m)}/P$ for the MMSE receiver. This is consistent with Figure 2.12 in that the MMSE

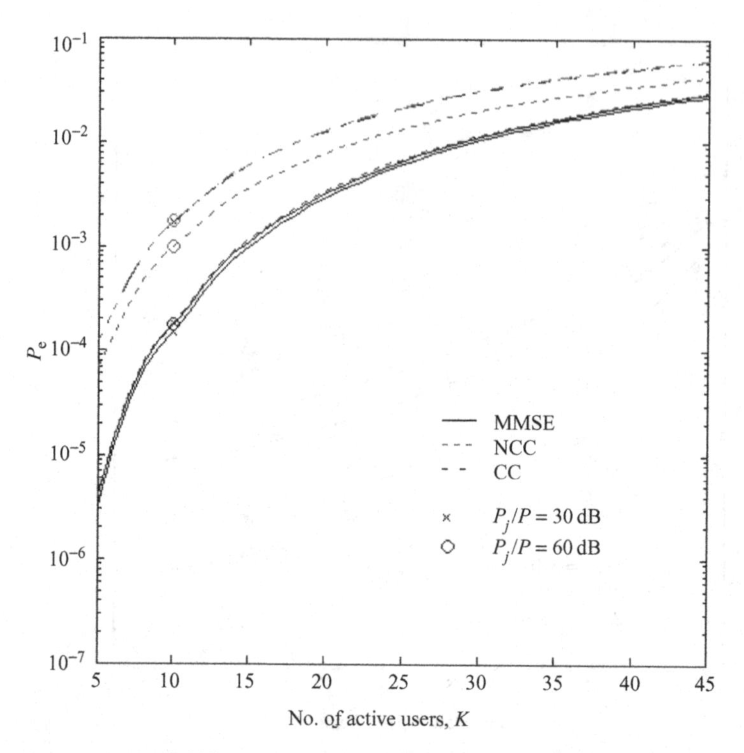

Figure 2.14 Probability of error against the number of active UWB system users.

receiver is more capable of suppressing strong narrowband interference than either the NCC or CC receivers, because the greater the narrowband interference suppression ability, the greater the allowable tolerance to other types of interference.

The relationship between error probability and transmitted power in terms of the signal-to-noise ratio (SNR) P/η_o is shown in Figure 2.15, for narrowband interferers of $P_j^{(m)}/P = 30$ dB and $P_j^{(m)}/P = 60$ dB, with the assumption of $P/\eta_o = 10$ dB, $M_i = 5$, $p_m = 0.05$ and $q_m = 0$. Notice that the performance curves for all types of receiver are of similar shape. The error probability shows a heavy drop as P/η_o increases but soon levels off for large P/η_o. This means that the overlay system is still capable of working under a low P/η_o operating environment, but it can be observed that narrowband interference exhibits a different degree of impact on different receivers. For the CC receiver, system performance is heavily downgraded in the presence of NBI. For the NCC receiver, the error probability is less affected when the NBI is weak (i.e., $P_j^{(m)}/P = 30$ dB) as the notch filter is still effective for suppressing the narrowband interference, but the system performance deteriorates for strong NBI. For the MMSE receiver, however, the overlay UWB system performance is much improved.

Frequency diversity can be explored under the multi-carrier system, but the presence of the narrowband system brings an adverse effect to the exploration of diversity. The effect of the number of sub-bands with narrowband interference M_i on the performance of the multicarrier CDMA system is shown in Figure 2.16, for $P_j^{(m)}/P = 30$ dB and

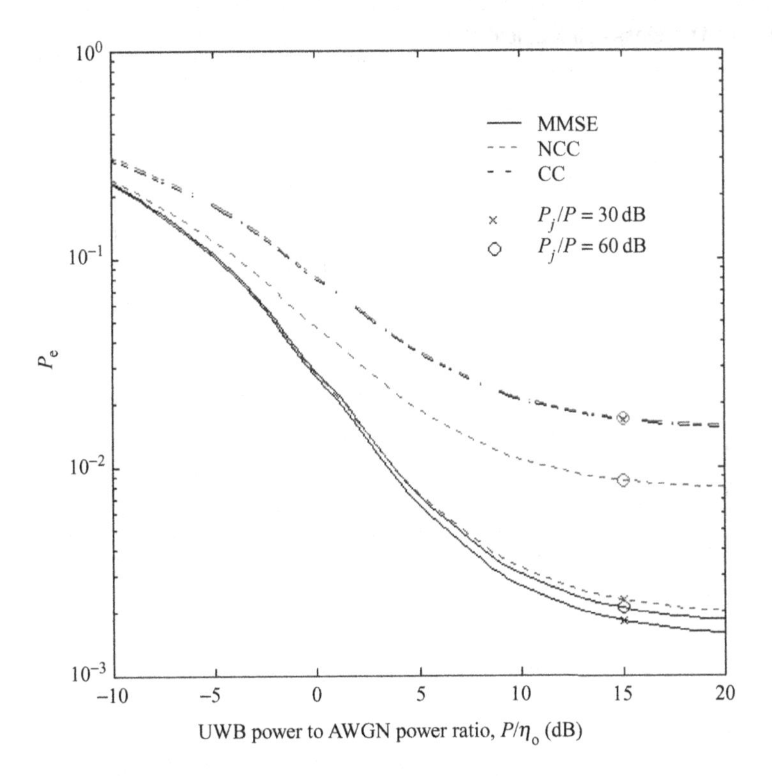

Figure 2.15 Probability of error versus the signal power-to-noise ratio.

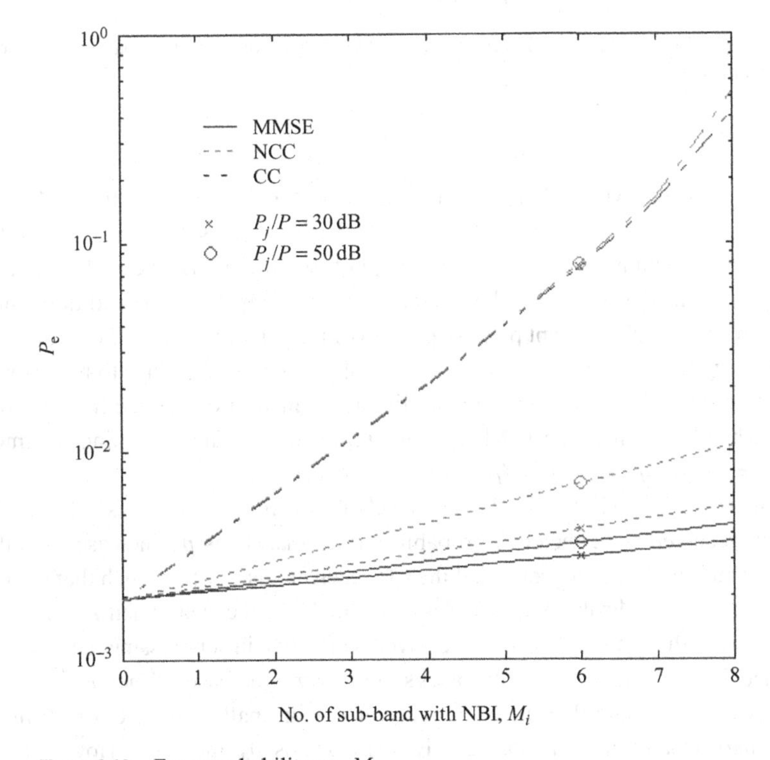

Figure 2.16 Error probability vs. M_i.

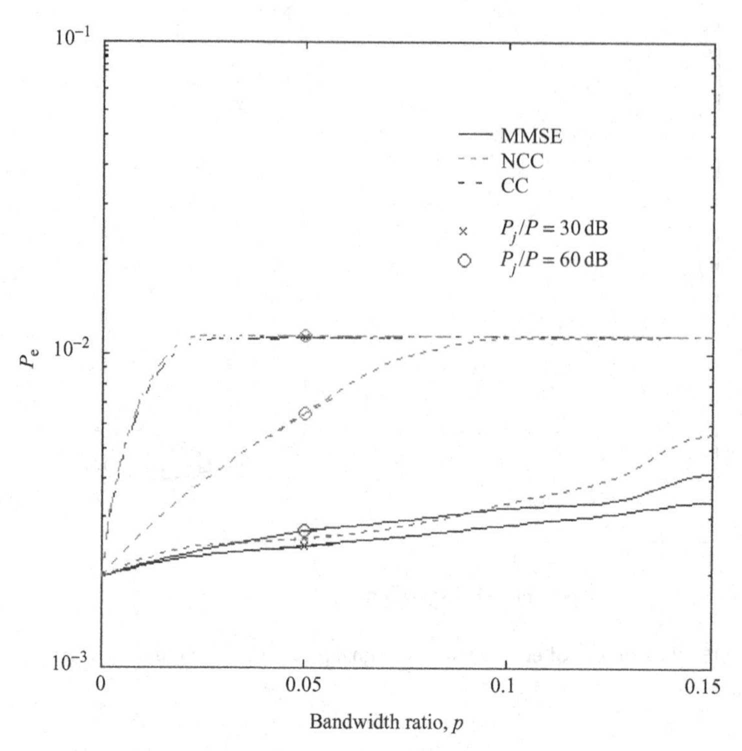

Figure 2.17 Probability of error versus the bandwidth ratio of the narrowband system to the UWB sub-band p_m.

$P_j^{(m)}/P = 50$ dB, where $P/\eta_o = 10$ dB, $p_m = 0.05$ and $q_m = 0$. For the MMSE receiver, the system performance experiences a slight degradation as M_i increases. For the NCC receiver, protection is valid for weak NBI, but for strong NBI the performance drop is obvious when M_i is large. For the CC receiver, which represents no additional protective measure against NBI except processing gain of the spreading code, the error probability rises drastically from losing frequency diversity since the tolerance to NBI is small.

Figure 2.17 shows the error probability as a function of the bandwidth ratio of the narrowband system to the UWB system sub-band p_m. The remaining parameters are $P/\eta_o = 10$ dB, $q_m = 0$ and $M_i = 3$.

With reference to Figure 2.12, look closely for the following curves in Figure 2.17: for the CC receiver, since the error probability rises sharply as p_m increases and the curve flattens out quickly, it indicates that the CC receiver has no effect such that the contribution of the polluted sub-band is useless. For the NCC receiver, when $P_j^{(m)}/P = 30$ dB, it represents the state that the NCC receiver is effective in suppressing narrowband interference. The error probability increases slightly as p_m increases. When $P_j^{(m)}/P = 60$ dB, it represents the state that the NCC receiver is marginally effective. Its ability to suppress narrowband interference steadily decreases as p_m increases. However, it can be seen that when the MMSE receiver is used, the system performance degrades slowly as

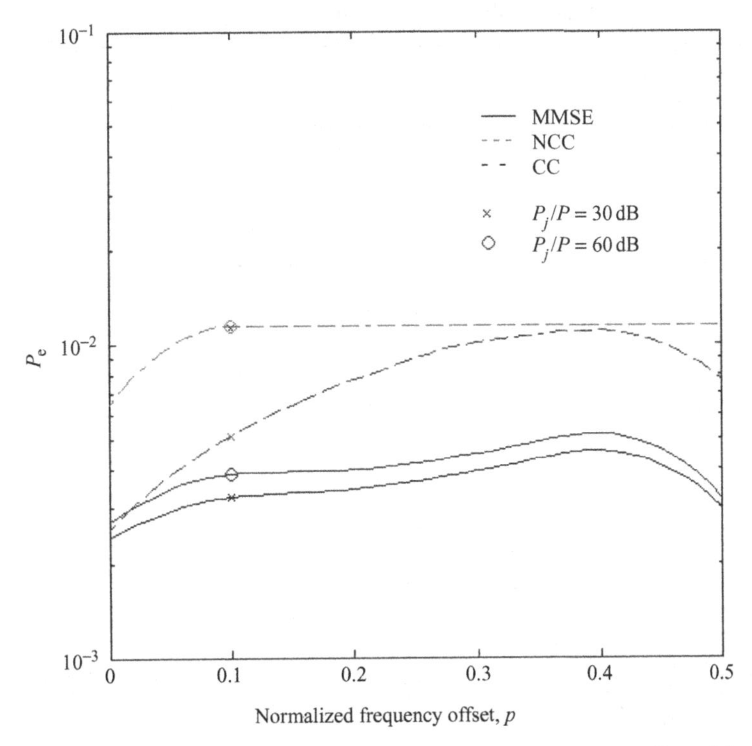

Figure 2.18 Probability of error versus the normalized frequency offset of the narrowband system from the sub-carrier center frequency q_m.

p_m increases regardless of the change in $P_j^{(m)}/P$, in contrast to both the NCC and CC receivers, whose system error probability increases dramatically for large $P_j^{(m)}/P$.

In Figure 2.18, error probability is plotted against the normalized frequency offset q_m of the narrowband system from the sub-carrier center, with $P/\eta_o = 10$ dB, $P_j^{(m)}/P = 60$ dB, $p_m = 0.05$ and $M_i = 3$. It can be seen that the MMSE receiver provides a more stable performance than either the NCC or CC receivers as q_m varies. The findings are in accordance with the discussion of the notch filtered chip shape in the previous sub-section: as the deepness of the notch produced decreases when q_m is non-zero, the narrowband rejection capability also decreases.

2.4.3 Other aspects

In this section, the results presented are based on the assumption that all sub-carriers are used to convey the same transmitted signal. Actually, each sub-carrier is operating independently of each another before decision statistic combining takes place. This suggests that, in addition to the approach described previously, each sub-carrier can be used to convey an independent transmitted signal, or divided into groups to transmit a different data stream. In comparison to the original scheme, the benefit of frequency diversity is traded for a higher data rate. In the proposed system, the UWB system has

Table 2.1 Sub-carrier grouping under different transmission rates

Transmission rate	1	2	4	8
Grouping of sub-carriers	8	4,4	2,2,2,2	1,1,1,1,1,1,1,1

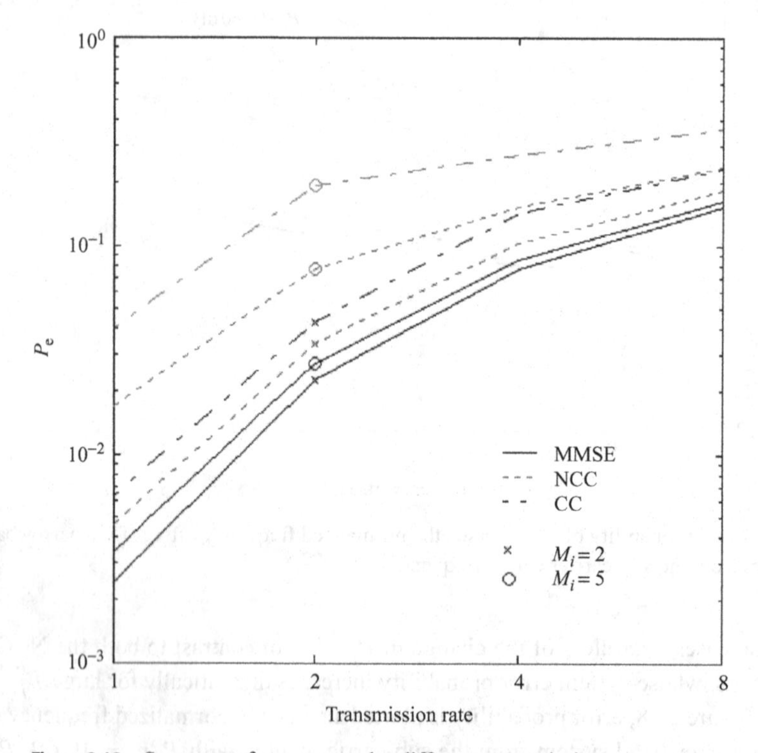

Figure 2.19 System performance under different transmission rates.

a sub-carrier M of 8, and it is assumed that the number of sub-carriers per group only takes values of powers of two, so there are four possible transmission rates, as illustrated in Table 2.1.

The transmission rate is defined as the number of independent data conveyed by the sub-carrier. For example, a transmission rate of 4 means that the system operates at four times its basic transmission rate (all sub-carriers are used to transmit the same data), eight sub-carriers are divided into four groups and the two sub-carriers of the same group are transmitting the same data stream.

In Figure 2.19, error probability is plotted against transmission rate, and the other system parameters are $P/\eta_o = 10\,\text{dB}$, $P_j^{(m)}/P = 60\,\text{dB}$, $p_m = 0.05$ and $q_m = 0$. $M_i = 2$ and $M_i = 5$ are selected for comparison in the performance evaluation. The possible distribution of the narrowband system over the sub-bands is shown in Table 2.2. All cases are assumed to be equally probable.

Table 2.2 Narrowband system and grouped sub-carriers

Transmission rate	1	2	4	8
$M_i = 2$	2	(2,0) or (1,1)	(2,0,0,0) or (1,1,0,0)	(1,1,0,0,0,0,0,0)
$M_i = 5$	5	(4,1) or (3, 2)	(2,1,1,1) or (2,2,1,0)	(1,1,1,1,1,0,0,0)

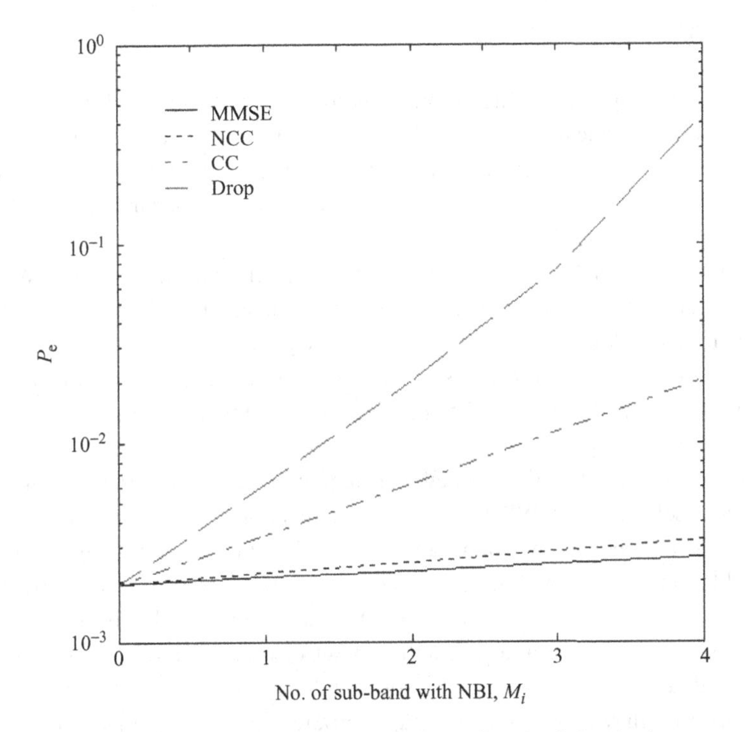

Figure 2.20 Comparison with the discarding polluted sub-band scheme.

From the graph, it can be seen that the MMSE receiver is best able to combat the effects of the reduction in frequency diversity when the transmission rate is increased. Meanwhile, the performance of the NCC receiver is acceptable at low transmission rate, but the performance of the NCC receiver drops drastically when the transmission rate increases. Once the NBI can be suppressed, the sub-band overlaid with the narrowband system is still useful, so there is capacity to exchange diversity for data rate.

To tackle the problem of overlaid narrowband systems, apart from applying the NBI rejection technique, a simpler way is to discard the polluted sub-band so that the decision statistic relies on the contributions from "clean" sub-carriers. For UWB devices, however, the bandwidth of each sub-carrier signal is extremely wide when compared with that of NBI. Simply discarding a sub-carrier that is experiencing NBI is too inefficient and even non-beneficial. Figure 2.20 illustrates the inefficiency of simply discarding the polluted sub-band in comparison to the preserving approach, with the system parameters as $P/\eta_o = 10$ dB, $p_m = 0.05$ and $q_m = 0$. The performance difference between the discarding approach and the preserving approach (MMSE, NCC and CC) is slim when M_i is small. However, the performance gap enlarges significantly for large values of M_i, where the NBI suppression (MMSE and NCC) for the preserving approach shows its effectiveness, since the jammed sub-carrier signals still make positive contributions to the net frequency diversity for the system.

2.5 Discussion

In this chapter, a spectrally filtered transmission approach is considered instead of the conventional carrier-less design for UWB communications. The application environment is assumed to be a short-range high-speed system with multiple users. Multicarrier CDMA overlay is proposed as a potential candidate. As described in Section 1.1, interference reduction and interference suppression are the key issues for harmoniously sharing the spectrum between the established narrowband systems and the overlaid UWB system. An investigation is carried out on the use of compromising measures incorporated in the multicarrier CDMA overlay system to meet these goals, as well as their impacts on the parties involved. Numerical expressions are derived in Section 2.4 for the performance evaluations presented in Section 2.5. In this section, we review the proposed design and discuss some related issues.

Multicarrier CDMA, although now considered to be the best modulation and multiple access scheme for UWB wireless communication, was previously thought to be spectrally inefficient in conventional applications, but this is relieved when an ultra-wide bandwidth is made available. As mentioned in Section 1.1, by using multicarrier CDMA, better spectral control can be achieved. CDMA has been proved to be an effective multiple access scheme under a large number of active users. In addition to the instinctive frequency diversity, the flexibility of resource allocation increases. Since the sub-bands can be disjoint, they can be shifted to appropriate spectral locations and therefore the system spectra can bypass certain frequency regions that require additional protection.

The processing speed requirement for both transmitting and receiving ends is lowered, because the 7.5 GHz ultra-wide bandwidth is sliced into a number of wideband portions, allowing devices with a lower dynamic range to be used. Furthermore, the transformation of ultra-wide bandwidth to several wideband spectra also suggests the possibility of applying technologies developed for wideband communications to ultra-wideband communications. All of these are found to be a problem for carrier-less UWB, but IR is attractive due to the absence of the costly analog RF mixer component, which is the major drawback of the spectrally filtered UWB approach, yet the production cost for analog components can be lower or even avoided due to the breakthrough in the development of software radio and advanced digital-signal-processing technologies.

Next the focus moved to the design of the transceiver. At the transmitter, interference reduction to established narrowband systems is achieved via chip-shape modification. It is assumed that the filter adapting information is accurate. Results demonstrate that the notch is formed at the target spectral position, and this measure is shown to be capable of providing significant protection to lower operating power narrowband systems.

One could argue that the narrowband devices should have a much higher operating power relative to any future UWB devices so that transmitter filtering would be unnecessary. However, one of the criteria of UWB wireless communication is that it should cause little interruption to established services. There would be a lack of coordination among the overlaid systems, since they are expected to exist independently and the UWB device would operate without licensing [19]. The actual power difference between the parties involved would not be as large as expected because this depends on the different operating ranges and the relative locations of the devices involved. Under the situation with a number of unknown factors or undetermined variables, the UWB devices should take the responsibility to have active avoidance in signal transmission and passive tolerance in signal reception.

In the use of notch filtering at the transmitter, for the UWB devices themselves, the benefits of cosine roll-off filtering, the band-limiting effect remains but the first Nyquist criterion in the time domain cannot be totally preserved. It is shown that the chip-shaping modification of the UWB signal slightly alters the original correlation property of the chip-shaping filter at the transmitter and the chip-matching filter at the receiver. As inter-chip interference increases, the MAI and MPI components of the decision statistic also increase, but its significance is less in comparison to the bad effect from NBI, as illustrated in the plot of the ratio of the average SINR of the correlator output with and without transmitter filtering. This indicates that the introduction of notch filtering at the transmitter has a limited influence on the performance of any UWB device, while it is effective in protecting established narrowband services, so there is no need to discard or turn off those overlaid sub-bands in the UWB device.

At the receiver, the data recovery process is carried out with the suppression of the interference from those narrowband systems, and it is shown that the use of the NBI suppression technique is more efficient than simply discarding the sub-bands with NBI. A comparison is made between the receiver using the minimum mean square error detection technique in a correlator, the receiver of a notch filter in cascade with a code-matched correlator, and the receiver of a conventional code-matched correlator. On the basis of the results obtained, the following remarks can be made:

(1) The conventional CC receiver cannot sustain strong narrowband interference. The NCC receiver can only provide sufficient suppression for narrowband interference when the narrowband system power to UWB system power ratio is moderate, but the MMSE receiver offers valid protection when the narrowband interference is much stronger.
(2) The MMSE receiver can support a higher number of active users than the NCC receiver as well as the CC receiver under strong narrowband interference from established services due to the reason stated above.
(3) For the same probability of error requirement, a lower transmission power is allowed when the MMSE receiver is used. The MMSE receiver also allows the potential of selective-maximal combining to be better exercised in comparison to either the NCC

or CC receivers. The MMSE receiver also shows a higher tolerance in the reduction of frequency diversity when more narrowband systems fall within the operating spectra of UWB devices.

(4) The MMSE receiver can sustain a narrowband system of wider bandwidth than the NCC receiver. The MMSE receiver is also found to have low susceptibility to the variations due to the chip-shape modification in the receiver. This allows interference reduction measures to the established services to be adopted with limited impact on the overlaid UWB multicarrier CDMA system performance.

Theoretical results show that pre-combining MMSE with selective-maximal combining provides the UWB system with much better performance than either the NCC or CC receivers can supply. A discussion of the implementation issues of the MMSE receiver for a CDMA system can be found in [18]. Nevertheless, since cost is also a major factor in design formulation, it is incorrect to reach the conclusion that the NCC and CC receivers are totally inferior to the MMSE receiver. For the structure of a notch filter in cascade with a code-matched correlator, its NBI rejection capability can be increased by increasing the number of taps of the transversal filter. The performance of any structure using a code-matched correlator can be improved by increasing the processing gain of the system. However, this requires further analysis of the trade-off between the improvement in performance and the growth in complexity before a decision is made.

Multicarrier CDMA has the potential to be used in UWB wireless communications. In this chapter, the major problems that would be encountered when using multicarrier CDMA for UWB applications are considered. The analytical framework formulated takes the possible problem of interference from narrowband systems and the large number of propagation paths in frequency selective fading into consideration. Study on the impacts of various system parameters is presented by numerical results. When freedom is provided, any possibility should not be underestimated or ignored. In Section 1.1, it was generally thought that ultra-wideband communication was equivalent to IR, because IR had played a dominant role in UWB. As time goes by, however, researchers are starting to seek alternatives due to the shortcomings of IR UWB system. With the recent development of UWB technology [15], it seems that more researchers are looking favorably at spectrally filtered UWB systems.

References

[1] T. T. Wong and J. Wang, "MMSE receiver for multicarrier CDMA overlay in ultra-wideband communications," *IEEE Trans. Vehicular Techn.*, vol. VT-54, pp. 603–614, March 2005.

[2] B. J. Rainbolt and S. L. Miller, "The necessity for and use of CDMA transmitter filtering in overlay systems," *IEEE J. Select. Areas Commun.*, vol. 16, no. 9, pp. 1756–1764, Dec. 1998.

[3] L. B. Milstein, "Interference rejection techniques in spread-spectrum communications," *Proc. IEEE*, vol. 76, no. 6, pp. 657–671, June 1988.

[4] J. G. Proakis, *Digital Communications*, 4th edition. New York: McGraw-Hill, 2001.

[5] D. Cassioli, M. Z. Win and A. F. Molisch, "The ultra-wide bandwidth indoor channel: from statistical model to simulations," *IEEE J. Select. Areas Commun.*, vol. 20, no. 6, pp. 1247–1257, Aug. 2002.

[6] J. R. Forester, "Ultra-wideband technology enabling low-power, high rate connectivity," paper presented at 2002 IEEE CAS Workshop on Wireless Communications and Networking, Sept. 2002.
[7] M. Nakagami, "The m-distribution – a general formula of intensity distribution of rapid fading," in W. C. Hoffman, ed., *Statistical Methods in Radio Wave Propagation*. Oxford: Pergamon Press, 1960.
[8] Y. Sanada, A. Kajiwara and M. Nakagawa, "Adaptive rake system for mobile communications," in *Proc. IEEE International Conference on Selected Topics in Wireless Communications 1992*, pp. 227–230, June 1992.
[9] C. N. Pateros and G. J. Saulnier, "An adaptive correlator receiver for direct-sequence spread-spectrum communication," *IEEE Trans. Commun.*, vol. 44, no. 11, pp. 1543–1552, Nov. 1996.
[10] H. V. Poor and X. Wang, "Code-aided interference suppression for DS/CDMA communications – part I: interference suppression capability," *IEEE Trans. Commun.*, vol. 45, no. 9, pp. 1101–1111, Sept. 1997.
[11] W. Xu and L. B. Milstein, "MMSE interference suppression for multicarrier DS-CDMA in frequency selective channels," in *Proc. IEEE Global Telecommunications Conference 1998, The Bridge to Global Integration, GLOBECOM 98*, vol. 1, pp. 259–264, Nov. 1998.
[12] M. Latva-aho and M. J. Juntti, "LMMSE detection for DS-CDMA systems in fading channels," *IEEE Trans. Commun.*, vol. 48, no. 2, pp 194–199, Feb. 2000.
[13] A. Papoulis and S. U. Pillai, *Probability, Random Variables and Stochastic Processes*, 4th edition. New York: McGraw-Hill, 2002.
[14] N. Kong and L. B. Milstein, "Average SNR of a generalized diversity selection combining scheme," *IEEE Commun. Lett.*, vol. 3, no. 3, pp. 57–59, March 1999.
[15] J. Wang and L. B. Milstein, "Multicarrier CDMA overlay for ultra-wideband communications," *IEEE Trans. Commun.*, vol. COM-52, pp. 1664–1669, Oct. 2004.
[16] T. T. Wong, "Multicarrier CDMA overlay for ultra-wideband wireless communications," M.Phil. Thesis, University of Hong Kong, 2003.
[17] A. Annamalai and C. Tellambura, "A new approach to performance evaluation of generalized selection diversity receivers in wireless channels," in *Proc. IEEE VTS 54th Vehicular Technology Conference, VTC 2001 Fall*, vol. 4, pp. 2309–2313, Oct. 2001.
[18] H. V. Poor and X. Wang, "Code-aided interference suppression for DS/CDMA communications – part II: parallel blind adaptive implementations," *IEEE Trans. Commun.*, vol. 45, No. 9, pp. 1112–1122, Sept. 1997.
[19] IEEE 802.15 Working Group for Wireless Personal Area Networks (WPANs), http://ieee802.org/15/index.html.

3 Impulse radio overlay in UWB communications

Interference suppression is important to allow UWB devices to operate over the spectrum occupied by narrowband systems. In this chapter, the use of a notch filter in TH-IR for UWB communication is considered, where a Gaussian monopulse is employed with pulse position modulation (PPM). A lognormal fading channel is assumed and a complete analytical framework is provided for performance evaluation using a transversal-type notch filter to reject narrowband interference. A closed-form expression of BER is derived, and the numerical results show that the use of a notch filter can significantly improve system performance. Furthermore, a performance comparison between TH-IR and multicarrier CDMA UWB systems is also made for the same transmit power, the same data rate and the same bandwidth. It is shown that, in the presence of narrowband interference, the TH-IR system with a notch filter can achieve a performance similar to multicarrier CDMA.

3.1 Introduction

Over the last decade, there has been a great interest in ultra-wideband (UWB) time-hopping (TH) impulse radio (IR) communication systems [1–8]. These systems make use of ultra-short duration pulses (monocycles), which yield ultra-wide bandwidth signals characterized by extremely low power densities. UWB systems are particularly promising for short-range high-speed wireless communications as they potentially combine reduced complexity with low power consumption, low probability of intercept, high accuracy positioning and immunity to multipath fading due to discontinuous transmission. Recently, the Federal Communications Commission (FCC) has defined the −10 dB emission mask between 3.1 GHz and 10.6 GHz for unlicensed UWB systems with a bandwidth of at least 500 MHz [9]. However, in this frequency band there are a variety of existing narrowband interfering signals, such as public safety band and wireless LAN (IEEE 802.11a) operating at frequencies of 4.9 GHz and 5.2 GHz, respectively. This means that UWB signals and narrowband signals must coexist and that there must be minimal mutual interference between them. Therefore, the successful deployment of UWB technology depends not only on the development of efficient multiple access techniques, but also on narrowband interference suppression techniques. Since the TH-IR receiver is operated by time-gating matched to the pulse duration [1–3], this time gating reduces the power of continuous-time interference to the duty cycle of IR. Therefore,

TH-IR inherently has the capability of narrowband interference suppression. However, when narrowband interference is very strong, we may need to use a notch filter [10] to help reject it.

Suppression schemes based on MMSE Rake combining were proposed in [11] and [12], while the computation complexity of the tap weight would be increased with the addition of branches within the Rake receiver as it is driven by the decision statistic. In this chapter, we study the use of a notch filter to suppress narrowband interference for TH-IR. A notch filter has been widely studied to reject narrowband interference for CDMA overlay systems when narrowband interference is very strong [13], [14]. Apart from the requirement for a higher sampling speed in comparison to the solutions proposed in [11] and [12], since the computation of the tap weights for notch filter is dependent on the nature of the narrowband interferer, the same set of weights can be adopted among the branches with the Rake receiver. Moreover, the notch filter also offers flexibility as its narrowband rejection ability can be enhanced by the addition of more taps. In this chapter, we will provide a complete performance analysis of error probability in addition to the simulation result.

The UWB concept can be based on several techniques, such as TH-IR or multicarrier CDMA techniques. In the proposed multicarrier CDMA system for UWB shown in Chapter 2, a notch filter has been shown to be very effective in rejecting narrowband signals. For the two different systems (TH-IR and multicarrier CDMA), we will compare their performances with a notch filter, assuming that they have the same transmit power, the same data rate and the same bandwidth.

3.2 System models

Suppose the time scale is divided into frames with duration T_f, and each frame is composed of N_h slots of duration T_c, the transmitted signal of the kth UWB user employing TH-IR with pulse position modulation (PPM) is given by:

$$s_k(t) = \sum_{n=-\infty}^{\infty} \sqrt{E_k}\,\Xi\big(t - nT_f - c_k^{(n)} T_c - \varepsilon\, d_k^{\lfloor n/N_s \rfloor}\big) \tag{3.1}$$

where E_k is the energy of a pulse and $\Xi(t)$ is the shape of the transmitted pulse. $\{c_k^{(n)}\}$ is the time hopping code of the kth user, where $c_k^{(n)} \in \{0, 1, \ldots, N_h - 1\}$, such that an additional time shift of $c_k^{(n)} T_c$ is introduced when the nth pulse of the kth user is transmitted. The code sequences of all users are assumed to be mutually independent. N_s is the number of pulses transmitted per symbol, i.e. the processing gain, the modulating data symbol changes every N_s hops (frames) and the index of the data symbol is $\lfloor n/N_s \rfloor$ ($\lfloor x \rfloor$ is the integer part of x). $\{d_k^{\lfloor n/N_s \rfloor}\}$ is the binary data sequence of the kth user and composed of equally likely symbols (or bits). A symbol has duration $T_s = N_s T_f$ and the symbol rate is $R_s = 1/T_s = 1/(N_s T_f)$. ε is the modulation index, where it is assumed that $\varepsilon > 0$, and $T_p + \varepsilon \le T_c$, where T_p denotes the pulse width, and the time shift added to a pulse by data is $\varepsilon\, d_k^{\lfloor n/N_s \rfloor}$. The optimal value of ε is around 20% of a pulse width [7].

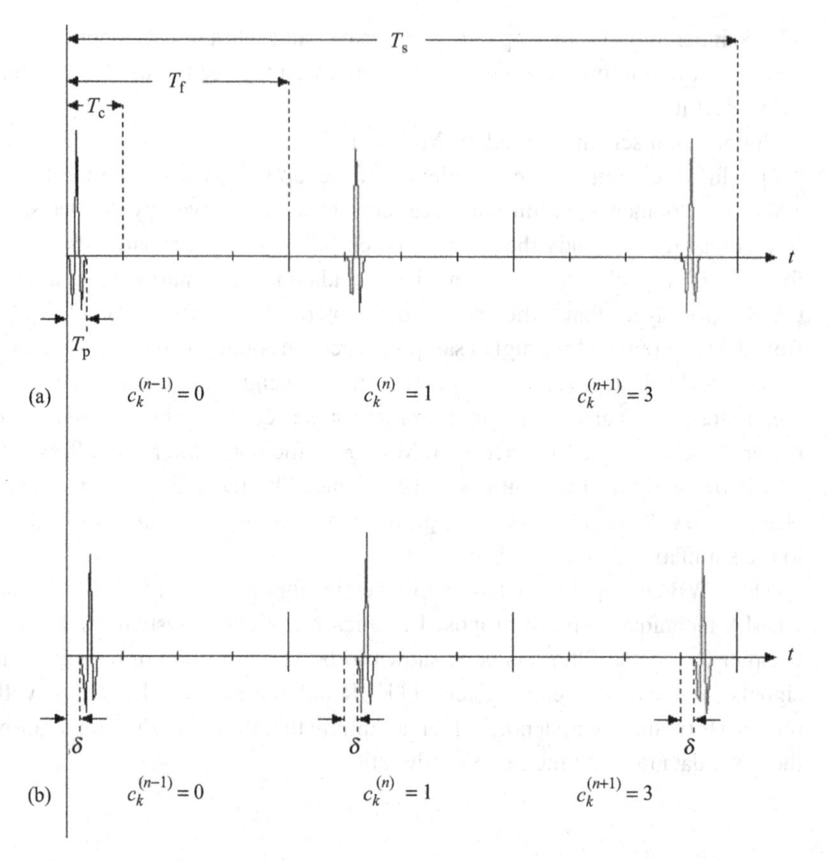

Figure 3.1 Sample of UWB TH-IR waveform ($N_s = 3$ frames, $N_h = 4$ bins). (a) Data: $d_k^{\lfloor n/n_s \rfloor} = 0$, (b) $d_k^{\lfloor n/n_s \rfloor} = 1$.

In this study, the monocycle denoted by $\Xi(t)$ is assumed to be a scaled second derivative of the Gaussian function with unit energy, i.e. $\int_{-\infty}^{+\infty} \Xi^2(t)\mathrm{d}t = 1$, and one form of a Gaussian monocycle in [8] is adopted as

$$\Xi(t + T_p/2) = \sqrt{\frac{28}{3T_p\sqrt{\pi}}}\left[1 - \left(\frac{7t}{T_p}\right)^2\right] \exp\left[-\frac{1}{2}\left(\frac{7t}{T_p}\right)^2\right] \tag{3.2}$$

and the -10 dB bandwidth of the Gaussian pulse is approximately given by

$$B_{\mathrm{UWB}} \approx 3.185/T_p \tag{3.3}$$

Figure 3.1 shows a sample of a UWB transmitted signal, where $N_s = 3$ and $N_h = 4$.

The narrowband interferer $j(t)$ is assumed to be a passband Gaussian with center frequency f_j and bandwidth B_j. The double-sided flat power spectral density of the narrowband interferer is given by

$$S_j(f) = \begin{cases} J/(2B_j), & |f \pm f_j| \leq B_j/2 \\ 0, & \text{otherwise} \end{cases} \tag{3.4}$$

where J is the power of the narrowband interference, and the autocorrelation function of $j(t)$ is

$$R_j(\tau) = \int_{-\infty}^{\infty} S_j(f) \exp(j2\pi f\tau) df = J \cdot \bar{R}_j(\tau) \tag{3.5}$$

where $\bar{R}_j(\tau)$ denotes the normalized autocorrelation function of $j(t)$, given by

$$\bar{R}_j(\tau) = \frac{\sin(\pi B_j \tau)}{\pi B_j \tau} \cos(2\pi f_j \tau) \tag{3.6}$$

For consistency with the analysis of multicarrier CDMA in later sections, the parameter p is defined as a ratio of the bandwidth of the narrowband signal to the -10 dB bandwidth, B_{UWB}, of the Gaussian pulse, which can be written as

$$p = B_j / B_{\text{UWB}} \tag{3.7}$$

For a typical narrowband system, the usual range of the ratio should be $0 \leq p \ll 0.1$. Another important parameter q is defined as the ratio of the difference in center frequencies (between the narrowband interferer and the Gaussian pulse) to the bandwidth of the Gaussian pulse, given by

$$q = (f_j - f_{\text{UWB}}) / B_{\text{UWB}} \tag{3.8}$$

where $-0.5 \leq q \leq 0.5$. In (3.8), f_{UWB} is the center frequency of the spectrum of the Gaussian pulse and $f_{\text{UWB}} \approx B_{\text{UWB}}/2$. Therefore, the parameter q can be written approximately as

$$q = \left(f_j - \frac{B_{\text{UWB}}}{2}\right) \Big/ B_{\text{UWB}} = f_j / B_{\text{UWB}} - \frac{1}{2} \tag{3.9}$$

The realistic UWB channel should be a dense multipath fading channel [7]. More than a dozen resolvable paths should exist. In the multiple access system with K users, the received signal can be expressed as

$$r(t) = \sum_{k=1}^{K} \sum_{l=1}^{L} \psi_{k,l} \beta_{k,l} s_k(t - \tau_{k,l}) + j(t) + n(t) \tag{3.10}$$

where $\psi_{k,l}$, $\beta_{k,l}$ and $\tau_{k,l}$ represent the phase, amplitude attenuation and delay, respectively, of the lth arrival path of the kth user. Independent fading is assumed for each path as well as for each user [15]. The phases are independent variables and take the value 1 or -1 with equal probability to account for signal inversion due to reflection. The delay is assumed to be uniformly distributed over $[0, T_s)$, i.e. the symbol duration. The amplitude attenuation is lognormally distributed with $\mu_{k,l}$ and $\sigma_{k,l}$ being the mean and standard deviation, respectively; its probability density function takes the form

$$p(\beta_{k,l}) = \frac{1}{\beta_{k,l} \sigma_{k,l} \sqrt{2\pi}} \exp\left\{ \frac{-[\ln(\beta_{k,l}) - \mu_{k.l}]^2}{2\sigma_{k,l}^2} \right\} \tag{3.11}$$

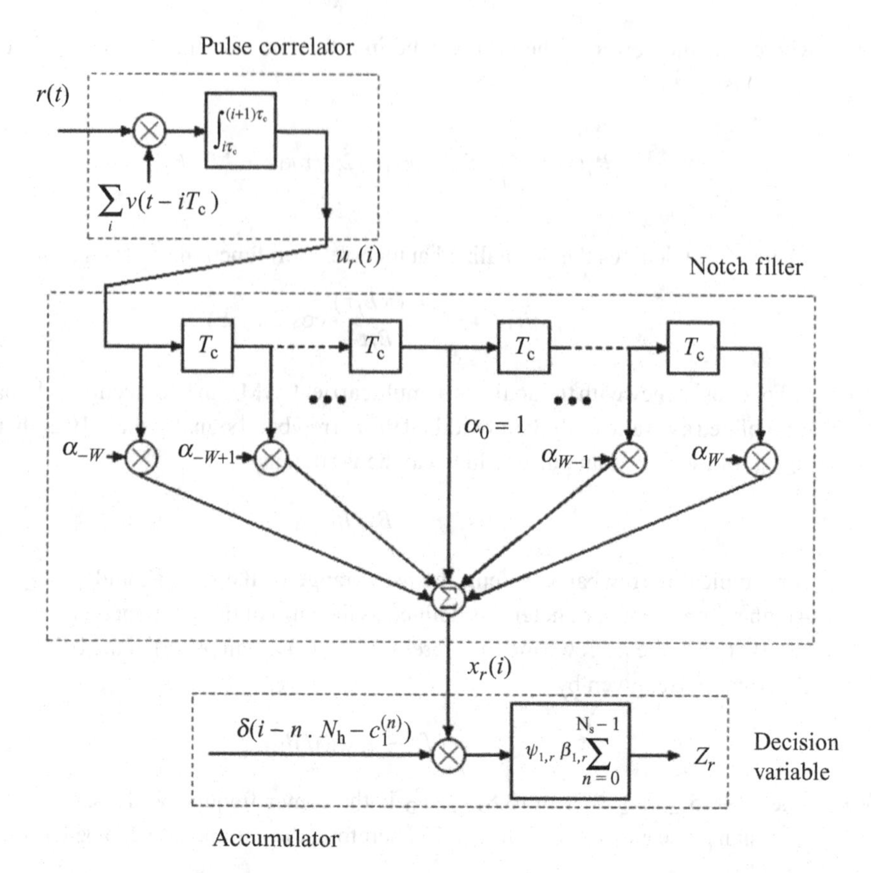

Figure 3.2 The rth branch of the receiver.

and its normalized intensity profile is given by

$$\Omega_{k,l} = E\left[\beta_{k,l}^2\right] = \frac{1 - \exp(-\nu)}{1 - \exp(-L\nu)} \exp[-(l-1)\nu] \tag{3.12}$$

where ν is the decay rate. Note that the normalization implies that the sum of the attenuation power of all paths is one $\left(\sum_{l=1}^{L} \Omega_{k,l} = 1\right)$. In (3.10), $n(t)$ is an AWGN with a double-sided power spectral density of $\eta_0/2$.

3.3 Performance evaluation

Selective maximal combination (SMC) is considered for the proposed receiver design, with the R highest ($R < L$) power paths out of all resolvable paths chosen for decision making. Perfect power control is also assumed. As shown in Figure 3.2, each branch of the Rake receiver consists of a pulse correlator, a notch filter and an accumulator. Assuming that the first user is the desired user, the output of the pulse correlator for the

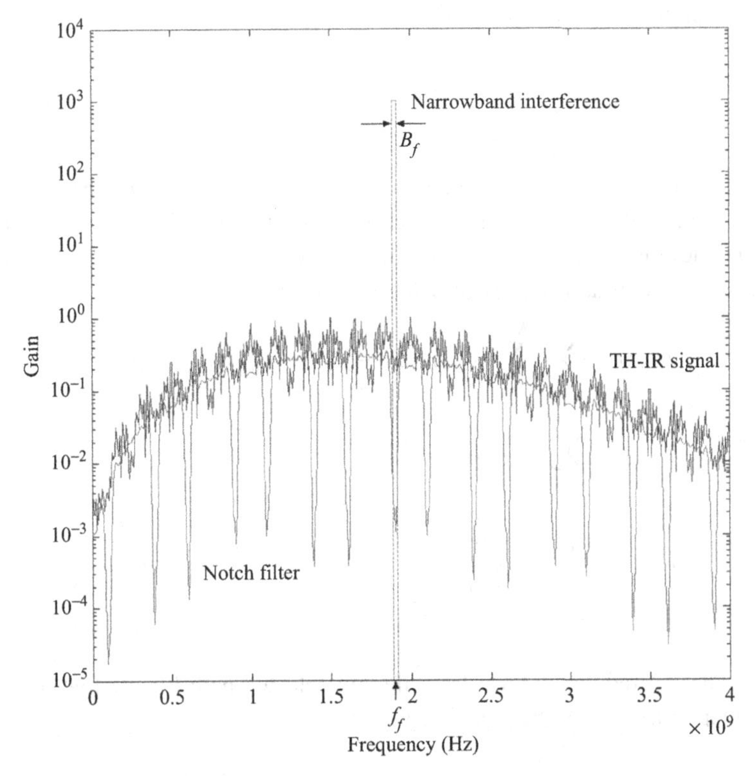

Figure 3.3 Frequency responses of the transmitted TH-IR signal and the notch filter at receiver.

rth selected path is given by

$$u_r(i) = \int_{iT_c}^{(i+1)T_c} r(t) \sum_i v(t - iT_c - \tau_{1,r})\mathrm{d}t = \int_0^{T_c} r(t + iT_c + \tau_{1,r})v(t)\mathrm{d}t \tag{3.13}$$

where $v(t)$ is the template pulse function, defined as

$$v(t) = \Xi(t) - \Xi(t - \varepsilon) \tag{3.14}$$

Note that the pulse correlator output is sampled at the rate of one sample per bin period T_c. The notch filter (Wiener filter) is used to predict and notch out the narrowband interference [10]. The number of taps on each side of the notch filter is W and thus the total number of taps is $2W + 1$. The coefficients of the filter are $\{\alpha_w\}$ with $w = [-W, \ldots, W]$ and $\alpha_0 = 1$. Note that when there is no narrowband interference the filter reduces to an all-pass filter, i.e. $\alpha_w = 0$ for $w \neq 0$. The conceptual frequency response of the notch filter along with the spectra of TH-IR signal and narrowband interference are shown in Figure 3.3, which shows that the narrowband interference can be suppressed. The output of the notch filter can be expressed as

$$x_r(i) = \sum_{w=-W}^{W} \alpha_w u_r(i-w) = \sum_{w=-W}^{W} \alpha_w \int_0^{T_c} r(t - wT_c + iT_c + \tau_{1,r})v(t)\mathrm{d}t \tag{3.15}$$

This output is passed to the accumulator using a delta function $\delta(i - nN_h - c_1^{(n)})$, where $\delta(i) = 1$ and 0 for $i = 0$ and $i \neq 0$, respectively. Thus, with the multiplication phase $\psi_{1,r}$ and the amplitude attenuation $\beta_{1,r}$ perfectly estimated from the channel, the random variable at the output of the accumulator is given by

$$\begin{aligned}
Z_r &= \beta_{1,r}\psi_{1,r} \sum_{i=-\infty}^{\infty} \sum_{n=0}^{N_s-1} x_r(i) \cdot \delta\left(i - nN_h - c_1^{(n)}\right) \\
&= \beta_{1,r}\psi_{1,r} \sum_{n=0}^{N_s-1} x_r\left(nN_h + c_1^{(n)}\right) \\
&= \beta_{1,r}\psi_{1,r} \sum_{n=0}^{N_s-1} \left[\sum_{w=-W}^{W} \alpha_w \int_0^{T_c} r\left(t - wT_c + nT_f + c_1^{(n)}T_c + \tau_{1,r}\right)v(t)\mathrm{d}t \right] \\
&= Z_{D|r} + Z_{I|r}
\end{aligned} \tag{3.16}$$

$Z_{D|r}$ is the desired signal term from the rth selected path for the first user, from the central tap of the notch filter, and is given by

$$\begin{aligned}
Z_{D|r} &= \sum_{n=0}^{N_s-1} \beta_{1,r}^2 \psi_{1,r}^2 \int_0^{T_c} s_1\left(t + nT_f + c_1^{(n)}T_c\right)v(t)\mathrm{d}t \\
&= \sqrt{E_1}\beta_{1,r}^2 \sum_{n=0}^{N_s-1} \int_0^{T_c} \Xi\left(t - \varepsilon d_1^{\lfloor n/N_s \rfloor}\right)v(t)\mathrm{d}t \\
&= \begin{cases} \sqrt{E_1}\beta_{1,r}^2 N_s \rho, & \text{for } d_1^{\lfloor n/N_s \rfloor} = 0 \\ -\sqrt{E_1}\beta_{1,r}^2 N_s \rho, & \text{for } d_1^{\lfloor n/N_s \rfloor} = 1 \end{cases}
\end{aligned} \tag{3.17}$$

where ρ stands for the correlation between the transmitted pulse and the template function:

$$\rho = \int_0^{T_c} \Xi(t)v(t)\mathrm{d}t \approx -\int_0^{T_c} \Xi(t)v(t-\varepsilon)\mathrm{d}t \tag{3.18}$$

$Z_{I|r}$ in (3.16) is the total interference for the rth selected path given by

$$Z_{I|r} = Z_{\text{MPI}|r} + Z_{\text{MAI}|r} + Z_{\text{NBI}|r} + Z_{\text{AWGN}|r} \tag{3.19}$$

where $Z_{\text{MPI}|r}$ is the multipath interference (MPI) from other paths of the desired user, $Z_{\text{MAI}|r}$ is the multiple access interference (MAI) from all other $K-1$ interfering users, $Z_{\text{NBI}|r}$ is the narrowband interference (NBI), and $Z_{\text{AWGN}|r}$ is the AWGN. Note that the disturbance (self-interference) caused by the notch filter to the desired user (or interference of the desired user caused by the non-central taps of the notch filter) should be very small when the bandwidth ratio p is small [14]. Including this interference makes the analysis very complicated since this term contains the same fading factors as the desired term, so self-interference is neglected in (3.19) for simple analysis. Since the terms in (3.19) are independent when the number of multipaths is large, the total interference can be approximated as Gaussian with variance

$$\mathrm{Var}[Z_{I|r}] = \mathrm{Var}[Z_{\mathrm{MPI}|r}] + \mathrm{Var}[Z_{\mathrm{MAI}|r}] + \mathrm{Var}[Z_{\mathrm{NBI}|r}] + \mathrm{Var}[Z_{\mathrm{AWGN}|r}] \tag{3.20}$$

where $\mathrm{Var}[Z_{\mathrm{MPI}|r}]$, $\mathrm{Var}[Z_{\mathrm{MAI}|r}]$ and $\mathrm{Var}[Z_{\mathrm{NBI}|r}]$ stand for the variances of the MPI, MAI and the NBI, respectively, derived in the Appendices 3A, 3B and 3C, respectively, and are given by (3A.11), (3B.10) and (3C.6), respectively. $\mathrm{Var}[Z_{\mathrm{AWGN}|r}]$ stands for the variance of the noise term, given by

$$\mathrm{Var}(Z_{\mathrm{AWGN}|r}) = \eta_{\mathrm{o}} \beta_{1,r}^2 \sigma_{\mathrm{AWGN}|r}^2 \tag{3.21}$$

where

$$\sigma_{\mathrm{AWGN}|r}^2 = \frac{N_{\mathrm{s}}}{2} \left(\sum_{w=-W}^{W} \alpha_w^2 \right) \left(\int_0^{T_{\mathrm{c}}} v^2(t) \mathrm{d}t \right) \tag{3.22}$$

Therefore, after weighting by amplitude attenuation, the SINR in Z_r is given by

$$\gamma_r = \frac{E^2[Z_r]}{\mathrm{Var}[Z_r]} = \beta_{1,r}^2 \left\{ \frac{\sigma_{\mathrm{MPI}|r}^2}{N_{\mathrm{s}}^2 \rho^2} + \frac{(K-1)\sigma_{\mathrm{MAI}|r}^2}{N_{\mathrm{s}}^2 \rho^2} + \left(\frac{J}{P} \right) \frac{\sigma_{\mathrm{NBI}|r}^2}{N_{\mathrm{s}}^2 \rho^2 N_{\mathrm{h}} T_{\mathrm{c}}} + \left(\frac{E_{\mathrm{b}}}{\eta_{\mathrm{o}}} \right)^{-1} \frac{\sigma_{\mathrm{AWGN}|r}^2}{N_{\mathrm{s}} \rho^2} \right\}^{-1} = \beta_{1,r}^2 \bar{\gamma}_r \tag{3.23}$$

where $\bar{\gamma}_r$ is the average SINR excluding the amplitude attenuation $\beta_{1,r}$, $E_{\mathrm{b}} = E_1 N_{\mathrm{s}}$ denotes the received bit energy, and $P = E_1/(N_{\mathrm{h}} T_{\mathrm{c}})$ represents the average transmitted power of the UWB signal. The outputs from the R selected branches of the Rake receiver are combined, so the decision variable is given by

$$Z = \sum_{r=1}^{R} Z_r \tag{3.24}$$

with SINR

$$\gamma = \sum_{r=1}^{R} \beta_{1,r}^2 \bar{\gamma}_r \tag{3.25}$$

Since the square of a lognormal random variable is still lognormally distributed, by Schwartz and Yeh's method [16], the distribution of the sum of lognormal random variables can be approximated by another lognormal distribution. Its mean and variance can be obtained by a recursive approach from the individual mean and standard derivation of the attenuation factors. If we define $\bar{\gamma}_{\mathrm{ave}} = \frac{1}{R} \sum_{r=1}^{R} \bar{\gamma}_r$ as the mean of the average SINR excluding the attenuation factor for all selected paths, and $\zeta = \sum_{r=1}^{R} \beta_r^2$ as the sum of the square of the attenuation factors, then the error probability can be approximated as

$$P_{\mathrm{e}} = \int_0^{\infty} Q(\sqrt{\gamma}) p(\gamma) \mathrm{d}\gamma \approx \int_0^{\infty} Q\left(\sqrt{\zeta \bar{\gamma}_{\mathrm{ave}}}\right) p(\zeta) \mathrm{d}\zeta \tag{3.26}$$

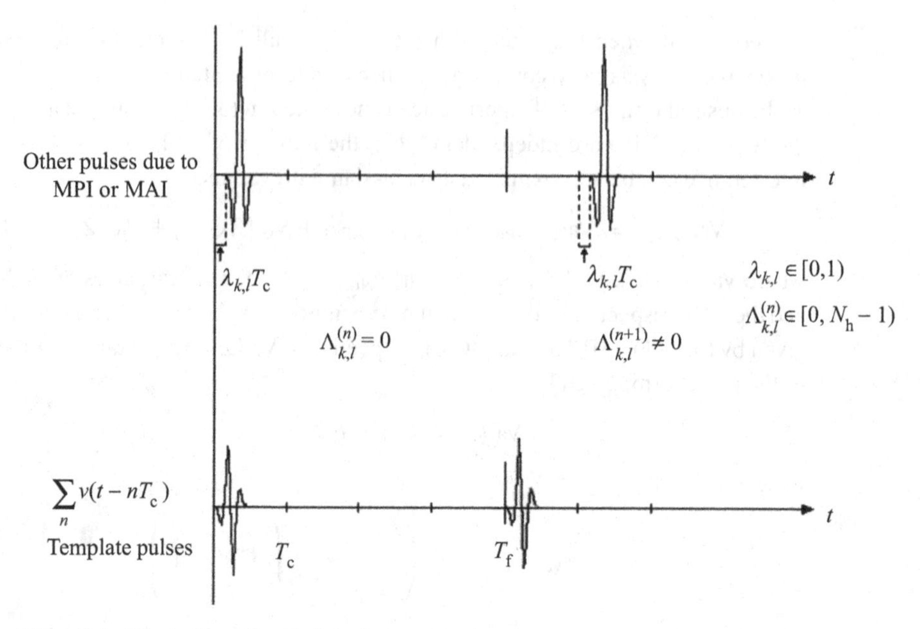

Figure 3.4 Illustration of pulse interference.

where the Q-function $Q(x)$ is defined as

$$Q(x) = \frac{1}{\sqrt{2\pi}} \int_x^{\infty} \exp(-\Theta^2/2)\mathrm{d}\Theta \tag{3.27}$$

3.4 Comparison of time-hopping and multicarrier CDMA

In order to compare the time-hopping system with the multicarrier (MC) CDMA in the presence of narrowband interference, we briefly describe the multicarrier CDMA, which was studied in Chapter 2. Multicarrier CDMA modulates different sub-carriers using the same data. All sub-carrier spectra are disjoint. The MC-CDMA has an inherent frequency diversity capability by combining the outputs of the different sub-carrier signals at the receiver. Moreover, it yields effective narrowband interference rejection in an overlay mode. For example, when there is a strong narrowband interferer in one of the sub-bands, in the worst case the receiver can simply ignore the signal in that sub-carrier band. An effective way is to use a notch filter to suppress narrowband interference in each sub-carrier. Then, even a jammed sub-carrier signal can still make a positive contribution to the net frequency diversity.

The pulse interference is shown in Figure 3.4. Figure 3.5 illustrates the transmitter of the multicarrier CDMA system. The source binary data sequence is first spread by the random binary sequence. Then, the spread signal is shaped by a chip-waveform-shaping filter (a square-root raised cosine filter of roll-off factor ϑ) with

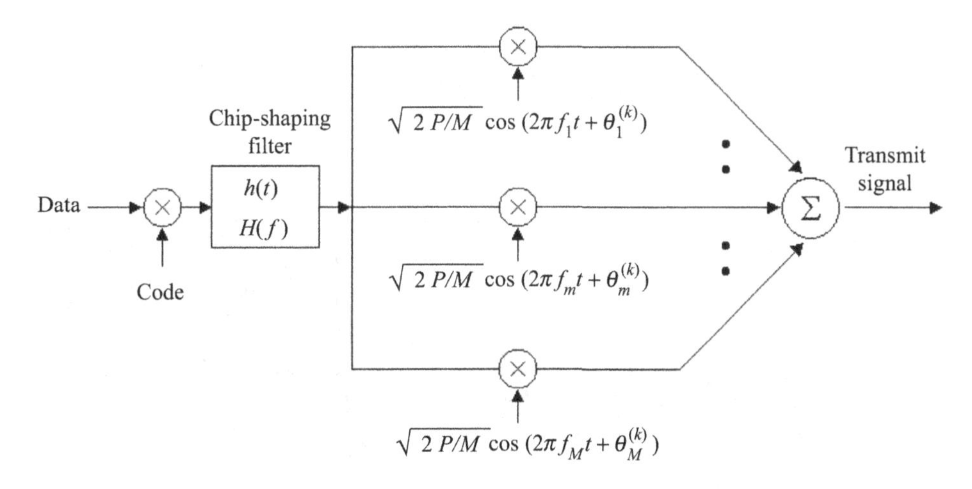

Figure 3.5 Transmitter for the kth user of the multicarrier CDMA UWB system.

frequency response $H(f)$. After that, the shaped signal modulates M different sub-carriers. Finally, the M sub-carrier modulated signals are summed to form a transmitter signal.

For fair comparison, the channel parameters have to be chosen appropriately as each sub-band of multicarrier CDMA occupies only a portion of the bandwidth of the UWB system. It is assumed that the number of paths in each sub-carrier is $L_{\mathrm{MC}} = L/M$. That is each sub-band of the multicarrier CDMA would have fewer multipaths but of greater power and with a larger decay rate than in the TH-IR system. In addition, the phase of each path is no longer a binary but is a random variable uniformly distributed over $[0, 2\pi)$, due to the introduction of the carrier.

As shown in Figure 3.6, the receiver consists of M parallel branches of the Rake receiver with R correlators, corresponding to M sub-carriers. In each branch, the received signal is input to a frequency down converter. After that, a baseband-matched filter with frequency response $H^*(f)$ is employed. Then, the output of the matched filter passes through a despreader. Similarly, the output of the despreader can be approximated as a Gaussian random variable with the SINR in the rth branch given by [13]

$$\gamma_{1,r}^{(m)} = \left\{ \frac{1}{2N^2} \left[\sum_{\substack{l=1 \\ l\neq r}}^{L_{\mathrm{MC}}} \Omega_{k,l} + (K-1) \right] \sum_{w_1=-W}^{W} \sum_{w_2=-W}^{W} \alpha_{w_1}^{(m)} \alpha_{w_2}^{(m)} \right.$$

$$\times \sum_{n=0}^{N-1} \sum_{\hat{n}=0}^{N-1} R(w_1 - \hat{n} + n, w_2 - \hat{n} + n) + \frac{M}{N\hat{p}(1+\vartheta)} \cdot \hat{\sigma}_J^2 \cdot \frac{J}{P}$$

$$\left. + M \left(\frac{E_{\mathrm{b}}}{\eta_{\mathrm{o}}} \right)^{-1} \sum_{w=-W}^{W} \left(\alpha_w^{(m)}\right)^2 \right\}^{-1} \tag{3.28}$$

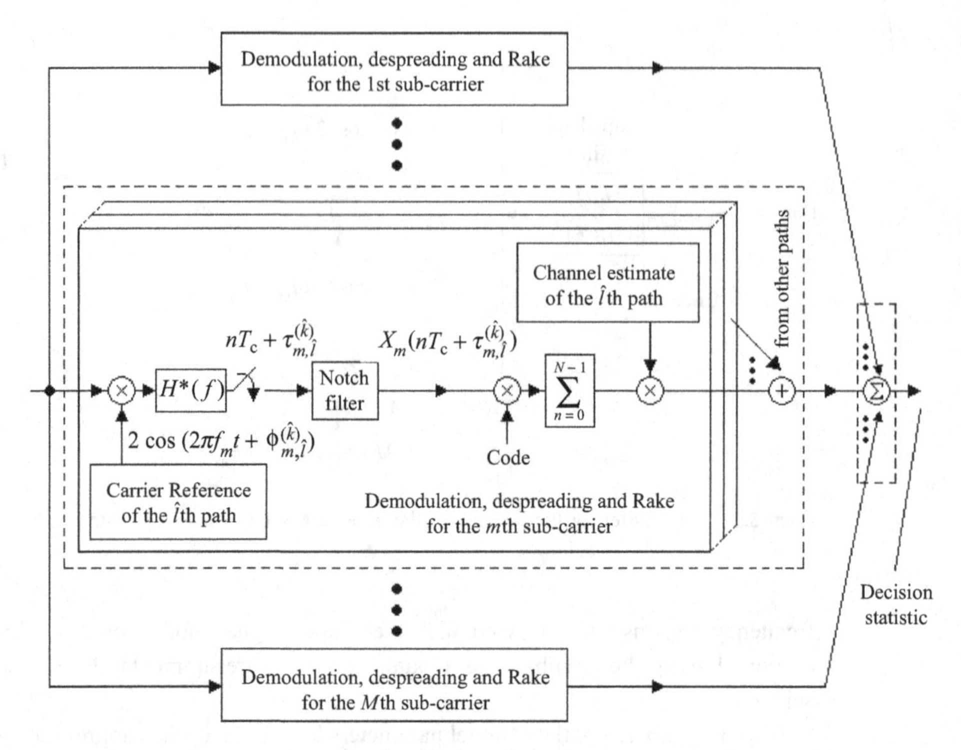

Figure 3.6 Receiver for the kth user of the multicarrier CDMA UWB system.

and in the absence of narrowband interference [13]

$$\gamma^{(m)}_{2,r} = \left\{ \frac{1}{2N^2} \left[\sum_{\substack{l=1 \\ l \neq r}}^{L_{MC}} \Omega_{k,l} + (K-1) \right] \sum_{n=0}^{N-1} \sum_{\hat{n}=0}^{N-1} R(n-\hat{n}, n-\hat{n}) + M \left(\frac{E_b}{\eta_o} \right)^{-1} \right\}^{-1} \tag{3.29}$$

where N is the spreading factor of one sub-carrier. $R(w_1 - \hat{n} + n, w_2 - \hat{n} + n)$ is defined as

$$R(n_1, n_2) = \int_0^1 x(\tau T_{ch} - n_1 T_{ch})\, x(\tau T_{ch} - n_2 T_{ch}) d\tau \tag{3.30}$$

where $x(t)$ is the impulse response of a raised cosine filter and T_{ch} is the chip rate. In (3.28), $\hat{\sigma}^2_J$ is given by

$$\hat{\sigma}^2_J = \int_{(1+\vartheta)(\hat{q}-\hat{p})/T_{ch}}^{(1+\vartheta)(\hat{q}+\hat{p})/T_{ch}} |H_w(f)|^2\, |H(f)|^2 df \tag{3.31}$$

$\hat{p}$ and $\hat{q}$ are defined, respectively, as the ratio of the bandwidth of the narrowband interferer to the bandwidth of one sub-band and the ratio of the difference in center frequencies (between the narrowband interferer and the corrupted sub-band) to the bandwidth of one sub-band.

The despreader output is weighted by a channel estimate, and the weighted outputs from all branches are summed to form the final test statistics:

$$\gamma = \sum_{m=1}^{M} \sum_{r=1}^{R} \beta_{m,r}^2 \gamma_r^{(m)} = \gamma_1 \beta_{M_1} + \gamma_2 \beta_{M-M_1} \tag{3.32}$$

where M_1 is the number of sub-bands with narrowband interference. In (3.32), γ_1 and γ_2 stand for the mean of the average SINR excluding the attenuation factor for all selected paths from all the sub-bands with and without narrowband interference, respectively. They are numerically expressed as

$$\gamma_1 = \frac{1}{M_1 R} \sum_{m=1}^{M_1} \sum_{r=1}^{R} \gamma_{1,r}^{(m)}, \tag{3.33}$$

$$\gamma_2 = \frac{1}{(M - M_1)R} \sum_{m=M_1+1}^{M} \sum_{r=1}^{R} \gamma_{2,r}^{(m)} \tag{3.34}$$

$\zeta_{M_1} = \sum_{m=1}^{M_1} \sum_{r=1}^{R} \beta_{m,r}^2$ is defined as the sum of the square of the attenuation factors for the sub-carriers with the narrowband interferers, and $\zeta_{M-M_1} = \sum_{m=M_1+1}^{M} \sum_{r=1}^{R} \beta_{m,r}^2$ denotes the sum of the remainders. The bit error rate (BER) for the multicarrier CDMA system can thus be expressed as

$$\begin{aligned} P_\mathrm{e} &= \int_0^\infty Q(\sqrt{\gamma})p(\gamma)\mathrm{d}\gamma \\ &= \int_0^\infty \int_0^\infty Q(\sqrt{\gamma_1 \zeta_{M_1} + \gamma_2 \zeta_{M-M_1}})p(\zeta_{M_1})p(\zeta_{M-M_1})\mathrm{d}\zeta_{M_1}\mathrm{d}\zeta_{M-M_1} \end{aligned} \tag{3.35}$$

where $p(x)$ denotes the lognormal probability density function.

The TH-IR and multicarrier CDMA systems are compared under the conditions of same signal power, same data rate and same bandwidth. The total bandwidth of the multicarrier CDMA UWB system is approximately

$$B_{\mathrm{UWB}} \approx M \cdot (1 + \vartheta)/T_{\mathrm{ch}} \tag{3.36}$$

Assuming that only one narrowband interferer is present, the parameters describing the narrowband interference for TH-IR UWB system and multicarrier CDMA system can be related as

$$\hat{p} = B_j/(B_{\mathrm{UWB}}/M) \approx p \cdot M \tag{3.37}$$

$$\hat{q} = \frac{f_j - f_m}{(B_{\mathrm{UWB}}/M)} \approx \frac{q}{|q|}\left[q \cdot M - \lfloor q \cdot M \rfloor - \frac{1}{2}\right] \tag{3.38}$$

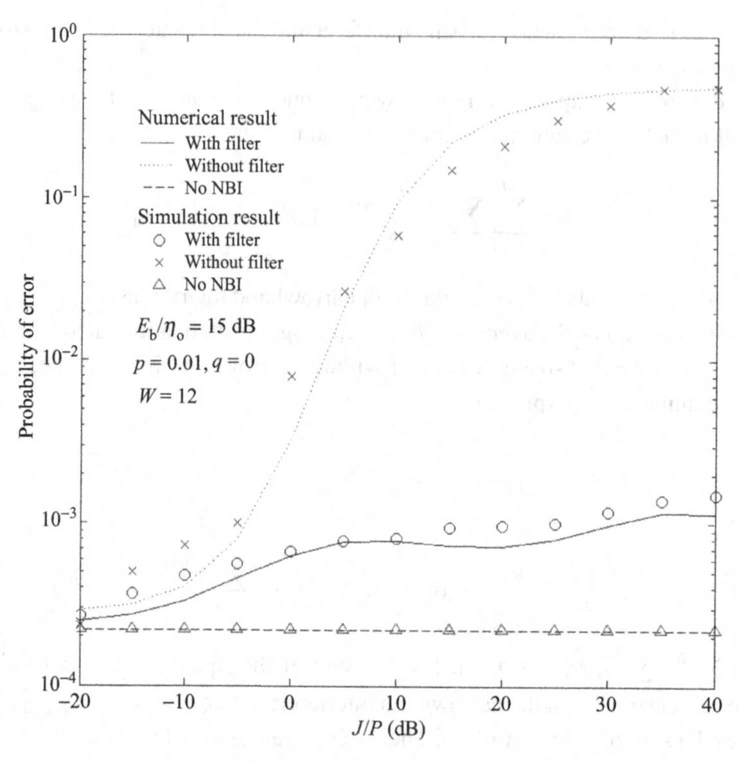

Figure 3.7 Error probability as a function of interference power to signal power ratio.

3.5 Numerical results

Some representative numerical results of the TH-IR UWB in the presence of narrowband interference are illustrated first in this section. The following system parameters are assumed unless explicitly specified: the pulse width of the Gaussian monocycle T_p, the modulation index ε, and the duration of the time bin T_c are 1.0 ns, 0.2 ns and 2.0 ns, respectively [7]. The number of pulses transmitted per symbol N_s is 5, the number of time slots per frame N_h is 8, and the number of active users K is 8. For the channel response, the number of multipaths L is 30, where the decay rate υ and the standard deviation σ are set at 0.15 and 0.5, respectively.

Figure 3.7 shows the error probabilities for a receiver with and without a notch filter as a function of J/P. For comparison, the case without narrowband interference is also shown. Simulation and analytical results are presented under these conditions: $E_b/\eta_o = 15$ dB, $p = 0.01$, $q = 0$, $R = 10$ and $W = 12$. It can be seen from the figure that the performance without a notch filter degrades dramatically as J/P increases, especially when J/P is larger than 0 dB. When J/P is small (less than 0 dB), performances with and without a notch filter are very close, so a notch filter is unnecessary. When J/P increases from 0 dB, the performance gap between the filter and no-filter conditions increases. When J/P is larger than 10 dB, the use of a notch filter gives an improvement

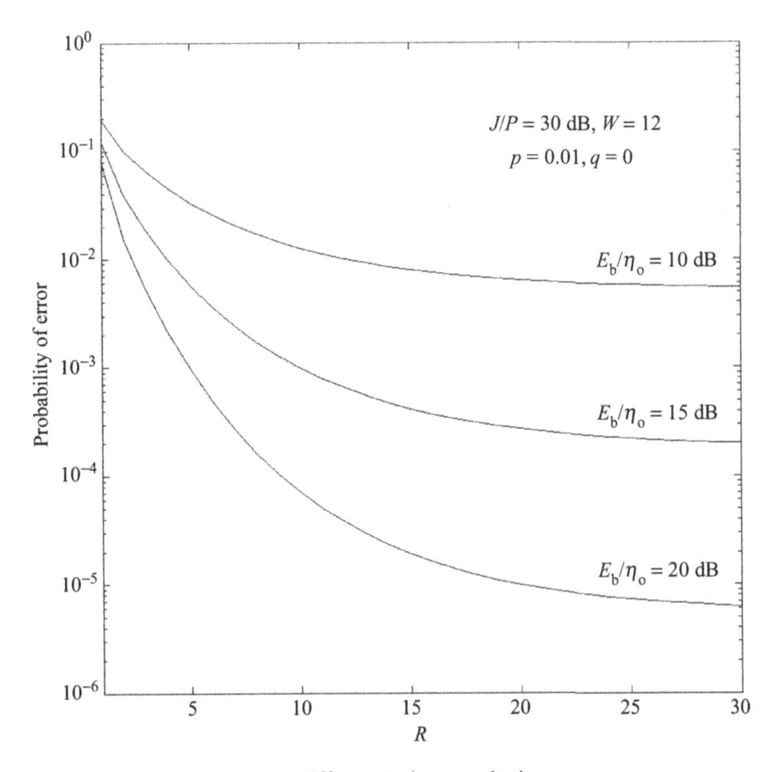

Figure 3.8 Bit error rate for different Rake complexity.

of two to three orders of magnitude. It can be observed that the analytic results and simulation results are close.

Figure 3.8 investigates the system performance by exploring multipath diversity (Rake receiver) for different values of signal-to-noise ratio ($E_b/\eta_o = 10$ dB, 15 dB and 20 dB). Other parameters are $J/P = 30$ dB, $p = 0.01$, $q = 0$ and $W = 12$. It can be seen that for a given E_b/η_o the error probability decreases sharply when the number of Rake fingers increases at the beginning ($R \leq 10$), with the performance improvement becoming small when R approaches L. This indicates that reliable symbol decision making can be achieved by using a limited number of significant paths, rather than all available paths.

The effect of the number of notch filter taps on system performance is presented in Figure 3.9, for $p = 0.01$, $E_b/\eta_o = 15$ dB, $J/P = 30$ dB, $q = 0$ and $R = 10$. It can be seen that the curves show a zig-zag shape in performance rather than a smooth fall. This is because the further addition of taps, although providing better narrowband interference suppression, introduces more pulse interference to the decision statistics, which is mainly due to the increased probability of capturing pulses from multipath and multiple access users. This phenomenon is serious when W is small, as the limited benefit of narrowband interference reduction is counterbalanced or exceeded by the worsening pulse interference. However, the overall trend of performance improvement continues as more taps are used. The curve levels off when $W > 8$, which indicates that

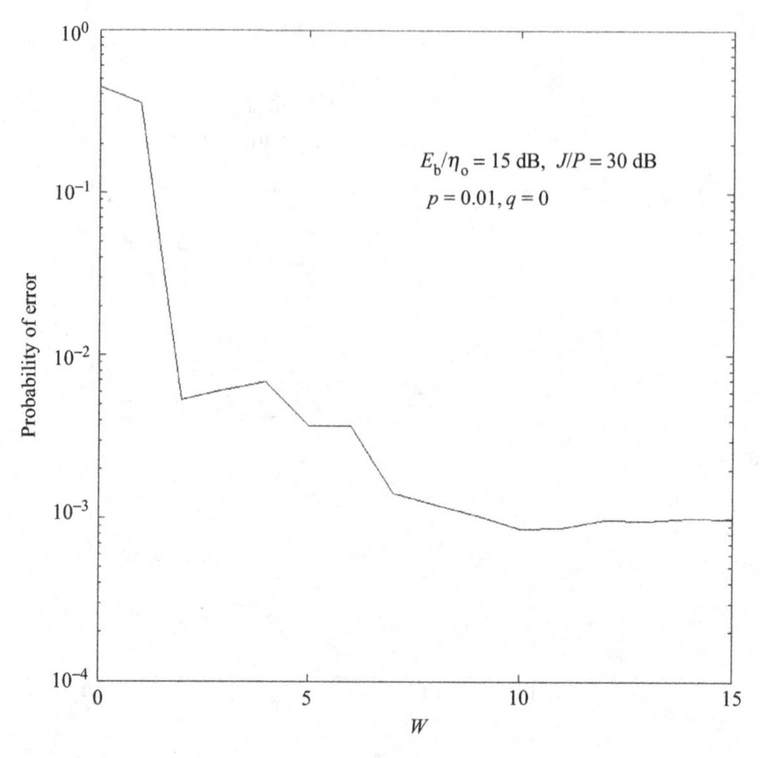

Figure 3.9 System performance corresponding to different of number of taps per side of the notch filter.

the marginal improvement is limited. The error rates converge to different levels for different bandwidth ratios.

Figures 3.10 and 3.11 are plotted against the bandwidth ratio p (for $q = 0$) and the ratio q (for $p = 0.02$) of the offset of the center frequency of the narrowband interferer to the bandwidth, respectively. The remaining system parameters are set at $J/P = 30$ dB with $E_b/\eta_o = 15$ dB. In accordance with the observation from Figure 3.9, a notch filter with larger W offers greater resistance to an increase in the bandwidth ratio. From Figure 3.11, it can be observed that the system performance varies with the normalized frequency offset. The frequency response of the notch filter can be written as

$$H_W(f) = \sum_{w=-W}^{W} \alpha_w \exp(-j2\pi f w T_c) \tag{3.39}$$

Such variations can be accounted by the periodicity and sinusoidal characteristics of the frequency response of the filter, as illustrated in Figure 3.3. Better jamming rejection can be achieved when the center frequency of the narrowband system coincides with certain harmonics of $1/T_c$. The degree of accuracy of the estimating function depends on the number of taps per side of the notch filter. A notch filter with $W = 6$ offers better protection than that with $W = 3$, but the reliability of the decision statistic is unstable as q varies. When W is increased to 12, however, the sensitivity of the error rate subjected

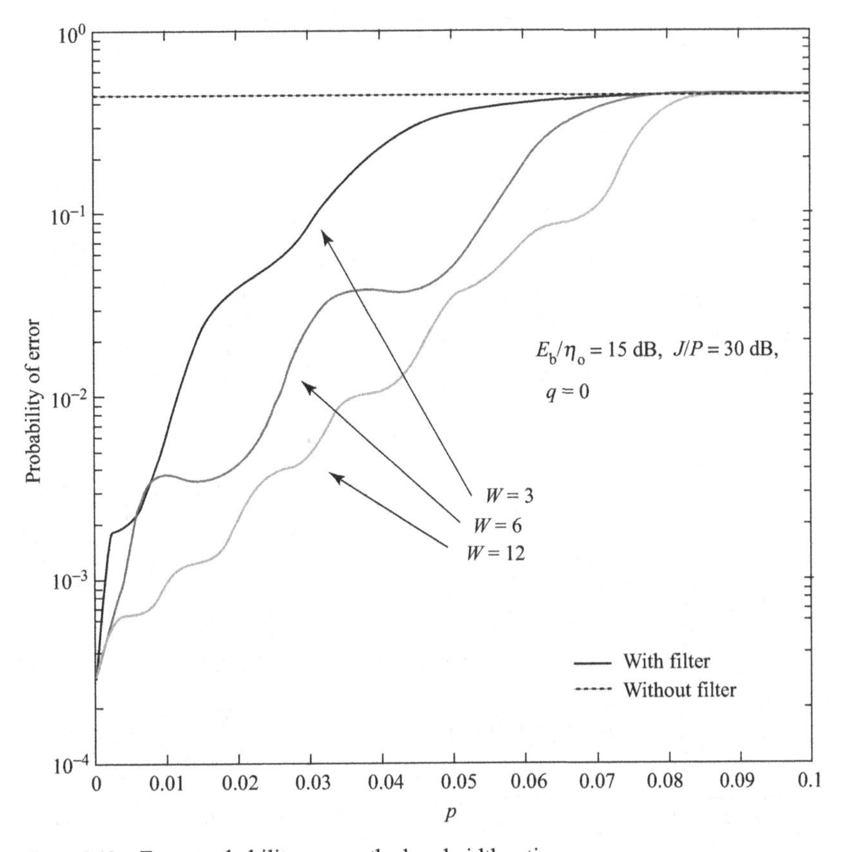

Figure 3.10 Error probability versus the bandwidth ratio.

to the change in q is reduced significantly. The rate and fluctuation of the bit error can be reduced by using a filter of larger W.

Figure 3.12 compares the performances of TH-IR and multicarrier CDMA UWB systems subject to the change in J/P. The symbol times for a TH-IR system and a multicarrier CDMA system, respectively, are given by $T_s = N_s N_h T_c$ and $T_s = NT_{ch}$. It is assumed that the roll-off factor ϑ of the raised-cosine filter is 0.3, and the chip period T_{ch} is 10 ns, the spreading gain per sub-carrier N is 8, and the number of sub-carriers M is 10. For the channel response, the number of multipaths L_{MC} is 3, and the decay rate ν_{MC} and the standard deviation σ_{MC} are 1.5 and 0.5, respectively. Only one narrowband signal is assumed within the spectrum of the system signal. When the bandwith and normalized offset ratios for the TH-IR system are $p = 0.01$ and $q = 0.05$, respectively, the corresponding ratios for the multicarrier CDMA system are $\hat{p} = 0.1$ (see (3.37)) and $\hat{q} = 0$ (see (3.38)), respectively. In the absence of the near–far effect, by assuming perfect power control for both systems, the signal-to-noise ratio E_b/η_o is 10 dB and the conditions for the number of active users K being 3 and 8 are illustrated. For the TH-IR system, the parameters W and R are kept unchanged at 12 and 10, respectively. For the multicarrier CDMA system, the number of taps per side of the notch filter W is also 12.

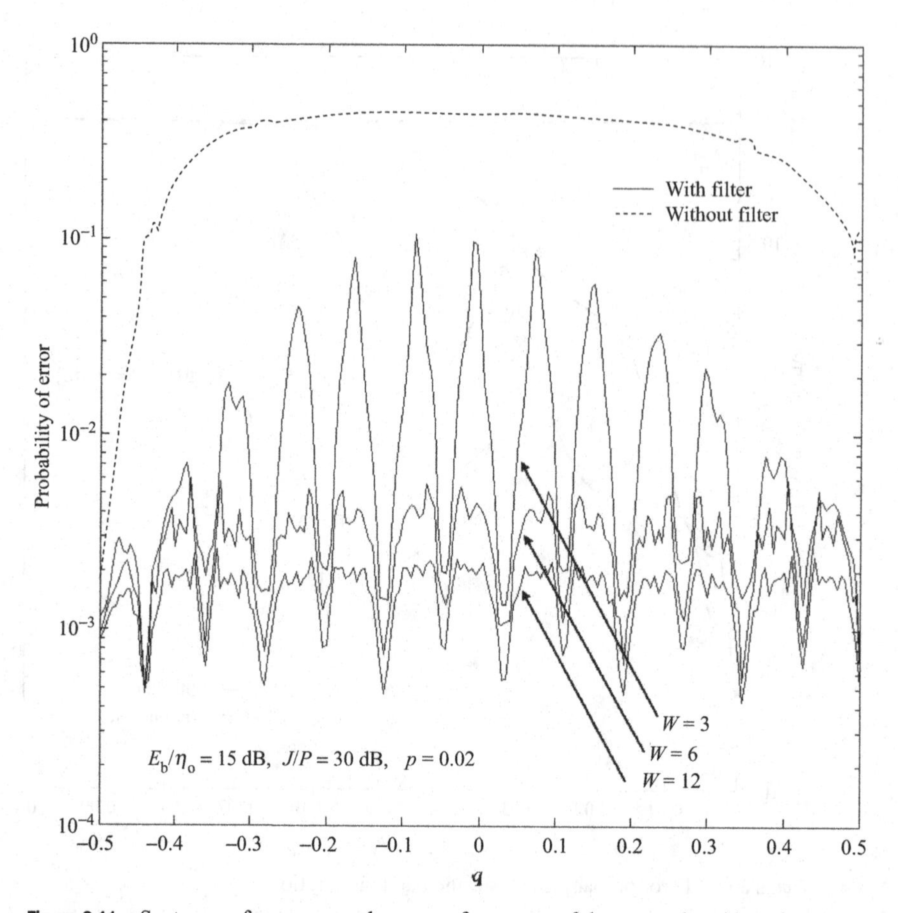

Figure 3.11 System performance as the center frequency of the narrowband interference varies.

The number of correlators per sub-carrier R is 1, so as to maintain the same degree of diversity as in the TH-IR system (10/30) in decision combining. Theoretically, the relative advantage of the multicarrier CDMA system over the TH-IR system is that the multicarrier CDMA system can enjoy frequency diversity of the branches unaffected by the narrowband system. For the multicarrier CDMA system, the impact of the jamming signal can be isolated within certain sub-bands. As observed, however, the use of a notch filter for TH-IR provides a good improvement in performance. Subject to a change in E_b/η_o, the performance difference between the TH-IR system and the multicarrier CDMA system is limited.

3.6 Conclusions

In this chapter, performance expressions are derived for the TH-IR UWB communication system with transversal type notch filtering at the receiver front end. Results show that the use of a notch filter can provide significant improvement in system performance in

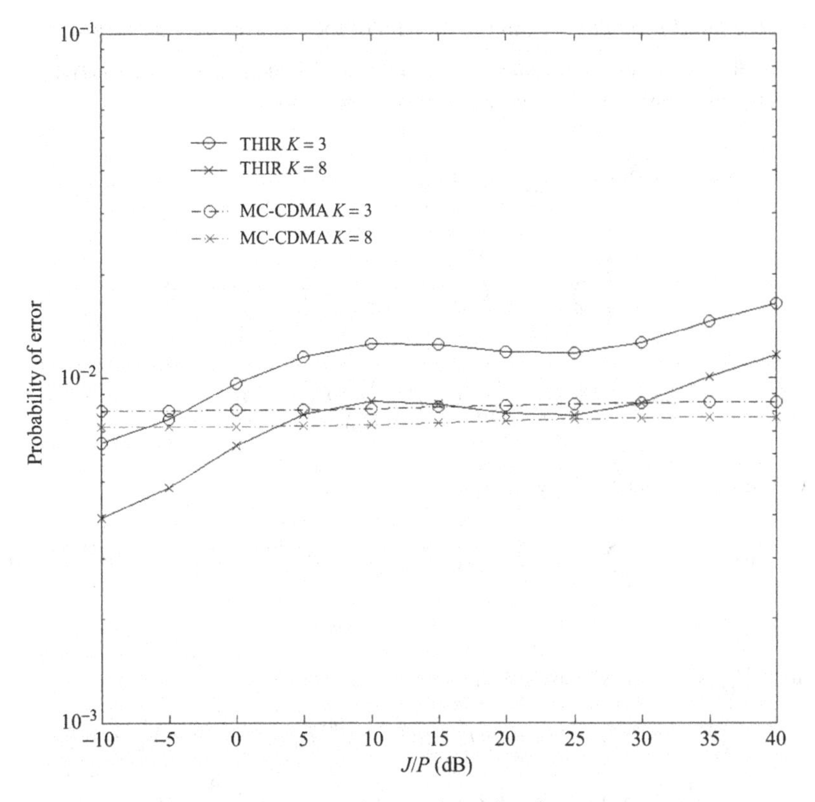

Figure 3.12 Comparison between TH-IR and multicarrier CDMA UWB systems.

various situations. Comparison between TH-IR and multicarrier CDMA UWB systems is also made. Under certain conditions, the relatively low complexity TH-IR UWB system with notch filtering at the receiver front end is capable of achieving a similar bit error rate as multicarrier CDMA.

Appendix 3A Derivation of the variances of multipath interference

Undesired pulses from multipath and multiple access can be collectively defined as an interfering pulse to the reference path of the first user. The possibility of sampling the interfering pulse is due to the combinational effect of random-coded time shifts, random propagation time delays and the sampling time at the receiver. If we consider the $\hat{n}$th sampling instant for the first user, the relative delay of the nth pulse from the lth path of the kth user with respect to the rth path of the first user can be expressed as

$$\tau_{k,l}^{(n)} = \tau_{1,r} + \hat{n}T_c - nT_c - \tau_{k,l} = \lambda_{k,l}T_c + \Lambda_{k,l}^{(n)}T_c + \Delta_{k,l}^{(n)}T_f \tag{3A.1}$$

where $\lambda_{k,l} \in [0, 1)$, $\Lambda_{k,l}^{(n)} \in [0, N_h - 1]$ and $\Delta_{k,l}^{(n)} \in \{0, 1, \ldots, \infty\}$. The pulse interference is shown in Figure 3.4. The template pulse for the rth correlator of the reference user and the interfering pulse from the lth path of the kth user may overlap when $\Lambda_{k,l}^{(n)} = 0$ or when the pulses are within the same time bin. When $\Lambda_{k,l}^{(n)} \neq 0$, no overlapping exists.

The multipath interference $Z_{\mathrm{MPI}|r}$ takes the form

$$\begin{aligned} Z_{\mathrm{MPI}|r} &= \beta_{1,r}\psi_{1,r}\sum_{\substack{l=1\\ l\neq r}}^{L}\psi_{1,l}\beta_{1,l}\sum_{n=0}^{N_s-1} \\ &\quad\times\left\{\sum_{w=-W}^{W}\alpha_w\int_0^{T_c} s_1\big(t-\tau_{1,l}-wT_c+nT_f+c_1^{(n)}T_c+\tau_{1,r}\big)v(t)\mathrm{d}t\right\} \\ &= \sqrt{E_1}\beta_{1,r}\psi_{1,r}\sum_{\substack{l=1\\ l\neq r}}^{L}\psi_{1,l}\beta_{1,l}\sum_{n=0}^{N_s-1} I_{\mathrm{MPI}|r}(l,n) \end{aligned} \tag{3A.2}$$

where $I_{\mathrm{MPI}|r}(l,n)$ can be written as

$$I_{\mathrm{MPI}|r}(l,n) = \sum_{w=-W}^{W}\alpha_w\int_0^{T_c}\Xi\big(t-\lambda_{1,l}T_c-\Lambda_{1,l}^{(n)}T_c-\Delta_{1,l}^{(n)}T_f-\varepsilon d_1^{\lfloor (n)/N_s\rfloor}\big)v(t)\mathrm{d}t \tag{3A.3}$$

and $\lambda_{1,l}^{(n)}$, $\Lambda_{1,l}^{(n)}$ and $\Delta_{1,l}^{(n)}$ are given by

$$\lambda_{1,l} = \big(\tau_{1,l}-\tau_{1,r}+wT_c-nT_f-c_1^{(n)}T_c-\Delta_{1,l}^{(n)}T_f-\Lambda_{1,l}^{(n)}T_c\big)/T_c \tag{3A.4}$$

$$\Lambda_{1,l}^{(n)} = \big\lfloor\big(\tau_{1,l}-\tau_{1,r}+wT_c-nT_f-c_1^{(n)}T_c-\Delta_{1,l}^{(n)}T_f\big)/T_c\big\rfloor \tag{3A.5}$$

$$\Delta_{1,l}^{(n)} = \big\lfloor\big(\tau_{1,l}-\tau_{1,r}+wT_c-nT_f-c_1^{(n)}T_c\big)/T_f\big\rfloor \tag{3A.6}$$

The integral in (3A.3) is non-zero only when the pulses are overlapped. Obviously, $\Lambda_{1,l}^{(n)} = 0$ leads to $\lfloor(\tau_{1,l}-\tau_{1,r})/T_c\rfloor - c_1^{(n)} = -w$, where the sum on the left-hand side of the expression takes the same probability of $1/N_h$ for any possible value of arithmetic modulo N_h (i.e. $0, 1, \ldots, N_h-1$). The probability of $\lfloor(\tau_{1,l}-\tau_{1,r})/T_c\rfloor - c_1^{(n)} = -w$ is given by

$$\mathrm{Prob}\big(\lfloor(\tau_{1,l}-\tau_{1,r})/T_c\rfloor - c_1^{(n)} = -w\big) = \left(\frac{1}{N_h}\right)^2 (N_h-|w|) \tag{3A.7}$$

Thus the variance of the MPI term is

$$\begin{aligned} \mathrm{Var}[Z_{\mathrm{MPI}|r}] &= E\left[\left(\psi_{1,r}\beta_{1,r}\sum_{\substack{l=1\\ l\neq r}}^{L}\psi_{1,l}\beta_{1,l}\sqrt{E_1}\sum_{n=0}^{N_s-1} I_{\mathrm{MPI}|r}(l,n)\right)^2\right] \\ &= E_1\beta_{1,r}^2\left(\sum_{\substack{l=1\\ l\neq r}}^{L}\Omega_{1,l}\right)\left(\sum_{n_1=0}^{N_s-1}\sum_{n_2=0}^{N_s-1}E[I_{\mathrm{MPI}|r}(n_1)I_{\mathrm{MPI}|r}(n_2)]\right) \end{aligned} \tag{3A.8}$$

where $E[I_{\mathrm{MPI}|r}(n_1)I_{\mathrm{MPI}|r}(n_2)]$ can be expressed as different forms corresponding to the values of n_1 and n_2. When $n_1 = n_2$,

$$E[I_{\mathrm{MPI}|r}(n_1)I_{\mathrm{MPI}|r}(n_2)] = \left(\sum_{w=-W}^{W} \frac{N_{\mathrm{h}} - |w|}{N_{\mathrm{h}}^2}\alpha_w^2\right) \cdot \left\{\int_0^1 \left[\frac{1}{2}\left(\int_0^{T_{\mathrm{c}}} \Xi(t - \lambda T_{\mathrm{c}})v(t)\mathrm{d}t\right)^2 + \frac{1}{2}\left(\int_0^{T_{\mathrm{c}}} \Xi(t - \lambda T_{\mathrm{c}} - \varepsilon)v(t)\mathrm{d}t\right)^2\right]\mathrm{d}\lambda\right\} \tag{3A.9}$$

When $n_1 \neq n_2$,

$$E[I_{\mathrm{MPI}|r}(n_1)I_{\mathrm{MPI}|r}(n_2)] = \left(\sum_{w_1=-W}^{W}\sum_{w_2=-W}^{W}\left(\frac{N_{\mathrm{h}} - |w_1|}{N_{\mathrm{h}}^2}\right)\left(\frac{N_{\mathrm{h}} - |w_2|}{N_{\mathrm{h}}^2}\right)\alpha_{w_1}\alpha_{w_2}\right) \cdot \left\{\int_0^1 \left[\frac{1}{2}\left(\int_0^{T_{\mathrm{c}}} \Xi(t - \lambda T_{\mathrm{c}})v(t)\mathrm{d}t\right)^2 + \frac{1}{2}\left(\int_0^{T_{\mathrm{c}}} \Xi(t - \lambda T_{\mathrm{c}} - \varepsilon)v(t)\mathrm{d}t\right)^2\right]\mathrm{d}\lambda\right\} \tag{3A.10}$$

Therefore the variance of $Z_{\mathrm{MPI}|r}$ can be written as

$$\mathrm{Var}[Z_{\mathrm{MPI}|r}] = \beta_{1,r}^2 E_1 \sigma_{\mathrm{MPI}|r}^2 \tag{3A.11}$$

where

$$\sigma_{\mathrm{MPI}|r}^2 = \left(\sum_{\substack{l=1\\ l\neq r}}^{L} \Omega_{1,l}\right)\left\{\sum_{n_1=0}^{N_{\mathrm{s}}-1}\sum_{n_2=0}^{N_{\mathrm{s}}-1} E[I_{\mathrm{MPI}|r}(n_1)I_{\mathrm{MPI}|r}(n_2)]\right\} \tag{3A.12}$$

Appendix 3B Derivation of the variances of multiple access interference

The multiple access interference $Z_{\mathrm{MAI}|r}$ can be written as

$$\begin{aligned} Z_{\mathrm{MAI}|r} &= \beta_{1,r}\psi_{1,r}\sum_{k=2}^{K}\sum_{l=1}^{L}\psi_{k,l}\beta_{k,l}\sum_{n=0}^{N_{\mathrm{s}}-1} \times \left\{\sum_{w=-W}^{W}\alpha_w \int_0^{T_{\mathrm{c}}} s_k(t - \tau_{k,l} - wT_{\mathrm{c}} + nT_{\mathrm{f}} + c_1^{(n)}T_{\mathrm{c}})v(t)\mathrm{d}t\right\} \\ &= \beta_{1,r}\psi_{1,r}\sum_{k=2}^{K}\sqrt{E_k}\sum_{l=1}^{L}\psi_{k,l}\beta_{k,l}\sum_{n=0}^{N_{\mathrm{s}}-1} I_{\mathrm{MAI}|r}(k,l,n) \end{aligned} \tag{3B.1}$$

where $I_{\mathrm{MAI}|r}(k,l,n)$ is given by

$$I_{\mathrm{MAI}|r}(k,l,n) = \sum_{w=-W}^{W} \alpha_w \int_0^{T_c} \Xi\left(t - \lambda_{k,l} T_c - \Lambda_{k,l}^{(n)} T_c - \Delta_{k,l}^{(n)} T_f - \varepsilon d_k^{\lfloor n/N_s \rfloor}\right) v(t) \mathrm{d}t \tag{3B.2}$$

and $\lambda_{k,l}^{(n)}$, $\Lambda_{k,l}^{(n)}$ and $\Delta_{k,l}^{(n)}$ are given by

$$\Delta_{k,l}^{(n)} = \left\lfloor \left(\tau_{k,l} - \tau_{1,r} + wT_c - nT_f + c_k^{(n)} T_c - c_1^{(n)} T_c\right) / T_f \right\rfloor \tag{3B.3}$$

$$\Lambda_{k,l}^{(n)} = \left\lfloor \left(\tau_{k,l} - \tau_{1,r} + wT_c - nT_f + c_k^{(n)} T_c - c_1^{(n)} T_c - \Delta_{k,l}^{(n)} T_f\right) / T_c \right\rfloor \tag{3B.4}$$

$$\lambda_{k,l} = \left(\tau_{k,l} - \tau_{1,r} + wT_c - nT_f + c_k^{(n)} T_c - c_1^{(n)} T_c - \Delta_{k,l}^{(n)} T_f - \Lambda_{k,l}^{(n)} T_c\right) / T_c \tag{3B.5}$$

Similarly, the integral in (3B.2) has a non-zero value only when $\Lambda_{k,l}^{(n)} = 0$, or alternatively $\lfloor(\tau_{k,l} - \tau_{1,r})/T_c\rfloor + c_k^{(n)} - c_1^{(n)} = -w$. Again, the sum takes the same probability of $1/N_h$ for any value in the set $[0, 1, \ldots, N_h - 1]$. The probability of pulse overlapping is given by

$$\mathrm{Prob}\left(\lfloor(\tau_{k,l} - \tau_{1,r})/T_c\rfloor + c_k^{(n)} - c_1^{(n)} = -w\right) = \left(\frac{1}{N_h}\right)^2 (N_h - |w|) \tag{3B.6}$$

Thus, the variance of the MAI term is

$$\begin{aligned} \mathrm{Var}\left[Z_{\mathrm{MAI}|r}\right] &= E\left[\left(\beta_{1,r}\psi_{1,r} \sum_{k=2}^{K} \sum_{l=1}^{L} \psi_{k,l}\beta_{k,l}\sqrt{E_k} \sum_{n=0}^{N_s-1} I_{\mathrm{MAI}|r}(k,l,n)\right)^2\right] \\ &= \beta_{1,r}^2 \sum_{k=2}^{K} E_k \left(\sum_{l=1}^{L} \Omega_{k,l}\right) \left(\sum_{n_1=0}^{N_s-1} \sum_{n_2=0}^{N_s-1} E[I_{\mathrm{MAI}|r}(n_1) I_{\mathrm{MAI}|r}(n_2)]\right) \end{aligned} \tag{3B.7}$$

where $E[I_{\mathrm{MAI}|r}(n_1) I_{\mathrm{MAI}|r}(n_2)]$ has a different form for different n_1 and n_2. For $n_1 = n_2$,

$$\begin{aligned} E\lfloor I_{\mathrm{MAI}|r}(n_1) I_{\mathrm{MAI}|r}(n_2)\rfloor &= \left(\sum_{w=-W}^{W} \frac{N_h - |w|}{N_h^2} \alpha_w^2\right) \left\{ \int_0^1 \left[\frac{1}{2} \left(\int_0^{T_c} \Xi(t - \lambda T_c) v(t) \mathrm{d}t \right)^2 \right.\right. \\ &\qquad \left.\left. + \frac{1}{2} \left(\int_0^{T_c} \Xi(t - \lambda T_c - \varepsilon) v(t) \mathrm{d}t \right)^2 \right] \mathrm{d}\lambda \right\} \end{aligned} \tag{3B.8}$$

For $n_1 \neq n_2$,

$$E[I_{\mathrm{MAI}|r}(n_1) I_{\mathrm{MAI}|r}(n_2)] = \left(\sum_{w_1=-W}^{W} \sum_{w_2=-W}^{W} \left(\frac{N_\mathrm{h} - |w_1|}{N_\mathrm{h}^2} \right) \left(\frac{N_\mathrm{h} - |w_2|}{N_\mathrm{h}^2} \right) \alpha_{w_1} \alpha_{w_2} \right) \cdot \left\{ \int_0^1 \left[\frac{1}{2} \left(\int_0^{T_\mathrm{c}} \Xi(t - \lambda T_\mathrm{c}) v(t)\, \mathrm{d}t \right)^2 + \frac{1}{2} \left(\int_0^{T_\mathrm{c}} \Xi(t - \lambda T_\mathrm{c}) v(t) \mathrm{d}t \right) \times \left(\int_0^{T_\mathrm{c}} \Xi(t - \lambda T_\mathrm{c} - \varepsilon) v(t) \mathrm{d}t \right) \right] \mathrm{d}\lambda \right\} \tag{3B.9}$$

Therefore the variance of $Z_{\mathrm{MAI}|r}$ is

$$\mathrm{Var}(Z_{\mathrm{MAI}|r}) = \beta_{1,r}^2 \left(\sum_{k=2}^{K} E_k \right) \sigma_{\mathrm{MAI}|r}^2 \tag{3B.10}$$

where

$$\sigma_{\mathrm{MAI}|r}^2 = \left\{ \sum_{n_1=0}^{N_s-1} \sum_{n_2=0}^{N_s-1} E[I_{\mathrm{MAI}|r}(n_1) I_{\mathrm{MAI}|r}(n_2)] \right\} \tag{3B.11}$$

Appendix 3C Derivation of the variances of narrowband interference

The narrowband interference $Z_{\mathrm{NBI}|r}$ is given by

$$Z_{\mathrm{NBI}|r} = \beta_{1,r} \psi_{1,r} \sum_{n=0}^{N_\mathrm{s}-1} \left\{ \sum_{w=-W}^{W} \alpha_w \int_0^{T_\mathrm{c}} j\left(t + \tau_{1,r} - w T_\mathrm{c} + n T_\mathrm{f} + c_1^{(n)} T_\mathrm{c}\right) v(t) \mathrm{d}t \right\} \tag{3C.1}$$

with variance

$$\mathrm{Var}(Z_{\mathrm{NBI}|r}) = \beta_{1,r}^2 \sum_{n_1=0}^{N_\mathrm{s}-1} \sum_{n_2=0}^{N_\mathrm{s}-1} = \left\{ \sum_{w_1=-W}^{W} \sum_{w_2=-W}^{W} \alpha_{w_1} \alpha_{w_2} \times \int_0^{T_\mathrm{c}} \int_0^{T_\mathrm{c}} E\left[j\left(t_1 + \tau_{1,r} + n_1 T_\mathrm{f} + c_1^{(n_1)} T_\mathrm{c} - w_1 T_\mathrm{c}\right) \cdot j\left(t_2 + \tau_{1,r} + n_2 T_\mathrm{f} + c_1^{(n_2)} T_\mathrm{c} - w_2 T_\mathrm{c}\right) \right] \times v(t_1) v(t_2) \mathrm{d}t_1 \mathrm{d}t_2 \right\}$$

$$= J \cdot \beta_{1,r}^2 \sum_{n_1=0}^{N_\mathrm{s}-1} \sum_{n_2=0}^{N_\mathrm{s}-1}$$

$$\times \left\{ \sum_{w_1=-W}^{W} \sum_{w_2=-W}^{W} \alpha_{w_1}\alpha_{w_2} \right.$$

$$\times \left[\int_0^{T_c}\int_0^{T_c} E\{\bar{R}_j[t_1 - t_2 + (n_1 - n_2)T_f + (c_1^{(n_1)} - c_1^{(n_2)})T_c - (w_1 - w_2)T_c]\} \right.$$

$$\left. \left. \cdot v(t_1)v(t_2)\mathrm{d}t_1\mathrm{d}t_2 \right] \right\}$$

$$= J \cdot \beta_{1,r}^2 \sum_{n_1=0}^{N_s-1}\sum_{n_2=0}^{N_s-1} I_{\mathrm{NBI}|r}(n_1, n_2) \tag{3C.2}$$

Since $\{c_1^{(n)}\}$ is a random sequence and each element takes the same probability of $1/N_h$ for any value of $0, 1, \ldots, N_h - 1$, the probability of $c_1^{(n_1)} - c_1^{(n_2)} = m$, $m \in [-(N_h - 1), N_h - 1]$, is given by

$$\mathrm{Prob}(c_1^{(n_1)} - c_1^{(n_2)} = m) = \left(\frac{1}{N_h}\right)^2 (N_h - |m|) \tag{3C.3}$$

$I_{\mathrm{NBI}|r}(n_1, n_2)$ can be expressed as different forms based on the values of n_1 and n_2. For $n_1 = n_2$,

$$I_{\mathrm{NBI}|r}(n_1, n_2) = \sum_{w_1=-W}^{W}\sum_{w_2=-W}^{W} \alpha_{w_1}\alpha_{w_2} \int_0^{T_c}\int_0^{T_c} \bar{R}_j[t_1 - t_2 - (w_1 - w_2)T_c]v(t_1)v(t_2)\mathrm{d}t_1\mathrm{d}t_2 \tag{3C.4}$$

For $n_1 \neq n_2$,

$$I_{\mathrm{NBI}|r}(n_1, n_2) = \sum_{w_1=-W}^{W}\sum_{w_2=-W}^{W} \alpha_{w_1}\alpha_{w_2} \sum_{m=-(N_h-1)}^{N_h-1} \left\{ (N_h - |m|/N_h^2) \right.$$

$$\left. \cdot \int_0^{T_c}\int_0^{T_c} \bar{R}_j[t_1 - t_2 + (n_1 - n_2)T_f + mT_c - (w_1 - w_2)T_c]v(t_1)v(t_2)\mathrm{d}t_1\mathrm{d}t_2 \right\} \tag{3C.5}$$

Thus, $\mathrm{Var}(Z_{\mathrm{NBI}|r})$ is given by

$$\mathrm{Var}(Z_{\mathrm{NBI}|r}) = J\beta_{1,r}^2\sigma_{\mathrm{NBI}|r}^2 \tag{3C.6}$$

where

$$\sigma_{\mathrm{NBI}|r}^2 = \sum_{n_1=0}^{N_s-1}\sum_{n_2=0}^{N_s-1} I_{\mathrm{NBI}|r}(n_1, n_2) \tag{3C.7}$$

References

[1] R. A. Scholtz, "Multiple access with time-hopping impulse modulation," in *Proc. IEEE Milcom'93*, pp. 447–450, 1993.

[2] M. Win and R. A. Scholtz, "Ultra-wide bandwidth time-hopping spread-spectrum impulse radio for wireless multiple access communications," *IEEE Trans. Commun.*, vol. COM-48, pp. 679–691, April 2000.

[3] M. Win and R. A. Scholtz, "Impulse radio: how it works," *IEEE Commun. Lett.*, vol. 2, pp. 10–12, Jan. 1998.

[4] L. Zhao and A. Haimovich, "Performance of ultra-wideband communications in presence of interference," *IEEE J. Select. Areas Commun.*, vol. JSAC-20, pp. 1684–1691, Dec. 2002.

[5] M. Hamalainen, V. Hovinen, R. Tesi, J. Iinatti and M. Latva-aho, "On the UWB system coexistence with GSM900, UMTS/WCDMA and GPS," *IEEE J. Select. Areas Commun.*, vol. JSAC-20, pp. 1712–1721, Dec. 2002.

[6] D. Laney, G. M. Maggio, F. Lehmann and L. Larson, "Multiple access for UWB impulse radio with pseudochaotic time hopping," *Proc. IEEE J. Select. Areas Commun.*, vol. JSAC-20, pp. 1692–1700, Dec. 2002.

[7] F. Ramirez-Mireles, "Signal design for ultra-wideband communications in dense multipath," *IEEE Trans. Vehicular Tech.*, vol. VT-51, pp. 1517–1521, Nov. 2002.

[8] J. T. Conroy, J. L. LoCicero and D. R. Ucci, "Communication techniques using monopulse waveforms," in *Proc. IEEE Milcom'99*, pp. 1181–1185, 1999.

[9] FCC First Report and Order Regarding Ultra-Wideband Transmission Systems, Order (FCC 02–48), 22 April 2002.

[10] L. B. Milstein, "Interference rejection techniques in spread-spectrum communications," *Proc. IEEE*, vol. 76, no. 6, pp. 657–671, June 1988.

[11] I. Bergel, E. Fishler and H. Messer, "Narrowband interference suppression in time hopping impulse-radio systems," in *Proc. IEEE Conference on Ultra Wideband Systems and Technologies*, pp. 303–307, May 2002.

[12] T. T. Wong and J. Wang, "MMSE receiver for multicarrier CDMA overlay in ultra-wideband communications," *IEEE Trans. Vehicular Tech.*, vol. VT-54, pp. 603–614, March 2005.

[13] J. Wang and L. B. Milstein, "Multicarrier CDMA overlay for ultra-wideband communications," *IEEE Trans. Commun.*, vol. 52, pp. 1664–1669, Oct. 2004.

[14] J. Wang and L. B. Milstein, "CDMA overlay situations for microcellular mobile communications," *IEEE Trans. Commun.*, vol. 43, no. 2–4, pp. 603–614, Feb.–April 1995.

[15] J. Foerster, "Channel modeling sub-committee report (final)," IEEE P802.15 Working Group for Wireless Personal Area Networks, December 2002.

[16] S. C. Schwartz and Y. S. Yeh, "On the distribution and moments of power sums with lognormal components," *The Bell System Tech. J.*, vol. 61, no. 7, pp. 1441–1462, Sept. 1982.

4 Rapid acquisition

Ultra-wideband impulse radio is a promising radio technology for networks delivering extremely high data rates at short ranges. The use of extremely short duration pulses, however, makes the synchronization task more difficult. In this chapter a two-stage acquisition with serial search noncoherent correlator for time-hopping impulse radio is proposed. The proposed two-stage acquisition scheme gets chip timing synchronization, and aligns the phase of the local time-hopping code in two successive stages. With the aid of the flow-graph approach, analytical expressions are presented for the mean acquisition time and the probability of acquisition. Numerical results in a slow fading channel show that the proposed two-stage acquisition method can offer a much shorter mean acquisition time or much higher probability of acquisition than that delivered by conventional acquisition.

4.1 Introduction

As explained in Section 1.1, one of the design challenges provided by the wide bandwidth property of IR-UWB signals is timing acquisition, so a rapid acquisition algorithm is very important in IR-UWB communications. A few papers have focused on acquisition for TH IR-UWB signals. In [1] the authors analyze the acquisition performance of the IR-UWB signal. In [2] a generalized analysis of various serial search strategies is presented for reducing the mean acquisition time for IR-UWB signals in a dense multipath environment, and finally in [3] hybrid schemes for IR-UWB signal acquisition are proposed to trade off the speed of parallel schemes with the simplicity of serial search schemes.

For an IR-UWB system that occupies the wideband from 3.1–10.6 GHz, a pulse-matched filter is not practical, while an active correlator is suitable exclusively for its implementation simplicity [4]. In addition, parallel search outperforms serial search but the associated complexity is often prohibitive [3]. Therefore, we focus our discussion on the serial search noncoherent-correlator-based approach for acquisition of TH IR-UWB signals. In this chapter, a novel two-stage serial search acquisition is proposed, which gets chip timing synchronization, and aligns the phase of the local time-hopping code in two successive stages. This novel acquisition scheme can offer a much shorter mean acquisition time and much higher probability of acquisition than that delivered by conventional acquisition. Furthermore, the novel two-stage acquisition scheme proposed in this chapter can also be adopted for parallel search acquisition.

4.2 System model

4.2.1 Signal model

In a TH-IR UWB system, the transmitted signal of the kth user in acquisition mode is unmodulated, given by

$$s^{(k)}(t) = A \sum_{j=-\infty}^{\infty} w\left(t - jT_{\mathrm{f}} - c_j^{(k)} T_{\mathrm{c}}\right) \tag{4.1}$$

where A is the amplitude and T_{f} denotes the time frame or pulse repetition time. Following [5], the monocycle pulse waveform is chosen to be the second derivative of a Gaussian function as

$$w(t) = \sqrt{\frac{4}{3t_n\sqrt{\pi}}}\left[1 - \left(\frac{t}{t_n}\right)^2\right]\exp\left[-\frac{1}{2}\left(\frac{t}{t_n}\right)^2\right] \tag{4.2}$$

where $\sqrt{4/(3t_n\sqrt{\pi})}$ ensures the normalized condition $\int_{-\infty}^{\infty} w^2(t) = 1$. It is assumed that $w(t) = 0$ and $|t| > T_{\mathrm{w}}/2$, where T_{w} is the width of monocycle duration determined by the parameter t_n and, usually, equal to less than 1 nanosecond.

An additional time shift of every monocycle in the pulse train is determined by the chip duration T_{c} and the TH sequence $\{c_j^{(k)}\}$ with length N_{p}. TH sequences support multiple-access communication systems and smooth the IR power spectral density in order to lower interference to existing narrowband radio at the same time. The elements $c_j^{(k)}$ of the sequence are chosen from a finite set $\{0, 1, \ldots, N_{\mathrm{h}} - 1\}$, so that the signal is transmitted with N_{h} possible hops (chips) per frame. For system simplicity, the chip duration T_{c} is always set at several nanoseconds [6], so that $T_{\mathrm{w}} \ll T_{\mathrm{c}}$ [7]. Thus, $T_{\mathrm{f}} = N_{\mathrm{h}}T_{\mathrm{c}} + T_{\mathrm{g}}$, where T_{g} is a guard time introduced to account for the processing delay at the receiver between two successive received frames (see, e.g., [7], [8]). Here, for simplicity, it is assumed that $T_{\mathrm{g}} = 0$ and $T_{\mathrm{f}} = N_{\mathrm{h}}T_{\mathrm{c}}$, but the novel scheme can encompass the case $T_{\mathrm{g}} \neq 0$ as well, by setting $T_{\mathrm{g}} = N_{\mathrm{g}}T_{\mathrm{c}}$, with integer N_{g} and restricting the sequence $c_j^{(k)}$ to take its value in $\{0, 1, \ldots, N_{\mathrm{h}}' - 1\}$, where $N_{\mathrm{h}}' = N_{\mathrm{h}} - N_{\mathrm{g}}$.

The UWB channel model has been described in Chapters 2 and 3. In the case of very low mobility of indoor UWB communication systems, the coherence time of the channel is larger than the length of one packet [9], and therefore the slow fading channel model is adopted from the point of acquisition. In [9] an UWB channel model is derived from the Saleh–Valenzuela model [10] with modifications that include using a lognormal rather than a Rayleigh distribution for the multipath gain magnitude. Multipath components provide replicas of the pilot that are all valid candidates for acquisition, but to keep analysis tractable a single path channel model is frequently used.

Therefore the received signal from the kth user can be represented as

$$r(t) = \alpha A \sum_{j=-\infty}^{\infty} w\left(t - jT_{\mathrm{f}} - c_j^{(k)} T_{\mathrm{c}} - \tau\right) + n(t) \tag{4.3}$$

where α is a fading coefficient, $n(t)$ is a white Gaussian noise with double-sided power spectral density of $N_0/2$ and τ represents the time delay between the clock of the transmitter and the receiver.

In (4.3) the sign of the fading coefficient, α, is equiprobable $+1/-1$ to account for signal inversion due to reflection. The absolute value $|\alpha| = 10^{\beta/20}$, where β is a Gaussian random variable with mean of -2.6526 and standard deviation of 4.8 satisfying the condition $E(|\alpha|^2) = 1$, then the probability density function (PDF) of α can be written as

$$\phi(|\alpha|) = \frac{\exp[-(\ln(|\alpha|) + 0.30539)^2 / 0.61078]}{\sqrt{0.61078\pi}\,|\alpha|} \tag{4.4}$$

4.2.2 TH-code design criteria and bound on auto- and cross-correlation

Equation (4.1) can be rewritten as

$$s^{(k)}(t) = A \sum_{n=-\infty}^{\infty} a_n^{(k)} w(t - nT_c) \tag{4.5}$$

where $a_n^{(k)}$ is defined as

$$a_n^{(k)} = \begin{cases} 1, & n = jN_h + c_j^{(k)} \\ 0, & \text{otherwise} \end{cases} \tag{4.6}$$

Then, the normalized periodic correlation between the sequences k_1 and k_2 ($1 \le k_1, k_2 \le K_u$ and K_u denotes the number of TH sequences) is

$$\rho_{k_1k_2}(n_\tau) = \frac{R_{k_1k_2}(n_\tau)}{R_{k_1k_1}(0)} = \frac{1}{N_p} \sum_{n=0}^{N_pN_h-1} a_{n\oplus n_\tau}^{(k_1)} a_n^{(k_2)} \tag{4.7}$$

where $R_{k_1k_1}(0) = N_p$ and $\oplus$ denotes addition modulo N_pN_h, leading to the periodic nature of the computation.

The Johnson bound [11, page 527] gives an upper bound of $B(N, 2d, W)$, the maximum number of binary vectors with length N, Hamming distance at least $2d$ and constant-weight W (the number of '1's in a binary vector) as follows:

$$B(N, 2d, W) \le \left\lfloor \frac{N}{W} \left\lfloor \frac{N-1}{W-1} \cdots \left\lfloor \frac{N-W+d+1}{d+1} \left\lfloor \frac{N-W+d}{d} \right\rfloor \right\rfloor \cdots \right\rfloor \right\rfloor \tag{4.8}$$

where $\lfloor x \rfloor$ denotes the integer part of x. Similar to [12], the sequences $\{a_n^{(k)}\}_{k=1}^{K_u}$ and their respective $N_pN_h - 1$ cyclic shifts are viewed as constant-weight code words over the field of $\{0, 1\}$. Then, we can transform the Johnson bound to one on the number of TH sequences that can be designed with a prescribed upper bound on auto- and cross-correlation.

The design criterion of $\{a_n^{(k)}\}_{k=1}^{K_u}$ is to minimize the periodic correlations $N_p\rho_{k_1k_2}(n_\tau)$ except when $k_1 = k_2$ and $n_\tau = 0$, in which case $\rho_{k_1k_1}(0) = 1$. Thus, by specifying λ, we can impose the condition

$$0 \le N_p\rho_{k_1k_2}(n_\tau) = \sum_{n=0}^{N_pN_h-1} a_{n\oplus n_\tau}^{(k_1)} a_n^{(k_2)} \le \lambda, \quad \text{when } k_1 \ne k_2 \quad \text{or} \quad n_\tau \ne 0 \tag{4.9}$$

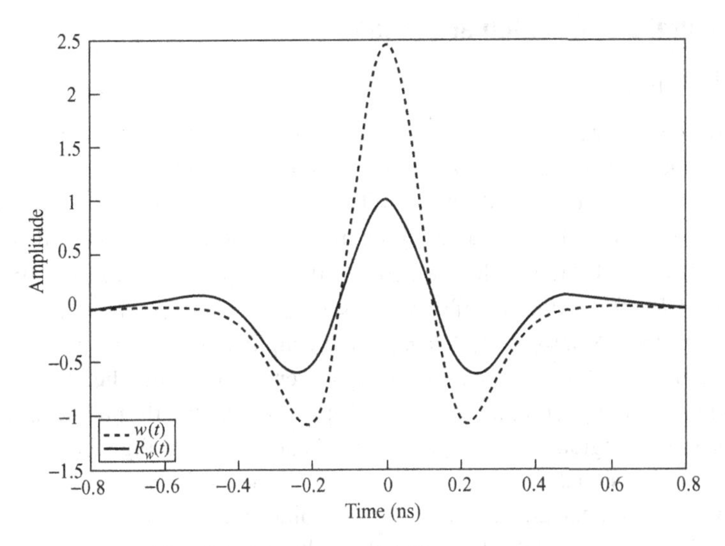

Figure 4.1 $w(t)$ and $R_{\mathrm{w}}(t)$ with $t_n = 0.125$ ns.

Substituting N by $N_{\mathrm{p}}N_{\mathrm{h}}$, W by N_{p}, and d by $N_{\mathrm{p}} - \lambda$ in (4.8), the Johnson bound states that

$$\begin{aligned} K_{\mathrm{u}} &\leq \frac{B(N_{\mathrm{p}}N_{\mathrm{h}}, 2(N_{\mathrm{p}} - \lambda), N_{\mathrm{p}})}{N_{\mathrm{p}}N_{\mathrm{h}}} \\ &= \frac{1}{N_{\mathrm{p}}N_{\mathrm{h}}}\left\lfloor \frac{N_{\mathrm{p}}N_{\mathrm{h}}}{N_{\mathrm{p}}} \left\lfloor \frac{N_{\mathrm{p}}N_{\mathrm{h}} - 1}{N_{\mathrm{p}} - 1} \cdots \left\lfloor \frac{N_{\mathrm{p}}N_{\mathrm{h}} - (\lambda - 1)}{N_{\mathrm{p}} - (\lambda - 1)} \left\lfloor \frac{N_{\mathrm{p}}N_{\mathrm{h}} - \lambda}{N_{\mathrm{p}} - \lambda} \right\rfloor \right\rfloor \cdots \right\rfloor \right\rfloor \end{aligned} \tag{4.10}$$

Equation (4.10) can be used to select parameters such as λ, N_{p} and N_{h} to support a reasonable number of TH sequences so that a multiple-access signal design is feasible.

For simple analysis, the upper bound of (4.9) is considered, i.e. $\rho_{k_1k_2}(n_\tau) = \lambda/N_{\mathrm{p}}$ when $k_1 \neq k_2$ or $n_\tau \neq 0$. Therefore the analysis is not too complicated and is independent of TH code design. Thus, the normalized periodic auto-correlation function (ACF) of $s^{(k)}(t)$ is

$$\rho_s(\tau) = \frac{R_s(\tau)}{R_s(0)} = \begin{cases} R_w(\tau), & |\tau| < T_w \\ \dfrac{\lambda}{N_{\mathrm{p}}} R_w(\tau - nT_{\mathrm{c}}), & n \neq 0 \quad \text{and} \quad |\tau - nT_{\mathrm{c}}| < T_w \\ 0, & \text{else} \end{cases} \tag{4.11}$$

where $R_w(\tau)$ is the autocorrelation of the monocycle pulse as

$$\begin{aligned} R_w(\tau) &= \int_{-\infty}^{\infty} w(t)w(t - \tau)\mathrm{d}t \\ &= \left[1 - 4\cdot\left(\frac{\tau}{2t_n}\right)^2 + \frac{4}{3}\cdot\left(\frac{\tau}{2t_n}\right)^4\right]\exp\left[-\left(\frac{\tau}{2t_n}\right)^2\right] \end{aligned} \tag{4.12}$$

Both $w(t)$ and $R_w(t)$ are plotted in Figure 4.1 with $t_n = 0.125$ ns in (4.2).

4.3 Conventional serial search acquisition

4.3.1 System description

In the absence of any a priori information regarding the phase τ of the incoming signal, the whole uncertainty region of τ is the full TH code period $T_p = N_p T_f$. The whole uncertainty region is quantized into a finite number of elements (cells), with the cell width being within the lock-in range of the tracking loop used for fine code alignment. From (4.12) it can be found that $R_w(\pm t_0) = 0,\ t_0 = (\sqrt{6 - 2\sqrt{6}})t_n \approx 1.05 t_n$, so the main-lobe of $R_w(t)$ is in the slot of $(-t_0, t_0) \approx (-t_n, t_n)$ where $R_w(t) > 0$. Therefore the cell width is set as $2\Delta t_n$, where $1/\Delta$ denotes the sampling time of every interval of length $2t_n$. The value of $1/\Delta$ should be large enough to ensure reducing the performance loss due to sampling timing error (e.g. $\Delta = 1, 1/2$ or $1/4$, etc.). Thus the number of cells in the whole uncertainty region is $N_c = N_p T_f/(2\Delta t_n)$. Acquisition is to align the unknown phase τ of the incoming signal with the known phase ζ of the local identical TH pulse train at the receiver to within the $2t_n$ interval, i.e. the absolute phase offset $|\Delta\tau| = |\zeta - \tau| < t_n$. The desired cell is denoted the H_1-cell, and the alternative hypothesis is denoted H_0-cell $|\Delta\tau| = |\zeta - \tau| \geq t_n$. In serial search acquisition, the search starts from a specific starting cell and serially examines the remaining cells until H_1-cell is found. If the acquisition is not achieved after the first round of searching, we start the next round.

To facilitate our analysis, the following situations are considered as in [13]: (1) all cell test statistics are independent; (2) the sampling occurs at the peak ($R_w(\Delta\tau) = 1$). Therefore, only one sample per $2t_n$ interval suffices ($\Delta = 1$, $N_c = N_p T_f/(2t_n)$), and there is one and only one H_1-cell in the uncertainty time region; (3) the acquisition can start from every cell in the uncertainty region with the uniform probability of $1/N_c$.

4.3.2 Noncoherent correlation detector

The basic unit in any acquisition receiver is the decision-making device (detector). In the noncoherent detector shown in Figure 4.2, the received signal is multiplied by the local pulse train, accumulated over the dwell time interval, passed through the square operator and compared with a threshold ξ to decide whether the cell is an H_1-cell for every cell test. Here full period correlation is adopted, i.e. the dwell time is equal to T_p. Under one static channel realization with given fading coefficient α, the output signal of the correlator is

$$Y = \frac{1}{AN_p}\sum_{j=0}^{N_p-1}\int_{jT_f}^{(j+1)T_f} r(t)w(t - jT_f - c_j T_c - \zeta)\mathrm{d}t = \alpha\rho_s(\Delta\tau) + \eta \tag{4.13}$$

where $\Delta\tau = \tau - \zeta$, and η is the noise component, given by

$$\eta = \frac{1}{AN_p}\sum_{j=0}^{N_p-1}\int_{jT_f}^{(j+1)T_f} n(t)w(t - jT_f - c_j T_c - \zeta)\mathrm{d}t \tag{4.14}$$

It can be shown that η is a Gaussian random variable with zero mean and variance of $N_0/(2N_p A^2)$. Therefore Y is a Gaussian random variable with mean $\alpha\rho_s(\Delta\tau)$ and

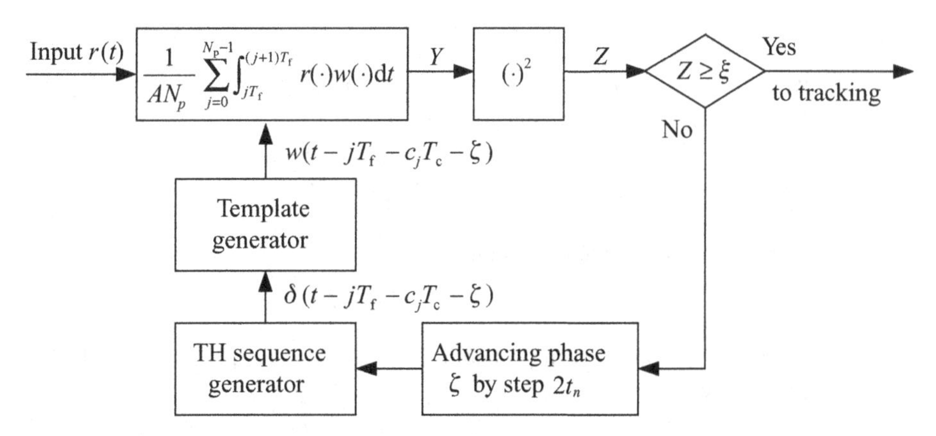

Figure 4.2 Block diagram of the conventional serial search acquisition scheme.

variance $\sigma_Y^2 = N_0/(2N_\mathrm{p}A^2)$. Thus the decision variable $Z = Y^2$ has a non-central chi-square PDF with one degree of freedom:

$$p(Z\,|\,\alpha) = \frac{1}{\sqrt{2\pi Z}\sigma_Y}\exp\left[\frac{-Z-\alpha^2\rho_s^2(\Delta\tau)}{2\sigma_Y^2}\right]\cosh\left[\frac{\sqrt{Z}\alpha\rho_s(\Delta\tau)}{\sigma_Y^2}\right],\ Z \geq 0 \tag{4.15}$$

where the hyperbolic cosine function is defined as $\cosh(x) = \dfrac{\mathrm{e}^x + \mathrm{e}^{-x}}{2}$.

It is assumed that for simplicity one T_c is composed of N_cn pulse main-lobes, i.e. $T_\mathrm{c} = 2N_\mathrm{cn}t_n$, and $\Delta\tau = 2nt_n$, $0 \leq n \leq N_\mathrm{c} - 1$. Since the maximum value of $R_w(t)$ out of the range of the main lobe is very small, its impact on the periodic ACF can be neglected. Therefore, one obtains

$$\rho_s(2nt_n) = \begin{cases} 1, & n = 0 \\ \dfrac{\lambda}{N_\mathrm{p}}, & n \bmod N_\mathrm{cn} = 0 \\ 0, & \text{else} \end{cases} \tag{4.16}$$

Thus the PDF of Z in the nth cell ($n = 0, 1, \ldots, N_\mathrm{c} - 1$) is

$$p^{(n)}(Z|\alpha) = \frac{1}{\sqrt{2\pi Z}\sigma_Y}\exp\left[\frac{-Z-\alpha^2\rho_s^2(2nt_n)}{2\sigma_Y^2}\right]\cosh\left[\frac{\sqrt{Z}\alpha\rho_s(2nt_n)}{\sigma_Y^2}\right],\quad Z \geq 0 \tag{4.17}$$

and the detection probability of an H_1-cell is

$$\begin{aligned} P_\mathrm{d}(\alpha) &= P_\mathrm{r}(Z > \xi\,|H_1) = \int_\xi^\infty p^{(0)}(Z|\alpha)\mathrm{d}Z \\ &= \int_\xi^\infty \frac{1}{\sqrt{2\pi Z}\sigma_Y}\exp\left[\frac{-Z-\alpha^2}{2\sigma_Y^2}\right]\cosh\left[\frac{\sqrt{Z}\alpha}{\sigma_Y^2}\right]\mathrm{d}Z \\ &= \int_{\overline{\xi}}^\infty \frac{1}{\sqrt{\pi\overline{Z}}}\exp\left(-\overline{Z}-\mu\alpha^2\right)\cosh\left(2\sqrt{\mu\alpha^2\overline{Z}}\right)\mathrm{d}\overline{Z} \end{aligned}$$

$$= C(\bar{\xi}, \mu\alpha^2) \tag{4.18}$$

where $C(x, y) = \int_x^{\infty} 1/\sqrt{\pi t}\exp(-t - y)\cosh(2\sqrt{yt})\mathrm{d}t$, μ is the average received SNR for the faded signal (after integration), given by

$$\mu = \frac{E(\alpha^2)}{2\sigma_Y^2} = \frac{N_p A^2}{N_0} \tag{4.19}$$

and $\bar{\xi}$ is the normalized threshold

$$\bar{\xi} = \mu\xi \tag{4.20}$$

Similarly, the false alarm probability $P_{\mathrm{FA}}^{(n)}(\alpha)(n = 1, 2, \ldots N_c - 1)$ of H_0-cells can be derived as

$$\begin{aligned} P_{\mathrm{FA}}^{(n)}(\alpha) &= P_r(Z > \xi | H_0) = \int_{\xi}^{\infty} p^{(n)}(Z|\alpha)\mathrm{d}Z \\ &= \begin{cases} C\left(\bar{\xi}, \dfrac{\mu\alpha^2\lambda^2}{N_p^2}\right), & n \bmod N_{\mathrm{cn}} = 0, \\ C\left(\bar{\xi}, 0\right), & \text{else} \end{cases} \end{aligned} \tag{4.21}$$

4.3.3 Flow diagram analysis

The discrete-time Markovian nature of the acquisition process allows it to be represented by a flow diagram with definite states [14], [15]. The flow diagram for the conventional serial search acquisition scheme for TH-IR signals is described in Figure 4.3, where the parameter z is used to mark time as one proceeds through the flow diagram and its power represents the number of time units (dwell time T_p). Each state in the diagram corresponds to a phase cell with uniform a priori probability $\pi_n = 1/N_c$, $n = 0, 1, 2, \ldots, N_c - 1$ of being the first state in the search. The two remaining states are the correct-acquisition and the false-alarm states. When a false alarm does occur, it can be recognized in the tracking system and the search for the desired cell will restart after a delay of the penalty time. Although the false alarm probability $P_{\mathrm{FA}}^{(n)}(\alpha)(n = 1, 2, \ldots N_c - 1)$ of H_0-cells is not identical, similar to [16] the generating function of the conventional serial search acquisition under one channel realization with given α can be derived as

$$\begin{aligned} F_{\mathrm{ACQ}}(z) &= \frac{H_{\mathrm{D}}(z)}{1 - H_{\mathrm{M}}(z)\prod_{n=1}^{N_c-1} H_0^{(n)}(z)} \left[\sum_{j=0}^{N_c-1} \pi_j \prod_{n=1}^{j} H_0^{(N_c-n)}(z)\right] \\ &= \frac{1}{N_c} \cdot \frac{P_{\mathrm{d}}(\alpha)z}{1 - (1 - P_{\mathrm{d}}(\alpha))z \prod_{n=1}^{N_c-1} \left[(1 - P_{\mathrm{FA}}^{(n)}(\alpha))z + P_{\mathrm{FA}}^{(n)}(\alpha)z^{K+1}\right]} \\ &\quad \times \left\{\sum_{j=0}^{N_c-1} \prod_{n=1}^{j} \left[(1 - P_{\mathrm{FA}}^{(n)}(\alpha))z + P_{\mathrm{FA}}^{(n)}(\alpha)z^{K+1}\right]\right\} \end{aligned} \tag{4.22}$$

where $H_{\mathrm{D}}(z) = P_{\mathrm{d}}(\alpha)z$ is a gain of the branch leading from the H_1 state (state 0) to acquisition state; $H_{\mathrm{M}}(z) = (1 - P_{\mathrm{d}}(\alpha))z$ is a gain of the branch connecting the H_1 state

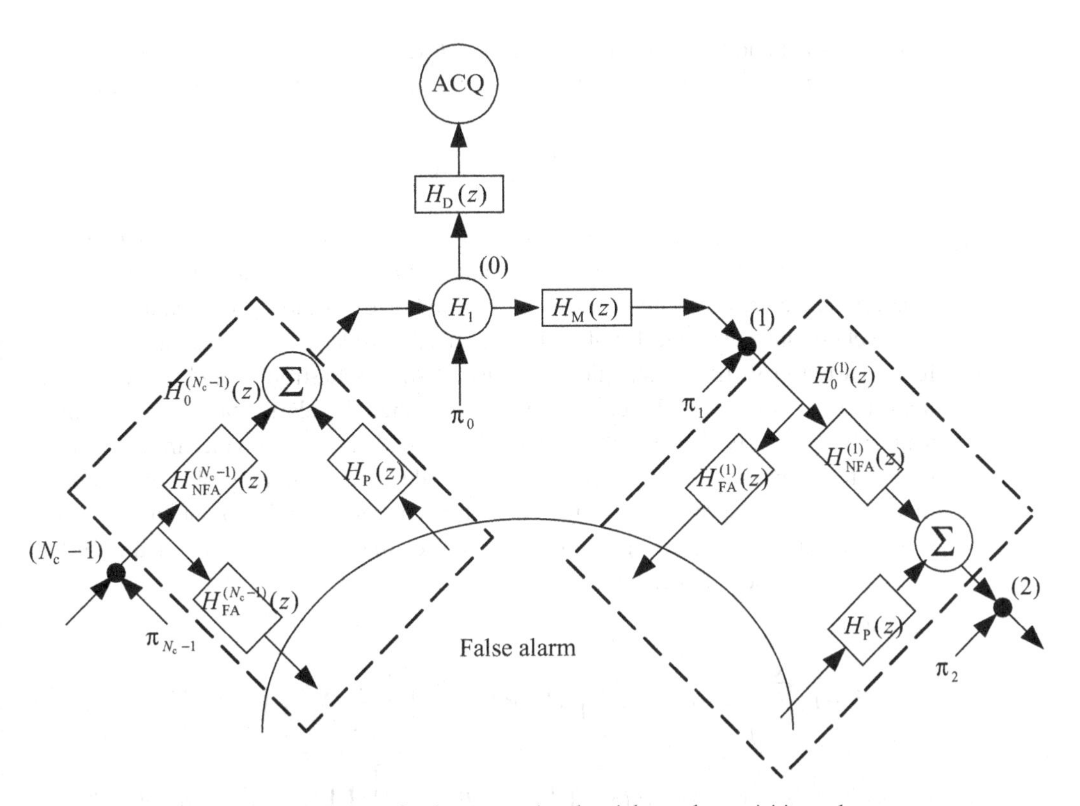

Figure 4.3 Flow diagram for the conventional serial search acquisition scheme.

(state 0) to state 1; $H_0^{(n)}(z) = H_{\text{NFA}}^{(n)}(z) + H_{\text{FA}}^{(n)}(z)H_{\text{P}}(z) = (1 - P_{\text{FA}}^{(n)}(\alpha))z + P_{\text{FA}}^{(n)}(\alpha)z^{K+1}$ is a gain of the branch connecting two successive states $(n, n+1)$, $n = 1, \ldots, N_c - 1$; $H_{\text{NFA}}^{(n)}(z) = (1 - P_{\text{FA}}^{(n)}(\alpha))z$ characterizes the path without a false alarm between two successive states $(n, n+1)$, $n = 1, \ldots, N_c - 1$; $H_{\text{FA}}^{(n)}(z) = P_{\text{FA}}^{(n)}(\alpha)z$ stands for a false alarm between two successive states $(n, n+1)$, $n = 1, \ldots, N_c - 1$; $H_{\text{P}}(z) = z^K$ is a gain of the transition from the false alarm state to the next cell; and $KT_{\text{p}} = KN_{\text{p}}T_{\text{f}}$ is the penalty time.

In the most common case of having a non-limited permitted acquisition time, which is typically employed in applications where data modulation is always present in the received waveform, the meaningful performance parameter generally used for serial search acquisition is the mean acquisition time. First, we get the mean acquisition time in one channel realization with given α as (after some algebra shown in Appendix 4A)

$$\begin{aligned} T_{\text{ACQ}}(\alpha) &= T_{\text{p}} \cdot \left.\frac{\mathrm{d}F_{\text{ACQ}}(z)}{dz}\right|_{z=1} \\ &= T_{\text{p}} \cdot \frac{(1-N_c)P_{\text{d}}(\alpha) + 2N_c + 2K\sum_{n=1}^{N_c-1} P_{\text{FA}}^{(n)}(\alpha) + 2KP_{\text{d}}(\alpha)\left[\frac{1}{N_c}\sum_{n=1}^{N_c-1} nP_{\text{FA}}^{(n)}(\alpha) - \sum_{n=1}^{N_c-1} P_{\text{FA}}^{(n)}(\alpha)\right]}{2P_{\text{d}}(\alpha)} \end{aligned} \tag{4.23}$$

From (4.18) and (4.21) it can be seen that $P_d(\alpha)$, $P_{FA}^{(n)}(\alpha)$ are irrelevant to the sign of α, then the mean acquisition time in a lognormal fading channel can be obtained by averaging as

$$T_{ACQ} = \int_0^{\infty} T_{ACQ}(\alpha)\phi(|\alpha|)d|\alpha| \tag{4.24}$$

In the case of a packetized transmission mode with a short preamble interval, if a false alarm occurs before the end of the preamble, the simple implementation is that the receiver must demodulate and decode for the duration of an entire packet and determine that cyclic-redundancy-check failure has occurred in order to detect that a false alarm has occurred [16]. In this situation the false alarm is catastrophic and causes a total miss of the correct code phase, which means that the false alarm state is an absorbing state and $H_P(z) = 0$. For packetized transmission mode the performance measure is the probability of acquisition in L or fewer dwells, while this cumulative probability requires first obtaining an expression for the PDF of the number of dwells to obtain successful synchronization [17]. Starting from (4.22), we rewrite the generating function in the form of a power series in z, namely

$$\begin{aligned}
F_{ACQ}(z) \\
&= P_d(\alpha)z\sum_{i=0}^{\infty}\left[(1-P_d(\alpha))z\prod_{n=1}^{N_c-1}(1-P_{FA}^{(n)}(\alpha))z\right]^i\left[\frac{1}{N_c}\sum_{j=0}^{N_c-1}\prod_{n=1}^{j}\left(1-P_{FA}^{(N_c-n)}(\alpha)\right)z\right] \\
&= \frac{P_d(\alpha)}{N_c}\sum_{i=0}^{\infty}\sum_{j=0}^{N_c-1}\left[(1-P_d(\alpha))\prod_{n=1}^{N_c-1}\left(1-P_{FA}^{(n)}(\alpha)\right)\right]^i\prod_{n=1}^{j}(1-P_{FA}^{(N_c-n)}(\alpha))z^{iN_c+j+1} \\
&= \sum_{l=0}^{\infty} f_l z^l
\end{aligned} \tag{4.25}$$

where f_l denotes the probability of the event that starting from some initial state the process will reach acquisition state in l dwells. After getting f_l, $l = 1, 2, \ldots, L$, the probability of acquisition in L or fewer dwells under one channel realization with given α can be obtained as

$$\begin{aligned}
P_{ACQ}(L,\alpha) &= \sum_{l=1}^{L} f_l \\
&= \frac{P_d(\alpha)}{N_c}\cdot\sum_{l=1}^{L}\left\{\left[(1-P_d(\alpha))\prod_{n=1}^{N_c-1}(1-P_{FA}^{(n)}(\alpha))\right]^{\left\lfloor\frac{l-1}{N_c}\right\rfloor}\right. \\
&\quad\times\left.\prod_{n=1}^{(l-1)\bmod N_c}(1-P_{FA}^{(N_c-n)}(\alpha))\right\}
\end{aligned} \tag{4.26}$$

Therefore, the probability of acquisition in L or fewer dwells in a lognormal fading channel is obtained by averaging

$$P_{ACQ}(L) = \int_0^{\infty} P_{ACQ}(L,\alpha)\phi(|\alpha|)d|\alpha| \tag{4.27}$$

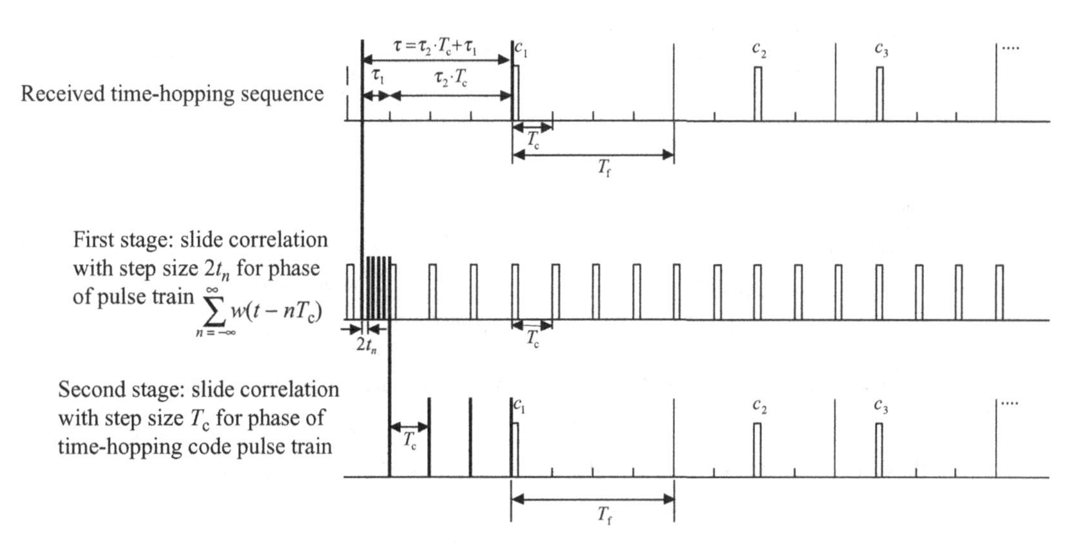

Figure 4.4 The novel two-stage serial search acquisition procedure.

4.4 Novel two-stage acquisition

4.4.1 System description

To overcome the drawback of the long time that is needed in conventional serial search acquisition with an advancing step size of $2t_n$, it is necessary to propose a novel scheme to improve the advancing step size. Assuming $\tau = \tau_2 \cdot T_c + \tau_1$, $\tau_2 = \lfloor \tau / T_c \rfloor$, $0 \le \tau_1 < T_c$, the novel acquisition method is based on the following idea: if we can estimate τ_1 (chip synchronization) correctly at first, then we can align the phase of the local TH sequence pulse train with the advancing step size enlarged to be T_c to search for the correct τ_2 (code synchronization). Without channel noise, this acquisition procedure can be done in two stages as shown in Figure 4.4. First, in order to estimate τ_1, the received signal is correlated by the regular spacing pulse train $\sum_{j=-\infty}^{\infty} w(t - jT_c)$ with an advancing step size of $2t_n$; second, once chip synchronization is achieved, the system proceeds with code synchronization and the received signal is correlated by the local TH sequence pulse train.

A block diagram of the novel two-stage acquisition method is shown in Figure 4.5. In the first stage, the output of the correlator is

$$Y_1 = \frac{1}{AN_pN_h} \sum_{j=0}^{N_p \cdot N_h^2 - 1} \int_{jT_c}^{(j+1)T_c} r(t) w(t - jT_c - \zeta_1) \mathrm{d}t \tag{4.28}$$

where ζ_1 is the phase of the regular spacing pulse train $\sum_{j=-\infty}^{\infty} w(t - jT_c)$ with an advancing step size of $2t_n$. The uncertainty region of τ_1 in the first stage is only T_c, which is quantized into a finite number of elements (sectors). The sector width is set as $2t_n$, therefore the number of sectors is $q_1 = T_c/(2t_n)$. In the whole uncertainty region, there exists only one H_1-sector that corresponds to the correct estimation of τ_1 ($|\Delta\tau_1| = |\zeta_1 - \tau_1| < t_n$). The other $q_1 - 1$ sectors are H_0-sectors ($|\Delta\tau_1| = |\zeta_1 - \tau_1| \ge t_n$) with PDFs identical to the decision variable $Z_1 = Y_1^2$. Once a "hit" ($Z_1 \ge \xi_1$) is observed by

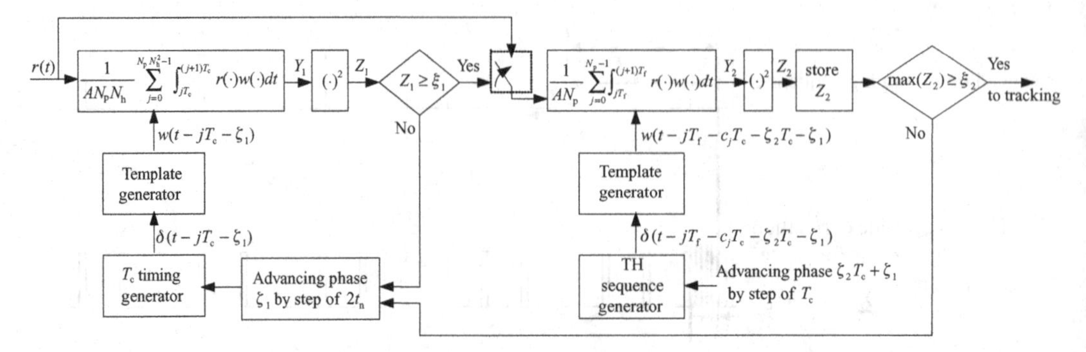

Figure 4.5 Block diagram of the novel two-stage serial search acquisition scheme.

the threshold detector, the current sector with phase of ζ_1 is considered to be an H_1-sector and the system goes into the second stage.

In the second stage, the output of the correlator is given by

$$Y_2 = \frac{1}{AN_{\rm p}} \sum_{j=0}^{N_{\rm p}-1} \int_{jT_{\rm f}}^{(j+1)T_{\rm f}} r(t)w(t - jT_{\rm f} - c_jT_{\rm c} - \zeta_2T_{\rm c} - \zeta_1){\rm d}t \tag{4.29}$$

There are a total of q_2 ($q_2 = N_{\rm p}N_{\rm h}$) unknown phase cells with the phases $\zeta_2T_{\rm c} + \zeta_1$, $\zeta_2 = 0, 1, 2, \ldots, q_2 - 1$ in the selected sector in the first stage. If the selected sector in the first stage is the H_1-sector, there exists one H_1-cell ($|\tau - (\zeta_2 \cdot T_{\rm f} + \zeta_1)| < t_n$), otherwise there exists no H_1-cell in this sector. There are two cases of H_0-cells: one is in the H_1-sector and the other is in the H_0-sector. The procedure of the second stage is similar to the MAX/TC criterion [18]. The serial decision variable (Z_2) for every cell test is stored. After a number of cell tests, q_2, the maximum variable is selected and compared to the threshold. If $\max(Z_2) \geq \xi_2$, the current cell is considered to be an H_1-cell and the system goes into the tracking mode, otherwise the first stage is resumed from the phase point next to ζ_1. If a false alarm does occur, it can be recognized by the tracking system and the first stage will restart after a delay.

4.4.2 Noncoherent correlation detector

To facilitate our analysis, the three situations mentioned in Sub-section 4.3.1 are also considered for the novel two-stage scheme. Under one-channel realization with given α, the output of the correlator Y_1 is given by

$$Y_1 = \begin{cases} \alpha + \eta_1, & |\Delta\tau_1| < t_n (H_1\text{-sector}) \\ \eta_1, & |\Delta\tau_1| \geq t_n (H_0\text{-sector}) \end{cases} \tag{4.30}$$

and the noise component

$$\eta_1 = \frac{1}{AN_{\rm p}N_{\rm h}} \sum_{j=0}^{N_{\rm p}\cdot N_{\rm h}^2-1} \int_{jT_{\rm c}}^{(j+1)T_{\rm c}} n(t)w(t - jT_{\rm c} - \zeta_1){\rm d}t \tag{4.31}$$

is a Gaussian random variable with zero mean and a variance of $N_0/(2N_pA^2)$. Therefore Y_1 is a Gaussian random variable:

$$Y_1 \sim \begin{cases} G\left(\alpha, \dfrac{N_0}{2N_pA^2}\right), & |\Delta\tau_1| < t_n(H_1\text{-sector}) \\ G\left(0, \dfrac{N_0}{2N_pA^2}\right), & |\Delta\tau_1| \geq t_n(H_0\text{-sector}) \end{cases} \tag{4.32}$$

The decision variable $Z_1 = Y_1^2$ has a noncentral chi-square distribution with one degree of freedom, and therefore the detection probability of the H_1-sector is given by

$$\begin{aligned} P_{d1}(\alpha) &= P_r(Z_1 > \xi_1|H_1) = \int_{\xi_1}^{\infty} p(Z_1|H_1)\mathrm{d}Z_1 \\ &= \int_{\xi_1}^{\infty} \frac{1}{\sqrt{\pi N_0/(N_pA^2)Z_1}} \exp\left[\frac{-Z_1-\alpha^2}{N_0/(N_pA^2)}\right] \cosh\left[\frac{2\sqrt{Z_1}\alpha}{N_0/(N_pA^2)}\right] \mathrm{d}Z_1 \\ &= \int_{\overline{\xi_1}}^{\infty} \frac{1}{\sqrt{\pi \overline{Z_1}}} \exp(-\overline{Z_1} - \mu\alpha^2) \cosh\left(2\sqrt{\mu\alpha^2\overline{Z_1}}\right) \mathrm{d}\overline{Z_1} \\ &= C(\overline{\xi_1}, \mu\alpha^2) \end{aligned} \tag{4.33}$$

where μ is given by (4.19) and $\overline{\xi_1}$ is the normalized threshold in the first stage as

$$\overline{\xi_1} = \mu\xi_1 \tag{4.34}$$

Thus, the false alarm probability of the H_0-sector is

$$\begin{aligned} P_{FA1} &= P_r(Z_1 > \xi_1|H_0) = \int_{\xi_1}^{\infty} p(Z_1|H_0)\mathrm{d}Z_1 \\ &= \int_{\xi_1}^{\infty} \frac{1}{\sqrt{\pi N_0/(N_pA^2)Z_1}} \exp\left[\frac{-Z_1}{N_0/(N_pA^2)}\right] \mathrm{d}Z_1 \\ &= \int_{\overline{\xi_1}}^{\infty} \frac{1}{\sqrt{\pi \overline{Z_1}}} \exp(-\overline{Z_1})\mathrm{d}\overline{Z_1} \\ &= C(\overline{\xi_1}, 0) \end{aligned} \tag{4.35}$$

For the second stage, the output of the correlator Y_2 is a Gaussian random variable:

$$Y_2 \sim \begin{cases} G\left(\alpha, \dfrac{N_0}{2N_pA^2}\right), & H_1\text{-cell} \\ G\left(\dfrac{\alpha\lambda}{N_p}, \dfrac{N_0}{2N_pA^2}\right), & H_0\text{-cell in } H_1\text{-sector} \\ G\left(0, \dfrac{N_0}{2N_pA^2}\right), & H_0\text{-cell in } H_0\text{-sector} \end{cases} \tag{4.36}$$

Under one-channel realization with given α, the decision variable $Z_2 = Y_2^2$ has a noncentral chi-square distribution with one degree of freedom, so that the PDF of Z_2 of the H_1-cell is

$$p_1(Z_2|\alpha) = \frac{1}{\sqrt{\pi N_0/(N_pA^2)Z_2}} \exp\left[\frac{-Z_2-\alpha^2}{N_0/(N_pA^2)}\right] \cosh\left[\frac{2\sqrt{\alpha^2 Z_2}}{N_0/(N_pA^2)}\right], \; Z_2 \geq 0 \tag{4.37}$$

The PDF of Z_2 of the H_0-cells in the H_1-sector is

$$p_0^{(1)}(Z_2|\alpha) = \frac{1}{\sqrt{\pi N_0/(N_\mathrm{p}A^2)Z_2}} \exp\left[\frac{-Z_2 - \alpha^2\lambda^2/N_\mathrm{p}^2}{N_0/(N_\mathrm{p}A^2)}\right] \cosh\left[\frac{2\sqrt{\alpha^2\lambda^2/N_\mathrm{p}^2 Z_2}}{N_0/(N_\mathrm{p}A^2)}\right], \quad Z_2 \geq 0 \tag{4.38}$$

and the PDF of Z_2 of the H_0-cells in the H_0-sector is

$$p_0^{(0)}(Z_2) = \frac{1}{\sqrt{\pi N_0/(N_\mathrm{p}A^2)Z_2}} \exp\left[\frac{-Z_2}{N_0/(N_\mathrm{p}A^2)}\right], \quad Z_2 \geq 0 \tag{4.39}$$

Three disjoint events are possible if the selected sector in the first stage is the H_1-sector:

(1) *Missed detection*: all of the q_2 test variables are below threshold ξ_2, with probability $P_{m2}(\alpha)$.
(2) *Correct detection*: the H_1-cell test variable is above threshold and is the maximum, with probability $P_{d2}(\alpha)$.
(3) *Error*: at least one test variable associated with an H_0-cell is above threshold and is greater than the H_1-cell test variable, with probability $P_{e2}(\alpha)$.

Another case is that the selected sector in the first stage is the H_0-sector, therefore there exists no H_1-cell in this sector for the search of the second stage. In this case two disjoint events are possible:

(1) *False alarm*: at least one test variable is above threshold ξ_2, with probability P_{FA2}.
(2) *Correct rejection*: all the q_2 test variables are below threshold, with probability $1 - P_{\mathrm{FA2}}$.

The probability of detection, error, and false alarm in the second stage can be written as

$$\begin{aligned} P_{\mathrm{d2}}(\alpha) &= \int_{\xi_2}^{\infty} p_1(Z_2|\,\alpha)\left[\int_{-\infty}^{Z_2} p_0^{(1)}(z|\,\alpha)\mathrm{d}z\right]^{N_\mathrm{p}N_\mathrm{h}-1} \mathrm{d}Z_2 \\ &= \int_{\overline{\xi_2}}^{\infty} \frac{1}{\sqrt{\pi\overline{Z_2}}} \exp(-\overline{Z_2} - \mu\alpha^2)\cosh\left(2\sqrt{\mu\alpha^2\overline{Z_2}}\right) \\ &\quad \times \left[1 - C\left(\overline{Z_2}, \frac{\mu\alpha^2\lambda^2}{N_\mathrm{p}^2}\right)\right]^{N_\mathrm{p}N_\mathrm{h}-1} \mathrm{d}\overline{Z_2} \end{aligned} \tag{4.40}$$

$$\begin{aligned} P_{\mathrm{m2}}(\alpha) &= \int_{-\infty}^{\xi_2} p_1(Z_2|\,\alpha)\mathrm{d}Z_2 \left[\int_{-\infty}^{\xi_2} p_0^{(1)}(Z_2|\,\alpha)\mathrm{d}Z_2\right]^{N_\mathrm{p}N_\mathrm{h}-1} \\ &= [1 - C(\overline{\xi_2}, \mu\alpha^2)]\left[1 - C\left(\overline{\xi_2}, \frac{\mu\alpha^2\lambda^2}{N_\mathrm{p}^2}\right)\right]^{N_\mathrm{p}N_\mathrm{h}-1} \end{aligned} \tag{4.41}$$

$$P_{\mathrm{e2}}(\alpha) = 1 - P_{\mathrm{d2}}(\alpha) - P_{\mathrm{m2}}(\alpha) \tag{4.42}$$

$$\begin{aligned} P_{\mathrm{FA2}} &= 1 - \left[\int_{-\infty}^{\xi_2} p_0^{(0)}(Z_2)\mathrm{d}Z_2\right]^{N_\mathrm{p}N_\mathrm{h}} \\ &= 1 - [1 - C(\overline{\xi_2}, 0)]^{N_\mathrm{p}N_\mathrm{h}} \end{aligned} \tag{4.43}$$

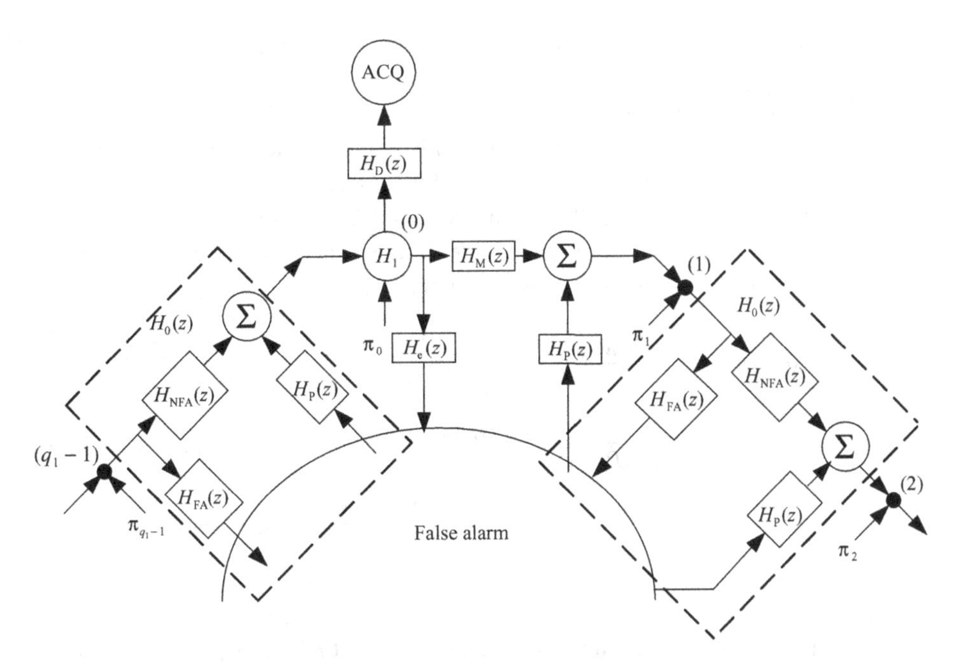

Figure 4.6 Flow diagram for the novel two-stage serial search acquisition scheme.

where $\overline{\xi_2}$ is the normalized threshold in the second stage as

$$\overline{\xi_2} = \mu\xi_2 \tag{4.44}$$

4.4.3 Flow diagram analysis

The flow diagram for the novel two-stage serial search acquisition scheme is shown in Figure 4.6. Each state in the flow diagram now corresponds to a sector in the uncertainty region of the first stage, and the H_1-sector is designated as state 0. A priori information of being the first state, uniformly distributed, in the search makes $\pi_n = 1/q_1,\ n = 0, 1, 2, \ldots, q_1 - 1$. Besides the correspondence between states and sectors, the other fundamental difference with a flow diagram of the conventional scheme is the presence of an additional branch out of the H_1 state to the false alarm state, corresponding to the error event. Note that here correlation length in the first stage is $N_p N_h^2 T_c = N_h T_p$, and the whole time needed for one search round of the second stage is $N_p N_h T_p$.

As in [16], the flow-diagram generating function of the novel two-stage acquisition under one-channel realization with given α can be derived as

$$F_{ACQ}(z) = \frac{1}{q_1} \cdot \frac{H_D(z)}{1 - [H_M(z) + H_e(z)H_P(z)]\, H_0^{q_1-1}(z)} \left[\sum_{j=0}^{q_1-1} H_0^j(z)\right] \tag{4.45}$$

where $H_D(z) = P_{d1}(\alpha)P_{D2}(\alpha)z^{(N_p N_h + N_h)}$ is a gain of the branch leading from the H_1 state (state 0) to the acquisition state; $H_M(z) = [(1 - P_{d1}(\alpha)) + P_{d1}(\alpha)P_{m2}(\alpha)z^{N_p N_h}]z^{N_h}$ is a gain of the branch connecting the H_1 state (state 0) to state 1 without an error event; $H_e(z) = P_{d1}(\alpha)(1 - P_{d2}(\alpha) - P_{m2}(\alpha))z^{N_p N_h + N_h}$ is a gain of the branch out of the H_1 state

(state 0) to the false alarm state, corresponding to the error event; $H_0(z) = H_{\mathrm{NFA}}^{(n)}(z) + H_{\mathrm{FA}}^{(n)}(z)H_{\mathrm{P}}(z) = [(1 - P_{\mathrm{FA1}}) + P_{\mathrm{FA1}}(1 - P_{\mathrm{FA2}})z^{N_{\mathrm{p}}N_{\mathrm{h}}} + P_{\mathrm{FA1}}P_{\mathrm{FA2}}z^{N_{\mathrm{p}}N_{\mathrm{h}}+K}]z^{N_{\mathrm{h}}}$ is a gain of the branch connecting two successive states; $H_{\mathrm{NFA}}(z) = [(1 - P_{\mathrm{FA1}}) + P_{\mathrm{FA1}}(1 - P_{\mathrm{FA2}})z^{N_{\mathrm{p}}N_{\mathrm{h}}}]z^{N_{\mathrm{h}}}$ characterizes the path going to the next state without false alarm; $H_{\mathrm{FA}}(z) = P_{\mathrm{FA1}}P_{\mathrm{FA2}}z^{N_{\mathrm{p}}N_{\mathrm{h}}+N_{\mathrm{h}}}$ stands for a false alarm; and $H_{\mathrm{p}}(z) = z^K$ is the gain of the transition from the false alarm state to the next cell.

The mean acquisition time under one-channel realization with given α is given by (after some algebra shown in Appendix 4B)

$$
\begin{aligned}
T_{\mathrm{ACQ}}(\alpha) &= T_{\mathrm{p}} \cdot \left.\frac{\mathrm{d}F_{\mathrm{ACQ}}(z)}{\mathrm{d}z}\right|_{z=1} \\
&= \frac{T_{\mathrm{p}}}{P_{\mathrm{d1}}(\alpha)P_{\mathrm{d2}}(\alpha)}\Bigg[-KP_{\mathrm{d1}}(\alpha)P_{\mathrm{d2}}(\alpha) - KP_{\mathrm{d1}}(\alpha)P_{\mathrm{m2}}(\alpha) + (K + N_{\mathrm{p}}N_{\mathrm{h}})P_{\mathrm{d1}}(\alpha) \\
&\quad + N_{\mathrm{h}} + (q_1 - 1)\left(1 - \frac{P_{\mathrm{d1}}(\alpha)P_{\mathrm{d2}}(\alpha)}{2}\right)(N_{\mathrm{h}} + N_{\mathrm{p}}N_{\mathrm{h}}P_{\mathrm{FA1}}KP_{\mathrm{FA1FA2}})\Bigg]
\end{aligned}
\tag{4.46}
$$

From (4.33), (4.40) and (4.41) it can be seen that $P_{\mathrm{d1}}(\alpha)$, $P_{\mathrm{d2}}(\alpha)$ and $P_{\mathrm{m2}}(\alpha)$ are irrelevant to the sign of α, then the mean acquisition time in a lognormal fading channel can be obtained with the same form as (4.24).

With the packetized transmission situation that a false alarm leads to a catastrophic situation, i.e. $H_{\mathrm{P}}(z) = 0$, we rewrite the generating function in the form of a power series in z, namely

$$
\begin{aligned}
&F_{\mathrm{ACQ}}(z) \\
&= \frac{H_{\mathrm{D}}(z)}{q_1}\sum_{i=0}^{\infty}\left\{[H_{\mathrm{M}}(z) + H_{\mathrm{e}}(z)H_{\mathrm{P}}(z)]\,H_0^{q_1-1}(z)\right\}^i\left[\sum_{j=0}^{q_1-1}H_0^j(z)\right] \\
&= \frac{P_{\mathrm{d1}}(\alpha)P_{\mathrm{d2}}(\alpha)}{q_1}\sum_{i=0}^{\infty}\sum_{j=0}^{q_1-1}\sum_{h=0}^{i(q_1-1)+j}\sum_{m=0}^{i}\binom{i(q_1-1)+j}{h}\binom{i}{m}[P_{\mathrm{FA1}}(1 - P_{\mathrm{FA2}})]^h \\
&\quad \cdot (1 - P_{\mathrm{FA1}})^{i(q_1-1)+j-h}\,(P_{\mathrm{d1}}(\alpha)P_{\mathrm{m2}}(\alpha))^m\,(1 - P_{\mathrm{d1}}(\alpha))^{i-m}z^{N_{\mathrm{h}}(iq_1+j+1)+N_{\mathrm{p}}N_{\mathrm{h}}(h+m+1)} \\
&= \sum_{l=0}^{\infty} f_l z^l
\end{aligned}
\tag{4.47}
$$

As in [17], determination of these coefficients f_l, $l = 1, 2, \ldots$ is possible but quite tedious. To make matters more tractable, with (4.33)–(4.35) and (4.40)–(4.44) and for one arbitrary number of L, we can make $i_{\max} = \left\lfloor \dfrac{L - N_{\mathrm{p}}N_{\mathrm{h}} - N_{\mathrm{h}}}{q_1 N_{\mathrm{h}}} \right\rfloor + 1$ as the upper limit of index i, and find out f_l, $l = 1, 2, \ldots, L$, respectively, using numerical computation. After that, the probability of acquisition in L or fewer dwells of the novel scheme in one-channel realization with given α can be obtained as $P_{\mathrm{acq}}(L, \alpha) = \sum_{l=1}^{L} f_l$. Finally, the probability of acquisition in L or fewer dwells in a lognormal fading channel can be obtained with the same form of (4.27).

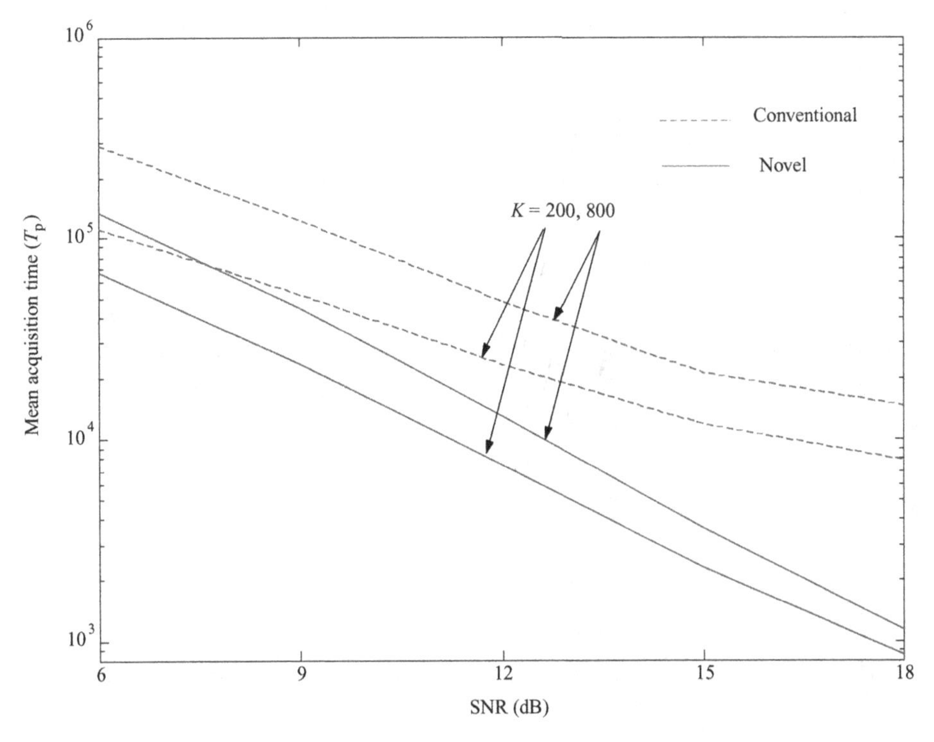

Figure 4.7 Mean acquisition time (T_p).

4.5 Numerical results

In this section, comparison of acquisition performance is investigated by numerical calculations on the equations in Sections 4.3 and 4.4 for the conventional and the novel two-stage acquisition schemes with different system parameters. The fading is assumed to be lognormally distributed, as noted in (4.4). The frame duration T_f is set to be 100 ns, and the chip duration T_c is chosen as 10 ns. Thus, we deduce $N_h = T_f/T_c = 10$. $t_n = 0.125$ ns is adopted to yield the monocycle pulse width $T_w < 1$ ns. The period of TH code, N_p, is set as 15, and $\lambda = 3$ is chosen to describe the ACF of the TH code. Therefore the number of cells in the whole time uncertainty region is $N_c = N_p T_f/(2t_n) = 6000$. There exist different thresholds setting rules in the decision-making device or detector, i.e. fixed thresholds, constant false alarm rate criteria, maximum selection (the case when no threshold is used), or optimum threshold setting [19], [20]. Here, in order to compare the performance of the two acquisition schemes and in order to avoid the effect of detailed threshold setting selection on acquisition performance, the thresholds ξ for the conventional scheme, and ξ_1 and ξ_2 for the novel two-stage scheme are adjusted to optimize acquisition performances such as the mean acquisition time and the probability of acquisition in L or fewer dwells for every value of L.

Figure 4.7 depicts the comparison of the mean acquisition time for the two acquisition schemes as a function of the average received SNR with two different values of the

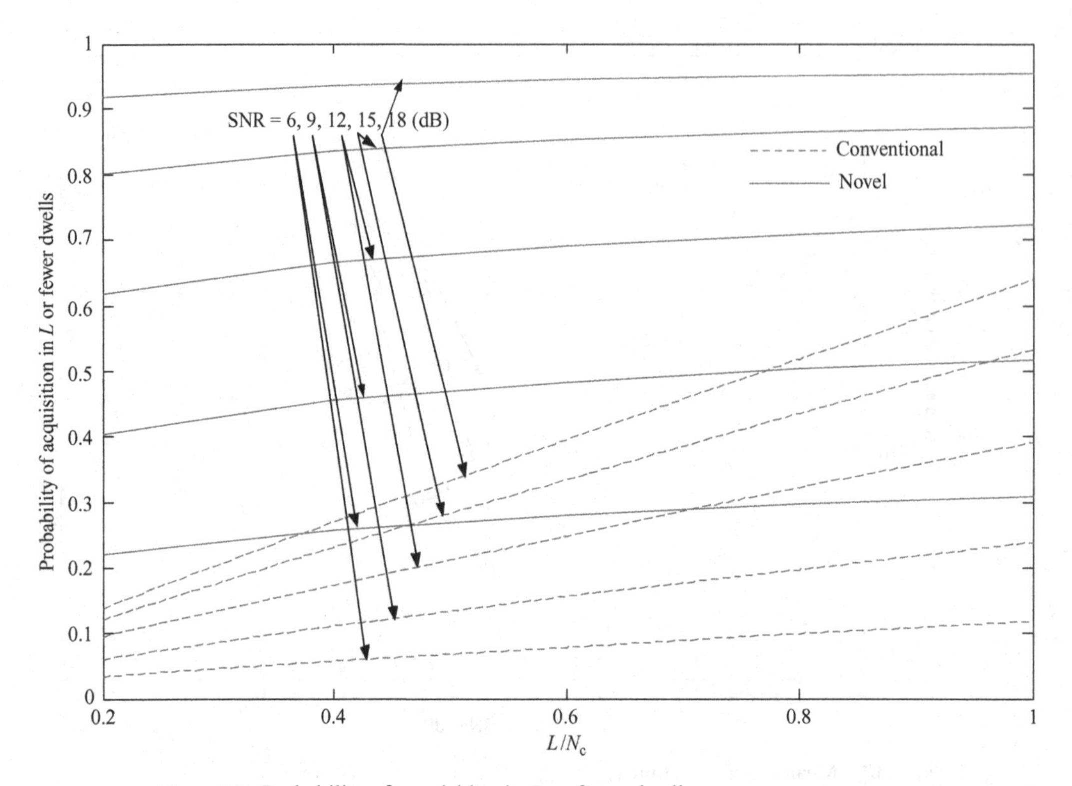

Figure 4.8 Probability of acquisition in L or fewer dwells.

penalty time, i.e. $K = 200,\ 800$. For both values of K, it can be seen that the mean acquisition time of the novel two-stage acquisition scheme is always much shorter than that of the conventional acquisition scheme, especially under low SNR. Furthermore, when the penalty time K increases, under low SNR the mean acquisition time of the conventional scheme increases significantly, whereas the increment of the mean acquisition time for the novel scheme is much smaller. When the SNR increases, for the both acquisition methods the false alarm probability of every H_0-cell (or H_0-sector) test can be controlled to lower level by adjusting the threshold, thereby the impact of penalty time K on the mean acquisition time becomes increasingly insignificant, especially negligible for the novel two-stage acquisition scheme with SNR larger than 15 dB.

Figure 4.8 compares the probability of acquisition in L or fewer dwells for the two acquisition schemes as a function of L/N_c for different values of SNR. It can be seen that for both schemes, and for a given L/N_c, the probability of acquisition increases as SNR increases. Furthermore, for a given SNR, the probability of acquisition of the novel scheme is larger than that of the conventional scheme especially when L/N_c is small. When L/N_c increases from 0.2 to 1, the probability of acquisition of the conventional scheme increases almost linearly, while the increment of the probability of acquisition for the novel scheme is not significant. The phenomenon implies that in order to reach large probability of acquisition under every SNR, the preamble length L needed in the

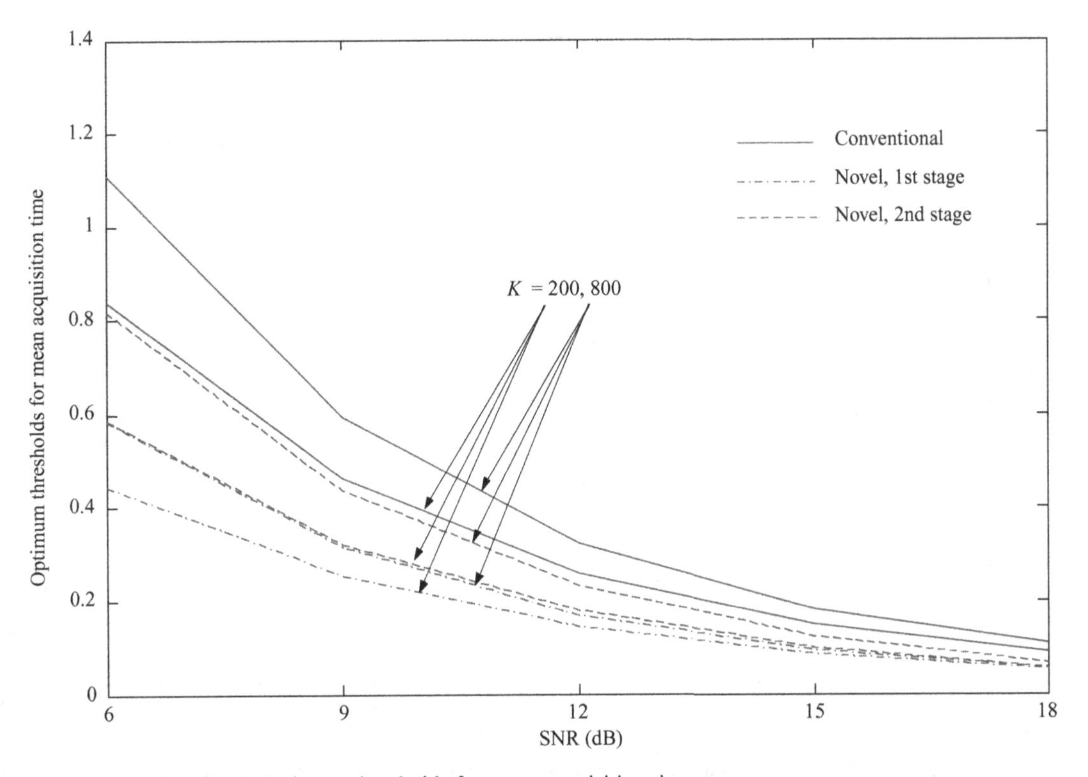

Figure 4.9 Optimum thresholds for mean acquisition time.

conventional acquisition scheme should be much larger than that in the novel acquisition scheme.

Figure 4.9 shows the optimum thresholds to minimize the mean acquisition time for different values of the penalty time K. It can be seen that, for a given K, the optimum thresholds ξ_1 and ξ_2 of the novel scheme are smaller than the optimum threshold ξ of the conventional scheme. For low SNR, the optimum thresholds of the two schemes are high to keep the false alarm probability in every detection of the H_0-cell (or H_0-sector) small enough and get a reasonable detection probability in the H_1-cell test, while these thresholds decrease when the SNR increases (or the variance of noise decreases). For example, with an SNR of 6 dB and K 800, the optimum threshold ξ of the conventional acquisition scheme is larger than 1 (the normalized amplitude of the useful signal after correlation in the H_1-cell test), which justifies the very poor detection probability in the H_1-cell test, whereas the optimum thresholds ξ_1 and ξ_2 of the novel acquisition scheme are much smaller than ξ of the conventional scheme. When the penalty time K increases, a lower false alarm probability is needed, and the optimum thresholds of the two schemes obviously increase.

Figure 4.10 illustrates the optimum thresholds in order to maximize the probability of acquisition in L or fewer dwells for two specific values of L ($L = 0.4N_c$ and N_c). It can be seen that when L increases in the conventional acquisition, the optimum threshold ξ

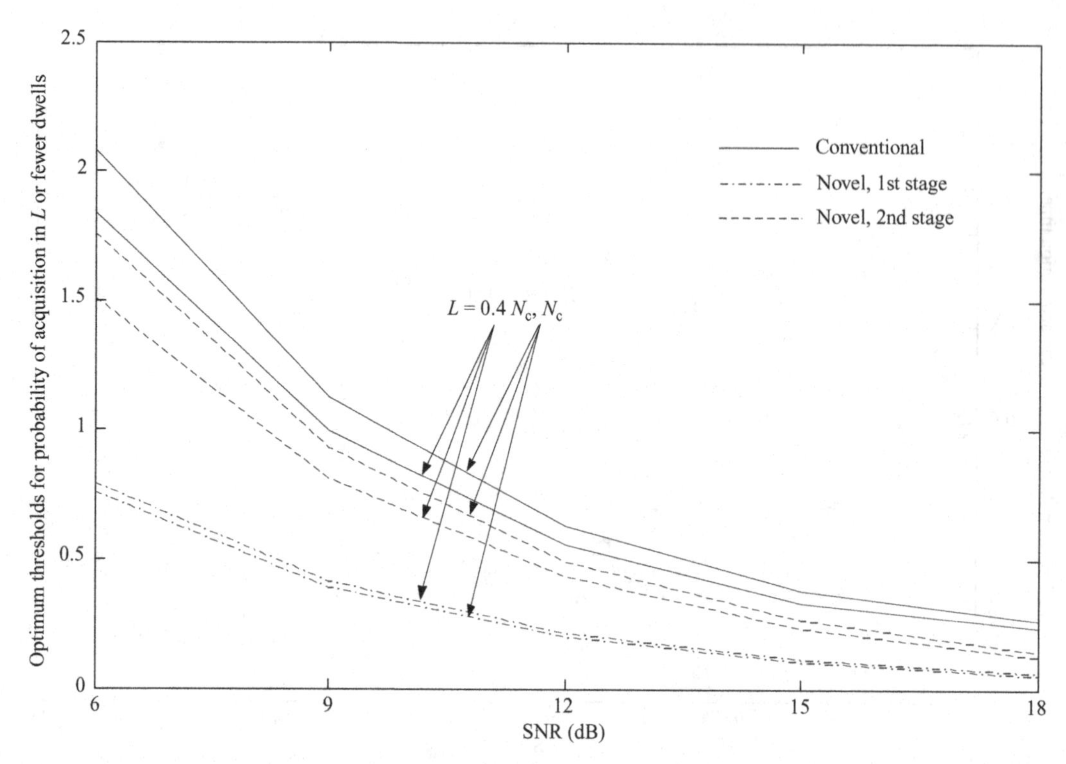

Figure 4.10 Optimum thresholds for probability of acquisition in L or fewer dwells.

must be significantly increased in order to reduce the catastrophic false alarm, while when L increases in the novel acquisition, ξ_2 in the second stage increases significantly to get a smaller false alarm probability in the H_0-cell test, whereas ξ_1 in the first stage is almost flat in order to get a higher detection probability in the H_1-sector test. Furthermore, for a given L, when the SNR increases the optimum threshold ξ in the conventional acquisition and ξ_2 in the novel acquisition decrease remarkably, while the decrease in ξ_1 in the novel acquisition is much slighter, which means that a higher detection probability in the H_1-sector test is more important in the first stage, while reducing false alarms in the H_0-cell test is more important in the second stage, to get the optimum probability of acquisition in L or fewer dwells.

One noticeable point is that the former analysis and corresponding numerical results are obtained in the single-user environment. In asynchronous multiuser systems, the false alarm probability in the first stage (chip acquisition) may be larger due to multiuser interference (MUI). The worst case is that there are many users without chip synchronization, and the false alarm probability remains 1 in every test of chip acquisition for the target user due to the severe MUI. In this situation, the flow diagrams show that due to chip acquisition more time is needed for the novel acquisition than for the conventional acquisition to complete one round of searching. However, it can easily be seen that such time consumed by chip acquisition is only a small fraction of one searching round time. Furthermore, in asynchronous multiuser systems this novel acquisition scheme can be combined with a multiuser detection receiver, such as the structure of interference cancellation (IC) followed by acquisition as proposed in [20]. Thus, after the interference from the synchronized users is cancelled by IC, the residual signal is sent to the novel acquisition scheme and the false alarm in the chip acquisition could be alleviated.

4.6 Summary

In this chapter, a novel two-stage serial search acquisition scheme has been proposed and analyzed for TH-IR code acquisition under a slow fading channel. It has been shown that the proposed novel scheme achieves much better acquisition performance than the conventional acquisition scheme:

(1) In the case of having non-limited permitted acquisition time, using the novel two-stage acquisition scheme reduces the mean acquisition time much more than using the conventional acquisition scheme, particularly under low SNR or with a large penalty time.
(2) In the case of a packetized transmission mode with a short preamble interval, the preamble length needed in the novel two-stage acquisition scheme is much shorter than that in the conventional acquisition scheme to reach the same probability of acquisition under every reasonable SNR.

Appendix 4A The derivation of T_{ACQ} in (4.23)

By using (4.22) and $H_{\mathrm{D}}(z) = P_{\mathrm{d}}(\alpha)z$, the derivative of $F_{\mathrm{ACQ}}(z)$ with respect to z is

$$\frac{\mathrm{d}F_{\mathrm{ACQ}}(z)}{\mathrm{d}z} = \frac{P_{\mathrm{d}}(\alpha)}{N_{\mathrm{c}}}[D(z) + zD'(z)] \tag{4A.1}$$

where

$$D(z) = \frac{\left[\sum_{j=0}^{N_{\mathrm{c}}-1}\prod_{n=1}^{j} H_0^{(N_{\mathrm{c}}-n)}(z)\right]}{1 - H_{\mathrm{M}}(z)\prod_{n=1}^{N_{\mathrm{c}}-1} H_0^{(n)}(z)} \tag{4A.2}$$

It is easy to get $H_{\mathrm{M}}(1) = H'_{\mathrm{M}}(z)\big|_{z=1} = (1 - P_{\mathrm{d}}(\alpha))$, $H_0^{(n)}(1) = 1$, and $[H_0^{(n)}(z)]'\big|_{z=1} = 1 + KP_{\mathrm{FA}}^{(n)}(\alpha)$. Therefore from (4A.1) we can obtain $D(1) = \dfrac{N_{\mathrm{c}}}{P_{\mathrm{d}}(\alpha)}$, and

$$D'(z)\Big|_{z=1}$$

$$= \frac{\left[1 - H_{\mathrm{M}}(1)\prod_{n=1}^{N_{\mathrm{c}}-1} H_0^{(n)}(1)\right]\cdot\left[\sum_{j=0}^{N_{\mathrm{c}}-1}\prod_{n=1}^{j} H_0^{(N_{\mathrm{c}}-n)}(z)\right]'\Bigg|_{z=1} + \left[H_{\mathrm{M}}(z)\prod_{n=1}^{N_{\mathrm{c}}-1} H_0^{(n)}(z)\right]'\Bigg|_{z=1}\cdot\left[\sum_{j=0}^{N_{\mathrm{c}}-1}\prod_{n=1}^{j} H_0^{(N_{\mathrm{c}}-n)}(1)\right]}{\left[1 - H_M(1)\prod_{n=1}^{N_{\mathrm{c}}-1} H_0^{(n)}(1)\right]^2}$$

$$= \frac{P_{\mathrm{d}}(\alpha)\cdot\left[\sum_{j=0}^{N_c-1}\prod_{n=1}^{j} H_0^{(N_c-n)}(z)\right]'\Bigg|_{z=1} + N_c\cdot\left[H_M(z)\prod_{n=1}^{N_c-1}H_0^{(n)}(z)\right]'\Bigg|_{z=1}}{P_{\mathrm{d}}^2(\alpha)} \tag{4A.3}$$

$$= \frac{P_{\mathrm{d}}(\alpha)\cdot\left[\sum_{j=1}^{N_c-1}\sum_{n=1}^{j}\left(1+KP_{\mathrm{FA}}^{(N_c-n)}(\alpha)\right)\right] + N_c(1-P_{\mathrm{d}}(\alpha))\left[1+\sum_{n=1}^{N_c-1}\left(1+KP_{\mathrm{FA}}^{(N_c-n)}(\alpha)\right)\right]}{P_{\mathrm{d}}^2(\alpha)}$$

$$= \frac{\frac{N_c(N_c-1)}{2}P_{\mathrm{d}}(\alpha) + \left[\sum_{n=1}^{N_c-1} nP_{\mathrm{FA}}^{(n)}(\alpha)\right]KP_{\mathrm{d}}(\alpha) + N_c(1-P_{\mathrm{d}}(\alpha))\left[N_c + K\sum_{n=1}^{N_c-1}P_{\mathrm{FA}}^{n}(\alpha)\right]}{P_{\mathrm{d}}^2(\alpha)}$$

Now the expression of T_{ACQ} in conventional serial search acquisition with no identical PDF of the H_0-cell can be written as

$$T_{\mathrm{ACQ}} = T_{\mathrm{p}}\cdot\frac{\mathrm{d}F_{\mathrm{ACQ}}(z)}{\mathrm{d}z}\Bigg|_{z=1}$$

$$= T_{\mathrm{p}}\cdot\frac{P_{\mathrm{d}}(\alpha)}{N_c}\left[D(z)+zD'(z)\right]\big|_{z=1}$$

$$= T_{\mathrm{p}}\cdot\frac{P_{\mathrm{d}}(\alpha)}{N_c}\left[\frac{N_c}{P_{\mathrm{d}}(\alpha)} + \frac{\frac{N_c(N_c-1)}{2}P_{\mathrm{d}}(\alpha) + \left[\sum_{n=1}^{N_c-1} nP_{\mathrm{FA}}^{(n)}(\alpha)\right]KP_{\mathrm{d}}(\alpha) + N_c(1-P_{\mathrm{d}}(\alpha))\left[N_c + K\sum_{n=1}^{N_c-1}P_{\mathrm{FA}}^{n}(\alpha)\right]}{P_{\mathrm{d}}^2(\alpha)}\right]$$

$$= T_{\mathrm{p}}\cdot\frac{\frac{N_c(N_c+1)}{2}P_{\mathrm{d}}(\alpha) + \left[\sum_{n=1}^{N_c-1} nP_{\mathrm{FA}}^{(n)}(\alpha)\right]KP_{\mathrm{d}}(\alpha) + N_c(1-P_{\mathrm{d}}(\alpha))\left[N_c + K\sum_{n=1}^{N_c-1}P_{\mathrm{FA}}^{n}(\alpha)\right]}{N_cP_{\mathrm{d}}(\alpha)}$$

$$= T_{\mathrm{p}}\cdot\frac{(1-N_c)P_{\mathrm{d}}(\alpha) + 2N_c + 2K\sum_{n=1}^{N_c-1} nP_{\mathrm{FA}}^{(n)}(\alpha) + 2KP_{\mathrm{d}}(\alpha)\left[\frac{1}{N_c}\sum_{n=1}^{N_c-1} nP_{\mathrm{FA}}^{(n)}(\alpha) - \sum_{n=1}^{N_c-1}P_{\mathrm{FA}}^{n}(\alpha)\right]}{2P_{\mathrm{d}}(\alpha)} \tag{4A.4}$$

Appendix 4B The derivation of T_{ACQ} in (4.46)

From (4.45), the mean acquisition time of the two-stage acquisition is given by

$$T_{\mathrm{ACQ}} = T_{\mathrm{p}}\cdot\frac{\mathrm{d}F_{\mathrm{ACQ}}(z)}{\mathrm{d}z}\Bigg|_{z=1}$$

$$= \frac{T_{\mathrm{p}}}{q_1}\cdot\frac{\left[1-[H_{\mathrm{M}}(1)+H_{\mathrm{e}}(1)H_{\mathrm{P}}(1)]H_0^{q_1-1}(1)\right]\cdot\left[H_{\mathrm{D}}'(1)\sum_{j=0}^{q_1-1}H_0^j(1) + H_{\mathrm{D}}(1)H_0'(1)\sum_{j=0}^{q_1-1}jH_0^{j-1}(1)\right]}{\left[1-[H_{\mathrm{M}}(1)+H_{\mathrm{e}}(1)H_{\mathrm{P}}(1)]H_0^{q_1-1}(1)\right]^2}$$

$$+ \frac{T_{\mathrm{p}}}{q_1}\cdot\frac{\left[(q_1-1)[H_{\mathrm{M}}(1)+H_{\mathrm{e}}(1)H_{\mathrm{P}}(1)]H_0^{q_1-2}(1)H_0'(1) + [H_{\mathrm{M}}'(1)+H_{\mathrm{e}}'(1)H_{\mathrm{P}}(1)+H_{\mathrm{e}}(1)H_{\mathrm{P}}'(1)]H_0^{q_1-1}(1)\right]H_{\mathrm{D}}(1)\sum_{j=0}^{q_1-1}H_0^j(1)}{\left[1-[H_{\mathrm{M}}(1)+H_{\mathrm{e}}(1)H_{\mathrm{P}}(1)]H_0^{q_1-1}(1)\right]^2} \tag{4B.1}$$

It is easy to get $H_0(1) = 1$ and $1 - [H_M(1) + H_e(1)H_P(1)] = H_D(1)$, so that (4B.1) can be rewritten as

$$\begin{aligned} T_{ACQ} &= \frac{T_p}{q_1} \cdot \frac{[1-[H_M(1)+H_e(1)H_P(1)]]\cdot\left[H'_D(1)q_1 + H_D(1)H'_0(1)\frac{q_1(q_1-1)}{2}\right]}{[1-[H_M(1)+H_e(1)H_P(1)]]^2} \\ &+ \frac{T_p}{q_1}\cdot\frac{[(q_1-1)[H_M(1)+H_e(1)H_P(1)]\,H'_0(1)+q_1\left[H'_M(1)+H'_e(1)H_P(1)+H_e(1)H'_P(1)\right]]\,H_D(1)}{[1-[H_M(1)+H_e(1)H_P(1)]]^2} \\ &= T_p\cdot\frac{1}{H_D(1)}\left\{H'_D(1)+H'_M(1)+H_e(1)H'_P(1)+H'_e(1)H_P(1)+(q_1-1)H'_0(1)\left[1-\frac{H_D(1)}{2}\right]\right\} \end{aligned} \tag{4B.2}$$

We also can get

$$H_D(1) = P_{d1}(\alpha)P_{d2}(\alpha) \tag{4B.3}$$

$$H'_D(1) = P_{d1}(\alpha)P_{d2}(\alpha)(N_pN_h + N_h) \tag{4B.4}$$

$$H'_M(1) = (1 - P_{d1}(\alpha))N_h + P_{d1}(\alpha)P_{m2}(\alpha)(N_pN_h + N_h) \tag{4B.5}$$

$$H_e(1) = P_{d1}(\alpha)(1 - P_{d2}(\alpha) - P_{m2}(\alpha)) \tag{4B.6}$$

$$H'_e(1) = P_{d1}(\alpha)(1 - P_{d2}(\alpha) - P_{m2}(\alpha))(N_pN_h + N_h) \tag{4B.7}$$

$$H_P(1) = 1 \tag{4B.8}$$

$$H'_P(1) = K \tag{4B.9}$$

$$\begin{aligned} H'_0(1) &= (1 - P_{FA1})N_h + P_{FA1}(1 - P_{FA2})(N_pN_h + N_h) \\ &+ P_{FA1}P_{FA2}(N_pN_h + N_h + K) \end{aligned} \tag{4B.10}$$

Therefore substituting (4B.3)–(4B.10) into (4B.2) gives the mean acquisition time for the two-stage acquisition as (4.46).

References

[1] Y. Ma, F. Chin, B. Kannan and S. Pasupathy, "Acquisition performance of an ultra wide-band communications system over a multiple-access fading channel," in *Proc. IEEE UWBST 2002*, Baltimore, MD, pp. 99–103, May 2002.

[2] E. A. Homier and R. A. Scholtz, "Rapid acquisition of ultra-wideband signals in the dense multipath channel," in *Proc. IEEE UWBST 2002*, Baltimore, MD, pp. 105–109, May 2002.

[3] E. A. Homier and R. A. Scholtz, "Hybrid fixed-dwell-time search techniques for rapid acquisition of ultra-wideband signals," in *Proc. IWUWBS 2003*, Oulu, Finland, June 2003.

[4] V. S. Somayazulu, J. R. Foerster and S. Roy, "Design challenges for very high data rate UWB systems," in *Conference Record of the Thirty-Sixth Asilomar Conference on Signals, Systems and Computers*, vol. 1, pp. 717–721, Nov. 2002.

[5] F. Ramirez-Mireles, "On the performance of ultra-wide-band signals in Gaussian noise and dense multipath," *IEEE Trans. Veh. Technol.*, vol. 50, no. 1, pp. 244–249, Jan. 2001.

[6] A. Petroff and P. Withington, *PulsON Technology Overview*, July 2001. PulseONOverview7_01.PDF. [Online]. Available: www.timedomain.com/Files/downloads/.

[7] C. J. Le Martret and G. B. Giannakis, "All-digital impulse radio with multiuser detection for wireless cellular systems," *IEEE Trans. Commun.*, vol. 50, no. 9, pp. 1440–1450, Sept. 2002.
[8] M. Z. Win, X. Qiu, R. A. Scholtz and V. O. K. Li, "ATM-based TH-SSMA network for multimedia PCS," *IEEE J. Select. Areas Commun.*, vol. 17, no. 5, pp. 824–836, May 1999.
[9] J. R. Foerster, "Channel modeling sub-committee report final," IEEE 802.15.SG3a Study Group, Dec. 2002.
[10] Y. Shin, H. Lee, B. Han and S. Im, "Multipath characteristics of impulse radio channel," in *Proc. of IEEE VTC 2000*, Boston, MA, pp. 2487–2491, Oct. 2000.
[11] F. J. MacWilliams and N. J. A. Sloane, *The Theory of Error Correcting Codes*. New York: North-Holland, 1986.
[12] O. Moreno, Z. Zhang, P. V. Kumar and V. Zinoviev, "New constructions of optimal cyclically permutable constant weight code," *IEEE Trans. Inform. Theory*, vol. 41, no. 2, pp. 448–456, March 1995.
[13] A. J. Viterbi, *CDMA: Principles of Spread-Spectrum Communication*. Massachusetts: Addison-Wesley, 1995.
[14] J. K. Holmes and C. C. Chen, "Acquisition time performance of PN spread-spectrum systems," *IEEE Trans. Commun.*, vol. 25, no. 8, pp. 778–784, Aug. 1977.
[15] A. Polydoros and M. K. Simon, "Generalized serial search code acquisition: the equivalent circular state diagram approach," *IEEE Trans. Commun.*, vol. 32, no. 12, pp. 1260–1268, Dec. 1984.
[16] A. Polydoros and C. L. Weber, "A unified approach to serial search spread-spectrum code acquisition–parts I and II," *IEEE Trans. Commun.*, vol. COM-32, pp. 542–561, May 1984.
[17] D. M. DiCarlo and C. L. Weber, "Statistical performance of single dwell serial synchronization systems," *IEEE Trans. Commun.*, vol. 28, no. 8, pp. 1382–1388, Aug. 1980.
[18] D. L. Noneaker, A. R. Raghavan and C. W. Baum, "The effect of automatic gain control on serial, matched-filter acquisition in direct-sequence packet radio communications," *IEEE Trans. Veh. Technol.*, vol. 50, no. 4, pp. 1140–1150, July 2001.
[19] J. H. J. Iinatti, "On the threshold setting principles in code acquisition of DS-SS signals," *IEEE J. Select. Areas Commun.*, vol. 18, no. 1, pp. 62–72, Jan. 2000.
[20] R. Cameron and B. Woerner, "Synchronization of CDMA systems employing interference cancellation," in *Proc. of IEEE VTC 1996*, pp. 178–182, April 1996.

Part III

Evolved 3G mobile communications

5 TD receiver with ideal channel state information

In this chapter, TD performance is studied by assuming that the ideal channel state information is available at the receiver side. Under a frequency-selective Rayleigh fading environment, the performance of EGC and GSC 2D-Rake receivers is analyzed for either mutually independent or mutually correlated pairs of channels. The spatial diversity gain provided by TD-STBC is compared with that of a system deploying only one transmit antenna.

5.1 Introduction

It is assumed that the ideal channel state information (CSI) is available at the receiver, in other words, the channel estimation is perfect. The performance of the TD-STBC scheme is investigated in terms of the BER (bit error rate). In the next chapter, a common pilot signal transmission is employed to assist receivers in estimating the channel fading coefficients; hence, the impact of imperfect channel estimation on system performance can be investigated through comparing the results obtained in Chapters 5 and 6. The effect of correlation between the pair of channels from two transmit antennas to receive antenna is studied. In order to emphasize the spatial diversity gain from using TD-STBC, the performance of the corresponding Rake receiver of the conventional CDMA system with only one transmit antenna is evaluated and compared.

The rest of this chapter is organized as follows. In Section 5.2, the transmitter, channel and receiver models are described. The performance of coherent reception for downlink of the CDMA system with and without TD-STBC is analyzed in Section 5.3, and the closed-form expressions of BER are obtained. In Section 5.4 the numerical results for various system scenarios and discussions are presented. Finally, Section 5.5 summarizes the chapter and draws a number of conclusions.

5.2 System model

5.2.1 Transmitter model in CDMA downlink

Two transmit antennas are deployed at a base station while only one receive antenna is employed at each mobile station. It is assumed that there are K simultaneously active CDMA users. Hence, K totally different spreading code sequences are required, which

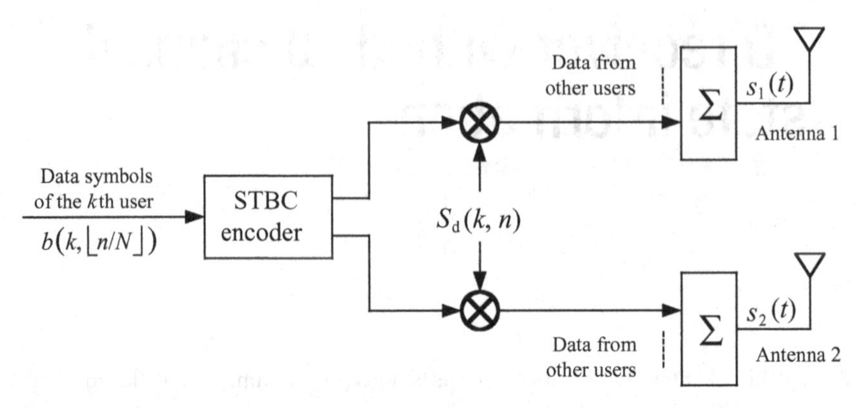

Figure 5.1 The block diagram of the transmitter.

are mutually orthogonal within one symbol interval. Furthermore, it is assumed that all spreading codes are unit-norm complex simple random binary sequences with value $(\pm 1 \pm \mathrm{j})/\sqrt{2}$, where $\mathrm{j} = \sqrt{-1}$. As shown in Figure 5.1, the lowpass equivalent transmitted signal at the first antenna is

$$s_1(t) = \sum_{n=-\infty}^{\infty} \left[\sqrt{E_c} \cdot \sum_{k=1}^{K} b(k, \lfloor n/N \rfloor) \cdot S_d(k, n) \right] \cdot g(t - nT_c) \tag{5.1}$$

This is compatible with the conventional system with only one transmit antenna, which is investigated and compared in Section 5.3.4.

Meanwhile, the signal transmitted from the second antenna is

$$s_2(t) = \sum_{n=-\infty}^{\infty} \left[\sqrt{E_c} \cdot \sum_{k=1}^{K} b'(k, \lfloor n/N \rfloor) \cdot S_d(k, n) \right] \cdot g(t - nT_c) \tag{5.2}$$

where E_c is the chip energy of the data channel and $S_d(k, n)$ represents the spreading code sequence of the data channel for the kth user. The operation $\lfloor \cdot \rfloor$ stands for the integer part of the operand. N is the spread factor (SF) and T_c is the chip interval. The chip waveform function $g(t)$ is assumed to be rectangular with unit energy. For the sake of simple analysis, the data symbol $b(k, n)$ of the kth user is assumed to be a real value. In (5.2), $b'(k, n)$ is the STBC encoded data symbol, given by

$$b'(k, \lfloor n/N \rfloor) = \begin{cases} -b(k, \lfloor n/N \rfloor + 1), & \lfloor n/N \rfloor \text{ is even} \\ b(k, \lfloor n/N \rfloor - 1), & \lfloor n/N \rfloor \text{ is odd} \end{cases} \tag{5.3}$$

5.2.2 Channel model

The discrete tap-delay-line channel model with time-varying tap coefficients is assumed and the channel from the ith $(i = 1, 2)$ transmit antenna to the receive antenna comprises L discrete resolvable paths, expressed through the channel coefficients $h_{i,l}$ $(l = 1, \ldots, L)$. The two sets of temporal multipaths corresponding to two transmit antennas experience independent but identical distributed (i.i.d.) Rayleigh fading with a constant multipath intensity profile (MIP). Therefore, all the channel coefficients $h_{i,l}$ are complex-valued Gaussian random variables with zero-mean and with the same variance of σ^2

in both real and imaginary parts, i.e. $h_{i,l} \in CN(0, \sigma^2)$. The lowpass equivalent complex impulse response of the channel between the ith transmit antenna and the receive antenna is

$$h_i(t) = \sum_{l=1}^{L} h_{i,l} \cdot \delta(t - \tau_l) \tag{5.4}$$

where τ_l is the delay of the lth resolvable path, irrespective of the antenna index i, and $\delta(\cdot)$ is the Kronecker delta function. For one given temporal path index l, $h_{1,l}$ and $h_{2,l}$ are identically Rayleigh distributed but can be correlated with a constant correlation coefficient ρ, defined as

$$\rho = \frac{E\{h_{1,l} \cdot h_{2,l}^*\}}{\sqrt{\mathrm{Var}\{h_{1,l}\}} \cdot \sqrt{\mathrm{Var}\{h_{2,l}\}}} \tag{5.5}$$

where the superscript * represents the complex conjugate operation and $E\{\cdot\}$ stands for the expected or mean value of the variable. When $\rho = 0$, $h_{1,l}$ and $h_{2,l}$ are independent from each other. Moreover, for a given spatial antenna index i, $h_{i,l}$ and $h_{i,l'}$ $(l \neq l')$ are mutually independent random variables. Finally, $h_{i,l}$ is assumed invariant over at least one STBC block, which refers to every two consecutive symbol periods in which two original data symbols are STBC encoded.

5.2.3 Receiver model

For any one mobile station, downlink signals from K synchronous data channels experience the same frequency selective fading and arrive at the receiver as

$$r(t) = \sum_{l=1}^{L} [h_{1,l} \cdot s_1(t - \tau_l) + h_{2,l} \cdot s_2(t - \tau_l)] + \eta(t) \tag{5.6}$$

where $\eta(t)$ represents the AWGN with double-sided power spectral density (PSD) of $\eta_0/2$. It is assumed that the coherent detection and chip synchronization at the receiver are perfect. Thus, the signal at each path can be resolved by a matched filter (MF) with a local delayed spreading code sequence that is properly locked to the concerned path. Within the context of wideband CDMA, it is assumed that multipath delays are approximately a few chips in duration and much smaller than the symbol period so that the inter-symbol interference (ISI) can be neglected.

Without loss of generality, let us focus on the mth STBC block. By sampling the output of the pulse MF, for any mobile station the received signal in the first and second symbol intervals of the mth STBC block can be given by, respectively,

$$u^{(1)}(m, n) = \sum_{k=1}^{K} \sum_{l=1}^{L} \{\sqrt{E_c} \cdot S_d(k, 2mN + n - \lfloor \tau_l / T_c \rfloor) \cdot [b_1(k, m) \cdot h_{1,l}(m) - b_2(k, m) \cdot h_{2,l}(m)]\} + \eta^{(1)}(m, n) \tag{5.7}$$

$$u^{(2)}(m, n) = \sum_{k=1}^{K} \sum_{l=1}^{L} \{\sqrt{E_c} \cdot S_d[k, (2m + 1) \cdot N + n - \lfloor \tau_l / T_c \rfloor] \cdot [b_2(k, m) \cdot h_{1,l}(m) + b_1(k, m) \cdot h_{2,l}(m)]\} + \eta^{(2)}(m, n) \tag{5.8}$$

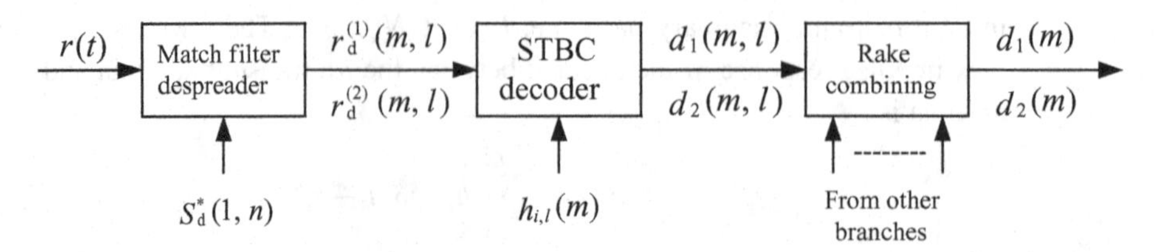

Figure 5.2 The block diagram of the receiver.

where $n = 0, 1, \ldots, N-1$ and $b_j(k, m)$ $(j = 1, 2)$ stands for the jth original data bit of the kth active user transmitted in the mth STBC block. $h_{i,l}(m)$ $(i = 1, 2)$ represents the fading coefficient of the lth resolvable path of the channel between the ith transmit antenna and the receive antenna in the mth STBC block. $\eta^{(j)}(m, n)$ $(j = 1, 2)$ is the sampled AWGN at the receiver in the jth symbol period of the mth STBC block.

5.3 Performance analysis of coherent reception

5.3.1 STBC decoding at each branch

Without loss of generality, the first user $(k = 1)$ is assumed to be the desired user. Hence, the following discussion is implicitly concerned with the first user. The receiver structure is shown in Figure 5.2.

Assuming perfect chip timing synchronization and that the local despreading code sequence is locked to the $\hat{l}$th $(\hat{l} = 1, 2, \ldots, L)$ resolvable path, the data channel of the first user at the $\hat{l}$th path is despread during the first and second symbol intervals of the mth STBC block as follows, respectively,

$$\begin{aligned} r_{\mathrm{d}}^{(1)}(m, \hat{l}) &= \sum_{n=0}^{N-1} u^{(1)}(m, n) \cdot S_{\mathrm{d}}^*(1, 2mN + n - \lfloor \tau_{\hat{l}} / T_{\mathrm{c}} \rfloor) \\ &= \sqrt{E_{\mathrm{c}}} \cdot N \cdot [b_1(1, m) \cdot h_{1,\hat{l}}(m) - b_2(1, m) \cdot h_{2,\hat{l}}(m)] + \eta_d^{(1)}(m, \hat{l}) \\ &\quad + \sum_{k=1}^{K} \sum_{\substack{l=1 \\ l \neq \hat{l}}}^{L} \sqrt{E_{\mathrm{c}}} \cdot R_{\mathrm{d,d}}^{(1)}(m, k; l, \hat{l}) \cdot [b_1(k, m) \cdot h_{1,l}(m) - b_2(k, m) \cdot h_{2,l}(m)] \end{aligned} \tag{5.9}$$

$$\begin{aligned} r_{\mathrm{d}}^{(2)}(m, \hat{l}) &= \sum_{n=0}^{N-1} u^{(2)}(m, n) \cdot S_{\mathrm{d}}^*[1, (2m+1) \cdot N + n - \lfloor \tau_{\hat{l}} / T_{\mathrm{c}} \rfloor] \\ &= \sqrt{E_{\mathrm{c}}} \cdot N \cdot [b_2(1, m) \cdot h_{1,\hat{l}}(m) + b_1(1, m) \cdot h_{2,\hat{l}}(m)] + \eta_{\mathrm{d}}^{(2)}(m, \hat{l}) \\ &\quad + \sum_{k=1}^{K} \sum_{\substack{l=1 \\ l \neq \hat{l}}}^{L} \sqrt{E_{\mathrm{c}}} \cdot R_{\mathrm{d,d}}^{(2)}(m, k; l, \hat{l}) \cdot [b_2(k, m) \cdot h_{1,l}(m) + b_1(k, m) \cdot h_{2,l}(m)] \end{aligned} \tag{5.10}$$

where the background AWGN component $\eta_{\mathrm{d}}^{(j)}(m, \hat{l})$ $(j = 1, 2)$ is given by

$$\eta_{\mathrm{d}}^{(j)}(m, \hat{l}) = \sum_{n=0}^{N-1} \eta^{(j)}(m, n) \cdot S_{\mathrm{d}}^*[1, (2m + j - 1) \cdot N + n - \lfloor \tau_{\hat{l}} / T_{\mathrm{c}} \rfloor] \tag{5.11}$$

and $R^{(j)}_{\mathrm{d,d}}(m,k;l,\hat{l})$ $(j=1,2)$ is the N-length discrete aperiodic correlation function [1] of two time-delayed data channel spreading code sequences used in the jth symbol period of the mth STBC block, given by

$$\begin{aligned} R^{(j)}_{\mathrm{d,d}}(m,k;l,\hat{l}) &= \sum_{n=0}^{N-1} S_{\mathrm{d}}[k,(2m+j-1)\cdot N+n-\lfloor \tau_l/T_{\mathrm{c}}\rfloor] \\ &\quad \cdot S^*_{\mathrm{d}}[1,(2m+j-1)\cdot N+n-\lfloor \tau_{\hat{l}}/T_{\mathrm{c}}\rfloor] \end{aligned} \tag{5.12}$$

Since two different spreading code sequences with the same time delay are orthogonal over one symbol interval, i.e. $R^{(j)}_{\mathrm{d,d}}(m,k;l,\hat{l})=0$ when $l=\hat{l}$ and $k\neq 1$. Otherwise, when $l\neq\hat{l}$, the two code sequences are nonorthogonal and $R^{(j)}_{\mathrm{d,d}}(m,k;l,\hat{l})$ is modeled as an i.i.d. random variable with zero-mean and variance $\mathrm{Var}\{R^{(j)}_{\mathrm{d,d}}(m,k;l,\hat{l})\}=N$.

With respect to the $\hat{l}$th resolvable path, it is assumed that the ideal CSI $h_{1,\hat{l}}(m)$ and $h_{2,\hat{l}}(m)$ are available at the receiver; thus, the branch random variable of the first estimated data bit in the mth STBC block can be constructed through STBC decoding as follows:

$$\begin{aligned} d_1(m,\hat{l}) &= r^{(1)}_{\mathrm{d}}(m,\hat{l})\cdot h^*_{1,\hat{l}}(m)+r^{(2)^*}_{\mathrm{d}}(m,\hat{l})\cdot h_{2,\hat{l}}(m) \\ &= \sqrt{E_{\mathrm{c}}}\cdot N\cdot b_1(1,m)\cdot[|h_{1,\hat{l}}(m)|^2+|h_{2,\hat{l}}(m)|^2] \\ &\quad +\eta^{(1)}_{\mathrm{d}}(m,\hat{l})\cdot h^*_{1,\hat{l}}(m)+\eta^{(2)^*}_{\mathrm{d}}(m,\hat{l})\cdot h_{2,\hat{l}}(m) \\ &\quad +\sum_{k=1}^{K}\sum_{\substack{l=1\\ l\neq\hat{l}}}^{L}\{\sqrt{E_{\mathrm{c}}}\cdot R^{(1)}_{\mathrm{d,d}}(m,k;l,\hat{l})\cdot[b_1(k,m)\cdot h_{1,l}(m)-b_2(k,m)\cdot h_{2,l}(m)] \\ &\quad\quad \cdot h^*_{1,\hat{l}}(m)+\sqrt{E_{\mathrm{c}}}\cdot R^{(2)*}_{\mathrm{d,d}}(m,k;l,\hat{l})\cdot[b_1(k,m)\cdot h^*_{2,l}(m)+b_2(k,m) \\ &\quad\quad \cdot h^*_{1,l}(m)]\cdot h_{2,\hat{l}}(m)\} \\ &= E_1(m,\hat{l})+N_1(m,\hat{l})+I_1(m,\hat{l}) \end{aligned} \tag{5.13}$$

It is shown that the above branch decision random variable consists of three distinct components: $E_1(m,\hat{l})$ is the desired signal component; $N_1(m,\hat{l})$ is the background AWGN component; and $I_1(m,\hat{l})$ is the multiple access and multipath interference component that results from the desired user as well as other simultaneously active users. They are defined as follows:

$$E_1(m,\hat{l})=\sqrt{E_{\mathrm{c}}}\cdot N\cdot b_1(1,m)\cdot\lfloor|h_{1,\hat{l}}(m)|^2+|h_{2,\hat{l}}(m)|^2\rfloor \tag{5.14}$$

$$N_1(m,\hat{l})=\eta^{(1)}_{\mathrm{d}}(m,\hat{l})\cdot h^*_{1,\hat{l}}(m)+\eta^{(2)^*}_{\mathrm{d}}(m,\hat{l})\cdot h_{2,\hat{l}}(m) \tag{5.15}$$

$$\begin{aligned} I_1(m,\hat{l}) &= \sum_{k=1}^{K}\sum_{\substack{l=1\\ l\neq\hat{l}}}^{L}\{\sqrt{E_{\mathrm{c}}}\cdot R^{(1)}_{\mathrm{d,d}}(m,k;l,\hat{l})\cdot[b_1(k,m)\cdot h_{1,l}(m)-b_2(k,m)\cdot h_{2,l}(m)]\cdot h^*_{1,\hat{l}}(m) \\ &\quad +\sqrt{E_{\mathrm{c}}}\cdot R^{(2)*}_{\mathrm{d,d}}(m,k;l,\hat{l})\cdot[b_1(k,m)\cdot h^*_{2,l}(m) \\ &\quad +b_2(k,m)\cdot h^*_{1,l}(m)]\cdot h_{2,\hat{l}}(m)\} \end{aligned} \tag{5.16}$$

Similarly, the random branch decision variable of the second estimated data bit can also be constructed through STBC decoding as follows:

$$d_2(m,\hat{l}) = r_{\rm d}^{(2)}(m,\hat{l}) \cdot h_{1,\hat{l}}^{*}(m) - r_{\rm d}^{(1)^*}(m,\hat{l}) \cdot h_{2,\hat{l}}(m)$$
$$= E_2(m,\hat{l}) + N_2(m,\hat{l}) + I_2(m,\hat{l}) \quad (5.17)$$

where $E_2(m,\hat{l})$ is the desired signal component, $N_2(m,\hat{l})$ is the background AWGN component, and $I_2(m,\hat{l})$ is the multiple access and multipath interference component given by, respectively,

$$E_2(m,\hat{l}) = \sqrt{E_{\rm c}} \cdot N \cdot b_2(1,m) \cdot \lfloor |h_{1,\hat{l}}(m)|^2 + |h_{2,\hat{l}}(m)|^2 \rfloor \quad (5.18)$$

$$N_2(m,\hat{l}) = \eta_{\rm d}^{(2)}(m,\hat{l}) \cdot h_{1,\hat{l}}^{*}(m) - \eta_{\rm d}^{(1)^*}(m,\hat{l}) \cdot h_{2,\hat{l}}(m) \quad (5.19)$$

$$I_2(m,\hat{l})$$
$$= \sum_{k=1}^{K} \sum_{\substack{l=1 \\ l\neq\hat{l}}}^{L} \{\sqrt{E_c} \cdot R_{d,d}^{(2)}(m,k;l,\hat{l}) \cdot [b_2(k,m) \cdot h_{1,l}(m) + b_1(k,m) \cdot h_{2,l}(m)] \cdot h_{1,\hat{l}}^{*}(m)$$
$$- \sqrt{E_{\rm c}} \cdot R_{\rm d,d}^{(1)*}(m,k;l,\hat{l}) \cdot \left[b_1(k,m) \cdot h_{1,l}^{*}(m) - b_2(k,m) \cdot h_{2,l}^{*}(m)\right] \cdot h_{2,\hat{l}}(m)\}$$
$$(5.20)$$

For a different user k and a different mth STBC block, the data bit $b_j(k,m)$ $(j = 1,2)$ is an i.i.d. binary random variable taking the value ± 1, and $R_{\rm d,d}^{(j)}(m,k;l,\hat{l})$ is an i.i.d. random variable with zero-mean and variance N, which is thus conditioned on the channel fading coefficients $h_{1,\hat{l}}(m)$ and $h_{2,\hat{l}}(m)$. $I_j(m,l)$ $(j = 1,2)$ is a random variable with zero mean, and the variance is given by

$$\text{Var}\{I_j(m,\hat{l})\} = E_{\rm c} \cdot N \cdot \sum_{k=1}^{K} \sum_{\substack{l=1 \\ l\neq\hat{l}}}^{L} E\{|h_{1,l}(m)|^2 + |h_{2,l}(m)|^2\} \cdot [|h_{1,\hat{l}}(m)|^2 + |h_{2,\hat{l}}(m)|^2]$$
$$= E_{\rm c} \cdot N \cdot K \cdot (L-1) \cdot 4\sigma^2 \cdot [|h_{1,\hat{l}}(m)|^2 + |h_{2,\hat{l}}(m)|^2] \quad (5.21)$$

where $\sigma^2 = (1/2) \cdot E\{|h_{i,l}(m)|^2\}$. The background AWGN component $N_j(m,\hat{l})$ $(j = 1,2)$ is a zero-mean Gaussian variable with variance

$$\text{Var}\{N_j(m,\hat{l})\} = N \cdot \frac{\eta_0}{2} \cdot [|h_{1,\hat{l}}(m)|^2 + |h_{2,\hat{l}}(m)|^2] \quad (5.22)$$

Since the data bits from K active users and the channel fading coefficients along L resolvable paths are independent random variables, $d_j(m,\hat{l})$ $(j = 1,2)$ in (5.13) or (5.17) is the sum of many independent random variables and hence can be approximated as a conditional Gaussian variable. Therefore, conditioned on the channel fading coefficients $h_{i,\hat{l}}(m)$ $(i = 1,2)$, the branch decision variable $d_j(m,\hat{l})$ is a Gaussian random variable with different mean and the same variance as follows, respectively,

$$E\{d_j(m,\hat{l})\} = E_j(m,\hat{l})$$
$$= \sqrt{E_{\rm c}} \cdot N \cdot b_j(1,m) \cdot [|h_{1,\hat{l}}(m)|^2 + |h_{2,\hat{l}}(m)|^2] \quad (5.23)$$

$$\text{Var}\{d_j(m,\hat{l})\} = \text{Var}\{N_j(m,\hat{l})\} + \text{Var}\{I_j(m,\hat{l})\}$$
$$= \left[E_{\rm c} \cdot N \cdot K \cdot (L-1) \cdot 4\sigma^2 + N \cdot \frac{\eta_0}{2}\right] \cdot [|h_{1,\hat{l}}(m)|^2 + |h_{2,\hat{l}}(m)|^2]$$
$$(5.24)$$

After the despreading and STBC decoding at each path, a Rake-type receiver is exploited to combine the signal energy from $L_c (L_c \leq L)$ selected paths. This can be regarded as a two-dimensional Rake (2D-Rake) combiner that collects both spatial diversity gain and path diversity gain, thus benefiting from both TD-STBC and DS-CDMA, respectively. However, this 2D-Rake is a bit different from the conventional coherent MRC Rake combiner employed in the conventional DS-CDMA system deployed with only one transmit antenna. It is proven [2] that overall TD-STBC processing exploits the coherent MRC combining along a pair of channels that correspond to two distinct transmit antennas and have the same time delay. Some novel combing strategies, which will be investigated in the following sections, add up the branch decision variables derived from multiple distinct resolvable paths, which might be regarded as the distinct fingers of the 2D-Rake receiver.

5.3.2 Equal gain combining 2D-Rake receiver

In the equal gain combining (EGC) receiver, the signals from L_c first arriving paths among L resolvable paths are selected and combined. Assuming that the fading of each path is independent and that the random decision variables in distinct branches of the 2D-Rake are independent from each other, the output of the 2D-Rake combiner can be represented as

$$d_j(m) = \sum_{l=1}^{L_c} d_j(m, l) \tag{5.25}$$

When $L_c = L$, the signals from all resolvable paths are combined. The decision variable $d_j(m)$ $(j = 1, 2)$ is a Gaussian variable with the conditional mean and variance given by, respectively,

$$E\{d_j(m)\} = \sum_{l=1}^{L_c} E\{d_j(m, l)\} = \sqrt{E_c} \cdot N \cdot b_j(1, m) \cdot \zeta \tag{5.26}$$

$$\begin{aligned} \mathrm{Var}\{d_j(m)\} &= \sum_{l=1}^{L_c} \mathrm{Var}\{d_j(m, l)\} \\ &= \left[E_c \cdot N \cdot K \cdot (L-1) \cdot 4\sigma^2 + N \cdot \frac{\eta_0}{2}\right] \cdot \zeta \end{aligned} \tag{5.27}$$

where ζ is defined as

$$\zeta = \sum_{l=1}^{L_c} [|h_{1,l}(m)|^2 + |h_{2,l}(m)|^2] \tag{5.28}$$

Therefore, conditioned on the instantaneous fading channel amplitudes of the multipaths, the BER is the same for $d_j(m)$ $(j = 1, 2)$ and can be obtained by [3]

$$P_{\mathrm{e,EGC}}(\gamma) = Q\left\{\left[\frac{|E\{d_j(m)\}|^2}{\mathrm{Var}\{d_j(m)\}}\right]^{1/2}\right\} = Q(\sqrt{2\gamma}) \tag{5.29}$$

where the Q-function is defined as $Q(x) = (1/\sqrt{2\pi}) \cdot \int_x^{\infty} \exp(-t^2/2) \cdot dt$, and by definition,

$$\gamma = \frac{E_c \cdot N \cdot \zeta}{E_c \cdot K \cdot (L-1) \cdot 8\sigma^2 + \eta_0} \tag{5.30}$$

Independent pair of channels

It is assumed that two transmit antennas are separated with enough distance and the pair of TD-STBC channels $h_{1,l}(m)$ and $h_{2,l}(m)$ are fading independently from each other; thus the summed random variable ζ in (5.28) follows a chi-square or gamma distribution with $4L_c$ degrees of freedom. Consequently, the probability density function (PDF) of γ in (5.30) is given by

$$p_{\gamma,\text{ind}}(\gamma) = \frac{1}{\bar{\gamma}_c^{2L_c} \cdot \Gamma(2L_c)} \cdot \gamma^{2L_c-1} \cdot \exp\left(-\frac{\gamma}{\bar{\gamma}_c}\right) \tag{5.31}$$

where, by definition,

$$\bar{\gamma}_c = \frac{E_c \cdot N \cdot (2\sigma^2)}{E_c \cdot K \cdot (L-1) \cdot 8\sigma^2 + \eta_0} \tag{5.32}$$

The resultant BER can be obtained through averaging the conditional BER $P_{e,\text{EGC}}(\gamma)$ in (5.29) over the PDF of γ in (5.31) [3], i.e.,

$$\begin{aligned} P_{e,\text{ind}} &= \int_0^\infty P_{e,\text{EGC}}(\gamma) \cdot p_{\gamma,\text{ind}}(\gamma) \cdot d\gamma \\ &= \left(\frac{1-\mu}{2}\right)^{2L_c} \cdot \sum_{l=0}^{2L_c-1} \binom{2L_c-1+l}{l} \cdot \left(\frac{1+\mu}{2}\right)^l \end{aligned} \tag{5.33}$$

where, by definition,

$$\mu = \sqrt{\frac{\bar{\gamma}_c}{1+\bar{\gamma}_c}} = \left[1 + \frac{4K \cdot (L-1)}{N} + \frac{1}{\bar{\gamma}_b}\right]^{-\frac{1}{2}} \tag{5.34}$$

where $\bar{\gamma}_b$ is the average signal-to-noise ratio (SNR) per bit, defined as

$$\bar{\gamma}_b = \frac{2E_c \cdot N \cdot \sigma^2}{\eta_0} \tag{5.35}$$

and note that $\bar{\gamma}_c$ is related to $\bar{\gamma}_b$ by

$$\bar{\gamma}_c^{-1} = \frac{4K \cdot (L-1)}{N} + \bar{\gamma}_b^{-1} \tag{5.36}$$

In order to elaborate more clearly the performance improvement from the spatial diversity gain due to TD-STBC, and the path diversity gain due to multipath Rake combining, we define the average SNR per bit per antenna per path as follows:

$$\bar{\gamma}_p = \frac{\bar{\gamma}_b}{N_T \cdot L} = \frac{2E_c \cdot N \cdot \sigma^2}{\eta_0 \cdot N_T \cdot L} \tag{5.37}$$

where N_T is the number of transmit antennas and $N_T = 2$ for TD-STBC, while $N_T = 1$ for the conventional DS-CDMA system with only one transmit antenna. Essentially, this implies that the total transmit power is restricted as a constant for different system scenarios in order to evaluate and compare the performance improvement from spatial diversity gain and multipath diversity gain.

Correlated pair of channels

It is assumed that the two transmit antennas are not adequately separated from each other and that the corresponding pair of channels $h_{1,l}(m)$ and $h_{2,l}(m)$ are mutually correlated by a spatial correlation coefficient ρ along two distinct spatial paths with the same index l ($l = 1, 2, \ldots, L$). When $\rho = 0$, the pair of channel coefficients $h_{1,l}(m)$ and $h_{2,l}(m)$ are mutually independent and the BER performance is investigated. When $\rho = 1$, the BER performance is the same as that of the conventional CDMA system without TD-STBC under the constraint of equivalent total transmit power, which is explored in Section 5.3.4.

Based on the conditional mean and variance in (5.23) and (5.24), the instantaneous signal-to-interference-plus-noise ratio (SINR) of the branch decision variable $d_j(m,l)$ ($l = 1, 2, \ldots, L$) at the lth resolvable path can be represented as

$$\frac{|E\{d_j(m,l)\}|^2}{\mathrm{Var}\{d_j(m,l)\}} = 2\gamma_l \tag{5.38}$$

where γ_l is defined by

$$\gamma_l = \frac{E_c \cdot N}{E_c \cdot K \cdot (L-1) \cdot 8\sigma^2 + \eta_0} \cdot [|h_{1,l}(m)|^2 + |h_{2,l}(m)|^2] \tag{5.39}$$

When $0 < \rho < 1$, the characteristic function of γ_l in (5.39) can be derived based on Appendix 5A and is given by

$$\phi_{\gamma_l}(jt) = \frac{1}{[1 - jt \cdot (1+\rho) \cdot \bar{\gamma}_l] \cdot [1 - jt \cdot (1-\rho) \cdot \bar{\gamma}_l]} \tag{5.40}$$

where $\bar{\gamma}_l$ is defined as in (5.32), i.e. $\bar{\gamma}_l = \bar{\gamma}_c$.

Since the fading along distinct resolvable paths are mutually independent and the combined signal γ in (5.30) is the sum of L_c statistically i.i.d. components γ_l ($l = 1, \ldots, L_c$) given in (5.39), therefore, the characteristic function of γ is the product of the individual characteristic functions, as follows:

$$\begin{aligned}
\phi_\gamma(jt) &= \prod_{l=1}^{L_c} \frac{1}{[1 - jt \cdot (1+\rho) \cdot \bar{\gamma}_l] \cdot [1 - jt \cdot (1-\rho) \cdot \bar{\gamma}_l]} \\
&= \frac{1}{[1 - jt \cdot (1+\rho) \cdot \bar{\gamma}_c]^{L_c} \cdot [1 - jt \cdot (1-\rho) \cdot \bar{\gamma}_c]^{L_c}} \\
&= \sum_{l=1}^{L_c} \binom{2L_c - l - 1}{L_c - 1} \cdot \frac{(\rho^2 - 1)^{L_c - l}}{(2\rho)^{2L_c - l}} \\
&\quad \cdot \left\{ \frac{(1+\rho)^l}{[1 - jt \cdot (1+\rho) \cdot \bar{\gamma}_c]^l} + \frac{(\rho - 1)^l}{[1 - jt \cdot (1-\rho) \cdot \bar{\gamma}_c]^l} \right\}
\end{aligned} \tag{5.41}$$

where the product in the last equation is decomposed by partial fractions [6].

The inverse Fourier transform of the characteristic function above yields a PDF of γ in the form [5]

$$\begin{aligned} p_{\gamma,\text{cor}}(\gamma) &= \sum_{l=1}^{L_c} \binom{2L_c - l - 1}{L_c - 1} \cdot \frac{(\rho^2 - 1)^{L_c - l} \cdot \gamma^{l-1}}{(2\rho)^{2L_c - l} \cdot \Gamma(l) \cdot \bar{\gamma}_c^l} \\ &\cdot \left\{ \exp\left[-\frac{\gamma}{(1+\rho)\cdot\bar{\gamma}_c} \right] + (-1)^l \cdot \exp\left[-\frac{\gamma}{(1-\rho)\cdot\bar{\gamma}_c} \right] \right\} \end{aligned} \tag{5.42}$$

Therefore, the resultant BER can be obtained by averaging the conditional BER $P_{e,\text{EGC}}(\gamma)$ in (5.29) over the PDF of γ in (5.42), i.e.

$$\begin{aligned} P_{e,\text{cor}} &= \int_0^\infty P_{e,\text{EGC}}(\gamma) \cdot p_{\gamma,\text{cor}}(\gamma) \cdot d\gamma \\ &= \frac{1}{2} \cdot \int_0^\infty \text{erfc}(\sqrt{\gamma}) \cdot p_{\gamma,\text{cor}}(\gamma) \cdot d\gamma \\ &= \sum_{l=1}^{L_c} \binom{2L_c - l - 1}{L_c - 1} \cdot \frac{(\rho^2 - 1)^{L_c - l}}{2 \cdot (2\rho)^{2L_c - l} \cdot \Gamma(l)} \\ &\cdot \left[(\rho + 1)^l \cdot \text{I}_\text{A}(l-1, q_1) + (\rho - 1)^l \cdot \text{I}_\text{A}(l-1, q_2) \right] \end{aligned} \tag{5.43}$$

where by the definitions

$$\text{I}_\text{A}(p, \alpha) = p! \cdot \left[1 - \sum_{k=0}^{p} \frac{\alpha^k}{k! \cdot \sqrt{\pi}} \cdot \frac{\Gamma(k + 1/2)}{(1+\alpha)^{k+1/2}} \right] \tag{5.44}$$

and

$$q_1 = \frac{1}{(1+\rho)\cdot\bar{\gamma}_c} \tag{5.45}$$

$$q_2 = \frac{1}{(1-\rho)\cdot\bar{\gamma}_c} \tag{5.46}$$

where $\bar{\gamma}_c$ is defined in (5.32).

5.3.3 Generalized selection combining the 2D-Rake receiver

In the context of conventional spread-spectrum communications with Rake reception, the complexity of MRC and EGC receivers depends on the number of resolvable paths available, which can be quite high, especially for the multipath diversity of wideband spread-spectrum signals due to the existence of many more resolvable paths. In addition, MRC and EGC are sensitive to channel estimation errors, and these errors tend to be more important when the instantaneous SNR is low. On the other hand, SC makes use of only one path out of the L resolvable (available) multipaths and hence does not fully exploit the amount of diversity offered by the multipath channel. Therefore, the gap between these two extremes, MRC/EGC and SC, is bridged by proposing generalized selection combining (GSC), which adaptively selects and combines the L_c strongest paths, or say, the L_c resolvable paths with highest SINR, among the L available multipaths. In the context of wideband CDMA systems, GSC schemes offer less complex receivers than the conventional EGC and MRC Rake receivers since they have a fixed number of fingers independent of the number of resolvable multipaths. In addition, GSC receivers are expected to be more robust toward channel estimation errors since the weakest SINR paths (and hence the ones that are the most exposed to these errors) are excluded from the combining process.

As is well known, in the context of DS-CDMA the number of resolvable multipaths depends on the spread-spectrum bandwidth as well as the maximum time delay spread under the particular radio propagation environment. From the implementation point of view, the Rake receiver with a fixed number of fingers is favorable since it offers less complexity and is independent from the number of resolvable multipaths available, which might vary from place to place. Therefore, it is expected to adaptively select and combine the signals from the L_c paths with highest SINR among the L available resolvable paths. Accordingly, a GSC 2D-Rake receiver that selects and combines the branch decision variables of the $L_c \leq L$ strongest paths is exploited next.

For the branch decision variable $d_j(m,l)$ ($j = 1, 2$) in the lth ($l = 1, 2, \ldots, L$) resolvable path, its instantaneous SINR is given in (5.38) and rewritten here:

$$\mathrm{SINR}\{d_j(m,l)\} = \frac{|E\{d_j(m,l)\}|^2}{\mathrm{Var}\{d_j(m,l)\}} = 2\gamma_l \tag{5.47}$$

where γ_l is defined in (5.39) and rewritten here:

$$\gamma_l = \frac{E_c \cdot N}{E_c \cdot K \cdot (L-1) \cdot 8\sigma^2 + \eta_0} \cdot [|h_{1,l}(m)|^2 + |h_{2,l}(m)|^2] \tag{5.48}$$

Moreover, $\gamma_{1:L} \geq \gamma_{2:L} \geq \cdots \geq \gamma_{L:L}$ are defined as the order statistics of instantaneous SINR variables that are obtained by arranging $\{\gamma_l \mid l = 1, 2, \ldots, L\}$ in the descending order of magnitude. For L_c strongest paths selected from L resolvable paths, the joint PDF of the order statistic variables [7], [8] $\{\gamma_{l:L} \mid l = 1, 2, \ldots, L_c\}$ is given by

$$p_{\gamma_{1:L}, \gamma_{2:L}, \ldots, \gamma_{L_c:L}}(\gamma_{1:L}, \gamma_{2:L}, \ldots, \gamma_{L_c:L}) = L_c! \cdot \binom{L}{L_c} \cdot [F_{\gamma_l}(\gamma_{L_c:L})]^{L-L_c} \cdot \prod_{l=1}^{L_c} p_{\gamma_l}(\gamma_{l:L}) \tag{5.49}$$

where $F(\cdot)$ is the cumulative distribution function (CDF) and $p(\cdot)$ is the relevant PDF.

After the L_c strongest paths are selected and combined by a GSC 2D-Rake receiver, the overall output SINR can be represented as [4]

$$\gamma_{\mathrm{GSC}} = 2 \cdot \sum_{l=1}^{L_c} \gamma_{l:L} \tag{5.50}$$

Thus, conditioned on $\{\gamma_{l:L} \mid l = 1, 2, \ldots, L_c\}$, the BER can be obtained by

$$P_{\mathrm{e,GSC}}(\gamma_{\mathrm{GSC}}) = Q(\sqrt{\gamma_{\mathrm{GSC}}}) = \frac{1}{2}\mathrm{erfc}\left(\sqrt{\sum_{l=1}^{L_c} \gamma_{l:L}}\right) \tag{5.51}$$

where the complementary error function is defined by $\mathrm{erfc}(x) = (2/\sqrt{\pi}) \int_x^\infty e^{-t^2}\, dt$. Therefore, the resultant BER can be achieved by averaging the conditional BER in (5.51) over the joint PDF of the order statistics of the selected paths in (5.49), as follows:

$$P_{\mathrm{e,GSC}}^{<L_c>} = \int_0^\infty \int_{\gamma_{L_c:L}}^\infty \cdots \int_{\gamma_{2:L}}^\infty Q(\sqrt{\gamma_{\mathrm{GSC}}}) \cdot p_{\gamma_{1:L},\gamma_{2:L},\ldots,\gamma_{L_c:L}}(\gamma_{1:L},\gamma_{2:L},\ldots,\gamma_{L_c:L}) \cdot \mathrm{d}\gamma_{1:L} \cdot \mathrm{d}\gamma_{2:L} \cdots \mathrm{d}\gamma_{L_c:L} \tag{5.52}$$

Independent pair of channels

It is assumed that the pair of TD-STBC channels $h_{1,l}(m)$ and $h_{2,l}(m)$ are independent from each other. Because the variable γ_l $(l = 1, 2, \ldots, L)$ in (5.48) follows a chi-square distribution with four degrees of freedom, its PDF and relevant CDF are given by

$$p_{\gamma_l,\mathrm{ind}}(\gamma_l) = \frac{\gamma_l}{\bar{\gamma}_l^2} \exp\left(-\frac{\gamma_l}{\bar{\gamma}_l}\right) \tag{5.53}$$

$$F_{\gamma_l,\mathrm{ind}}(\gamma_l) = 1 - \left(1 + \frac{\gamma_l}{\bar{\gamma}_l}\right) \exp\left(-\frac{\gamma_l}{\bar{\gamma}_l}\right) \tag{5.54}$$

where $\bar{\gamma}_l$ is defined as in (5.32), i.e. $\bar{\gamma}_l = \bar{\gamma}_c$.

Substituting (5.53) and (5.54) into (5.49) to get the joint PDF, and applying the general integral formula (5.52), we further obtain the closed-form BER for several special cases. The detailed derivations are presented in the Appendix 5B.2. When $L_c = 1$, only the strongest path is selected so that the closed-form BER can be manipulated as follows:

$$\begin{aligned} P_{\mathrm{e,ind}}^{<1>} &= \binom{L}{1} \cdot \int_0^\infty [F_{\gamma_l,\mathrm{ind}}(\gamma_{1:L})]^{L-1} \cdot p_{\gamma_l,\mathrm{ind}}(\gamma_{1:L}) \cdot Q(\sqrt{2\cdot\gamma_{1:L}}) \cdot \mathrm{d}\gamma_{1:L} \\ &= \frac{L}{2} \cdot \sum_{m=0}^{L-1} \sum_{n=0}^{m} \binom{L-1}{m} \cdot \binom{m}{n} \cdot \frac{(-1)^m}{(m+1)^{n+2}} \cdot I_{\mathrm{A}}\left(n+1, \frac{m+1}{\bar{\gamma}_c}\right) \end{aligned} \tag{5.55}$$

where $\bar{\gamma}_c$ is given by (5.32) and the function $I_{\mathrm{A}}(\cdot,\cdot)$ is defined in (5.44).

When $L_c = 2$, the two strongest paths are selected and combined so that the closed-form BER can be manipulated through the two-fold integral in (5.52) as

$$\begin{aligned} P_{\mathrm{e,ind}}^{<2>} &= 2 \cdot \binom{L}{2} \cdot \int_0^\infty \int_{\gamma_{2:L}}^\infty [F_{\gamma_l,\mathrm{ind}}(\gamma_{2:L})]^{L-2} \cdot \prod_{l=1}^{2} p_{\gamma_l,\mathrm{ind}}(\gamma_{l:L}) \\ &\quad \cdot Q\left(\sqrt{2 \cdot \sum_{l=1}^{2} \gamma_{l:L}}\right) \cdot \mathrm{d}\gamma_{1:L} \cdot \mathrm{d}\gamma_{2:L} \\ &= \frac{L(L-1)}{2} \cdot \left[P_{\mathrm{e1}}^{<2>} + \sum_{m=1}^{L-2} \sum_{n=0}^{m} \binom{L-2}{m} \cdot \binom{m}{n} \cdot (-1)^m \cdot (P_{\mathrm{e2}}^{<2>} + P_{\mathrm{e3}}^{<2>})\right] \end{aligned} \tag{5.56}$$

where, by the definitions,

$$P_{\mathrm{e1}}^{<2>} = \frac{1}{2} - \frac{16\bar{\gamma}_{\mathrm{c}}^3 + 56\bar{\gamma}_{\mathrm{c}}^2 + 70\bar{\gamma}_{\mathrm{c}} + 35}{32 \cdot (1+\bar{\gamma}_{\mathrm{c}})^3 \cdot \sqrt{1+1/\bar{\gamma}_{\mathrm{c}}}} \tag{5.57}$$

$$P_{\mathrm{e2}}^{<2>} = -\frac{(2m+2)\cdot\bar{\gamma}_{\mathrm{c}} + m^2 + 4m + 2}{m \cdot (m+2) \cdot (1+\bar{\gamma}_{\mathrm{c}})} \cdot \sqrt{\frac{2}{\pi}} \cdot \frac{\Gamma(n+5/2)\cdot \bar{\gamma}_{\mathrm{c}}^{1/2}}{(m+2+2\bar{\gamma}_{\mathrm{c}})^{n+5/2}} \tag{5.58}$$

$$P_{\mathrm{e3}}^{<2>} = \frac{2(n-m+2)(1+\bar{\gamma}_{\mathrm{c}}) - m}{2 \cdot m^{n+3} \cdot (1+\bar{\gamma}_{\mathrm{c}}) \cdot \sqrt{1+1/\bar{\gamma}_{\mathrm{c}}}} \cdot \mathrm{I_A}\left(n+1, \frac{m}{2+2\bar{\gamma}_{\mathrm{c}}}\right) + \frac{n+m+4}{(m+2)^{n+3}} \cdot \mathrm{I_A}\left(n+1, \frac{m+2}{2\bar{\gamma}_{\mathrm{c}}}\right) \tag{5.59}$$

When $L_{\mathrm{c}} = 3$, the three strongest paths are selected and combined so that the closed-form BER can be manipulated through the three-fold integral in (5.52) as

$$\begin{aligned} P_{\mathrm{e,ind}}^{<3>} &= 3! \cdot \binom{L}{3} \cdot \int_0^{\infty} \int_{\gamma_{3:L}}^{\infty} \int_{\gamma_{2:L}}^{\infty} [F_{\gamma_l,\mathrm{ind}}(\gamma_{3:L})]^{L-3} \cdot \prod_{l=1}^{3} p_{\gamma_l,\mathrm{ind}}(\gamma_{l:L}) \\ &\quad \cdot Q\left(\sqrt{2 \cdot \sum_{l=1}^{3} \gamma_{l:L}}\right) \cdot \mathrm{d}\gamma_{1:L} \cdot \mathrm{d}\gamma_{2:L} \cdot \mathrm{d}\gamma_{3:L} \\ &= \frac{3!}{2} \cdot \binom{L}{3} \cdot \left[P_{\mathrm{e1}}^{<3>} + \sum_{m=1}^{L-3} \sum_{n=0}^{m} \binom{L-3}{m} \cdot \binom{m}{n} \cdot (-1)^m \cdot P_{\mathrm{e2}}^{<3>} \right. \\ &\quad \left. + \sum_{m=0}^{L-3} \sum_{n=0}^{m} \binom{L-3}{m} \cdot \binom{m}{n} \cdot (-1)^m \cdot (P_{\mathrm{e3}}^{<3>} + P_{\mathrm{e4}}^{<3>}) \right] \end{aligned} \tag{5.60}$$

where, by the definitions,

$$P_{\mathrm{e1}}^{<3>} = -\frac{288\bar{\gamma}_{\mathrm{c}}^3 + 848\bar{\gamma}_{\mathrm{c}}^2 + 790\bar{\gamma}_{\mathrm{c}} + 477}{13824 \cdot (1+\bar{\gamma}_{\mathrm{c}})^5 \cdot \sqrt{1+1/\bar{\gamma}_{\mathrm{c}}}} \tag{5.61}$$

$$\begin{aligned} P_{\mathrm{e2}}^{<3>} &= -\frac{3}{4 \cdot m^{n+5} \cdot \sqrt{1+1/\bar{\gamma}_{\mathrm{c}}}} \cdot I_{\mathrm{A}}\left(n+4, \frac{m}{3+3\bar{\gamma}_{\mathrm{c}}}\right) \\ &\quad + \frac{2\bar{\gamma}_{\mathrm{c}} + 3}{8 \cdot m^{n+4} \cdot (1+\bar{\gamma}_{\mathrm{c}}) \cdot \sqrt{1+1/\bar{\gamma}_{\mathrm{c}}}} \cdot I_{\mathrm{A}}\left(n+3, \frac{m}{3+3\bar{\gamma}_{\mathrm{c}}}\right) \\ &\quad + \frac{8\bar{\gamma}_{\mathrm{c}}^2 + 20\bar{\gamma}_{\mathrm{c}} + 15}{16 \cdot m^{n+3} \cdot (1+\bar{\gamma}_{\mathrm{c}})^2 \cdot \sqrt{1+1/\bar{\gamma}_{\mathrm{c}}}} \cdot I_{\mathrm{A}}\left(n+2, \frac{m}{3+3\bar{\gamma}_{\mathrm{c}}}\right) \\ &\quad - \frac{16\bar{\gamma}_{\mathrm{c}}^3 + 56\bar{\gamma}_{\mathrm{c}}^2 + 70\bar{\gamma}_{\mathrm{c}} + 35}{32 \cdot m^{n+2} \cdot (1+\bar{\gamma}_{\mathrm{c}})^3 \cdot \sqrt{1+1/\bar{\gamma}_{\mathrm{c}}}} \cdot I_{\mathrm{A}}\left(n+1, \frac{m}{3+3\bar{\gamma}_{\mathrm{c}}}\right) \end{aligned} \tag{5.62}$$

$$\begin{aligned} P_{\mathrm{e3}}^{<3>} &= \frac{1}{2 \cdot (m+3)^{n+4}} \cdot I_{\mathrm{A}}\left(n+3, \frac{m+3}{3\bar{\gamma}_{\mathrm{c}}}\right) + \frac{1}{(m+3)^{n+3}} \\ &\quad \cdot \mathrm{I_A}\left(n+2, \frac{m+3}{3\bar{\gamma}_{\mathrm{c}}}\right) + \frac{1}{2 \cdot (m+3)^{n+2}} \cdot I_{\mathrm{A}}\left(n+1, \frac{m+3}{3\bar{\gamma}_{\mathrm{c}}}\right) \end{aligned} \tag{5.63}$$

$$P_{\rm e4}^{<3>} = \sqrt{\frac{3\bar{\gamma}_{\rm c}}{\pi}} \cdot \left[\frac{1}{4\cdot(1+\bar{\gamma}_{\rm c})} \cdot \frac{\Gamma(n+9/2)}{(m+3+3\bar{\gamma}_{\rm c})^{n+9/2}} - \frac{2+\bar{\gamma}_{\rm c}}{4\cdot(1+\bar{\gamma}_{\rm c})^2} \cdot \frac{\Gamma(n+7/2)}{(m+3+3\bar{\gamma}_{\rm c})^{n+7/2}} - \frac{19+22\bar{\gamma}_{\rm c}+8\bar{\gamma}_{\rm c}^2}{16\cdot(1+\bar{\gamma}_{\rm c})^3} \cdot \frac{\Gamma(n+5/2)}{(m+3+3\bar{\gamma}_{\rm c})^{n+5/2}} \right] \quad (5.64)$$

Correlated pair of channels

It is assumed that the pair of channels $h_{1,l}(m)$ and $h_{2,l}(m)$ are mutually correlated by a spatial correlation coefficient ρ along two distinct spatial paths with the same temporal index. When $\rho = 0$, the pair of channel coefficients $h_{1,l}(m)$ and $h_{2,l}(m)$ are independent from each other and the resultant BER performance is investigated. When $\rho = 1$, the BER performance is the same as that of the conventional CDMA system without TD-STBC under the constraint of equivalent total transmit power, which is explored in the Section 5.3.4.

When $0 < \rho < 1$, the characteristic function of the variable γ_l $(l = 1, 2, \ldots, L)$ in (5.48) is given in (5.40) and rewritten here as

$$\phi_{\gamma_l}(jt) = \frac{1}{[1 - jt\cdot(1+\rho)\cdot\bar{\gamma}_l]\cdot[1 - jt\cdot(1-\rho)\cdot\bar{\gamma}_l]} \quad (5.65)$$

where $\bar{\gamma}_l$ is defined as in (5.32), i.e. $\bar{\gamma}_l = \bar{\gamma}_{\rm c}$.

Through the inverse Fourier transform of its characteristic function, the PDF of γ_l can be attained as

$$p_{\gamma_l,{\rm cor}}(\gamma_l) = \frac{1}{2\rho\cdot\bar{\gamma}_l}\left\{\exp\left[-\frac{\gamma_l}{(1+\rho)\bar{\gamma}_l}\right] - \exp\left[-\frac{\gamma_l}{(1-\rho)\bar{\gamma}_l}\right]\right\} \quad (5.66)$$

Furthermore, its CDF can be integrated as

$$F_{\gamma_l,{\rm cor}}(\gamma_l) = 1 - \frac{1+\rho}{2\rho}\cdot\exp\left[-\frac{\gamma_l}{(1+\rho)\bar{\gamma}_l}\right] + \frac{1-\rho}{2\rho}\cdot\exp\left[-\frac{\gamma_l}{(1-\rho)\bar{\gamma}_l}\right] \quad (5.67)$$

For the scenario of independent pairs of channels, substituting (5.66) and (5.67) into (5.49) to get the joint PDF, and applying the general integral formula of (5.52), we further obtain the closed-form BER for several special cases. The detailed derivations are presented in the Appendix 5B.3. When $L_{\rm c} = 1$, only the strongest path is selected so that the closed-form BER can be manipulated, as follows:

$$\begin{aligned} P_{\rm e,cor}^{<1>} &= \binom{L}{1}\cdot\int_0^{\infty}[F_{\gamma_l,{\rm cor}}(\gamma_{1:L})]^{L-1}\cdot p_{\gamma_l,{\rm cor}}(\gamma_{1:L})\cdot Q(\sqrt{2\cdot\gamma_{1:L}})\cdot {\rm d}\gamma_{1:L} \\ &= \frac{L}{2}\cdot\frac{1}{2\rho\cdot\bar{\gamma}_{\rm c}}\cdot\sum_{m=0}^{L-1}\sum_{n=0}^{m}\left\{\binom{L-1}{m}\cdot\binom{m}{n}\cdot\left(-\frac{1+\rho}{2\rho}\right)^n\cdot\left(\frac{1-\rho}{2\rho}\right)^{m-n} \cdot\left[\frac{1}{p_1} - \frac{1}{p_1}\sqrt{\frac{1}{1+p_1}} - \frac{1}{p_2} + \frac{1}{p_2}\sqrt{\frac{1}{1+p_2}}\right]\right\} \end{aligned} \quad (5.68)$$

where, by the definitions,

$$p_1 = \frac{n+1}{(1+\rho)\cdot\bar{\gamma}_{\rm c}} + \frac{m-n}{(1-\rho)\cdot\bar{\gamma}_{\rm c}} \quad (5.69)$$

$$p_2 = \frac{n}{(1+\rho)\cdot\bar{\gamma}_{\rm c}} + \frac{m-n+1}{(1-\rho)\cdot\bar{\gamma}_{\rm c}} \quad (5.70)$$

When $L_c = 2$, the two strongest paths are selected and combined so that the closed-form BER can be manipulated through the two-fold integral in (5.52) as

$$\begin{aligned}
P_{\mathrm{e,cor}}^{<2>} &= 2 \cdot \binom{L}{2} \cdot \int_0^{\infty} \int_{\gamma_{2:L}}^{\infty} [F_{\gamma_l,\mathrm{cor}}(\gamma_{2:L})]^{L-2} \cdot \prod_{l=1}^{2} p_{\gamma_l,\mathrm{cor}}(\gamma_{l:L}) \\
&\quad \cdot Q\left(\sqrt{2 \cdot \sum_{l=1}^{2} \gamma_{l:L}}\right) \cdot \mathrm{d}\gamma_{1:L} \cdot \mathrm{d}\gamma_{2:L} \\
&= \frac{L \cdot (L-1)}{2} \cdot \left(\frac{1}{2\rho \cdot \bar{\gamma}_c}\right)^2 \\
&\quad \cdot \sum_{m=0}^{L-2} \sum_{n=0}^{m} \left\{ \binom{L-2}{m} \cdot \binom{m}{n} \cdot \left(-\frac{1+\rho}{2\rho}\right)^n \cdot \left(\frac{1-\rho}{2\rho}\right)^{m-n} \right. \\
&\quad \left. \cdot \left[I_B(p_1, q_1) - I_B(p_1, q_2) - I_B(p_2, q_1) + I_B(p_2, q_2) \right] \right\}
\end{aligned} \tag{5.71}$$

where p_1 and p_2 are defined in (5.69) and (5.70), and q_1 and q_2 are defined in (5.45) and (5.46). In addition, by the definition,

$$I_B(a, b) = \begin{cases} \dfrac{1}{b \cdot (a+b)} - \dfrac{1}{b \cdot (a-b) \cdot \sqrt{1+b}} + \dfrac{2}{a^2 - b^2} \cdot \sqrt{\dfrac{2}{a+b+2}}, & a \neq b \\[2ex] \dfrac{1}{2b^2} - \dfrac{2+3b}{4b^2 \cdot (1+b)^{3/2}}, & a = b \end{cases} \tag{5.72}$$

When $L_c = 3$, the three strongest paths are selected and combined so that the closed-form BER can be manipulated through the three-fold integral in (5.52) as

$$\begin{aligned}
P_{\mathrm{e,cor}}^{<3>} &= 3! \cdot \binom{L}{3} \cdot \int_0^{\infty} \int_{\gamma_{3:L}}^{\infty} \int_{\gamma_{2:L}}^{\infty} [F_{\gamma_l,\mathrm{cor}}(\gamma_{3:L})]^{L-3} \cdot \prod_{l=1}^{3} p_{\gamma_l,\mathrm{cor}}(\gamma_{l:L}) \\
&\quad \cdot Q\left(\sqrt{2 \cdot \sum_{l=1}^{3} \gamma_{l:L}}\right) \cdot \mathrm{d}\gamma_{1:L} \cdot \mathrm{d}\gamma_{2:L} \cdot \mathrm{d}\gamma_{3:L} \\
&= \frac{L \cdot (L-1) \cdot (L-2)}{2} \cdot \left(\frac{1}{2\rho \cdot \bar{\gamma}_c}\right)^3 \\
&\quad \cdot \sum_{m=0}^{L-3} \sum_{n=0}^{m} \left\{ \binom{L-3}{m} \cdot \binom{m}{n} \cdot \left(-\frac{1+\rho}{2\rho}\right)^n \cdot \left(\frac{1-\rho}{2\rho}\right)^{m-n} \right. \\
&\quad \cdot \left[I_C(p_1, q_1) - I_C(p_2, q_1) + I_C(p_1, q_2) - I_C(p_2, q_2) \right. \\
&\quad \left. \left. - I_D(p_1, q_2, q_1) + I_D(p_2, q_2, q_1) - I_D(p_1, q_1, q_2) + I_D(p_2, q_1, q_2) \right] \right\}
\end{aligned} \tag{5.73}$$

where p_1 and p_2 are defined in (5.69) and (5.70), and q_1 and q_2 are defined in (5.45) and (5.46). In addition, $I_C(\cdot,\cdot)$ is defined by

$$I_C(a,b) = \begin{cases} \frac{1}{2b^2\cdot(a+2b)} + \frac{9b^2+8b-3ab-2a}{4b^2\cdot(a-b)^2\cdot(1+b)^{3/2}} \\ \quad - \frac{9}{2\cdot(a+2b)\cdot(a-b)^2}\cdot\sqrt{\frac{3}{a+2b+3}}, & a \neq b \\ \frac{1}{6b^3} - \frac{1}{6b^3\cdot(1+b)^{1/2}} - \frac{1}{12b^2\cdot(1+b)^{3/2}} \\ \quad - \frac{1}{16\cdot b\cdot(1+b)^{5/2}}, & a = b \end{cases} \tag{5.74}$$

and $I_D(\cdot,\cdot,\cdot)$ is defined by

$$I_D(a,b,c) = I_{D,1}(a,b,c) + I_{D,2}(a,b,c) \tag{5.75}$$

where, by the definitions,

$$I_{D,1}(a,b,c) = \begin{cases} \frac{1}{c\cdot(b+c)\cdot(a+b+c)} + \frac{4}{(b^2-c^2)\cdot(2a-b-c)} \\ \quad \cdot\sqrt{\frac{2}{2+b+c}} - \frac{2a-b+5c}{c\cdot(b-c)\cdot(2a-b-c)\cdot(a+b+c)} \\ \quad \cdot\sqrt{\frac{3}{3+a+b+c}}, & 2a-b-c \neq 0 \\ \frac{1}{c\cdot(b+c)}\cdot\frac{1}{a+b+c}\cdot\left(1-\sqrt{\frac{3}{3+a+b+c}}\right) \\ \quad + \frac{2\sqrt{2}}{3\cdot(b^2-c^2)\cdot(2+b+c)^{3/2}}, & 2a-b-c = 0 \end{cases} \tag{5.76}$$

$$I_{D,2}(a,b,c) = \begin{cases} -\frac{1}{c\cdot(b-c)\cdot(a+b-2c)\cdot\sqrt{1+c}}\cdot\left[1-\sqrt{\frac{3(1+c)}{3+a+b+c}}\right], \\ \qquad a+b-2c \neq 0 \\ -\frac{1}{6c\cdot(b-c)\cdot(1+c)^{3/2}}, \qquad a+b-2c = 0 \end{cases} \tag{5.77}$$

5.3.4 Conventional Rake receiver without TD-STBC

In order to elaborate explicitly the spatial diversity gain provided by TD-STBC, in this section a conventional DS-CDMA system with only one transmit antenna is studied, so that the BER performance of the system with and without TD-STBC can easily be observed and compared.

Since only one transmit antenna is deployed at the base station, the structure of the transmitter is the same as the branch of the first transmit antenna shown in Figure 5.1, thus the transmit signal is similar to that given by (5.1), i.e.

$$s_{1\text{Tx}}(t) = \sum_{n=-\infty}^{\infty} \left[\sqrt{E_{\text{c},1\text{Tx}}} \cdot \sum_{k=1}^{K} b(k, \lfloor n/N \rfloor) \cdot S_{\text{d}}(k, n) \right] \cdot g(t - nT_{\text{c}}) \tag{5.78}$$

where it is also assumed that there are K active CDMA users, and hence K different spreading code sequences are needed. $E_{\text{c},1\text{Tx}}$ is the chip energy of the data channel in the system with only one transmit antenna and without TD-STBC. Unless noted otherwise, all other identical symbols and operators in this section are defined as previously in this chapter and have the same properties.

The involved discrete tap-delay-line channel model here can be characterized by

$$h_{1\text{Tx}}(t) = \sum_{l=1}^{L} h_{1,l} \cdot \delta(t - \tau_l) \tag{5.79}$$

and the received signal at any one mobile station can be represented as

$$r_{1\text{Tx}}(t) = \sum_{l=1}^{L} h_{1,l} \cdot s_{1\text{Tx}}(t - \tau_l) + \eta(t) \tag{5.80}$$

By sampling the output of the pulse MF, the received signal during the mth symbol interval can be obtained as

$$\begin{aligned} u_{1\text{Tx}}(m, n) = &\sum_{k=1}^{K} \sum_{l=1}^{L} \sqrt{E_{\text{c},1\text{Tx}}} \cdot S_{\text{d}}(k, mN + n - \lfloor \tau_l / T_{\text{c}} \rfloor) \\ &\cdot b(k, m) \cdot h_{1,l}(m) + \eta(m, n) \end{aligned} \tag{5.81}$$

where $n = 0, 1, \ldots, N-1$ and $b(k, m)$ stands for the data bit of the kth active user transmitted in the mth symbol interval; $h_{1,l}(m)$ represents the fading coefficient of the lth resolvable path of the channel between the transmit and receive antenna in the mth symbol period; and $\eta(m, n)$ is the sampled AWGN in the mth symbol period.

Similarly, the first user ($k = 1$) is assumed to be the desired user. The receiver structure is the same as shown in Figure 5.2 regardless of the STBC decoder. Assuming perfect chip timing synchronization and that the local despreading code sequence is locked to the $\hat{l}$th ($\hat{l} = 1, 2, \ldots, L$) resolvable path, the data channel of the first user at the $\hat{l}$th path is despread during the mth symbol period, as follows:

$$\begin{aligned} r_{\text{d},1\text{Tx}}(m, \hat{l}) &= \sum_{n=0}^{N-1} u_{1\text{Tx}}(m, n) \cdot S_{\text{d}}^{*}(1, mN + n - \lfloor \tau_{\hat{l}} / T_{\text{c}} \rfloor) \\ &= \sqrt{E_{\text{c},1\text{Tx}}} \cdot N \cdot b(1, m) \cdot h_{1,\hat{l}}(m) + \eta_{\text{d}}(m, \hat{l}) \\ &\quad + \sum_{k=1}^{K} \sum_{\substack{l=1 \\ l \neq \hat{l}}}^{L} \sqrt{E_{\text{c},1\text{Tx}}} \cdot R_{\text{d,d}}(m, k; l, \hat{l}) \cdot b(k, m) \cdot h_{1,l}(m) \end{aligned} \tag{5.82}$$

where the background AWGN component $\eta_{\text{d}}(m, \hat{l})$ is given by

$$\eta_{\text{d}}(m, \hat{l}) = \sum_{n=0}^{N-1} \eta(m, n) \cdot S_{\text{d}}^{*}(1, mN + n - \lfloor \tau_{\hat{l}} / T_{\text{c}} \rfloor) \tag{5.83}$$

$R_{\mathrm{d,d}}(m,k;l,\hat{l})$ is the N-length discrete aperiodic correlation function [1] of two time-delayed data channel spreading code sequences used in the mth symbol period, given by

$$R_{\mathrm{d,d}}(m,k;l,\hat{l}) = \sum_{n=0}^{N-1} S_{\mathrm{d}}[k, mN+n-\lfloor \tau_l/T_{\mathrm{c}} \rfloor] \cdot S_{\mathrm{d}}^*[1, mN+n-\lfloor \tau_{\hat{l}}/T_{\mathrm{c}} \rfloor] \quad (5.84)$$

since two different spreading code sequences with the same time delay are orthogonal over one symbol interval, i.e. $R_{\mathrm{d,d}}(m,k;l,\hat{l}) = 0$ when $l = \hat{l}$ and $k \neq 1$. Otherwise, when $l \neq \hat{l}$, the two code sequences are non-orthogonal and $R_{\mathrm{d,d}}(m,k;l,\hat{l})$ is modeled as an i.i.d. random variable with zero mean and variance $\mathrm{Var}\{R_{\mathrm{d,d}}(m,k;l,\hat{l})\} = N$.

It is assumed that the ideal CSI in the $\hat{l}$th resolvable path, $h_{1,\hat{l}}(m)$, is available at the receiver and, thus, the branch random variable of the estimated data bit in the mth symbol interval is constructed through coherent reception as

$$\begin{aligned} d_{1\mathrm{Tx}}(m,\hat{l}) &= r_{\mathrm{d},1\mathrm{Tx}}(m,\hat{l}) \cdot h_{1,\hat{l}}^*(m) \\ &= \sqrt{E_{\mathrm{c},1\mathrm{Tx}}} \cdot N \cdot b(1,m) \cdot |h_{1,\hat{l}}(m)|^2 + \eta_{\mathrm{d}}(m,\hat{l}) \cdot h_{1,\hat{l}}^*(m) \\ &\quad + \sum_{k=1}^{K} \sum_{\substack{l=1 \\ l \neq \hat{l}}}^{L} \sqrt{E_{\mathrm{c},1\mathrm{Tx}}} \cdot R_{\mathrm{d,d}}(m,k;l,\hat{l}) \cdot b(k,m) \cdot h_{1,l}(m) \cdot h_{1,\hat{l}}^*(m) \\ &= E_{1\mathrm{Tx}}(m,\hat{l}) + N_{1\mathrm{Tx}}(m,\hat{l}) + I_{1\mathrm{Tx}}(m,\hat{l}) \end{aligned} \quad (5.85)$$

It is shown that the above branch decision random variable consists of three distinct components: $E_{1\mathrm{Tx}}(m,\hat{l})$ is the desired signal component; $N_{1\mathrm{Tx}}(m,\hat{l})$ is the background AWGN component; and $I_{1\mathrm{Tx}}(m,\hat{l})$ is the multiple access and multipath interference component that results from the desired user as well as other simultaneously active users. They are defined as follows:

$$E_{1\mathrm{Tx}}(m,\hat{l}) = \sqrt{E_{\mathrm{c},1\mathrm{Tx}}} \cdot N \cdot b(1,m) \cdot |h_{1,\hat{l}}(m)|^2 \quad (5.86)$$

$$N_{1\mathrm{Tx}}(m,\hat{l}) = \eta_{\mathrm{d}}(m,\hat{l}) \cdot h_{1,\hat{l}}^*(m) \quad (5.87)$$

$$I_{1\mathrm{Tx}}(m,\hat{l}) = \sum_{k=1}^{K} \sum_{\substack{l=1 \\ l \neq \hat{l}}}^{L} \sqrt{E_{\mathrm{c},1\mathrm{Tx}}} \cdot R_{\mathrm{d,d}}(m,k;l,\hat{l}) \cdot b(k,m) \cdot h_{1,l}(m) \cdot h_{1,\hat{l}}^*(m) \quad (5.88)$$

For a different user k and during a different mth symbol period, the data bit $b(k,m)$ is an i.i.d. binary random variable taking the value ± 1, and $R_{\mathrm{d,d}}(m,k;l,\hat{l})$ is an i.i.d. random variable with zero-mean and variance N. Therefore, conditioned on the channel fading coefficients $h_{1,\hat{l}}(m)$, $I_{1\mathrm{Tx}}(m,l)$ is a random variable with zero mean and variance given by

$$\begin{aligned} \mathrm{Var}\{I_{1\mathrm{Tx}}(m,\hat{l})\} &= E_{\mathrm{c},1\mathrm{Tx}} \cdot N \cdot \sum_{k=1}^{K} \sum_{\substack{l=1 \\ l \neq \hat{l}}}^{L} E\{|h_{1,l}(m)|^2\} \cdot |h_{1,\hat{l}}(m)|^2 \\ &= E_{\mathrm{c},1\mathrm{Tx}} \cdot N \cdot K \cdot (L-1) \cdot 2\sigma^2 \cdot |h_{1,\hat{l}}(m)|^2 \end{aligned} \quad (5.89)$$

The background AWGN component $N_{1\text{Tx}}(m, \hat{l})$ is a zero-mean Gaussian variable with variance

$$\text{Var}\{N_{1\text{Tx}}(m, \hat{l})\} = N \cdot \frac{\eta_0}{2} \cdot |h_{1,\hat{l}}(m)|^2 \tag{5.90}$$

Since the data bits from K active users and the channel fading coefficients along L resolvable paths are independent random variables, $d_{1\text{Tx}}(m, \hat{l})$ in (5.85) is the sum of many independent random variables and hence can be approximated as a conditional Gaussian variable. Therefore, conditioned on the channel fading coefficients $h_{1,\hat{l}}(m)$, the branch decision variable $d_{1\text{Tx}}(m, \hat{l})$ is Gaussian random variable with mean and variance as follows:

$$E\{d_{1\text{Tx}}(m, \hat{l})\} = E_{1\text{Tx}}(m, \hat{l}) = \sqrt{E_{\text{c},1\text{Tx}}} \cdot N \cdot b(1, m) \cdot |h_{1,\hat{l}}(m)|^2 \tag{5.91}$$

$$\begin{aligned}\text{Var}\{d_{1\text{Tx}}(m, \hat{l})\} &= \text{Var}\{N_{1\text{Tx}}(m, \hat{l})\} + \text{Var}\{I_{1\text{Tx}}(m, \hat{l})\} \\ &= \left[E_{\text{c},1\text{Tx}} \cdot N \cdot K \cdot (L-1) \cdot 2\sigma^2 + N \cdot \frac{\eta_0}{2}\right] \cdot |h_{1,\hat{l}}(m)|^2\end{aligned} \tag{5.92}$$

After despreading at each path, a conventional Rake receiver is used to combine the signal energy from L_c ($L_\text{c} \leq L$) selected paths. This can obtain the path diversity gain benefiting from DS-CDMA merits.

EGC Rake receiver

The EGC Rake receiver selects and combines signals of the first L_c arriving paths among the L available resolvable paths. Assuming that the fading of each path is independent and that the decision random variables in distinct branches are independent from each other, the output of the Rake combiner can be represented as

$$d_{1\text{Tx}}(m) = \sum_{l=1}^{L_\text{c}} d_{1\text{Tx}}(m, l) \tag{5.93}$$

When $L_\text{c} = L$, the signals from all resolvable paths are combined. The decision variable $d_{1\text{Tx}}(m)$ is a Gaussian variable with conditional mean and variance given by, respectively:

$$E\{d_{1\text{Tx}}(m)\} = \sum_{l=1}^{L_\text{c}} E\{d_{1\text{Tx}}(m, l)\} = \sqrt{E_{\text{c},1\text{Tx}}} \cdot N \cdot b(1, m) \cdot \zeta_{1\text{Tx}} \tag{5.94}$$

$$\begin{aligned}\text{Var}\{d_{1\text{Tx}}(m)\} &= \sum_{l=1}^{L_\text{c}} \text{Var}\{d_{1\text{Tx}}(m, l)\} \\ &= \left[E_{\text{c},1\text{Tx}} \cdot N \cdot K \cdot (L-1) \cdot 2\sigma^2 + N \cdot \frac{\eta_0}{2}\right] \cdot \zeta_{1\text{Tx}}\end{aligned} \tag{5.95}$$

where $\zeta_{1\text{Tx}}$ is defined as

$$\zeta_{1\text{Tx}} = \sum_{l=1}^{L_\text{c}} |h_{1,l}(m)|^2 \tag{5.96}$$

Therefore, conditioned on the instantaneous fading channel amplitudes of the multipaths, the BER can be obtained by [3], [4]

$$P_{\mathrm{e,EGC}}(\gamma_{1\mathrm{Tx}}) = Q\left\{\left[\frac{|E\{d_{1\mathrm{Tx}}(m)\}|^2}{\mathrm{Var}\{d_{1\mathrm{Tx}}(m)\}}\right]^{1/2}\right\} = Q(\sqrt{2\gamma_{1\mathrm{Tx}}}) \tag{5.97}$$

where, by definition,

$$\gamma_{1\mathrm{Tx}} = \frac{E_{\mathrm{c,1Tx}} \cdot N \cdot \zeta_{1\mathrm{Tx}}}{E_{\mathrm{c,1Tx}} \cdot K \cdot (L-1) \cdot 4\sigma^2 + \eta_0} \tag{5.98}$$

Let us review and compare this with the EGC 2D-Rake receiver discussed in Section 5.3.2. When the spatial correlation coefficient between the pair of channels $\rho = 1, h_{1,l}(m)$ and $h_{2,l}(m)$ are always identical, hence $h_{1,l}(m) = h_{2,l}(m)$ and $\zeta = 2\zeta_{1\mathrm{Tx}}$. On the other hand, since the total transmit power is constrained as constant in both systems with and without TD-STBC, it is $E_{c,1\mathrm{Tx}} = 2E_c$. Therefore, when $\rho = 1$, γ defined in (5.30) is equivalent to $\gamma_{1\mathrm{Tx}}$ defined in (5.98) and the BER performance of the EGC 2D-Rake receiver given by (5.29) degrades to the BER of the EGC Rake receiver of the conventional CDMA system without TD-STBC given by (5.97).

Since the summed random variable $\zeta_{1\mathrm{Tx}}$ in (5.96) follows a chi-square distribution with $2L_{\mathrm{c}}$ degrees of freedom, the PDF of $\gamma_{1\mathrm{Tx}}$ in (5.98) is given by

$$p_{\gamma,1\mathrm{Tx}}(\gamma_{1\mathrm{Tx}}) = \frac{1}{\bar{\gamma}_{\mathrm{c,1Tx}}^{L_{\mathrm{c}}} \cdot \Gamma(L_{\mathrm{c}})} \cdot \gamma_{1\mathrm{Tx}}^{L_{\mathrm{c}}-1} \cdot \exp\left(-\frac{\gamma_{1\mathrm{Tx}}}{\bar{\gamma}_{\mathrm{c,1Tx}}}\right) \tag{5.99}$$

where, by definition,

$$\bar{\gamma}_{\mathrm{c,1Tx}} = \frac{E_{\mathrm{c,1Tx}} \cdot N \cdot (2\sigma^2)}{E_{\mathrm{c,1Tx}} \cdot K \cdot (L-1) \cdot 4\sigma^2 + \eta_0} \tag{5.100}$$

The resultant BER can be obtained by averaging the conditional BER $P_{\mathrm{e,EGC}}(\gamma_{1\mathrm{Tx}})$ in (5.97) over the PDF of $\gamma_{1\mathrm{Tx}}$ in (5.99) [3], i.e.,

$$\begin{aligned} P_{\mathrm{e,1Tx}} &= \int_0^{\infty} P_{\mathrm{e,EGC}}(\gamma_{1\mathrm{Tx}}) \cdot p_{\gamma,1\mathrm{Tx}}(\gamma_{1\mathrm{Tx}}) \cdot \mathrm{d}\gamma_{1\mathrm{Tx}} \\ &= \left(\frac{1-\mu_{1\mathrm{Tx}}}{2}\right)^{L_{\mathrm{c}}} \cdot \sum_{l=0}^{L_{\mathrm{c}}-1} \binom{L_c - 1 + l}{l} \cdot \left(\frac{1+\mu_{1\mathrm{Tx}}}{2}\right)^l \end{aligned} \tag{5.101}$$

where, by definition,

$$\mu_{1\mathrm{Tx}} = \sqrt{\frac{\bar{\gamma}_{\mathrm{c,1Tx}}}{1+\bar{\gamma}_{\mathrm{c,1Tx}}}} = \left[1 + \frac{2K \cdot (L-1)}{N} + \frac{1}{\bar{\gamma}_{\mathrm{b,1Tx}}}\right]^{-\frac{1}{2}} \tag{5.102}$$

where $\bar{\gamma}_{\mathrm{b,1Tx}}$ is the average SNR per bit, defined as

$$\bar{\gamma}_{\mathrm{b,1Tx}} = \frac{2E_{\mathrm{c,1Tx}} \cdot N \cdot \sigma^2}{\eta_0} \tag{5.103}$$

and noting that $\bar{\gamma}_{\mathrm{c,1Tx}}$ is related to $\bar{\gamma}_{\mathrm{b,1Tx}}$ by

$$\bar{\gamma}_{\mathrm{c,1Tx}}^{-1} = 4K \cdot (L-1)/N + \bar{\gamma}_{\mathrm{b,1Tx}}^{-1} \tag{5.104}$$

GSC Rake receiver

The GSC Rake receiver is exploited to adaptively select and combine the signals from the L_c paths with highest SINR among the L available resolvable paths. For the branch decision variable $d_{1\mathrm{Tx}}(m,l)$ $(l = 1, 2, \ldots, L)$ in the lth resolvable path, its instantaneous SINR can be deduced from (5.91) and (5.92), as follows:

$$\frac{|E\{d_{1\mathrm{Tx}}(m,l)\}|^2}{\mathrm{Var}\{d_{1\mathrm{Tx}}(m,l)\}} = 2\gamma_{l,1\mathrm{Tx}} \tag{5.105}$$

where $\gamma_{l,1\mathrm{Tx}}$ is defined by

$$\gamma_{l,1\mathrm{Tx}} = \frac{E_{\mathrm{c},1\mathrm{Tx}} \cdot N}{E_{\mathrm{c},1\mathrm{Tx}} \cdot K \cdot (L-1) \cdot 4\sigma^2 + \eta_0} \cdot |h_{1,l}(m)|^2 \tag{5.106}$$

Moreover, the order statistics of the instantaneous SINR $\gamma_{1:L,1\mathrm{Tx}} \geq \gamma_{2:L,1\mathrm{Tx}} \geq \cdots \geq \gamma_{L:L,1\mathrm{Tx}}$ are defined and obtained by arranging $\{\gamma_{l,1\mathrm{Tx}} \mid l = 1, 2, \ldots, L\}$ in a descending order of magnitude. For the L_c strongest paths selected from L resolvable paths, the joint PDF of the order statistic [7], [8] variables $\{\gamma_{l:L,1\mathrm{Tx}} \mid l = 1,\ 2, \ldots,\ L_c\}$ is given by

$$\begin{aligned} & p_{\gamma_{1:L,1\mathrm{Tx}},\gamma_{2:L,1\mathrm{Tx}},\ldots,\gamma_{L_c:L,1\mathrm{Tx}}}(\gamma_{1:L,1\mathrm{Tx}}, \gamma_{2:L,1\mathrm{Tx}}, \ldots, \gamma_{L_c:L,1\mathrm{Tx}}) \\ & = L_c! \cdot \binom{L}{L_c} \cdot [F_{\gamma_{l,1\mathrm{Tx}}}(\gamma_{L_c:L,1\mathrm{Tx}})]^{L-L_c} \cdot \prod_{l=1}^{L_c} p_{\gamma_{l,1\mathrm{Tx}}}(\gamma_{l:L,1\mathrm{Tx}}) \end{aligned} \tag{5.107}$$

where the relevant PDF and CDF of the variable $\gamma_{l,1\mathrm{Tx}}$ in (5.106) are given by

$$p_{\gamma_{l,1\mathrm{Tx}}}(\gamma_{l,1\mathrm{Tx}}) = \frac{1}{\bar{\gamma}_{l,1\mathrm{Tx}}} \exp\left(-\frac{\gamma_{l,1\mathrm{Tx}}}{\bar{\gamma}_{l,1\mathrm{Tx}}}\right) \tag{5.108}$$

$$F_{\gamma_{l,1\mathrm{Tx}}}(\gamma_{l,1\mathrm{Tx}}) = 1 - \exp\left(-\frac{\gamma_{l,1\mathrm{Tx}}}{\bar{\gamma}_{l,1\mathrm{Tx}}}\right) \tag{5.109}$$

where $\bar{\gamma}_{l,1\mathrm{Tx}}$ is defined as in (5.100), i.e. $\bar{\gamma}_{l,1\mathrm{Tx}} = \bar{\gamma}_{\mathrm{c},1\mathrm{Tx}}$.

After the L_c strongest paths are selected and combined by the GSC Rake receiver, the total output SINR can be represented as [4]

$$\gamma_{\mathrm{GSC},1\mathrm{Tx}} = 2 \cdot \sum_{l=1}^{L_c} \gamma_{l:L,1\mathrm{Tx}} \tag{5.110}$$

Thus, conditioned on $\{\gamma_{l:L,1\mathrm{Tx}} \mid l = 1,\ 2, \ldots,\ L_c\}$, the BER can be obtained by

$$P_{\mathrm{e,GSC}}(\gamma_{\mathrm{GSC},1\mathrm{Tx}}) = Q(\sqrt{\gamma_{\mathrm{GSC},1\mathrm{Tx}}}) = \frac{1}{2}\mathrm{erfc}\left(\sqrt{\sum_{l=1}^{L_c} \gamma_{l:L,1\mathrm{Tx}}}\right) \tag{5.111}$$

Therefore, the resultant BER can be achieved by averaging the conditional BER in (5.111) over the joint PDF of the order statistics of selected paths in (5.107), as follows:

$$\begin{aligned} P_{\mathrm{e},1\mathrm{Tx}}^{<L_c>} = & \int_0^\infty \int_{\gamma_{L_c:L,1\mathrm{Tx}}}^\infty \cdots \int_{\gamma_{2:L,1\mathrm{Tx}}}^\infty Q(\sqrt{\gamma_{\mathrm{GSC},1\mathrm{Tx}}}) \cdot p_{\gamma_{1:L,1\mathrm{Tx}},\gamma_{2:L,1\mathrm{Tx}},\ldots,\gamma_{L_c:L,1\mathrm{Tx}}} \\ & \cdot (\gamma_{1:L,1\mathrm{Tx}}, \gamma_{2:L,1\mathrm{Tx}}, \ldots, \gamma_{L_c:L,1\mathrm{Tx}}) \cdot d\gamma_{1:L,1\mathrm{Tx}} \cdot d\gamma_{2:L,1\mathrm{Tx}} \cdots d\gamma_{L_c:L,1\mathrm{Tx}} \end{aligned} \tag{5.112}$$

Substituting (5.108) and (5.109) into (5.107) to get the joint PDF, and applying the general integral formula (5.112), we further obtain the closed-form BER for several special cases. The detailed derivations are presented in the Appendix 5B.4. When $L_c = 1$, only the strongest path is selected so that the closed-form BER can be manipulated, as follows:

$$\begin{aligned} P_{e,1Tx}^{<1>} &= \binom{L}{1} \cdot \int_0^{\infty} [F_{\gamma_{l,1Tx}}(\gamma_{1:L,1Tx})]^{L-1} \cdot p_{\gamma_{l,1Tx}}(\gamma_{1:L,1Tx}) \cdot Q(\sqrt{2 \cdot \gamma_{1:L,1Tx}}) \cdot d\gamma_{1:L,1Tx} \\ &= \frac{L}{2} \cdot \sum_{m=0}^{L-1} \binom{L-1}{m} \cdot (-1)^m \cdot \frac{1}{m+1} \cdot \left(1 - \sqrt{\frac{\bar{\gamma}_{c,1Tx}}{1+m+\bar{\gamma}_{c,1Tx}}}\right) \end{aligned} \tag{5.113}$$

where $\bar{\gamma}_{c,1Tx}$ is given by (5.100).

When $L_c = 2$, the two strongest paths are selected and combined so that the closed-form BER can be manipulated through the two-fold integral in (5.112) as

$$\begin{aligned} P_{e,1Tx}^{<2>} &= 2 \cdot \binom{L}{2} \cdot \int_0^{\infty} \int_{\gamma_{2:L,1Tx}}^{\infty} [F_{\gamma_{l,1Tx}}(\gamma_{2:L,1Tx})]^{L-2} \cdot \prod_{l=1}^{2} p_{\gamma_{l,1Tx}}(\gamma_{l:L,1Tx}) \\ &\quad \cdot Q\left(\sqrt{2 \cdot \sum_{l=1}^{2} \gamma_{l:L,1Tx}}\right) \cdot d\gamma_{1:L,1Tx} \cdot d\gamma_{2:L,1Tx} \\ &= \frac{L(L-1)}{2} \cdot \left[P_{e1,1Tx}^{<2>} + \sum_{m=1}^{L-2} \binom{L-2}{m} \cdot (-1)^m \cdot P_{e2,1Tx}^{<2>}\right] \end{aligned} \tag{5.114}$$

where, by the definitions,

$$P_{e1,1Tx}^{<2>} = \frac{1}{2} - \frac{3+2\bar{\gamma}_{c,1Tx}}{4 \cdot (1+\bar{\gamma}_{c,1Tx})} \cdot \sqrt{\frac{\bar{\gamma}_{c,1Tx}}{1+\bar{\gamma}_{c,1Tx}}} \tag{5.115}$$

$$P_{e2,1Tx}^{<2>} = \frac{1}{m+2} - \frac{1}{m} \cdot \sqrt{\frac{\bar{\gamma}_{c,1Tx}}{1+\bar{\gamma}_{c,1Tx}}} + \frac{2}{m \cdot (m+2)} \cdot \sqrt{\frac{2\bar{\gamma}_{c,1Tx}}{2+m+2\bar{\gamma}_{c,1Tx}}} \tag{5.116}$$

When $L_c = 3$, the three strongest paths are selected and combined so that the closed-form BER can be manipulated through the three-fold integral in (5.112) as

$$\begin{aligned} P_{e,1Tx}^{<3>} &= 3! \cdot \binom{L}{3} \cdot \int_0^{\infty} \int_{\gamma_{3:L,1Tx}}^{\infty} \int_{\gamma_{2:L,1Tx}}^{\infty} [F_{\gamma_{l,1Tx}}(\gamma_{3:L,1Tx})]^{L-3} \cdot \prod_{l=1}^{3} p_{\gamma_{l,1Tx}}(\gamma_{l:L,1Tx}) \\ &\quad \cdot Q\left(\sqrt{2 \cdot \sum_{l=1}^{3} \gamma_{l:L,1Tx}}\right) \cdot d\gamma_{1:L,1Tx} \cdot d\gamma_{2:L,1Tx} \cdot d\gamma_{3:L,1Tx} \\ &= \frac{3!}{2} \cdot \binom{L}{3} \cdot \left[P_{e1,1Tx}^{<3>} + \sum_{m=1}^{L-3} \binom{L-3}{m} \cdot (-1)^m \cdot P_{e2,1Tx}^{<3>}\right] \end{aligned} \tag{5.117}$$

where, by the definitions,

$$P_{\mathrm{e1,1Tx}}^{<3>} = \frac{1}{6} - \frac{15 + 20\bar{\gamma}_{\mathrm{c,1Tx}} + 8\bar{\gamma}_{\mathrm{c,1Tx}}^2}{48 \cdot (1 + \bar{\gamma}_{\mathrm{c,1Tx}})^2} \cdot \sqrt{\frac{\bar{\gamma}_{\mathrm{c,1Tx}}}{1 + \bar{\gamma}_{\mathrm{c,1Tx}}}} \tag{5.118}$$

$$P_{\mathrm{e2,1Tx}}^{<3>} = \frac{1}{2(m+3)} + \frac{6 \cdot (1 + \bar{\gamma}_{\mathrm{c,1Tx}}) - m \cdot (3 + 2\bar{\gamma}_{\mathrm{c,1Tx}})}{4m^2 \cdot (1 + \bar{\gamma}_{\mathrm{c,1Tx}})} \cdot \sqrt{\frac{\bar{\gamma}_{\mathrm{c,1Tx}}}{1 + \bar{\gamma}_{\mathrm{c,1Tx}}}} - \frac{9\sqrt{3}}{2m^2 \cdot (m+3)} \cdot \sqrt{\frac{\bar{\gamma}_{\mathrm{c,1Tx}}}{3 + m + 3\bar{\gamma}_{\mathrm{c,1Tx}}}} \tag{5.119}$$

5.4 Numerical results and discussion

In this section, the system performance of TD-STBC and the resultant spatial diversity gain over the conventional DS-CDMA system are numerically evaluated in terms of BER under different system configurations and parameters. In order to demonstrate the spatial and path diversity gain provided by the 2D-Rake receiver, the total transmit power is restricted as constant, irrespective of the number of transmit antennas and resolvable multipaths. Therefore, the parameter of the SNR per bit per antenna per path $\bar{\gamma}_{\mathrm{p}}$ in (5.37) is used to calculate BER while the SNR per bit $\bar{\gamma}_{\mathrm{b}}$ in (5.35) is used to plot the performance figures. Unless noted otherwise, the number of resolvable multipaths $L = 6$, the number of active users $K = 10$ and the spreading factor $N = 128$.

5.4.1 Performance of the EGC 2D-Rake receiver

In Figure 5.3, the BER of the EGC 2D-Rake receiver with an independent pair of channels is illustrated versus the average SNR per bit $\bar{\gamma}_{\mathrm{b}}$ for different numbers of Rake fingers, i.e. $L_{\mathrm{c}} = 1 \sim 6$, respectively. It is clearly seen that the 2D-Rake receiver takes advantage of the multipath and improves the performance when the number of fingers increases. For the combined path $L_{\mathrm{c}} = 4$, BER $= 10^{-2}$ can be obtained when the average SNR per bit $\bar{\gamma}_{\mathrm{b}} \approx 12$ dB. If the channel coding is further utilized, an acceptable BER performance can thus be achieved.

Figure 5.4 illustrates the BER of the full EGC or GSC 2D-Rake receiver, which combines the signals from all resolvable multipaths, versus the correlation coefficient ρ between the pair of channels $h_{1,l}$ and $h_{2,l}$. It is assumed that the number of resolvable multipaths varies from $L = 1$ to $L = 6$ and that the concerned average SNR per bit inspected is $\bar{\gamma}_{\mathrm{b}} = 10$ dB. It can be seen that the BER monotonously increases when the correlation of the pair of channels increases. As the number of resolvable paths increases, however, the degradation of BER due to the increasing correlation of the pair of channels becomes less. In other words, in the rich scattering propagation environment the increased multipath diversity gain is somehow able to compensate for the reduction of spatial diversity gain so that the overall system performance is more robust.

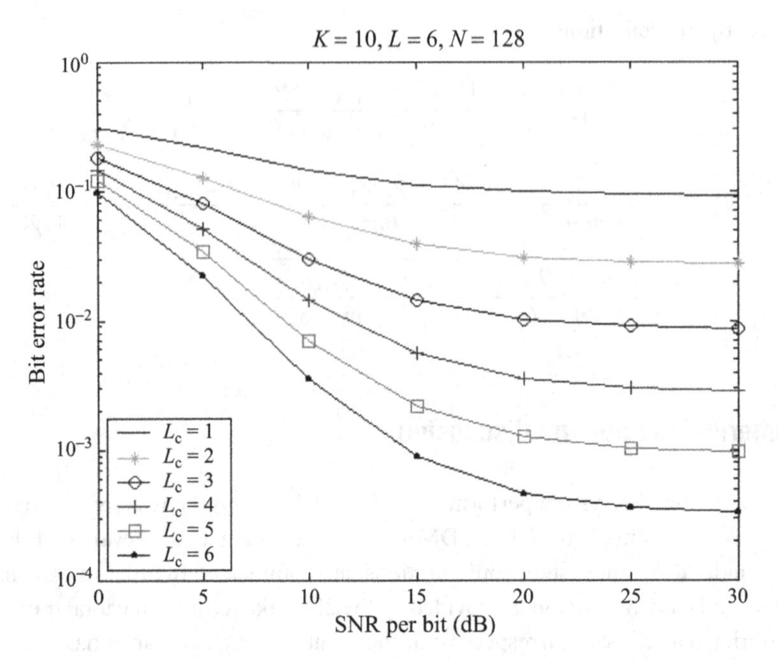

Figure 5.3 BER of EGC 2D-Rake receiver with ideal CSI available and an independent pair of channels.

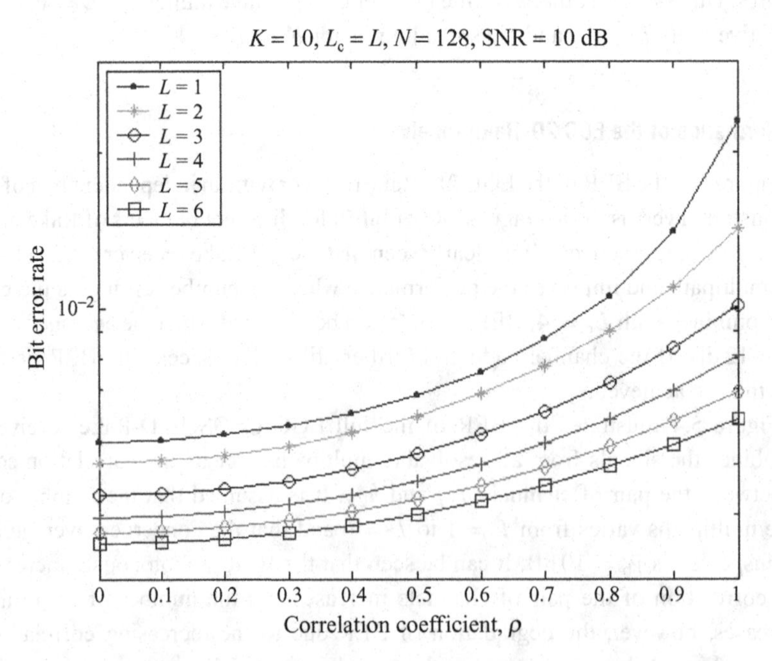

Figure 5.4 BER of full EGC/GSC 2D-RAKE receiver versus the correlation coefficient with ideal CSI available.

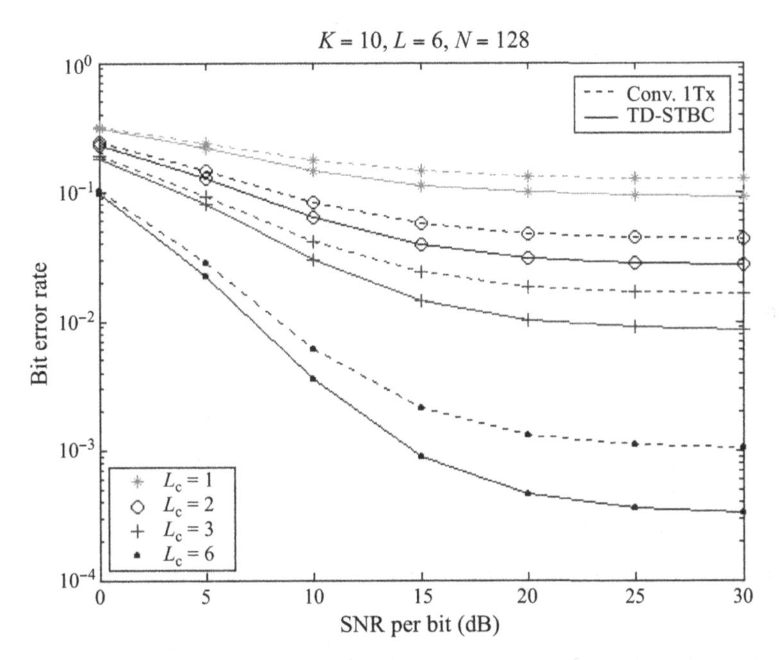

Figure 5.5 BER comparison of EGC 2D-Rake receiver with TD-STBC and conventional EGC Rake receiver with only one transmit antenna, assuming ideal CSI available.

Figure 5.5 demonstrates the spatial diversity gain benefiting from TD-STBC by comparing the BER of the EGC 2D-Rake receiver with the BER of the conventional EGC Rake receiver with only one transmit antenna. It is assumed that the pair of channels $h_{1,l}$ and $h_{2,l}$ are independent from each other and the different number of Rake fingers is $L_c = 1, 2, 3, 6$. The solid curves are for the 2D-Rake receiver with TD-STBC while the dotted curves are for the conventional Rake receiver of the DS-CDMA system with only one transmit antenna and without TD-STBC. It can be seen that TD-STBC can provide significant spatial diversity gain with respect to BER when the ideal CSI is assumed available at the receiver. For example, as the combined path $L_c = 6$ and BER $= 10^{-3}$ the obtained diversity gain is about 10 dB.

5.4.2 Performance of the GSC 2D-Rake receiver

In Figure 5.6, the BER of the GSC 2D-Rake receiver with an independent pair of channels is illustrated versus the average SNR per bit $\bar{\gamma}_b$ for different numbers of Rake fingers, i.e. $L_c = 1, 2, 3, 6$, respectively. It is also seen that the 2D-Rake receiver improves the performance when the number of fingers increases. For the combined path $L_c = 3$, BER $= 10^{-2}$ can be obtained when the average SNR per bit $\bar{\gamma}_b \approx 10\,\text{dB}$. Compared with receiving only the strongest path, it is shown that the GSC receiver with $L_c = 3$ can achieve more than half of the diversity gain achieved by the full 2D-Rake receiver that combines signals of all resolvable multipaths. On the other hand, by comparison of the corresponding curves in Figures 5.6 and 5.3, it is seen that the GSC receiver outperforms

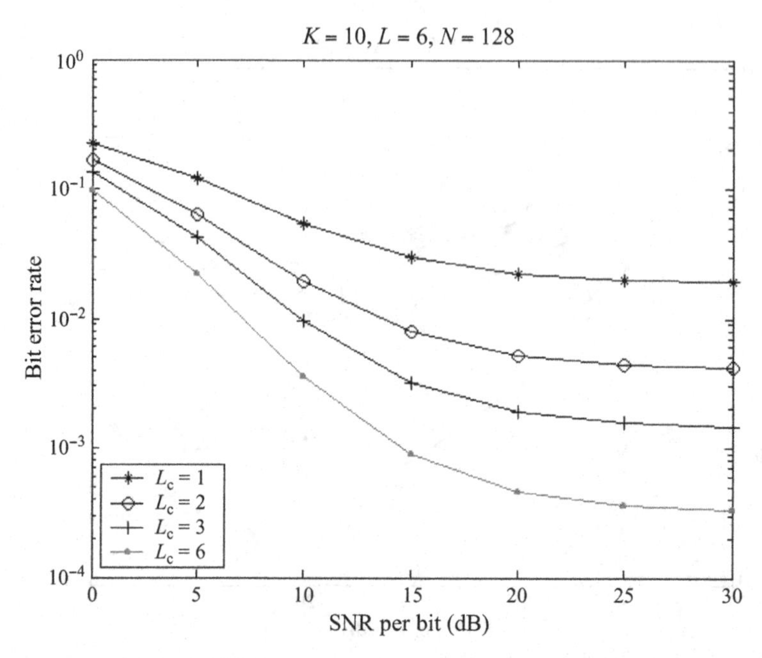

Figure 5.6 BER of GSC 2D-Rake receiver with ideal CSI available and an independent pair of channels.

the EGC receiver. This proves the feasibility of implementation of GSC receivers that select and combine limited numbers of strongest paths in practice.

When the pair of channels $h_{1,l}$ and $h_{2,l}$ are correlated, it is predicted that the BER increases due to the lack of spatial diversity promised by TD-STBC. In Figure 5.7, the BERs of both the EGC and GSC 2D-Rake receivers are illustrated versus the correlation coefficient ρ between the pair of channels $h_{1,l}$ and $h_{2,l}$. The concerned number of Rake fingers is $L_c = 1, 2, 3, 6$, respectively, and the average SNR per bit is $\bar{\gamma}_b = 30$ dB. The dotted-line curves are for the EGC 2D-Rake receiver while the solid-line curves are for the GSC 2D-Rake receiver. It can be seen that the BER performance is just slightly degraded when the correlation of the pair of channels increases. Thus the DS-CDMA system with TD-STBC performs robustly even under the environment of severely correlated pairs of channels because the system benefits from the spatial diversity gain as well as the multipath diversity gain. It can also be seen that for the same number of fingers to combine the GSC receiver outperforms the EGC receiver with respect to the BER performance as well as the robustness property toward the correlation between the pair of channels.

Figure 5.8 demonstrates the spatial diversity gain of TD-STBC by comparing the BER of the GSC 2D-Rake receiver and the BER of the conventional GSC Rake receiver with only one transmit antenna. It is assumed that the pair of channels $h_{1,l}$ and $h_{2,l}$ are independent from each other and that the number of Rake fingers to combine is $L_c = 1, 2, 3$, respectively. Similarly to Figure 5.5, under the assumption of ideal CSI available at the receiver, it is seen that the GSC 2D-Rake receiver with TD-STBC can

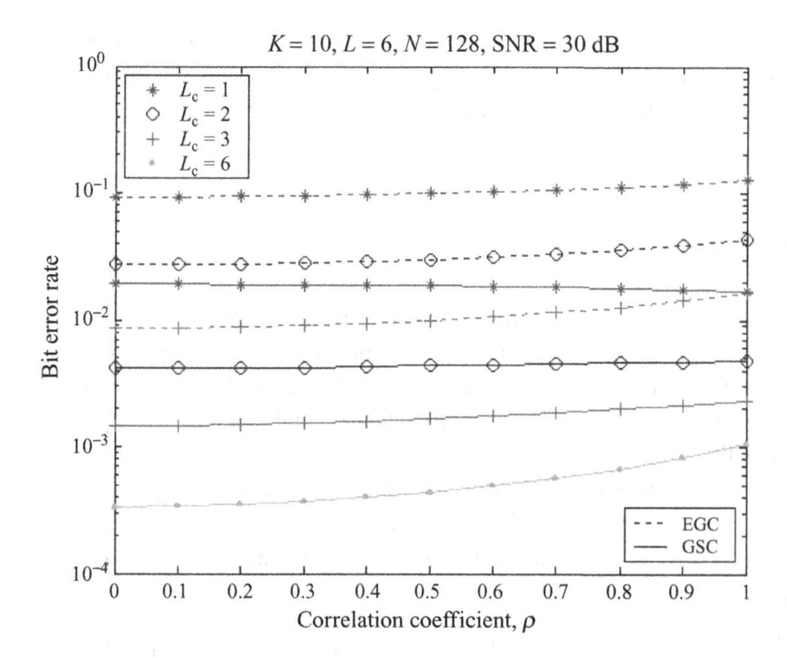

Figure 5.7 BER of EGC and GSC 2D-Rake receivers versus the correlation coefficient with ideal CSI available.

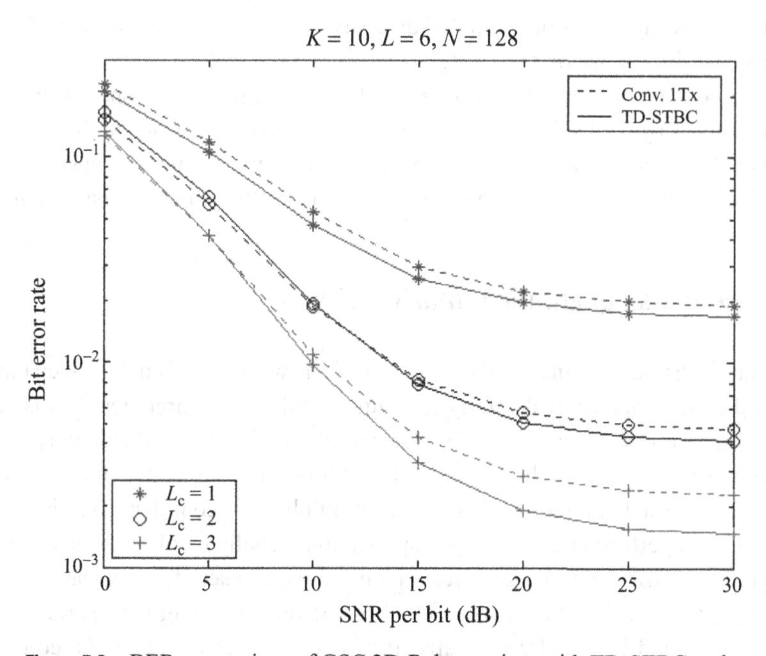

Figure 5.8 BER comparison of GSC 2D-Rake receiver with TD-STBC and conventional GSC Rake receiver with only one transmit antenna, assuming ideal CSI available.

still achieve spatial diversity gain in BER over the conventional Rake receiver without TD-STBC when the concerned average SNR per bit is relatively high. For example, as the combined path $L_c = 3$ and BER $= 2 \times 10^{-3}$, the obtained diversity gain is nearly 4 dB.

5.5 Summary

In this chapter, the downlink performance of the DS-CDMA system with and without TD-STBC is investigated in terms of BER under frequency-selective fading channels. It is assumed that the ideal CSI is available at the receiver and the pair of channels $h_{1,l}$ and $h_{2,l}$ corresponding to two transmit antennas might be independent from each other or mutually correlated by a common coefficient.

For the system with TD-STBC, the 2D-Rake receiver is exploited to collect both spatial diversity gain and multipath diversity gain. Two different combining methods are analyzed and some closed-form BER results are obtained for both EGC and GSC Rake receivers in DS-CDMA systems.

(1) Both EGC and GSC 2D-Rake receivers improve the system performance when the number of fingers increases. However, for the same fingers, the GSC receiver outperforms the EGC receiver.
(2) In comparison with the conventional system deploying only one transmit antenna, it is observed that significant spatial diversity gain on BER is achieved by the 2D-Rake receiver by the use of TD-STBC.
(3) The correlation between the pair of channels degrades the spatial diversity gain promised by TD-STBC; however, the BER performance of the 2D-Rake receiver is rather robust in a rich scattering propagation environment because the multipath diversity gain can compensate for a reduction in the spatial diversity gain.

Appendix 5A Hermitian quadratic forms in CGRV

The Rayleigh fading channel is the most typical and widely used analytical channel model when the performance of diversity techniques in digital communications is studied. It is well known that the magnitude of the zero-mean complex-valued Gaussian random variable (CGRV) follows the Rayleigh distribution. It is sometimes more convenient to represent the Rayleigh fading channel as a multiplicative complex Gaussian distributed channel. The performance analysis in applications usually involves the evaluation of the probability distribution of a generic quadratic form of a set of zero-mean CGRVs [3].

Let $z_n (n = 1, \ldots, N)$ be a set of CGRVs, the real and imaginary parts of which are independent and identically normally distributed, and denote as $\mathbf{Z}$ the column vector formed from the z_n. A generic quadratic form of an $N \times 1$ CGRV $\mathbf{Z}$ is a real-valued random variable given by

$$Q_Z = \mathbf{Z}^{\mathrm{H}} \cdot \mathbf{W} \cdot \mathbf{Z} \tag{5A.1}$$

where **W** is any arbitrary $N \times N$ Hermitian matrix and the superscript H stands for the conjugate transpose. Then it is found that the general characteristic function of Q_Z is given by

$$\Phi_{Q_Z}(t) = |\mathbf{I}_N - jt \cdot \mathbf{RW}|^{-1} \cdot \exp\{-\overline{\mathbf{Z}}^{\mathrm{H}}\, \mathbf{R}^{-1}[\mathbf{I}_N - (\mathbf{I}_N - jt \cdot \mathbf{RW})^{-1}]\,\overline{\mathbf{Z}}\} \quad (5\mathrm{A}.2)$$

where $\mathbf{I}_N$ is the $N \times N$ identity matrix, $\overline{\mathbf{Z}} = E\{\mathbf{Z}\}$ is the column vector of complex means, and $\mathbf{R} = E\{(\mathbf{Z} - \overline{\mathbf{Z}})(\mathbf{Z} - \overline{\mathbf{Z}})^{\mathrm{H}}\}$ is the $N \times N$ covariance matrix of **Z**, which is assumed to be non-singular. The operation $|\cdot|$ stands for the determinant of the matrix. According to the relationship of the Fourier transform pair between the characteristic function and the PDF, the PDF of Q_Z can be derived through the standard characteristic function inversion technique, i.e.

$$p_{Q_Z}(q_Z) = \frac{1}{2\pi}\int_{-\infty}^{\infty} \Phi_{Q_Z}(t) \cdot \exp\left(-jt{\cdot}q_Z\right) \cdot \mathrm{d}t \quad (5\mathrm{A}.3)$$

The density function of Q_Z in (5A.3) is generally difficult to solve precisely because of the complicated nature of the exponential factor in $\Phi_{Q_Z}(t)$. However, with the special case that often arises in performance analysis of digital communications, in which the variables z_n have zero means, thus $\overline{\mathbf{Z}} = \mathbf{0}$, and the characteristic function in (5A.2) reduces to

$$\Phi_{Q_Z}(t) = |\mathbf{I}_N - jt \cdot \mathbf{RW}|^{-1} = \prod_{1}^{N} \frac{1}{1 - jt\lambda_n} \quad (5\mathrm{A}.4)$$

where the λ_n are the eigenvalues of the matrix **RW**. By noting that **W** is Hermitian and **R** is nonnegative definite and Hermitian, it can be shown that **RW** has real-valued eigenvalues [4]. Hence, the singularities of $\Phi_{Q_Z}(t)$ in this case consist solely of a finite number of finite-order poles; once the eigenvalues of **RW** are obtained, the PDF $p_{Q_Z}(q_Z)$ can be therefore readily determined by taking advantage of the residue theorem [5]. In particular, the quadratic form composed of two mutually independent or correlated CGRVs can be represented as

$$Q = A \cdot |X|^2 + B \cdot |Y|^2 + CXY^* + C^*X^*Y \quad (5\mathrm{A}.5)$$

where A, B, C are constants, and X and Y are a pair of independent or correlated CGRVs with zero means. With respect to (5A.4), the two eigenvalues concerned (which are not necessarily distinct) can be generally derived as

$$\lambda_{1,2} = \frac{\omega_2 \pm \sqrt{\omega_2^2 + 4\omega_1}}{2} \quad (5\mathrm{A}.6)$$

where, by definitions,

$$\omega_1 = (|C|^2 - AB) \cdot (\xi_{xx}\xi_{yy} - |\xi_{xy}|^2) \quad (5\mathrm{A}.7)$$

$$\omega_2 = A\xi_{xx} + B\xi_{yy} + C\xi_{xy} + C^*\xi_{xy}^* \quad (5\mathrm{A}.8)$$

where ξ_{xx} and ξ_{yy} are the variances of CGRV X and Y, respectively, and ξ_{xy} is the covariance of X and Y defined by

$$\xi_{xy} = \xi_{yx}^* = E\{(X - \overline{X})(Y - \overline{Y})^*\} \tag{5A.9}$$

Further considering two correlated CGRVs, the correlation coefficient of X and Y can be defined as

$$\rho_{x,y} = \frac{\text{COV}(x, y)}{\sigma_x \cdot \sigma_y} = \frac{\xi_{xy}}{\sqrt{\xi_{xx}} \cdot \sqrt{\xi_{yy}}} \tag{5A.10}$$

When the eigenvalues in (5A.6) are two distinct values, the PDF can be achieved through inverse-Fourier transformation (5A.3), or alternatively using the residue theorem [5], as follows:

$$p_Q(q) = \frac{1}{\lambda_1 - \lambda_2} \cdot \exp\left(-\frac{q}{\lambda_1}\right) - \frac{1}{\lambda_1 - \lambda_2} \cdot \exp\left(-\frac{q}{\lambda_2}\right) \tag{5A.11}$$

Alternatively, when two eigenvalues in (5A.6) are single twice repeated values $\lambda = \lambda_1 = \lambda_2$, the PDF can be derived as

$$p_Q(q) = \frac{q}{\lambda^2} \cdot \exp\left(-\frac{q}{\lambda}\right) \tag{5A.12}$$

Appendix 5B Involved integral derivations

5B.1 Some basic integral formulas

Based on formula (7.4.19) of [9] and (2-1-95) of [3], it can be obtained that

$$\int_0^\infty \mathrm{e}^{-at} \cdot \operatorname{erfc}(\sqrt{bt}) \cdot \mathrm{d}t = \frac{1}{a} \cdot \left(1 - \sqrt{\frac{b}{a+b}}\right) \tag{5B.1}$$

where $\operatorname{Re}(a + b) > 0$. By setting $b = 1$ and taking the derivative of both sides of (5B.1) with respect to a continuously n times, the following can be obtained:

$$\int_0^\infty x^n \cdot \mathrm{e}^{-ax} \cdot \operatorname{erfc}(\sqrt{x}) \cdot \mathrm{dx} = \frac{n!}{a^{n+1}} \cdot \left\{1 - \sum_{q=0}^{n} \frac{a^q}{q! \cdot \sqrt{\pi}} \cdot \frac{\Gamma\left(q + \frac{1}{2}\right)}{(1+a)^{q+\frac{1}{2}}}\right\} \tag{5B.2}$$

where $n \geq 0,\ a > 0$. Another approach to prove (5B.2) is to utilize the method of integration by parts and mathematical induction.

An alternative integral result can be deduced from [3], as follows:

$$\int_0^\infty x^n \cdot \mathrm{e}^{-ax} \cdot \operatorname{erfc}(\sqrt{x}) \cdot \mathrm{dx}$$
$$= \frac{2 \cdot n!}{a^{n+1}} \cdot \left[\frac{1}{2}\left(1 - \frac{1}{\sqrt{1+a}}\right)\right]^{n+1} \cdot \sum_{k=0}^{n} \binom{n+k}{k} \cdot \left(\frac{1}{2}\right)^{n+k} \cdot \left[\frac{1}{2}\left(1 + \frac{1}{\sqrt{1+a}}\right)\right]^k \tag{5B.3}$$

From the binomial theorem [3], [9],

$$(a+x)^n = \sum_{k=0}^{n} \binom{n}{k} \cdot x^k \cdot a^{n-k} \tag{5B.4}$$

$$\begin{aligned}(a+b+c)^L &= \sum_{m=0}^{L}\sum_{n=0}^{m} \binom{L}{m} \cdot \binom{m}{n} \cdot a^n \cdot b^{m-n} \cdot c^{L-m} \\ &= \sum_{m=0}^{L}\sum_{n=0}^{m} \frac{L!}{(L-m)! \cdot (m-n)! \cdot n!} \cdot a^n \cdot b^{m-n} \cdot c^{L-m}\end{aligned} \tag{5B.5}$$

From the definition of the complementary error function (erfc),

$$\operatorname{erfc}(x) = \frac{2}{\sqrt{\pi}} \int_x^{\infty} e^{-t^2} \cdot \mathrm{d}t \tag{5B.6}$$

its derivative is given by

$$[\operatorname{erfc}(\sqrt{x})]' = -\frac{1}{\sqrt{\pi}} \cdot \frac{\mathrm{e}^{-x}}{\sqrt{x}} \cdot \mathrm{d}x \tag{5B.7}$$

When $\alpha + \beta > 0$, and for any n, it is readily proven that

$$\lim_{x\to\infty} \mathrm{e}^{-\alpha x} \cdot \operatorname{erfc}(\sqrt{\beta x}) = 0 \tag{5B.8}$$

$$\lim_{x\to\infty} x^n \cdot \operatorname{erfc}(\sqrt{x}) = 0 \tag{5B.9}$$

Using the method of integration by parts, it can be deduced that

$$\begin{aligned}\int_x^{\infty} \operatorname{erfc}(\sqrt{t}) \cdot \mathrm{d}t &= t \cdot \operatorname{erfc}(\sqrt{t})\Big|_x^{\infty} - \int_x^{\infty} t \cdot \frac{-1}{\sqrt{\pi}} \cdot \frac{\mathrm{e}^{-t}}{\sqrt{t}} \cdot \mathrm{d}t \\ &= -x \cdot \operatorname{erfc}(\sqrt{x}) + \frac{1}{\sqrt{\pi}} \cdot \int_x^{\infty} \sqrt{t} \cdot \mathrm{e}^{-t} \cdot \mathrm{d}t\end{aligned} \tag{5B.10}$$

Furthermore, by variant substitution, $\sqrt{t} = y$ and $\mathrm{d}t = 2y\mathrm{d}y$, the integral in (5B.10) can be deduced:

$$\begin{aligned}\frac{1}{\sqrt{\pi}} \cdot \int_x^{\infty} \sqrt{t} \cdot \mathrm{e}^{-t} \cdot \mathrm{d}t &= \frac{2}{\sqrt{\pi}} \cdot \int_{\sqrt{x}}^{\infty} y^2 \cdot \mathrm{e}^{-y^2} \cdot \mathrm{d}y \\ &= \frac{1}{\sqrt{\pi}} \cdot \sqrt{x} \cdot \mathrm{e}^{-x} + \frac{1}{2} \cdot \operatorname{erfc}(\sqrt{x})\end{aligned} \tag{5B.11}$$

Therefore, it is obtained that

$$\int_x^{\infty} \operatorname{erfc}(\sqrt{t}) \cdot \mathrm{d}t = \frac{1}{\sqrt{\pi}} \cdot \sqrt{x} \cdot \mathrm{e}^{-x} + \left(\frac{1}{2} - x\right) \cdot \operatorname{erfc}(\sqrt{x}) \tag{5B.12}$$

Similarly, through the method of integration by parts and variant substitution, the followed integral formulas can be derived:

$$\begin{aligned}&\int_x^{\infty} e^{-\alpha t} \cdot \operatorname{erfc}(\sqrt{t}) \cdot \mathrm{d}t \\ &= \frac{1}{\alpha} \cdot \mathrm{e}^{-\alpha x} \cdot \operatorname{erfc}(\sqrt{x}) - \frac{1}{\alpha\sqrt{1+\alpha}} \cdot \operatorname{erfc}[\sqrt{(1+\alpha) \cdot x}] \qquad \alpha > -1\end{aligned} \tag{5B.13}$$

$$\int_x^\infty \sqrt{t}\cdot \mathrm{e}^{-\alpha t}\cdot \mathrm{d}t = \frac{1}{\alpha}\sqrt{x}\cdot \mathrm{e}^{-\alpha x} + \frac{\sqrt{\pi}}{2\cdot \alpha^{3/2}}\cdot \mathrm{erfc}(\sqrt{\alpha x}) \qquad \alpha > 0 \tag{5B.14}$$

$$\int_x^\infty t\cdot \mathrm{erfc}(\sqrt{t})\cdot \mathrm{d}t$$
$$= -\frac{1}{2}x^2\cdot \mathrm{erfc}(\sqrt{x}) + \frac{1}{2\sqrt{\pi}}\cdot x^{\frac{3}{2}}\cdot \mathrm{e}^{-x} + \frac{3}{4\sqrt{\pi}}\cdot \sqrt{x}\cdot \mathrm{e}^{-x} + \frac{3}{8}\cdot \mathrm{erfc}(\sqrt{x}) \tag{5B.15}$$

$$\int_x^\infty t^2\cdot \mathrm{erfc}(\sqrt{t})\cdot \mathrm{d}t$$
$$= -\frac{1}{3}x^3\cdot \mathrm{erfc}(\sqrt{x}) + \frac{1}{3\sqrt{\pi}}\cdot x^{\frac{5}{2}}\cdot \mathrm{e}^{-x} + \frac{5}{6\sqrt{\pi}}\cdot x^{\frac{3}{2}}\cdot \mathrm{e}^{-x}$$
$$+\frac{5}{4\sqrt{\pi}}\cdot \sqrt{x}\cdot \mathrm{e}^{-x} + \frac{5}{8}\cdot \mathrm{erfc}(\sqrt{x}) \tag{5B.16}$$

$$\int_x^\infty t\cdot \mathrm{e}^{-\alpha t}\cdot \mathrm{erfc}(\sqrt{t})\cdot \mathrm{d}t$$
$$= \frac{1}{\alpha}\cdot x\cdot \mathrm{e}^{-\alpha x}\cdot \mathrm{erfc}(\sqrt{x}) + \frac{1}{\alpha^2}\cdot \mathrm{e}^{-\alpha x}\cdot \mathrm{erfc}(\sqrt{x})$$
$$-\frac{1}{\alpha\cdot(1+\alpha)\cdot\sqrt{\pi}}\cdot \sqrt{x}\cdot \mathrm{e}^{-(1+\alpha)x} - \frac{2+3\alpha}{2\alpha^2\cdot(1+\alpha)^{\frac{3}{2}}}\cdot \mathrm{erfc}[\sqrt{(1+\alpha)\cdot x}] \tag{5B.17}$$

$$\int_x^\infty t^2\cdot \mathrm{e}^{-\alpha t}\cdot \mathrm{erfc}(\sqrt{t})\cdot \mathrm{d}t$$
$$= \frac{1}{\alpha}\cdot x^2\cdot \mathrm{e}^{-\alpha x}\cdot \mathrm{erfc}(\sqrt{x}) + \frac{2}{\alpha^2}\cdot x\cdot \mathrm{e}^{-\alpha x}\cdot \mathrm{erfc}(\sqrt{x}) + \frac{2}{\alpha^3}\cdot \mathrm{e}^{-\alpha x}\cdot \mathrm{erfc}(\sqrt{x})$$
$$-\frac{15\alpha^2+20\alpha+8}{4\alpha^3\cdot(1+\alpha)^{\frac{5}{2}}}\cdot \mathrm{erfc}[\sqrt{(1+\alpha)\cdot x}] - \frac{4+7\alpha}{2\alpha^2\cdot(1+\alpha)^2\sqrt{\pi}}$$
$$\cdot\sqrt{x}\cdot \mathrm{e}^{-(1+\alpha)x} - \frac{1}{\alpha\cdot(1+\alpha)\cdot\sqrt{\pi}}\cdot x^{\frac{3}{2}}\cdot \mathrm{e}^{-(1+\alpha)x} \tag{5B.18}$$

5B.2 Independent pair of channels

In Sections 5B.2 and 5B.3, we set $\alpha = 1/\overline{\gamma}_\mathrm{c}$ to simplify the notation. For the integral in (5.55), by substituting (5.53) and (5.54) into (5.55) and applying the binomial theorem to the power series, it is obtained that

$$P_{\mathrm{e,ind}}^{<1>} = \binom{L}{1}\cdot\int_0^\infty [1-(1+\alpha y_1)\cdot\exp(-\alpha y_1)]^{L-1}$$
$$\cdot\alpha^2\cdot y_1\cdot\exp(-\alpha y_1)\cdot Q(\sqrt{2y_1})\cdot \mathrm{d}y_1$$
$$= \binom{L}{1}\cdot\frac{1}{2}\cdot\sum_{m=0}^{L-1}\sum_{n=0}^{m}\binom{L-1}{m}\cdot\binom{m}{n}\cdot(-1)^m\cdot\alpha^{n+2}$$
$$\cdot\int_0^\infty y_1^{n+1}\cdot\exp[-\alpha(m+1)y_1]\cdot\mathrm{erfc}(\sqrt{y_1})\cdot\mathrm{d}y_1 \tag{5B.19}$$

By applying (5B.2), the above formula (5B.19) can be deduced as

$$
\begin{aligned}
P_{\text{e,ind}}^{<1>} &= \frac{L}{2}\cdot\sum_{m=0}^{L-1}\sum_{n=0}^{m}\binom{L-1}{m}\cdot\binom{m}{n}\cdot(-1)^m\cdot\frac{(n+1)!}{(m+1)^{n+2}} \\
&\quad\cdot\left\{1-\sum_{q=0}^{n+1}\frac{[\alpha\,(m+1)]^q}{q!\cdot\sqrt{\pi}}\cdot\frac{\Gamma\,(q+1/2)}{(\alpha m+\alpha+1)^{q+1/2}}\right\} \\
&= \frac{L}{2}\cdot\sum_{m=0}^{L-1}\sum_{n=0}^{m}\binom{L-1}{m}\cdot\binom{m}{n}\cdot(-1)^m\cdot\frac{1}{(m+1)^{n+2}} \\
&\quad\cdot \mathrm{I_A}\,[n+1,\alpha\,(m+1)]
\end{aligned}
\qquad (5B.20)
$$

where $I_A(\cdot,\cdot)$ is defined in (5.44). This is the integral result presented in (5.55).

For the integral in (5.56), by substituting (5.53) and (5.54) into (5.56), it is obtained that

$$
\begin{aligned}
P_{\text{e,ind}}^{<2>} &= 2!\cdot\binom{L}{2}\cdot\int_0^{\infty}\int_{y_2}^{\infty}\sum_{m=0}^{L-2}\sum_{n=0}^{m}\binom{L-2}{m}\cdot\binom{m}{n} \\
&\quad\cdot(-1)^m\cdot\alpha^{n+4}\cdot y_1\cdot y_2^{n+1}\cdot\exp(-\alpha m y_2) \\
&\quad\cdot\exp[-\alpha(y_1+y_2)]\cdot\frac{1}{2}\cdot\mathrm{erfc}(\sqrt{y_1+y_2})\cdot dy_1\cdot dy_2 \\
&= 2\cdot\binom{L}{2}\cdot\sum_{m=0}^{L-2}\sum_{n=0}^{m}\binom{L-2}{m}\cdot\binom{m}{n}\cdot(-1)^m\cdot\alpha^{n+4}\cdot\frac{1}{2}\cdot\int_0^{\infty} y_2^{n+1} \\
&\quad\cdot\exp(-\alpha m y_2)\cdot\int_{y_2}^{\infty} y_1\cdot\exp[-\alpha(y_1+y_2)] \\
&\quad\cdot\mathrm{erfc}(\sqrt{y_1+y_2})\cdot \mathrm{d}y_1\cdot \mathrm{d}y_2
\end{aligned}
\qquad (5B.21)
$$

Assuming $t = y_1 + y_2$, and substituting $y_1 = t - y_2$ and $\mathrm{d}y_1 = \mathrm{d}t$ into the inner integral of (5B.21), it can be written as

$$
\begin{aligned}
&\int_{y_2}^{\infty} y_1\cdot\exp[-\alpha(y_1+y_2)]\cdot\mathrm{erfc}(\sqrt{y_1+y_2})\cdot\mathrm{d}y_1 \\
&= \int_{2y_2}^{\infty} t\cdot\mathrm{e}^{-\alpha t}\cdot\mathrm{erfc}(\sqrt{t})\cdot\mathrm{d}t - y_2\cdot\int_{2y_2}^{\infty}\mathrm{e}^{-\alpha t}\cdot\mathrm{erfc}(\sqrt{t})\cdot\mathrm{d}t
\end{aligned}
\qquad (5B.22)
$$

Applying the integration formula in (5B.17) and (5B.13), the integral in (5B.22) can be further resolved as

$$
\begin{aligned}
&\int_{y_2}^{\infty} y_1\cdot\exp[-\alpha(y_1+y_2)]\cdot\mathrm{erfc}(\sqrt{y_1+y_2})\cdot\mathrm{d}y_1 \\
&= \frac{1}{\alpha}\cdot y_2\cdot\mathrm{e}^{-2\alpha y_2}\cdot\mathrm{erfc}(\sqrt{2y_2}) + \frac{1}{\alpha^2}\cdot\mathrm{e}^{-2\alpha y_2}\cdot\mathrm{erfc}(\sqrt{2y_2}) - \frac{1}{\alpha^2\sqrt{1+\alpha}} \\
&\quad\cdot\mathrm{erfc}[\sqrt{2(1+\alpha)\cdot y_2}] + \frac{1}{\alpha\sqrt{1+\alpha}}\cdot y_2\cdot\mathrm{erfc}[\sqrt{2(1+\alpha)\cdot y_2}] \\
&\quad - \frac{1}{\alpha(1+\alpha)}\cdot\sqrt{\frac{2}{\pi}}\cdot\sqrt{y_2}\cdot\exp[-2\cdot(1+\alpha)\cdot y_2] - \frac{1}{2\alpha\cdot(1+\alpha)^{3/2}} \\
&\quad\cdot\mathrm{erfc}[\sqrt{2(1+\alpha)\cdot y_2}]
\end{aligned}
\qquad (5B.23)
$$

Substituting (5B.23) into the outer integral in (5B.21), it can be written as

$$\begin{aligned}
&\int_0^\infty y_2^{n+1}\cdot\exp(-\alpha m y_2)\cdot\int_{y_2}^\infty y_1\cdot\exp[-\alpha(y_1+y_2)]\cdot\mathrm{erfc}(\sqrt{y_1+y_2})\cdot dy_1\cdot dy_2\\
&=\frac{1}{\alpha}\cdot\int_0^\infty y_2^{n+2}\cdot\exp[-\alpha(m+2)y_2]\cdot\mathrm{erfc}(\sqrt{2y_2})\cdot dy_2\\
&+\frac{1}{\alpha^2}\cdot\int_0^\infty y_2^{n+1}\cdot\exp[-\alpha(m+2)y_2]\cdot\mathrm{erfc}(\sqrt{2y_2})\cdot dy_2\\
&-\frac{1}{\alpha^2\sqrt{1+\alpha}}\cdot\int_0^\infty y_2^{n+1}\cdot\exp(-\alpha m y_2)\cdot\mathrm{erfc}[\sqrt{2(1+\alpha)\cdot y_2}]\cdot dy_2\\
&+\frac{1}{\alpha\sqrt{1+\alpha}}\cdot\int_0^\infty y_2^{n+2}\cdot\exp(-\alpha m y_2)\cdot\mathrm{erfc}[\sqrt{2(1+\alpha)\cdot y_2}]\cdot dy_2\\
&-\frac{1}{\alpha(1+\alpha)}\cdot\sqrt{\frac{2}{\pi}}\cdot\int_0^\infty y_2^{n+\frac{3}{2}}\cdot\exp\{-[\alpha(m+2)+2]\cdot y_2\}\cdot dy_2\\
&-\frac{1}{2\alpha\cdot(1+\alpha)^{3/2}}\cdot\int_0^\infty y_2^{n+1}\cdot\exp(-\alpha m y_2)\cdot\mathrm{erfc}[\sqrt{2(1+\alpha)\cdot y_2}]\cdot dy_2
\end{aligned}\tag{5B.24}$$

For any m, the first and second integrals in (5B.24) can be obtained according to the generic formula (5B.2) as follows, respectively:

$$\begin{aligned}
&\frac{1}{\alpha}\cdot\int_0^\infty y^{n+2}\cdot\exp[-\alpha(m+2)y]\cdot\mathrm{erfc}\,(\sqrt{2y})\cdot dy\\
&=\frac{(n+2)!}{\alpha^{n+4}\cdot(m+2)^{n+3}}\cdot\left\{1-\sum_{q=0}^{n+2}\sqrt{\frac{2}{\pi}}\cdot\frac{[\alpha(m+2)]^q}{q!}\cdot\frac{\Gamma(q+1/2)}{[\alpha(m+2)+2]^{q+1/2}}\right\}
\end{aligned}\tag{5B.25}$$

$$\begin{aligned}
&\frac{1}{\alpha^2}\cdot\int_0^\infty y^{n+1}\cdot\exp[-\alpha(m+2)y]\cdot\mathrm{erfc}\,(\sqrt{2y})\cdot dy\\
&=\frac{(n+1)!}{\alpha^{n+4}\cdot(m+2)^{n+2}}\cdot\left\{1-\sum_{q=0}^{n+1}\sqrt{\frac{2}{\pi}}\cdot\frac{[\alpha(m+2)]^q}{q!}\cdot\frac{\Gamma(q+1/2)}{[\alpha(m+2)+2]^{q+1/2}}\right\}
\end{aligned}\tag{5B.26}$$

Based on the formula (3.381–4) of [9], the fifth integral in (5B.24) can be deduced as

$$\begin{aligned}
&-\frac{1}{\alpha(1+\alpha)}\cdot\sqrt{\frac{2}{\pi}}\cdot\int_0^\infty y^{n+\frac{3}{2}}\cdot\exp\{-[\alpha(m+2)+2]\cdot y\}\cdot dy\\
&=-\frac{1}{\alpha(1+\alpha)}\cdot\sqrt{\frac{2}{\pi}}\cdot\frac{\Gamma(n+5/2)}{[\alpha(m+2)+2]^{n+5/2}}
\end{aligned}\tag{5B.27}$$

When $m > 0$, the third, fourth and sixth integrals in (5B.24) can be derived with the help of the generic formula (5B.2) as follows, respectively:

$$-\frac{1}{\alpha^2\sqrt{1+\alpha}}\cdot\int_0^\infty y^{n+1}\cdot\exp(-\alpha m y)\cdot\text{erfc}[\sqrt{2(1+\alpha)\cdot y}]\cdot \mathrm{d}y$$
$$= -\frac{(n+1)!}{m^{n+2}\cdot\alpha^{n+4}\cdot\sqrt{1+\alpha}}\cdot\left\{1-\sum_{q=0}^{n+1}\sqrt{\frac{2(1+\alpha)}{\pi}}\cdot\frac{\alpha^q\cdot m^q}{q!}\cdot\frac{\Gamma(q+1/2)}{[\alpha(m+2)+2]^{q+1/2}}\right\} \tag{5B.28}$$

$$\frac{1}{\alpha\sqrt{1+\alpha}}\cdot\int_0^\infty y^{n+2}\cdot\exp(-\alpha m y)\cdot\text{erfc}[\sqrt{2(1+\alpha)\cdot y}]\cdot \mathrm{d}y$$
$$= \frac{(n+2)!}{m^{n+3}\cdot\alpha^{n+4}\cdot\sqrt{1+\alpha}}\cdot\left\{1-\sum_{q=0}^{n+2}\sqrt{\frac{2(1+\alpha)}{\pi}}\cdot\frac{\alpha^q\cdot m^q}{q!}\cdot\frac{\Gamma(q+1/2)}{[\alpha(m+2)+2]^{q+1/2}}\right\} \tag{5B.29}$$

$$-\frac{1}{2\alpha\cdot(1+\alpha)^{3/2}}\cdot\int_0^\infty y^{n+1}\cdot\exp(-\alpha m y)\cdot\text{erfc}[\sqrt{2(1+\alpha)\cdot y}]\cdot \mathrm{d}y$$
$$= -\frac{(n+2)!}{2m^{n+2}\cdot\alpha^{n+3}\cdot(1+\alpha)^{n+\frac{3}{2}}}\cdot\left\{1-\sum_{q=0}^{n+1}\sqrt{\frac{2(1+\alpha)}{\pi}}\cdot\frac{\alpha^q\cdot m^q}{q!}\cdot\frac{\Gamma(q+1/2)}{[\alpha(m+2)+2]^{q+1/2}}\right\} \tag{5B.30}$$

When $m = 0$, it is noted that $n = 0$ from (5B.21), thus the integral of (5B.24) can be written as

$$\int_0^\infty y_2\cdot\int_{y_2}^\infty y_1\cdot\exp[-\alpha(y_1+y_2)]\cdot\text{erfc}(\sqrt{y_1+y_2})\cdot \mathrm{d}y_1\cdot \mathrm{d}y_2$$
$$= \frac{1}{\alpha}\cdot\int_0^\infty y^2\cdot \mathrm{e}^{-2\alpha y}\cdot\text{erfc}(\sqrt{2y})\cdot \mathrm{d}y + \frac{1}{\alpha^2}\cdot\int_0^\infty y\cdot \mathrm{e}^{-2\alpha y}\cdot\text{erfc}(\sqrt{2y})\cdot \mathrm{d}y$$
$$-\frac{1}{\alpha^2\sqrt{1+\alpha}}\cdot\int_0^\infty y\cdot\text{erfc}[\sqrt{2(1+\alpha)\cdot y}]\cdot \mathrm{d}y$$
$$+\frac{1}{\alpha\sqrt{1+\alpha}}\cdot\int_0^\infty y^2\cdot\text{erfc}[\sqrt{2(1+\alpha)\cdot y}]\cdot \mathrm{d}y$$
$$-\frac{1}{\alpha(1+\alpha)}\cdot\sqrt{\frac{2}{\pi}}\cdot\int_0^\infty y^{\frac{3}{2}}\cdot \mathrm{e}^{-2(1+\alpha)y}\cdot \mathrm{d}y - \frac{1}{2\alpha\cdot(1+\alpha)^{3/2}}\cdot\int_0^\infty y\cdot\text{erfc}[\sqrt{2(1+\alpha)\cdot y}]\cdot \mathrm{d}y \tag{5B.31}$$

Based on the generic formula (5B.2), the first and second integrals in (5B.31) can be obtained:

$$\frac{1}{\alpha} \cdot \int_0^\infty y^2 \cdot e^{-2\alpha y} \cdot \mathrm{erfc}(\sqrt{2y}) \cdot dy = \frac{1}{4\alpha^4} \cdot \left[1 - \frac{15\alpha^2 + 20\alpha + 8}{8(1+\alpha)^2 \cdot \sqrt{1+\alpha}}\right] \tag{5B.32}$$

$$\frac{1}{\alpha^2} \cdot \int_0^\infty y \cdot e^{-2\alpha y} \cdot \mathrm{erfc}(\sqrt{2y}) \cdot dy = \frac{1}{4\alpha^4} \cdot \left[1 - \frac{2 + 3\alpha}{2(1+\alpha) \cdot \sqrt{1+\alpha}}\right] \tag{5B.33}$$

According to the formula (6.281–1) of [9], the third, fourth and sixth integrals in (5B.31) can be obtained as follows, respectively:

$$-\frac{1}{\alpha^2\sqrt{1+\alpha}} \cdot \int_0^\infty y \cdot \mathrm{erfc}[\sqrt{2(1+\alpha) \cdot y}] \cdot dy = -\frac{3}{32\alpha^2 \cdot (1+\alpha)^2 \cdot \sqrt{1+\alpha}} \tag{5B.34}$$

$$\frac{1}{\alpha\sqrt{1+\alpha}} \cdot \int_0^\infty y^2 \cdot \mathrm{erfc}[\sqrt{2(1+\alpha) \cdot y}] \cdot dy = \frac{5}{64\alpha \cdot (1+\alpha)^3 \cdot \sqrt{1+\alpha}} \tag{5B.35}$$

$$-\frac{1}{2\alpha \cdot (1+\alpha)^{3/2}} \cdot \int_0^\infty y \cdot \mathrm{erfc}[\sqrt{2(1+\alpha) \cdot y}] \cdot dy = -\frac{3}{64\alpha \cdot (1+\alpha)^3 \cdot \sqrt{1+\alpha}} \tag{5B.36}$$

Based on the formula (3.381–4) of [9], the fifth integral in (5B.31) can be deduced as

$$-\frac{1}{\alpha(1+\alpha)} \cdot \sqrt{\frac{2}{\pi}} \cdot \int_0^\infty y^{\frac{3}{2}} \cdot e^{-2(1+\alpha)y} \cdot dy = -\frac{3}{16\alpha \cdot (1+\alpha)^3 \cdot \sqrt{1+\alpha}} \tag{5B.37}$$

Therefore, when $m = n = 0$, substituting equations (5B.32)–(5B.37) into (5B.31), the double integral is achieved as

$$\int_0^\infty y_2 \cdot \int_{y_2}^\infty y_1 \cdot \exp[-\alpha(y_1 + y_2)] \cdot \mathrm{erfc}(\sqrt{y_1 + y_2}) \cdot dy_1 \cdot dy_2$$
$$= \frac{1}{\alpha^4} \cdot \left\{\frac{1}{2} - \frac{35\alpha^3 + 70\alpha^2 + 56\alpha + 16}{32 \cdot (1+\alpha)^{\frac{7}{2}}}\right\} \tag{5B.38}$$

Combining the results from (5B.25)–(5B.30) and (5B.38), the integral in (5.56) is achieved. For the integral in (5.60), by substituting (5.53) and (5.54) into (5.60), it is obtained that

$$P_{\mathrm{e,ind}}^{<3>} = 3! \cdot \binom{L}{3} \cdot \int_0^\infty \int_{y_3}^\infty \int_{y_2}^\infty \sum_{m=0}^{L-3} \sum_{n=0}^{m} \binom{L-3}{m} \cdot \binom{m}{n}$$
$$\cdot (-1)^m \cdot \alpha^{n+6} \cdot y_1 \cdot y_2 \cdot y_3^{n+1} \cdot \exp(-\alpha m y_3)$$
$$\cdot \exp[-\alpha(y_1 + y_2 + y_3)] \cdot \frac{1}{2} \cdot \mathrm{erfc}(\sqrt{y_1 + y_2 + y_3}) \cdot dy_1 \cdot dy_2 \cdot dy_3$$
$$= 3! \cdot \binom{L}{3} \cdot \sum_{m=0}^{L-3} \sum_{n=0}^{m} \binom{L-3}{m} \cdot \binom{m}{n} \cdot (-1)^m \cdot \frac{1}{2} \cdot \alpha^{n+6} \tag{5B.39}$$
$$\cdot \int_0^\infty y_3^{n+1} \cdot \exp(-\alpha m y_3) \cdot \int_{y_3}^\infty y_2 \cdot \int_{y_2}^\infty y_1$$
$$\cdot \exp[-\alpha(y_1 + y_2 + y_3)] \cdot \mathrm{erfc}(\sqrt{y_1 + y_2 + y_3}) \cdot dy_1 \cdot dy_2 \cdot dy_3$$

Assuming $t = y_1 + y_2 + y_3$, and substituting $y_1 = t - (y_2 + y_3)$ and $\mathrm{d}y_1 = \mathrm{d}t$ into the innermost integral of (5B.39), it can be written as

$$\begin{aligned}
&\int_{y_2}^{\infty} y_1 \cdot \exp[-\alpha(y_1 + y_2 + y_3)] \cdot \mathrm{erfc}(\sqrt{y_1 + y_2 + y_3}) \cdot \mathrm{d}y_1 \\
&\quad = \int_{2y_2+y_3}^{\infty} t \cdot \mathrm{e}^{-\alpha t} \cdot \mathrm{erfc}(\sqrt{t}) \cdot \mathrm{d}t - (y_2 + y_3) \cdot \int_{2y_2+y_3}^{\infty} \mathrm{e}^{-\alpha t} \cdot \mathrm{erfc}(\sqrt{t}) \cdot \mathrm{d}t
\end{aligned} \tag{5B.40}$$

Applying the integral formulas in (5B.17) and (5B.13), the integral in (5B.40) can be further resolved as

$$\begin{aligned}
&\int_{y_2}^{\infty} y_1 \cdot \exp[-\alpha(y_1 + y_2 + y_3)] \cdot \mathrm{erfc}(\sqrt{y_1 + y_2 + y_3}) \cdot \mathrm{d}y_1 \\
&\quad = \frac{1}{\alpha} \cdot y_2 \cdot \exp[-\alpha \cdot (2y_2 + y_3)] \cdot \mathrm{erfc}(\sqrt{2y_2 + y_3}) \\
&\qquad + \frac{1}{\alpha\sqrt{1+\alpha}} \cdot y_2 \cdot \mathrm{erfc}[\sqrt{(1+\alpha) \cdot (2y_2 + y_3)}] \\
&\qquad + \frac{[2\alpha(1+\alpha)]y_3 - (2 + 3\alpha)}{2\alpha^2 \cdot (1+\alpha)^{\frac{3}{2}}} \cdot \mathrm{erfc}[\sqrt{(1+\alpha) \cdot (2y_2 + y_3)}] \\
&\qquad + \frac{1}{\alpha^2} \cdot \exp[-\alpha \cdot (2y_2 + y_3)] \cdot \mathrm{erfc}(\sqrt{2y_2 + y_3}) \\
&\qquad - \frac{\sqrt{2y_2 + y_3}}{\alpha\sqrt{\pi} \cdot (1+\alpha)} \cdot \exp[-(1+\alpha) \cdot (2y_2 + y_3)]
\end{aligned} \tag{5B.41}$$

Therefore, the second inner integral of (5B.39) can be written as

$$\begin{aligned}
&\int_{y_3}^{\infty} y_2 \cdot \int_{y_2}^{\infty} y_1 \cdot \exp[-\alpha(y_1 + y_2 + y_3)] \cdot \mathrm{erfc}(\sqrt{y_1 + y_2 + y_3}) \cdot \mathrm{d}y_1 \cdot \mathrm{d}y_2 \\
&\quad = \frac{1}{\alpha} \cdot \int_{y_3}^{\infty} y^2 \cdot \exp[-\alpha \cdot (2y + y_3)] \cdot \mathrm{erfc}(\sqrt{2y + y_3}) \cdot \mathrm{d}y \\
&\qquad + \frac{1}{\alpha\sqrt{1+\alpha}} \cdot \int_{y_3}^{\infty} y^2 \cdot \mathrm{erfc}[\sqrt{(1+\alpha) \cdot (2y + y_3)}] \cdot \mathrm{d}y \\
&\qquad + \frac{[2\alpha(1+\alpha)]y_3 - (2 + 3\alpha)}{2\alpha^2 \cdot (1+\alpha)^{\frac{3}{2}}} \cdot \int_{y_3}^{\infty} y \cdot \mathrm{erfc}[\sqrt{(1+\alpha) \cdot (2y + y_3)}] \cdot \mathrm{d}y \\
&\qquad + \frac{1}{\alpha^2} \cdot \int_{y_3}^{\infty} y \cdot \exp[-\alpha \cdot (2y + y_3)] \cdot \mathrm{erfc}\,(\sqrt{2y + y_3}) \cdot \mathrm{d}y \\
&\qquad - \frac{1}{\alpha\sqrt{\pi} \cdot (1+\alpha)} \cdot \int_{y_3}^{\infty} y \cdot \sqrt{2y + y_3} \cdot \exp[-(1+\alpha) \cdot (2y + y_3)] \cdot \mathrm{d}y
\end{aligned} \tag{5B.42}$$

Through substitution of individual proper variants, the integrals in (5B.42) can be deduced as follows, respectively:

$$\int_{y_3}^{\infty} y^2 \cdot \exp[-\alpha \cdot (2y + y_3)] \cdot \mathrm{erfc}(\sqrt{2y + y_3}) \cdot \mathrm{d}y$$
$$= \frac{1}{8} \cdot \int_{3y_3}^{\infty} t^2 \cdot \mathrm{e}^{-\alpha t} \cdot \mathrm{erfc}(\sqrt{t}) \cdot \mathrm{d}t - \frac{1}{4} \cdot y_3 \cdot \int_{3y_3}^{\infty} t \cdot \mathrm{e}^{-\alpha t} \cdot \mathrm{erfc}(\sqrt{t}) \cdot \mathrm{d}t$$
$$+ \frac{1}{8} \cdot y_3^2 \cdot \int_{3y_3}^{\infty} \mathrm{e}^{-\alpha t} \cdot \mathrm{erfc}(\sqrt{t}) \cdot \mathrm{d}t \quad (5B.43)$$

$$\int_{y_3}^{\infty} y^2 \cdot \mathrm{erfc}[\sqrt{(1+\alpha) \cdot (2y + y_3)}] \cdot \mathrm{d}y$$
$$= \frac{1}{8(1+\alpha)^3} \int_{3(1+\alpha)y_3}^{\infty} t^2 \cdot \mathrm{erfc}(\sqrt{t}) \cdot \mathrm{d}t - \frac{y_3}{4(1+\alpha)^2} \int_{3(1+\alpha)y_3}^{\infty} t \cdot \mathrm{erfc}(\sqrt{t}) \cdot \mathrm{d}t$$
$$+ \frac{y_3^2}{8(1+\alpha)} \int_{3(1+\alpha)y_3}^{\infty} \mathrm{erfc}(\sqrt{t}) \cdot \mathrm{d}t \quad (5B.44)$$

$$\int_{y_3}^{\infty} y \cdot \mathrm{erfc}[\sqrt{(1+\alpha) \cdot (2y + y_3)}] \cdot \mathrm{d}y$$
$$= \frac{1}{4(1+\alpha)^2} \cdot \int_{3(1+\alpha)y_3}^{\infty} t \cdot \mathrm{erfc}(\sqrt{t}) \cdot \mathrm{d}t - \frac{1}{4(1+\alpha)} \cdot y_3$$
$$\cdot \int_{3(1+\alpha)y_3}^{\infty} \mathrm{erfc}(\sqrt{t}) \cdot \mathrm{d}t \quad (5B.45)$$

$$\int_{y_3}^{\infty} y \cdot \exp[-\alpha \cdot (2y + y_3)] \cdot \mathrm{erfc}(\sqrt{2y + y_3}) \cdot \mathrm{d}y$$
$$= \frac{1}{4} \cdot \int_{3y_3}^{\infty} t \cdot \mathrm{e}^{-\alpha t} \cdot \mathrm{erfc}(\sqrt{t}) \cdot \mathrm{d}t - \frac{1}{4} \cdot y_3 \cdot \int_{3y_3}^{\infty} \mathrm{e}^{-\alpha t} \cdot \mathrm{erfc}(\sqrt{t}) \cdot \mathrm{d}t \quad (5B.46)$$

$$\int_{y_3}^{\infty} y \cdot \sqrt{2y + y_3} \cdot \exp[-(1+\alpha) \cdot (2y + y_3)] \cdot \mathrm{d}y$$
$$= \frac{1}{4(1+\alpha)} \cdot (3y_3)^{\frac{3}{2}} \cdot \exp[-3(1+\alpha)y_3]$$
$$+ \left[\frac{3}{8(1+\alpha)} - \frac{1}{4} \cdot y_3\right]$$
$$\cdot \left\{\frac{\sqrt{3y_3}}{1+\alpha} \cdot \exp[-3(1+\alpha)y_3] + \frac{\sqrt{\pi}}{2(1+\alpha)^{\frac{3}{2}}} \cdot \mathrm{erfc}[\sqrt{3(1+\alpha)y_3}]\right\} \quad (5B.47)$$

Therefore, substituting (5B.43)–(5B.47) into (5B.42) and applying the general formulas (5B.12)–(5B.18), the integral in (5B.42) can be further resolved as

$$\int_{y_3}^{\infty} y_2 \cdot \int_{y_2}^{\infty} y_1 \cdot \exp[-\alpha(y_1+y_2+y_3)] \cdot \mathrm{erfc}(\sqrt{y_1+y_2+y_3}) \cdot \mathrm{d}y_1 \cdot \mathrm{d}y_2$$
$$= \frac{1}{2\alpha^2} \cdot y_3^2 \cdot \exp(-3\alpha y_3) \cdot \mathrm{erfc}(\sqrt{3y_3})$$
$$+ \frac{1}{\alpha^3} \cdot y_3 \cdot \exp(-3\alpha y_3) \cdot \mathrm{erfc}(\sqrt{3y_3})$$
$$+ \frac{1}{2\alpha^4} \cdot \exp(-3\alpha y_3) \cdot \mathrm{erfc}(\sqrt{3y_3})$$
$$- \frac{3}{4\alpha \cdot \sqrt{1+\alpha}} \cdot y_3^3 \cdot \mathrm{erfc}[\sqrt{3(1+\alpha)y_3}]$$
$$+ \frac{2+3\alpha}{8\alpha^2 \cdot (1+\alpha)^{3/2}} \cdot y_3^2 \cdot \mathrm{erfc}[\sqrt{3(1+\alpha)y_3}]$$
$$+ \frac{15\alpha^2+20\alpha+8}{16\alpha^3 \cdot (1+\alpha)^{5/2}} \cdot y_3 \cdot \mathrm{erfc}[\sqrt{3(1+\alpha)y_3}]$$
$$- \frac{35\alpha^3+70\alpha^2+56\alpha+16}{32\alpha^4 \cdot (1+\alpha)^{7/2}} \cdot \mathrm{erfc}[\sqrt{3(1+\alpha)y_3}]$$
$$+ \frac{\sqrt{3}}{4\alpha \cdot (1+\alpha) \cdot \sqrt{\pi}} \cdot y_3^{5/2} \cdot \exp[-3(1+\alpha)y_3]$$
$$- \frac{(2\alpha+1) \cdot \sqrt{3}}{4\alpha^2 \cdot (1+\alpha)^2 \cdot \sqrt{\pi}} \cdot y_3^{3/2} \cdot \exp[-3(1+\alpha)y_3]$$
$$- \frac{(19\alpha^2+22\alpha+8) \cdot \sqrt{3}}{16\alpha^3 \cdot (1+\alpha)^3 \cdot \sqrt{\pi}} \cdot y_3^{1/2} \cdot \exp[-3(1+\alpha)y_3] \tag{5B.48}$$

Substituting (5B.48) into the outermost integral in (5B.39), it can be rewritten after substituting proper variants as follows:

$$\int_0^{\infty} y_3^{n+1} \cdot \exp(-\alpha m y_3) \cdot \int_{y_3}^{\infty} y_2 \cdot \int_{y_2}^{\infty} y_1 \cdot \exp[-\alpha(y_1+y_2+y_3)]$$
$$\cdot \mathrm{erfc}\left(\sqrt{y_1+y_2+y_3}\right) \cdot \mathrm{d}y_1 \cdot \mathrm{d}y_2 \cdot \mathrm{d}y_3$$
$$= \frac{1}{2\alpha^2} \cdot \left(\frac{1}{3}\right)^{n+4} \cdot \int_0^{\infty} t^{n+3} \cdot \exp\left(-\frac{m+3}{3}\alpha t\right) \cdot \mathrm{erfc}(\sqrt{t}) \cdot \mathrm{d}t$$
$$+ \frac{1}{\alpha^3} \cdot \left(\frac{1}{3}\right)^{n+3} \cdot \int_0^{\infty} t^{n+2} \cdot \exp\left(-\frac{m+3}{3}\alpha t\right) \cdot \mathrm{erfc}(\sqrt{t}) \cdot \mathrm{d}t$$
$$+ \frac{1}{2\alpha^4} \cdot \left(\frac{1}{3}\right)^{n+2} \cdot \int_0^{\infty} t^{n+1} \cdot \exp\left(-\frac{m+3}{3}\alpha t\right) \cdot \mathrm{erfc}(\sqrt{t}) \cdot \mathrm{d}t$$

$$
\begin{aligned}
& -\frac{3}{4\alpha\cdot\sqrt{1+\alpha}}\cdot\left[\frac{1}{3(1+\alpha)}\right]^{n+5}\cdot\int_0^\infty t^{n+4} \\
& \cdot\exp\left[-\frac{\alpha m}{3(1+\alpha)}\cdot t\right]\cdot\operatorname{erfc}(\sqrt{t})\cdot dt \\
& +\frac{2+3\alpha}{8\alpha^2\cdot(1+\alpha)^{3/2}}\cdot\left[\frac{1}{3(1+\alpha)}\right]^{n+4}\cdot\int_0^\infty t^{n+3} \\
& \cdot\exp\left[-\frac{\alpha m}{3(1+\alpha)}\cdot t\right]\cdot\operatorname{erfc}(\sqrt{t})\cdot dt \\
& +\frac{15\alpha^2+20\alpha+8}{16\alpha^3\cdot(1+\alpha)^{5/2}}\cdot\left[\frac{1}{3(1+\alpha)}\right]^{n+3}\cdot\int_0^\infty t^{n+2} \\
& \cdot\exp\left[-\frac{\alpha m}{3(1+\alpha)}\cdot t\right]\cdot\operatorname{erfc}(\sqrt{t})\cdot dt \\
& -\frac{35\alpha^3+70\alpha^2+56\alpha+16}{32\alpha^4\cdot(1+\alpha)^{7/2}}\cdot\left[\frac{1}{3(1+\alpha)}\right]^{n+2}\cdot\int_0^\infty t^{n+1} \\
& \cdot\exp\left[-\frac{\alpha m}{3(1+\alpha)}\cdot t\right]\cdot\operatorname{erfc}(\sqrt{t})\cdot dt \\
& +\frac{\sqrt{3}}{4\alpha\cdot(1+\alpha)\cdot\sqrt{\pi}}\cdot\int_0^\infty t^{n+\frac{7}{2}}\cdot\exp\{-[\alpha(m+3)+3]\cdot t\}\cdot dt \\
& -\frac{(2\alpha+1)\cdot\sqrt{3}}{4\alpha^2\cdot(1+\alpha)^2\cdot\sqrt{\pi}}\cdot\int_0^\infty t^{n+\frac{5}{2}}\cdot\exp\{-[\alpha(m+3)+3]\cdot t\}\cdot dt \\
& -\frac{19\alpha^2+22\alpha+8}{16\alpha^3\cdot(1+\alpha)^3\cdot\sqrt{\pi}}\cdot\sqrt{3}\cdot\int_0^\infty t^{n+\frac{3}{2}}\cdot\exp\{-[\alpha(m+3)+3]\cdot t\}\cdot dt
\end{aligned}
\tag{5B.49}
$$

As $m = n = 0$, the fourth to seventh integrals in (5B.49) can be obtained according to the formula (6.281–1) of [9] as follows, respectively:

$$
-\frac{3}{4\alpha\cdot\sqrt{1+\alpha}}\cdot\left[\frac{1}{3(1+\alpha)}\right]^5\cdot\int_0^\infty t^4\cdot\operatorname{erfc}(\sqrt{t})\cdot dt = -\frac{7}{384\alpha\cdot(1+\alpha)^{11/2}} \tag{5B.50}
$$

$$
\frac{2+3\alpha}{8\alpha^2\cdot(1+\alpha)^{3/2}}\cdot\left[\frac{1}{3(1+\alpha)}\right]^4\cdot\int_0^\infty t^3\cdot\operatorname{erfc}(\sqrt{t})\cdot dt = \frac{35\cdot(2+3\alpha)}{512\cdot 27\cdot\alpha^2\cdot(1+\alpha)^{11/2}} \tag{5B.51}
$$

$$
\frac{15\alpha^2+20\alpha+8}{16\alpha^3\cdot(1+\alpha)^{5/2}}\cdot\left[\frac{1}{3(1+\alpha)}\right]^3\cdot\int_0^\infty t^2\cdot\operatorname{erfc}(\sqrt{t})\cdot dt = \frac{5\cdot(15\alpha^2+20\alpha+8)}{128\cdot 27\cdot\alpha^3\cdot(1+\alpha)^{11/2}} \tag{5B.52}
$$

$$-\frac{35\alpha^3+70\alpha^2+56\alpha+16}{32\alpha^4\cdot(1+\alpha)^{7/2}}\cdot\left[\frac{1}{3(1+\alpha)}\right]^2\cdot\int_0^\infty t\cdot\operatorname{erfc}(\sqrt{t})\cdot dt$$
$$=-\frac{35\alpha^3+70\alpha^2+56\alpha+16}{256\cdot 3\cdot\alpha^4\cdot(1+\alpha)^{11/2}} \tag{5B.53}$$

The other integrals in (5B.49) have the same results for any value of m. As $m\neq 0$, the weighted triple integral in (5B.39) can be obtained according to the generic formula (5B.2) and the formulas (6.281–1) and (3.381–4) of [9] as follows:

$$\alpha^{n+6}\cdot\int_0^\infty y_3^{n+1}\cdot\exp(-\alpha m y_3)\cdot\int_{y_3}^\infty y_2\cdot\int_{y_2}^\infty y_1\cdot\exp[-\alpha(y_1+y_2+y_3)]$$
$$\cdot\operatorname{erfc}(\sqrt{y_1+y_2+y_3})\cdot dy_1\cdot dy_2\cdot dy_3$$
$$=\frac{(n+3)!}{2\cdot(m+3)^{n+4}}\cdot\left\{1-\sum_{q=0}^{n+3}\frac{[(m+3)\alpha]^q\cdot\sqrt{3}}{q!\cdot\sqrt{\pi}}\cdot\frac{\Gamma(q+1/2)}{[3+(m+3)\alpha]^{q+1/2}}\right\}$$
$$+\frac{(n+2)!}{(m+3)^{n+3}}\cdot\left\{1-\sum_{q=0}^{n+2}\frac{[(m+3)\alpha]^q\cdot\sqrt{3}}{q!\cdot\sqrt{\pi}}\cdot\frac{\Gamma(q+1/2)}{[3+(m+3)\alpha]^{q+1/2}}\right\}$$
$$+\frac{(n+1)!}{2\cdot(m+3)^{n+2}}\cdot\left\{1-\sum_{q=0}^{n+1}\frac{[(m+3)\alpha]^q\cdot\sqrt{3}}{q!\cdot\sqrt{\pi}}\cdot\frac{\Gamma(q+1/2)}{[3+(m+3)\alpha]^{q+1/2}}\right\}$$
$$-\frac{3\cdot(n+4)!}{4m^{n+5}\cdot\sqrt{1+\alpha}}\cdot\left\{1-\sum_{q=0}^{n+4}\frac{\alpha^q\cdot m^q\cdot\sqrt{3(1+\alpha)}}{q!\cdot\sqrt{\pi}}\cdot\frac{\Gamma(q+1/2)}{[\alpha(m+3)+3]^{q+1/2}}\right\}$$
$$+\frac{(2+3\alpha)\cdot(n+3)!}{8m^{n+4}\cdot(1+\alpha)^{3/2}}\cdot\left\{1-\sum_{q=0}^{n+3}\frac{\alpha^q\cdot m^q\cdot\sqrt{3(1+\alpha)}}{q!\cdot\sqrt{\pi}}\cdot\frac{\Gamma(q+1/2)}{[\alpha(m+3)+3]^{q+1/2}}\right\}$$
$$+\frac{(15\alpha^2+20\alpha+8)\cdot(n+2)!}{16m^{n+3}\cdot(1+\alpha)^{5/2}}$$
$$\cdot\left\{1-\sum_{q=0}^{n+2}\frac{\alpha^q\cdot m^q\cdot\sqrt{3(1+\alpha)}}{q!\cdot\sqrt{\pi}}\cdot\frac{\Gamma(q+1/2)}{[\alpha(m+3)+3]^{q+1/2}}\right\}$$
$$-\frac{35\alpha^3+70\alpha^2+56\alpha+16}{32m^{n+2}\cdot(1+\alpha)^{7/2}}\cdot(n+1)!$$
$$\cdot\left\{1-\sum_{q=0}^{n+1}\frac{\alpha^q\cdot m^q\cdot\sqrt{3(1+\alpha)}}{q!\cdot\sqrt{\pi}}\cdot\frac{\Gamma(q+1/2)}{[\alpha(m+3)+3]^{q+1/2}}\right\}$$
$$+\frac{\sqrt{3}}{4\cdot(1+\alpha)\cdot\sqrt{\pi}}\cdot\alpha^{n+5}\cdot\frac{\Gamma(n+9/2)}{[\alpha(m+3)+3]^{n+9/2}}$$
$$-\frac{(2\alpha+1)\cdot\sqrt{3}}{4\cdot(1+\alpha)^2\cdot\sqrt{\pi}}\cdot\alpha^{n+4}\cdot\frac{\Gamma(n+7/2)}{[\alpha(m+3)+3]^{n+7/2}}-\frac{19\alpha^2+22\alpha+8}{16\cdot(1+\alpha)^3\cdot\sqrt{\pi}}$$
$$\cdot\sqrt{3}\cdot\alpha^{n+3}\cdot\frac{\Gamma(n+5/2)}{[\alpha(m+3)+3]^{n+5/2}} \tag{5B.54}$$

Therefore, combining and substituting the results (5B.50)–(5B.54) into (5B.39), the integral representation in (5.60) can be achieved.

5B.3 Correlated pair of channels

For the integral in (5.68), by substituting (5.66) and (5.67) into (5.68) and applying the binomial theorem to the power series, it is obtained that

$$\begin{aligned}
P_{\mathrm{e,cor}}^{<1>} &= \binom{L}{1} \cdot \int_0^{\infty} \left[1 - \frac{1+\rho}{2\rho} \cdot \exp\left(-\frac{\alpha}{1+\rho} \cdot y_1\right)\right. \\
&\quad \left. + \frac{1-\rho}{2\rho} \cdot \exp\left(-\frac{\alpha}{1-\rho} \cdot y_1\right)\right]^{L-1} \\
&\quad \cdot \frac{\alpha}{2\rho} \cdot \left[\exp\left(-\frac{\alpha}{1+\rho} \cdot y_1\right) - \exp\left(-\frac{\alpha}{1-\rho} \cdot y_1\right)\right] \cdot Q\sqrt{2y_1} \cdot \mathrm{d}y_1 \\
&= \binom{L}{1} \cdot \frac{\alpha}{2\rho} \cdot \frac{1}{2} \cdot \sum_{m=0}^{L-1} \sum_{n=0}^{m} \binom{L-1}{m} \cdot \binom{m}{n} \cdot \left(-\frac{1+\rho}{2\rho}\right)^n \\
&\quad \cdot \left(\frac{1-\rho}{2\rho}\right)^{m-n} \cdot \int_0^{\infty} \left\{\exp\left[-\left(\frac{n+1}{1+\rho} + \frac{m-n}{1-\rho}\right) \cdot \alpha y_1\right]\right. \\
&\quad \left. - \exp\left[-\left(\frac{n}{1+\rho} + \frac{m-n+1}{1-\rho}\right) \cdot \alpha y_1\right]\right\} \cdot \mathrm{erfc}\sqrt{y_1} \cdot \mathrm{d}y_1
\end{aligned} \tag{5B.55}$$

Applying (5B.1) to the integral in (5B.55), the integral result of (5.68) can be obtained. For the integral in (5.71), by substituting (5.66) and (5.67) into (5.71) and applying the binomial theorem to the power series, it is obtained that

$$\begin{aligned}
&P_{\mathrm{e,cor}}^{<2>} \\
&= 2! \cdot \binom{L}{2} \cdot \left(\frac{\alpha}{2\rho}\right)^2 \cdot \frac{1}{2} \cdot \sum_{m=0}^{L-2} \sum_{n=0}^{m} \binom{L-2}{m} \cdot \binom{m}{n} \cdot \left(-\frac{1+\rho}{2\rho}\right)^n \cdot \left(\frac{1-\rho}{2\rho}\right)^{m-n} \\
&\times \left\{\int_0^{\infty} \int_{y_2}^{\infty} \exp\left[-\left(\frac{n+1}{1+\rho} + \frac{m-n}{1-\rho}\right) \cdot \alpha y_2 - \frac{1}{1+\rho} \cdot \alpha y_1\right]\right. \\
&\quad \cdot \mathrm{erfc}(\sqrt{y_1 + y_2}) \cdot \mathrm{d}y_1 \cdot \mathrm{d}y_2 \\
&\quad - \int_0^{\infty} \int_{y_2}^{\infty} \exp\left[-\left(\frac{n+1}{1+\rho} + \frac{m-n}{1-\rho}\right) \cdot \alpha y_2 - \frac{1}{1-\rho} \cdot \alpha y_1\right] \\
&\quad \cdot \mathrm{erfc}(\sqrt{y_1 + y_2}) \cdot \mathrm{d}y_1 \cdot \mathrm{d}y_2 \\
&\quad - \int_0^{\infty} \int_{y_2}^{\infty} \exp\left[-\left(\frac{n}{1+\rho} + \frac{m-n+1}{1-\rho}\right) \cdot \alpha y_2 - \frac{1}{1+\rho} \cdot \alpha y_1\right] \\
&\quad \cdot \mathrm{erfc}(\sqrt{y_1 + y_2}) \cdot \mathrm{d}y_1 \cdot \mathrm{d}y_2 \\
&\quad + \int_0^{\infty} \int_{y_2}^{\infty} \exp\left[-\left(\frac{n}{1+\rho} + \frac{m-n+1}{1-\rho}\right) \cdot \alpha y_2 - \frac{1}{1-\rho} \cdot \alpha y_1\right] \\
&\quad \left. \cdot \mathrm{erfc}(\sqrt{y_1 + y_2}) \cdot \mathrm{d}y_1 \cdot \mathrm{d}y_2\right\}
\end{aligned} \tag{5B.56}$$

Let us first consider the general integral $\int_{y_2}^{\infty} \exp(-ay_2 - by_1) \cdot \mathrm{erfc}(\sqrt{y_1 + y_2}) \cdot \mathrm{d}y_1$. Through setting $t = y_1 + y_2$ and substituting $y_1 = t - y_2$ and $\mathrm{d}y_1 = \mathrm{d}t$ into it, when $b > 0$, with the help of (5B.13) it can be solved as follows:

$$\begin{aligned}
&\int_{y_2}^{\infty} \exp(-ay_2 - by_1) \cdot \mathrm{erfc}(\sqrt{y_1 + y_2}) \cdot \mathrm{d}y_1 \\
&= \exp[-(a-b)y_2] \cdot \int_{2y_2}^{\infty} \exp(-bt) \cdot \mathrm{erfc}(\sqrt{t}) \cdot \mathrm{d}t \\
&= \frac{1}{b} \cdot \exp[-(a+b)y_2] \cdot \mathrm{erfc}(\sqrt{2y_2}) \\
&\quad - \frac{1}{b\sqrt{1+b}} \cdot \exp[-(a-b)y_2] \cdot \mathrm{erfc}[\sqrt{2 \cdot (1+b)y_2}]
\end{aligned} \tag{5B.57}$$

Next consider the double integral

$$\begin{aligned}
&\int_0^{\infty} \int_{y_2}^{\infty} \exp(-ay_2 - by_1) \cdot \mathrm{erfc}(\sqrt{y_1 + y_2}) \cdot \mathrm{d}y_1 \cdot \mathrm{d}y_2 \\
&= \frac{1}{b} \cdot \int_0^{\infty} \exp[-(a+b)y_2] \cdot \mathrm{erfc}(\sqrt{2y_2}) \cdot \mathrm{d}y_2 \\
&\quad - \frac{1}{b\sqrt{1+b}} \cdot \int_0^{\infty} \exp[-(a-b)y_2] \cdot \mathrm{erfc}[\sqrt{2 \cdot (1+b)y_2}] \cdot \mathrm{d}y_2
\end{aligned} \tag{5B.58}$$

Applying (5B.1) to the integrals in (5B.58), if $a \neq b$, it is obtained that

$$\begin{aligned}
&\int_0^{\infty} \int_{y_2}^{\infty} \exp(-ay_2 - by_1) \cdot \mathrm{erfc}(\sqrt{y_1 + y_2}) \cdot \mathrm{d}y_1 \cdot \mathrm{d}y_2 \\
&= \frac{1}{b \cdot (a+b)} - \frac{1}{b \cdot (a-b) \cdot \sqrt{1+b}} + \frac{2}{a^2 - b^2} \cdot \sqrt{\frac{2}{a+b+2}}
\end{aligned} \tag{5B.59}$$

If $a = b$, it is obtained that

$$\begin{aligned}
&\int_0^{\infty} \int_{y_2}^{\infty} \exp(-ay_2 - by_1) \cdot \mathrm{erfc}(\sqrt{y_1 + y_2}) \cdot \mathrm{d}y_1 \cdot \mathrm{d}y_2 \\
&= \frac{1}{2b^2} - \frac{2+3b}{4b^2 \cdot (1+b)^{3/2}}
\end{aligned} \tag{5B.60}$$

Therefore, applying the general equations (5B.59) and (5B.60) to the corresponding integrals in (5B.56), it is straightforward to obtain the result in (5.71). For the integral in (5.73), by substituting (5.66) and (5.67) into (5.73) and applying the binomial theorem to the power series, it is obtained that

$$\begin{aligned}
&P_{\mathrm{e,cor}}^{<3>} \\
&= 3! \cdot \binom{L}{3} \cdot \left(\frac{\alpha}{2\rho}\right)^3 \cdot \frac{1}{2} \cdot \sum_{m=0}^{L-3} \sum_{n=0}^{m} \binom{L-3}{m} \cdot \binom{m}{n} \cdot \left(-\frac{1+\rho}{2\rho}\right)^n \cdot \left(\frac{1-\rho}{2\rho}\right)^{m-n} \\
&\cdot \int_0^{\infty} \int_{y_3}^{\infty} \int_{y_2}^{\infty} \left\{ \exp\left[-\left(\frac{n+1}{1+\rho} + \frac{m-n}{1-\rho}\right) \cdot \alpha y_3 - \frac{\alpha}{1+\rho} \cdot (y_2 + y_1) \right] \right. \\
&- \exp\left[-\left(\frac{n}{1+\rho} + \frac{m-n+1}{1-\rho}\right) \cdot \alpha y_3 - \frac{\alpha}{1+\rho} \cdot (y_2 + y_1) \right]
\end{aligned}$$

$$+ \exp\left[-\left(\frac{n+1}{1+\rho}+\frac{m-n}{1-\rho}\right)\cdot \alpha y_3 - \frac{\alpha}{1-\rho}\cdot (y_2+y_1)\right]$$
$$- \exp\left[-\left(\frac{n}{1+\rho}+\frac{m-n+1}{1-\rho}\right)\cdot \alpha y_3 - \frac{\alpha}{1-\rho}\cdot (y_2+y_1)\right]$$
$$- \exp\left[-\left(\frac{n+1}{1+\rho}+\frac{m-n}{1-\rho}\right)\cdot \alpha y_3 - \frac{1}{1-\rho}\cdot \alpha y_2 - \frac{1}{1+\rho}\cdot \alpha y_1\right]$$
$$+ \exp\left[-\left(\frac{n}{1+\rho}+\frac{m-n+1}{1-\rho}\right)\cdot \alpha y_3 - \frac{1}{1-\rho}\cdot \alpha y_2 - \frac{1}{1+\rho}\cdot \alpha y_1\right]$$
$$- \exp\left[-\left(\frac{n+1}{1+\rho}+\frac{m-n}{1-\rho}\right)\cdot \alpha y_3 - \frac{1}{1+\rho}\cdot \alpha y_2 - \frac{1}{1-\rho}\cdot \alpha y_1\right]$$
$$+ \exp\left[-\left(\frac{n}{1+\rho}+\frac{m-n+1}{1-\rho}\right)\cdot \alpha y_3 - \frac{1}{1+\rho}\cdot \alpha y_2 - \frac{1}{1-\rho}\cdot \alpha y_1\right]\Bigg\}$$
$$\cdot \operatorname{erfc}(\sqrt{y_1+y_2+y_3})\cdot dy_1\cdot dy_2\cdot dy_3 \tag{5B.61}$$

Let us next consider the general integral $\int_{y_2}^{\infty}\exp[-ay_3 - b(y_2+y_1)]\cdot \operatorname{erfc}(\sqrt{y_1+y_2+y_3})\cdot dy_1$. Through setting $t = y_1+y_2+y_3$ and substituting $y_1 = t - y_2 - y_3$ and $dy_1 = dt$ into the integral, when $b > 0$, with the help of (5B.13) it can be solved as follows:

$$\begin{aligned}\int_{y_2}^{\infty} & \exp[-ay_3 - b(y_2+y_1)]\cdot \operatorname{erfc}(\sqrt{y_1+y_2+y_3})\cdot dy_1\\ &= \exp[-(a-b)y_3]\cdot \int_{2y_2+y_3}^{\infty}\exp(-bt)\cdot \operatorname{erfc}(\sqrt{t})\cdot dt\\ &= \frac{1}{b}\cdot \exp(-ay_3 - 2by_2)\cdot \operatorname{erfc}(\sqrt{2y_2+y_3})\\ &\quad - \frac{1}{b\sqrt{1+b}}\cdot \exp[-(a-b)y_3]\cdot \operatorname{erfc}[\sqrt{(1+b)(2y_2+y_3)}]\end{aligned} \tag{5B.62}$$

Next consider the double integral

$$\begin{aligned}\int_{y_3}^{\infty}\int_{y_2}^{\infty} & \exp[-ay_3 - b(y_2+y_1)]\cdot \operatorname{erfc}(\sqrt{y_1+y_2+y_3})\cdot dy_1\cdot dy_2\\ &= \frac{1}{b}\cdot \int_{y_3}^{\infty}\exp(-ay_3 - 2by_2)\cdot \operatorname{erfc}(\sqrt{2y_2+y_3})\cdot dy_2\\ &\quad - \frac{1}{b\sqrt{1+b}}\cdot \int_{y_3}^{\infty}\exp[-(a-b)y_3]\cdot \operatorname{erfc}[\sqrt{(1+b)(2y_2+y_3)}]\cdot dy_2\end{aligned} \tag{5B.63}$$

By substituting $t = 2y_2 + y_3$ and applying (5B.13), when $b > 0$, it can be deduced that

$$\begin{aligned}\frac{1}{b}\cdot \int_{y_3}^{\infty} & \exp(-ay_3 - 2by_2)\cdot \operatorname{erfc}(\sqrt{2y_2+y_3})\cdot dy_2\\ &= \frac{1}{2b}\cdot \exp[-(a-b)y_3]\cdot \int_{3y_3}^{\infty}\exp(-bt)\cdot \operatorname{erfc}(\sqrt{t})\cdot dt\end{aligned}$$

$$= \frac{1}{2b^2} \cdot \exp[-(a+2b)y_3] \cdot \mathrm{erfc}(\sqrt{3y_3})$$
$$- \frac{1}{2b^2 \cdot \sqrt{1+b}} \cdot \exp[-(a-b)y_3] \cdot \mathrm{erfc}[\sqrt{3(1+b)y_3}] \tag{5B.64}$$

By substituting $t = (1+b)(2y_2 + y_3)$ and applying (5B.12), it can be solved as

$$-\frac{1}{b\sqrt{1+b}} \cdot \exp[-(a-b)y_3] \cdot \int_{y_3}^{\infty} \mathrm{erfc}[\sqrt{(1+b)(2y_2+y_3)}] \cdot \mathrm{d}y_2$$
$$= -\frac{1}{2b \cdot (1+b)^{3/2}} \cdot \exp[-(a-b)y_3] \cdot \int_{3(1+b)y_3}^{\infty} \mathrm{erfc}(\sqrt{t}) \cdot \mathrm{d}t$$
$$= -\frac{1}{2b \cdot (1+b)^{3/2}} \cdot \exp[-(a-b)y_3]$$
$$\cdot \left\{ \frac{1}{\sqrt{\pi}} \cdot \sqrt{3(1+b)y_3} \cdot \exp[-3(1+b)y_3] \right.$$
$$\left. + \left[\frac{1}{2} - 3(1+b)y_3 \right] \cdot \mathrm{erfc}[\sqrt{3(1+b)y_3}] \right\} \tag{5B.65}$$

Substituting (5B.64) and (5B.65) into (5B.63) and further considering the triple integral

$$\int_0^{\infty} \int_{y_3}^{\infty} \int_{y_2}^{\infty} \exp[-ay_3 - b(y_2+y_1)] \cdot \mathrm{erfc}(\sqrt{y_1+y_2+y_3}) \cdot \mathrm{d}y_1 \cdot \mathrm{d}y_2 \cdot \mathrm{d}y_3$$
$$= I_{1,1}^{<3>} + I_{1,2}^{<3>} + I_{1,3}^{<3>} \tag{5B.66}$$

where, by applying (5B.1),

$$I_{1,1}^{<3>} = \frac{1}{2b^2} \cdot \int_0^{\infty} \exp[-(a+2b)y_3] \cdot \mathrm{erfc}(\sqrt{3y_3}) \cdot \mathrm{d}y_3$$
$$= \frac{1}{2b^2} \cdot \frac{1}{a+2b} \cdot \left(1 - \sqrt{\frac{3}{a+2b+3}}\right) \tag{5B.67}$$

When $a \neq b$,

$$I_{1,2}^{<3>} = -\frac{1}{2b^2 \cdot \sqrt{1+b}} \cdot \int_0^{\infty} \exp[-(a-b)y_3] \cdot \mathrm{erfc}[\sqrt{3(1+b)y_3}] \cdot \mathrm{d}y_3$$
$$= -\frac{1}{2b^2 \cdot \sqrt{1+b}} \cdot \frac{1}{a-b} \cdot \left[1 - \sqrt{\frac{3(1+b)}{a+2b+3}}\right] \tag{5B.68}$$

whereas when $a = b$,

$$I_{1,2}^{<3>} = -\frac{1}{2b^2 \cdot \sqrt{1+b}} \cdot \int_0^{\infty} \exp[-(a-b)y_3] \cdot \mathrm{erfc}[\sqrt{3(1+b)y_3}] \cdot \mathrm{d}y_3$$
$$= -\frac{1}{2b^2 \cdot \sqrt{1+b}} \cdot \int_0^{\infty} \mathrm{erfc}[\sqrt{3(1+b)y_3}] \cdot \mathrm{d}y_3$$
$$= -\frac{1}{12b^2 \cdot (1+b)^{3/2}} \tag{5B.69}$$

$$
\begin{aligned}
\mathrm{I}_{1,3}^{<3>} = & -\frac{\sqrt{3}}{2b\cdot(1+b)\cdot\sqrt{\pi}}\int_0^{\infty}\sqrt{y_3}\cdot\exp[-(a+2b+3)y_3]\cdot dy_3 \\
& -\frac{1}{4b\cdot(1+b)^{3/2}}\cdot\int_0^{\infty}\exp[-(a-b)y_3]\cdot\mathrm{erfc}[\sqrt{3(1+b)y_3}]\cdot dy_3 \\
& +\frac{3}{2b\cdot\sqrt{1+b}}\cdot\int_0^{\infty}y_3\cdot\exp[-(a-b)y_3]\cdot\mathrm{erfc}[\sqrt{3(1+b)y_3}]\cdot dy_3
\end{aligned} \tag{5B.70}
$$

Through applying the formula (3.381–4) of [9] to the first integral in (5B.70) and applying (5B.1) to the second integral in (5B.70), and by substituting $t = 3\cdot(1+b)\cdot y_3$ and applying (5B.2) to the third integral in (5B.70), it can be obtained when $a \neq b$ that

$$
\begin{aligned}
I_{1,3}^{<3>} = & -\frac{1}{4b\cdot(1+b)^{3/2}\cdot(a-b)}+\frac{3}{2b\cdot(a-b)^2\cdot\sqrt{1+b}} \\
& -\frac{3}{2b\cdot(a-b)^2}\cdot\sqrt{\frac{3}{3+a+2b}}
\end{aligned} \tag{5B.71}
$$

When $a = b$,

$$
\begin{aligned}
I_{1,3}^{<3>} & = -\frac{1}{12\cdot b\cdot(1+b)^{5/2}}-\frac{1}{24\cdot b\cdot(1+b)^{5/2}}+\frac{1}{16\cdot b\cdot(1+b)^{5/2}} \\
& = -\frac{1}{16\cdot b\cdot(1+b)^{5/2}}
\end{aligned} \tag{5B.72}
$$

When $a \neq b$, by substituting (5B.67), (5B.68) and (5B.71) into (5B.66), it can be found that

$$
\begin{aligned}
& \int_0^{\infty}\int_{y_3}^{\infty}\int_{y_2}^{\infty}\exp[-ay_3-b(y_2+y_1)]\cdot\mathrm{erfc}(\sqrt{y_1+y_2+y_3})\cdot dy_1\cdot dy_2\cdot dy_3 \\
& = \frac{1}{2b^2\cdot(a+2b)}+\frac{9b^2+8b-3ab-2a}{4b^2\cdot(a-b)^2\cdot(1+b)^{3/2}} \\
& \quad -\frac{9}{2(a+2b)\cdot(a-b)^2}\cdot\sqrt{\frac{3}{a+2b+3}}
\end{aligned} \tag{5B.73}
$$

When $a = b$, by substituting (5B.67), (5B.69) and (5B.72) into (5B.66), it can be found that

$$
\begin{aligned}
& \int_0^{\infty}\int_{y_3}^{\infty}\int_{y_2}^{\infty}\exp[-ay_3-b(y_2+y_1)]\cdot\mathrm{erfc}(\sqrt{y_1+y_2+y_3})\cdot dy_1\cdot dy_2\cdot dy_3 \\
& = \frac{1}{6b^3}-\frac{1}{6b^3\cdot(1+b)^{1/2}}-\frac{1}{12b^2\cdot(1+b)^{3/2}}-\frac{1}{16\cdot b\cdot(1+b)^{5/2}}
\end{aligned} \tag{5B.74}
$$

Next, for the general integral $\int_{y_2}^{\infty}\exp(-ay_3-by_2-cy_1)\cdot\mathrm{erfc}(\sqrt{y_1+y_2+y_3})\cdot dy_1$, through setting $t = y_1+y_2+y_3$ and substituting $y_1 = t-y_2-y_3$ and $dy_1 = dt$ into the integral, for $c > 0$, by the help of (5B.13) it can be solved as follows:

$$\begin{aligned}
&\int_{y_2}^{\infty} \exp(-ay_3 - by_2 - cy_1) \cdot \text{erfc}(\sqrt{y_1 + y_2 + y_3}) \cdot dy_1 \\
&\quad = \exp[-(a-c)y_3 - (b-c)y_2] \cdot \int_{2y_2+y_3}^{\infty} \exp(-ct) \cdot \text{erfc}(\sqrt{t}) \cdot dt \\
&\quad = \frac{1}{c} \cdot \exp[-ay_3 - (b+c)y_2] \cdot \text{erfc}(\sqrt{2y_2 + y_3}) \\
&\qquad - \frac{1}{c\sqrt{1+c}} \cdot \exp[-(a-c)y_3 - (b-c)y_2] \cdot \text{erfc}[\sqrt{(1+c)(2y_2 + y_3)}]
\end{aligned} \tag{5B.75}$$

Furthermore, considering the double integral

$$\begin{aligned}
&\int_{y_3}^{\infty}\int_{y_2}^{\infty} \exp(-ay_3 - by_2 - cy_1) \cdot \text{erfc}(\sqrt{y_1 + y_2 + y_3}) \cdot dy_1 \cdot dy_2 \\
&\quad = \frac{1}{c} \cdot \int_{y_3}^{\infty} \exp[-ay_3 - (b+c)y_2] \cdot \text{erfc}(\sqrt{2y_2 + y_3}) \cdot dy_2 \\
&\qquad - \frac{1}{c\sqrt{1+c}} \cdot \int_{y_3}^{\infty} \exp[-(a-c)y_3 - (b-c)y_2] \\
&\qquad \cdot \text{erfc}[\sqrt{(1+c)(2y_2 + y_3)}] \cdot dy_2
\end{aligned} \tag{5B.76}$$

By substituting $t = 2y_2 + y_3$ and applying (5B.13), for $b + c > 0$ it can be deduced that

$$\begin{aligned}
&\frac{1}{c} \cdot \int_{y_3}^{\infty} \exp[-ay_3 - (b+c)y_2] \cdot \text{erfc}(\sqrt{2y_2 + y_3}) \cdot dy_2 \\
&\quad = \frac{1}{2c} \cdot \exp\left[\left(-a + \frac{b+c}{2}\right) y_3\right] \cdot \int_{3y_3}^{\infty} \exp\left(-\frac{b+c}{2} t\right) \cdot \text{erfc}(\sqrt{t}) \cdot dt \\
&\quad = \frac{1}{c \cdot (b+c)} \cdot \exp\left[-(a+b+c)\, y_3\right] \cdot \text{erfc}(\sqrt{3y_3}) \\
&\qquad - \frac{1}{c \cdot (b+c) \cdot \sqrt{\frac{2+b+c}{2}}} \cdot \exp\left[-\left(a - \frac{b+c}{2}\right) y_3\right] \cdot \text{erfc}\left[\sqrt{3\left(1 + \frac{b+c}{2}\right) y_3}\right]
\end{aligned} \tag{5B.77}$$

By substituting $t = (1+c)(2y_2 + y_3)$ and applying (5B.13), for $b \neq c$ it can be deduced that

$$\begin{aligned}
&-\frac{1}{c\sqrt{1+c}} \cdot \int_{y_3}^{\infty} \exp[-(a-c)y_3 - (b-c)y_2] \cdot \text{erfc}[\sqrt{(1+c)(2y_2 + y_3)}] \cdot dy_2 \\
&\quad = -\frac{1}{2c \cdot (1+c)^{3/2}} \cdot \exp\left[-\left(a - \frac{b+c}{2}\right) y_3\right] \cdot \int_{3(1+c)y_3}^{\infty} \\
&\qquad \times \exp\left[-\frac{b-c}{2 \cdot (1+c)} \cdot t\right] \cdot \text{erfc}(\sqrt{t}) \cdot dt
\end{aligned}$$

$$
\begin{aligned}
&= -\frac{1}{c\cdot(b-c)\cdot\sqrt{1+c}}\cdot\exp[-(a+b-2c)y_3]\cdot\operatorname{erfc}[\sqrt{3(1+c)y_3}] \\
&\quad+\frac{1}{c\cdot(b-c)}\cdot\sqrt{\frac{2}{2+b+c}}\cdot\exp\left[-\left(a-\frac{b+c}{2}\right)y_3\right] \\
&\quad\cdot\operatorname{erfc}\left[\sqrt{\frac{3}{2}(2+b+c)y_3}\right]
\end{aligned} \tag{5B.78}
$$

Substituting (5B.77) and (5B.78) into (5B.76) and further considering the triple integral

$$
\begin{aligned}
&\int_0^\infty\int_{y_3}^\infty\int_{y_2}^\infty \exp(-ay_3-by_2-cy_1)\cdot\operatorname{erfc}(\sqrt{y_1+y_2+y_3})\cdot dy_1\cdot dy_2\cdot dy_3 \\
&= I_{2,1}^{<3>}+I_{2,2}^{<3>}+I_{2,3}^{<3>}+I_{2,4}^{<3>}
\end{aligned} \tag{5B.79}
$$

When $a+b+c>0$, and by applying (5B.1),

$$
\begin{aligned}
I_{2,1}^{<3>} &= \frac{1}{c\cdot(b+c)}\cdot\int_0^\infty \exp[-(a+b+c)y_3]\cdot\operatorname{erfc}(\sqrt{3y_3})\cdot dy_3 \\
&= \frac{1}{c\cdot(b+c)}\cdot\frac{1}{a+b+c}\cdot\left(1-\sqrt{\frac{3}{3+a+b+c}}\right)
\end{aligned} \tag{5B.80}
$$

As $2a-b-c\neq 0$, by applying (5B.1), it is obtained that

$$
\begin{aligned}
I_{2,2}^{<3>} &= -\frac{1}{c\cdot(b+c)\cdot\sqrt{\frac{2+b+c}{2}}}\cdot\int_0^\infty \exp\left[-\left(a-\frac{b+c}{2}\right)y_3\right] \\
&\quad\cdot\operatorname{erfc}\left[\sqrt{3\left(1+\frac{b+c}{2}\right)y_3}\right]\cdot dy_3 \\
&= -\frac{2}{c\cdot(b+c)\cdot(2a-b-c)}\cdot\sqrt{\frac{2}{2+b+c}}\cdot\left[1-\sqrt{\frac{3\cdot(2+b+c)}{2\cdot(3+a+b+c)}}\right]
\end{aligned} \tag{5B.81}
$$

When $2a-b-c=0$, by applying the formula (6.281–1) of [9], it is obtained that

$$
I_{2,2}^{<3>} = -\frac{1}{3c\cdot(b+c)\cdot(2+b+c)}\cdot\sqrt{\frac{2}{2+b+c}} \tag{5B.82}
$$

As $a+b-2c\neq 0$, by applying (5B.1), it is obtained that

$$
\begin{aligned}
I_{2,3}^{<3>} &= -\frac{1}{c\cdot(b-c)\cdot\sqrt{1+c}} \\
&\quad\cdot\int_0^\infty \exp[-(a+b-2c)y_3]\cdot\operatorname{erfc}[\sqrt{3(1+c)y_3}]\cdot dy_3 \\
&= -\frac{1}{c\cdot(b-c)\cdot(a+b-2c)\cdot\sqrt{1+c}}\cdot\left[1-\sqrt{\frac{3(1+c)}{3+a+b+c}}\right]
\end{aligned} \tag{5B.83}
$$

When $a + b - 2c = 0$, by applying the formula (6.281–1) of [9], it is obtained that

$$I_{2,3}^{<3>} = -\frac{1}{6c \cdot (b-c) \cdot (1+c)^{3/2}} \tag{5B.84}$$

When $2a - b - c \neq 0$, by applying (5B.1), one obtains that

$$\begin{aligned} I_{2,4}^{<3>} &= \frac{1}{c \cdot (b-c)} \cdot \sqrt{\frac{2}{2+b+c}} \cdot \int_0^\infty \exp\left[-\left(a - \frac{b+c}{2}\right) y_3\right] \\ &\quad \cdot \operatorname{erfc}\left[\sqrt{\frac{3}{2}(2+b+c)y_3}\right] \cdot dy_3 \\ &= \frac{2}{c \cdot (b-c) \cdot (2a-b-c)} \cdot \sqrt{\frac{2}{2+b+c}} \cdot \left[1 - \sqrt{\frac{3 \cdot (2+b+c)}{2 \cdot (3+a+b+c)}}\right] \end{aligned} \tag{5B.85}$$

When $2a - b - c = 0$, by applying the formula (6.281–1) of [9], it is obtained that

$$I_{2,4}^{<3>} = \frac{1}{3c \cdot (b-c) \cdot (2+b+c)} \cdot \sqrt{\frac{2}{2+b+c}} \tag{5B.86}$$

Substituting the results from (5B.80)–(5B.86) into (5B.79), the triple integral can be resolved with respect to different parameters. Therefore, applying the general equations of the triple integral in (5B.73), (5B.74) and (5B.79) to the integral equation (5B.61), the consequent expression (5.73) can be achieved.

5B.4 Conventional Rake receiver without TD-STBC

In this section, we set $\alpha = 1/\overline{\gamma}_{\mathrm{c,1Tx}}$ to simplify the notation. For the integral in (5.113), by substituting (5.108) and (5.109) into (5.113) and by applying the binomial theorem to the power series, with the help of (5B.1), it can be obtained that

$$\begin{aligned} P_{\mathrm{e,1Tx}}^{<1>} &= \binom{L}{1} \cdot \int_0^\infty [1 - \exp(-\alpha y_1)]^{L-1} \cdot \alpha \cdot \exp(-\alpha y_1) \cdot Q(\sqrt{2y_1}) \cdot dy_1 \\ &= \binom{L}{1} \cdot \frac{\alpha}{2} \cdot \sum_{m=0}^{L-1} \binom{L-1}{m} \cdot (-1)^m \cdot \int_0^\infty \exp[-\alpha(m+1) \cdot y_1] \cdot j(\sqrt{y_1}) \cdot dy_1 \\ &= \frac{L}{2} \cdot \sum_{m=0}^{L-1} \binom{L-1}{m} \cdot (-1)^m \cdot \frac{1}{m+1} \cdot \left(1 - \sqrt{\frac{1}{1+\alpha+\alpha m}}\right) \end{aligned} \tag{5B.87}$$

This is the integral result presented in (5.113).

For the integral in (5.114), by substituting (5.108) and (5.109) into (5.114), and by applying the binomial theorem to the power series, it can be found that

$$P_{\mathrm{e,1Tx}}^{<2>} = 2! \cdot \binom{L}{2} \cdot \frac{\alpha^2}{2} \cdot \sum_{m=0}^{L-2} \binom{L-2}{m} \cdot (-1)^m \cdot \int_0^{\infty} \exp(-\alpha m y_2) \cdot \int_{y_2}^{\infty} \exp[-\alpha(y_1+y_2)] \cdot \operatorname{erfc}(\sqrt{y_1+y_2}) \cdot dy_1 \cdot dy_2 \tag{5B.88}$$

Through substituting $t = y_1 + y_2$ and applying (5B.13), the double integral of (5B.88) can be deduced as follows:

$$\begin{aligned} &\int_0^{\infty} \exp(-\alpha m y_2) \cdot \int_{y_2}^{\infty} \exp[-\alpha(y_1+y_2)] \cdot \operatorname{erfc}(\sqrt{y_1+y_2}) \cdot dy_1 \cdot dy_2 \\ &= \frac{1}{\alpha} \cdot \int_0^{\infty} \exp[-\alpha(m+2)y_2] \cdot \operatorname{erfc}(\sqrt{2y_2}) \cdot dy_2 \\ &\quad - \frac{1}{\alpha\sqrt{1+\alpha}} \cdot \int_0^{\infty} \exp(-\alpha m y_2) \cdot \operatorname{erfc}[\sqrt{2(1+\alpha) \cdot y_2}] \cdot dy_2 \end{aligned} \tag{5B.89}$$

Through applying (5B.1), it is obtained that

$$\frac{1}{\alpha} \cdot \int_0^{\infty} \exp[-\alpha(m+2)y_2] \cdot \operatorname{erfc}(\sqrt{2y_2}) \cdot dy_2 = \frac{1}{(m+2) \cdot \alpha^2} \cdot \left(1 - \sqrt{\frac{2}{\alpha m + 2\alpha + 2}}\right) \tag{5B.90}$$

When $m = 0$, under the help of formula (6.281–1) of [9] it is obtained that

$$-\frac{1}{\alpha\sqrt{1+\alpha}} \cdot \int_0^{\infty} \exp(-\alpha m y_2) \cdot \operatorname{erfc}[\sqrt{2(1+\alpha) \cdot y_2}] \cdot dy_2 = -\frac{1}{4\alpha \cdot (1+\alpha)^{3/2}} \tag{5B.91}$$

When $m \neq 0$, by applying (5B.1), it is obtained that

$$-\frac{1}{\alpha\sqrt{1+\alpha}} \cdot \int_0^{\infty} \exp(-\alpha m y_2) \cdot \operatorname{erfc}[\sqrt{2(1+\alpha) \cdot y_2}] \cdot dy_2 = -\frac{1}{m\alpha^2 \cdot \sqrt{1+\alpha}} + \frac{1}{m\alpha^2} \cdot \sqrt{\frac{2}{\alpha m + 2\alpha + 2}} \tag{5B.92}$$

Therefore, substituting equations (5B.90)–(5B.92) into (5B.89) and then (5B.88), the results of (5.114) can be achieved. For the integral in (5.117), by substituting (5.108) and (5.109) into (5.117), and by applying the binomial theorem to the power series, it is obtained that

$$\begin{aligned} P_{\mathrm{e,1Tx}}^{<3>} &= 3! \cdot \binom{L}{3} \cdot \alpha^3 \cdot \frac{1}{2} \cdot \sum_{m=0}^{L-3} \binom{L-3}{m} \cdot (-1)^m \\ &\quad \cdot \int_0^{\infty} \exp(-\alpha m y_3) \cdot \int_{y_3}^{\infty} \int_{y_2}^{\infty} \exp[-\alpha(y_1+y_2+y_3)] \\ &\quad \cdot \operatorname{erfc}(\sqrt{y_1+y_2+y_3}) \cdot dy_1 \cdot dy_2 \cdot dy_3 \end{aligned} \tag{5B.93}$$

Through substituting $t = y_1 + y_2 + y_3$ and applying (5B.13), the inner integral of (5B.93) can be deduced as follows:

$$\int_{y_2}^{\infty} \exp[-\alpha(y_1 + y_2 + y_3)] \cdot \operatorname{erfc}(\sqrt{y_1 + y_2 + y_3}) \cdot dy_1$$
$$= \frac{1}{\alpha} \cdot \exp[-\alpha \cdot (2y_2 + y_3)] \cdot \operatorname{erfc}(\sqrt{2y_2 + y_3}) - \frac{1}{\alpha\sqrt{1+\alpha}} \cdot \operatorname{erfc}[\sqrt{(1+\alpha)\cdot(2y_2 + y_3)}] \quad (5B.94)$$

Substituting (5B.94) into the inner double integral in (5B.93), the first term can be conducted under the help of substitution $2y_2 + y_3 = t$ and (5B.13) as

$$\int_{y_3}^{\infty} \frac{1}{\alpha} \cdot \exp[-\alpha \cdot (2y_2 + y_3)] \cdot \operatorname{erfc}(\sqrt{2y_2 + y_3}) \cdot dy_2$$
$$= \frac{1}{2\alpha^2} \cdot \exp(-3\alpha y_3) \cdot \operatorname{erfc}(\sqrt{3y_3}) - \frac{1}{2\alpha^2 \cdot \sqrt{1+\alpha}} \cdot \operatorname{erfc}[\sqrt{3\cdot(1+\alpha)\cdot y_3}] \quad (5B.95)$$

Similarly, after substituting (5B.94) into the inner double integral in (5B.93), the second term can be conducted under the help of substitution $(1+\alpha)(2y_2 + y_3) = t$ and (5B.12) as

$$-\frac{1}{\alpha\sqrt{1+\alpha}} \cdot \int_{y_3}^{\infty} \operatorname{erfc}[\sqrt{(1+\alpha)\cdot(2y_2 + y_3)}] \cdot dy_2$$
$$= -\frac{1}{2\sqrt{\pi}\cdot\alpha\cdot(1+\alpha)^{3/2}} \cdot \sqrt{3(1+\alpha)y_3} \cdot \exp[-3(1+\alpha)y_3] - \frac{1}{4\alpha\cdot(1+\alpha)^{3/2}} \cdot \operatorname{erfc}[\sqrt{3(1+\alpha)y_3}] + \frac{3}{2\alpha\cdot\sqrt{1+\alpha}} \cdot y_3 \cdot \operatorname{erfc}[\sqrt{3(1+\alpha)y_3}] \quad (5B.96)$$

Substituting (5B.95), (5B.96) and (5B.94) into the triple integral of (5B.93), it is obtained that

$$\int_0^{\infty} \exp -\alpha m y_3) \cdot \int_{y_3}^{\infty} \int_{y_2}^{\infty} \exp[-\alpha(y_1 + y_2 + y_3)] \cdot \operatorname{erfc}(\sqrt{y_1 + y_2 + y_3}) \cdot dy_1 \cdot dy_2 \cdot dy_3$$
$$= \frac{1}{2\alpha^2} \cdot \int_0^{\infty} \exp[-(m+3)\alpha y_3] \cdot \operatorname{erfc}(\sqrt{3y_3}) \cdot dy_3 - \frac{1}{2\sqrt{\pi}\cdot\alpha\cdot(1+\alpha)} \cdot \int_0^{\infty} \sqrt{3y_3} \cdot \exp[-(\alpha m + 3\alpha + 3)\cdot y_3] \cdot dy_3$$
$$- \frac{2+3\alpha}{4\alpha^2\cdot(1+\alpha)^{3/2}} \cdot \int_0^{\infty} \exp(-\alpha m y_3) \cdot \operatorname{erfc}[\sqrt{3(1+\alpha)y_3}] \cdot dy_3 + \frac{3}{2\alpha\cdot\sqrt{1+\alpha}} \cdot \int_0^{\infty} y_3 \cdot \exp(-\alpha m y_3) \cdot \operatorname{erfc}[\sqrt{3(1+\alpha)y_3}] \cdot dy_3 \quad (5B.97)$$

Through applying (5B.1) and (3.381–4) of [9], it is obtained that

$$\frac{1}{2\alpha^2}\cdot\int_0^{\infty}\exp[-(m+3)\alpha y_3]\cdot\mathrm{erfc}(\sqrt{3y_3})\cdot dy_3 = \frac{1}{2(m+3)\cdot\alpha^3}\cdot\left(1-\sqrt{\frac{3}{\alpha m+3\alpha+3}}\right) \tag{5B.98}$$

$$-\frac{1}{2\sqrt{\pi}\cdot\alpha\cdot(1+\alpha)}\cdot\int_0^{\infty}\sqrt{3y_3}\cdot\exp[-(\alpha m+3\alpha+3)\cdot y_3]\cdot dy_3 = -\frac{\sqrt{3}}{4\alpha\cdot(1+\alpha)}\cdot\frac{1}{(\alpha m+3\alpha+3)^{3/2}} \tag{5B.99}$$

When $m = 0$, by applying the formula (6.281–1) of [9], it is obtained that

$$-\frac{2+3\alpha}{4\alpha^2\cdot(1+\alpha)^{3/2}}\cdot\int_0^{\infty}\exp(-\alpha m y_3)\cdot\mathrm{erfc}[\sqrt{3(1+\alpha)y_3}]\cdot dy_3 = -\frac{2+3\alpha}{24\alpha^2\cdot(1+\alpha)^{5/2}} \tag{5B.100}$$

$$\frac{3}{2\alpha\cdot\sqrt{1+\alpha}}\cdot\int_0^{\infty}y_3\cdot\exp(-\alpha m y_3)\cdot\mathrm{erfc}[\sqrt{3(1+\alpha)y_3}]\cdot dy_3 = \frac{1}{16\alpha\cdot(1+\alpha)^{5/2}} \tag{5B.101}$$

When $m \neq 0$, by applying (5B.1), it is obtained that

$$-\frac{2+3\alpha}{4\alpha^2\cdot(1+\alpha)^{3/2}}\cdot\int_0^{\infty}\exp(-\alpha m y_3)\cdot\mathrm{erfc}[\sqrt{3(1+\alpha)y_3}]\cdot dy_3 = -\frac{2+3\alpha}{4m\alpha^3\cdot(1+\alpha)^{3/2}}+\frac{2+3\alpha}{4m\alpha^3\cdot(1+\alpha)}\cdot\sqrt{\frac{3}{\alpha m+3\alpha+3}} \tag{5B.102}$$

By substituting $t = 3(1+\alpha)y_3$ and (5B.2), it is deduced that

$$\frac{3}{2\alpha\cdot\sqrt{1+\alpha}}\cdot\int_0^{\infty}y_3\cdot\exp(-\alpha m y_3)\cdot\mathrm{erfc}[\sqrt{3(1+\alpha)y_3}]\cdot dy_3 = \frac{3}{2m^2\cdot\alpha^3\cdot\sqrt{1+\alpha}}-\frac{9\sqrt{3}\cdot(\alpha m+2\alpha+2)}{4m^2\cdot\alpha^3\cdot(\alpha m+3\alpha+3)^{3/2}} \tag{5B.103}$$

Furthermore, when $m = 0$, through substituting (5B.98)–(5B.101) into (5B.97), the triple integral in (5B.93) can be derived as

$$\int_0^{\infty}\int_{y_3}^{\infty}\int_{y_2}^{\infty}\exp(-\alpha m y_3)\cdot\exp[-\alpha(y_1+y_2+y_3)]\cdot\mathrm{erfc}(\sqrt{y_1+y_2+y_3})\cdot dy_1\cdot dy_2\cdot dy_3 = \frac{1}{6\alpha^3}-\frac{8+20\alpha+15\alpha^2}{48\alpha^3\cdot(1+\alpha)^{5/2}} \tag{5B.104}$$

When $m \neq 0$, substituting (5B.98), (5B.99), (5B.102) and (5B.103) into (5B.97), the triple integral can be derived as

$$\int_0^\infty \int_{y_3}^\infty \int_{y_2}^\infty \exp(-\alpha m y_3) \cdot \exp[-\alpha(y_1 + y_2 + y_3)] \cdot \operatorname{erfc}(\sqrt{y_1 + y_2 + y_3}) \cdot dy_1 \cdot dy_2 \cdot dy_3 = \frac{1}{2(m+3) \cdot \alpha^3} + \frac{6 \cdot (1+\alpha) - m \cdot (2+3\alpha)}{4m^2 \cdot \alpha^3 \cdot (1+\alpha)^{3/2}} - \frac{9\sqrt{3}}{2m^2 \cdot (m+3) \cdot \alpha^3 \cdot (\alpha m + 3\alpha + 3)^{1/2}} \tag{5B.105}$$

Therefore, substituting the triple integral results of (5B.104) and (5B.105) into (5B.93), the consequent expression (5.117) can finally be achieved.

References

[1] J. Wang and J. Chen, "Performance of wideband CDMA systems with complex spreading and imperfect channel estimation," *IEEE J. Select. Areas Comm.*, vol. 19, no. 1, pp. 152–163, Jan. 2001.

[2] S. M. Alamouti, "A simple transmit diversity technique for wireless communications," *IEEE J. Select. Areas Comm.*, vol. 16, no. 8, pp. 1451–1458, Oct. 1998.

[3] J. G. Proakis, *Digital Communications*, 4th edition. New York: McGraw Hill, 2001.

[4] X. Y. Wang, "Transmit diversity in CDMA for wireless communications," Ph.D. Thesis, University of Hong Kong, 2003.

[5] E. B. Saff and A. D. Snider, *Fundamentals of Complex Analysis for Mathematics, Science, and Engineering*, 2nd edition. Englewood Cliffs, NJ: Prentice Hall, 1993.

[6] I. S. Gradshteyn and I. M. Ryzhik, *Table of Integrals, Series, and Products*, 6th edition. San Diego: Academic Press, 2000.

[7] H. A. David, *Order Statistics*, 2nd edition. New York: John Wiley & Sons, 1981.

[8] B. C. Arnold, N. Balakrishnan and H. N. Nagaraja, *A First Course in Order Statistics*. New York: John Wiley & Sons, 1992.

[9] M. Abramowitz and I. A. Stegun, *Handbook of Mathematical Functions with Formulas, Graphs and Mathematical Tables*. New York: Dover Publications, 1965.

6 TD receiver with imperfect channel estimation

In this chapter, a common pilot signal transmission scheme is utilized to assist receivers to estimate the channel fading coefficients; hence, the effect of imperfect channel estimation on the TD-STBC system performance is investigated for the independent pair of channels. The power ratio of pilot to data channels and the lowpass filter used to improve the channel estimation are addressed.

6.1 Introduction

In this chapter, the TD-STBC in the DS-CDMA system with an imperfect channel estimation scheme based on the 3GPP standard [1] is studied. In the downlink of the WCDMA system, two common pilot channel signals are transmitted simultaneously from two antennas, which are employed to assist mobile stations to estimate the channel fading coefficients. In this chapter, however, it is merely assumed that the pair of channels corresponding to two transmit antennas are independent from each other. The impact of imperfect channel estimation on the system performance can be investigated through comparing the results obtained in Chapters 5 and 6. In terms of the resultant BER and system capacity, the effect of some important parameters on the system performance is also evaluated.

The rest of this chapter is organized as follows. In Section 6.2, the transmitter, channel and receiver models are described. The performance of coherent reception in the downlink of the CDMA system with and without TD-STBC are analyzed in Section 6.3. In Section 6.4, the numerical results of the system with various parameters and consequent discussions are presented. Finally, Section 6.5 summarizes and draws some conclusions.

6.2 System model

6.2.1 Transmitter model in CDMA downlink

Two transmit antennas are deployed at a base station while only one receive antenna is utilized at each mobile station. It is assumed that there are K active CDMA users and a common pilot channel is transmitted simultaneously from two transmit antennas for channel estimation [1]. Therefore, $K + 1$ different spreading code sequences are

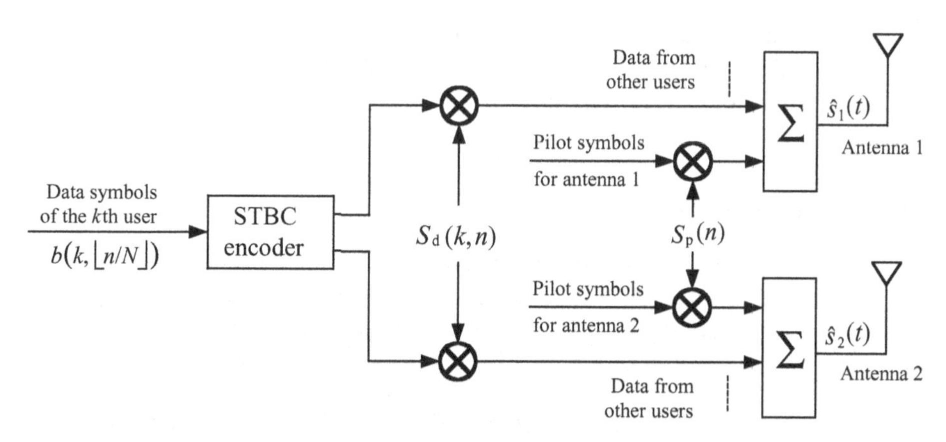

Figure 6.1 Block diagram of the transmitter.

required, which are mutually orthogonal within one symbol interval. Furthermore, it is assumed that all spreading codes have the same properties as presented in Chapter 5.

As shown in Figure 6.1, the lowpass equivalent transmitted signals from those two antennas are represented as follows, respectively:

$$\hat{s}_1(t) = \sum_{n=-\infty}^{\infty} \left[\sqrt{E_c} \cdot \sum_{k=1}^{K} b(k, \lfloor n/N \rfloor) \cdot S_d(k,n) + \sqrt{gE_c} \cdot S_p(n) \right] \cdot g(t - nT_c) \tag{6.1}$$

$$\hat{s}_2(t) = \sum_{n=-\infty}^{\infty} \left[\sqrt{E_c} \cdot \sum_{k=1}^{K} b'(k, \lfloor n/N \rfloor) \cdot S_d(k,n) + \sqrt{gE_c} \cdot (-1)^{\lfloor n/N \rfloor} \cdot S_p(n) \right] \cdot g(t - nT_c) \tag{6.2}$$

where g is the transmit power ratio of the pilot channel to data channels, and $S_p(n)$ stands for the spreading code sequence of the pilot channel. All other identical symbols and operators in this chapter are defined as in Chapter 5.

6.2.2 Channel model

The discrete tap-delay-line channel model is assumed to be as presented in Chapter 5 except that only the scenario with an independent pair of channels is taken into account. Furthermore, the lowpass equivalent complex impulse response of the channel between the ith ($i = 1, 2$) transmit antenna and the receive antenna is rewritten from (5.4) here:

$$h_i(t) = \sum_{l=1}^{L} h_{i,l} \cdot \delta(t - \tau_l) \tag{6.3}$$

For one given temporal path index l ($l = 1, 2, \ldots, L$), the channel fading coefficients $h_{1,l}$ and $h_{2,l}$ are identically Rayleigh distributed and independent from each other. Moreover, for a given spatial antenna index i, $h_{i,l}$ and $h_{i,l'}$ ($l \neq l'$) are mutually independent random variables. Similarly, $h_{i,l}$ is assumed invariant over at least one STBC block, which refers to every two consecutive symbol intervals during which two original data symbols are STBC encoded.

6.2.3 Receiver model

For any one mobile station, downlink signals from K synchronous data channels as well as one pilot channel experience the same frequency selective fading and arrive at the receiver as

$$\hat{r}(t) = \sum_{l=1}^{L} [h_{1,l} \cdot \hat{s}_1(t - \tau_l) + h_{2,l} \cdot \hat{s}_2(t - \tau_l)] + \eta(t) \tag{6.4}$$

where $\eta(t)$ represents the AWGN with double-sided power spectral density of $\eta_0/2$.

It is assumed that the chip synchronization at the receiver is perfect and that the received signal at each path can be resolved by an MF with a local delayed spreading code sequence. Similarly, it is assumed that the maximum multipath delay is approximately a few chips in duration and is much smaller than the symbol period so that the inter-symbol interference can be neglected. Without loss of generality, let us focus on the mth STBC block. By sampling the output of the pulse MF, the received signal in the first and second symbol intervals of the mth STBC block can be given by, respectively,

$$\begin{aligned} \hat{u}^{(1)}(m,n) = & \sum_{k=1}^{K}\sum_{l=1}^{L} \sqrt{E_c} \cdot S_d\,(k, 2mN + n - \lfloor \tau_l \,/\, T_c \rfloor) \\ & \cdot [b_1(k,m) \cdot h_{1,l}(m) - b_2(k,m) \cdot h_{2,l}(m)] \\ & + \sum_{l=1}^{L} \sqrt{gE_c} \cdot S_p\,(2mN + n - \lfloor \tau_l \,/\, T_c \rfloor) \\ & \cdot [h_{1,l}(m) + h_{2,l}(m)] + \eta^{(1)}(m,n) \end{aligned} \tag{6.5}$$

$$\begin{aligned} \hat{u}^{(2)}(m,n) = & \sum_{k=1}^{K}\sum_{l=1}^{L} \sqrt{E_c} \cdot S_d\,[k, (2m+1) \cdot N + n - \lfloor \tau_l \,/\, T_c \rfloor] \\ & \cdot [b_2(k,m) \cdot h_{1,l}(m) + b_1(k,m) \cdot h_{2,l}(m)] \\ & + \sum_{l=1}^{L} \sqrt{gE_c} \cdot S_p\,[(2m+1) \cdot N + n - \lfloor \tau_l \,/\, T_c \rfloor] \\ & \cdot [h_{1,l}(m) - h_{2,l}(m)] + \eta^{(2)}(m,n) \end{aligned} \tag{6.6}$$

where $n = 0, 1, \ldots, N-1$; $b_j(k,m)$ $(j = 1, 2)$ stands for the jth data bit of the kth active user transmitted in the mth STBC block; $h_{i,l}(m)$ $(i = 1, 2)$ represents the fading coefficient of the lth resolvable path of the channel between the ith transmit antenna and the receiving antenna in the mth STBC block; and $\eta^{(j)}(m,n)$ $(j = 1, 2)$ is the sampled AWGN in the jth symbol period of the mth STBC block.

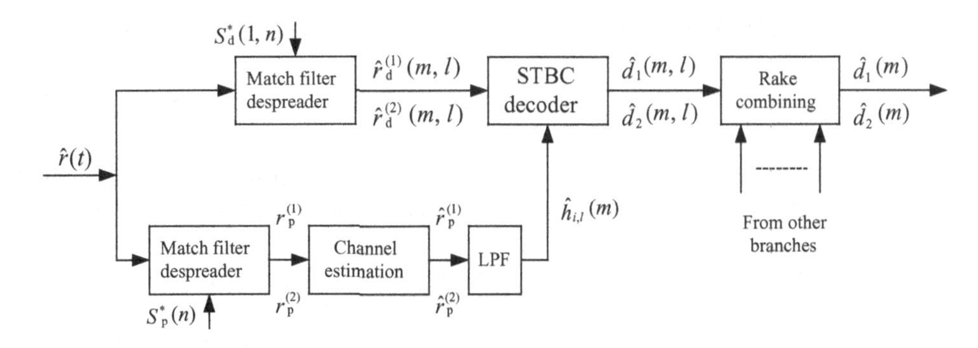

Figure 6.2 Block diagram of the receiver.

6.3 Performance analysis of coherent reception

6.3.1 STBC decoding at each branch

The first user ($k = 1$) is assumed to be the desired user. The 2D-Rake receiver structure is shown in Figure 6.2.

It is assumed that the chip-timing synchronization is perfect and the local despreading code sequence is locked to the $\hat{l}$th ($\hat{l} = 1, 2, \ldots, L$) resolvable path. During the first and second symbol periods of the mth STBC block, the pilot channel at the $\hat{l}$th path can be despread from (6.5) and (6.6) as follows, respectively:

$$\begin{aligned}
r_{\mathrm{p}}^{(1)}(m, \hat{l}) &= \sum_{n=0}^{N-1} \hat{u}^{(1)}(m, n) \cdot S_{\mathrm{p}}^*(2mN + n - \lfloor \tau_{\hat{l}} / T_{\mathrm{c}} \rfloor) \\
&= \sqrt{gE_{\mathrm{c}}} \cdot N \cdot [h_{1,\hat{l}}(m) + h_{2,\hat{l}}(m)] + \sum_{\substack{l=1 \\ l \neq \hat{l}}}^{L} \sqrt{gE_{\mathrm{c}}} \\
&\quad \cdot R_{\mathrm{p,p}}^{(1)}(m; l, \hat{l}) \cdot [h_{1,l}(m) + h_{2,l}(m)] \\
&\quad + \sum_{k=1}^{K} \sum_{\substack{l=1 \\ l \neq \hat{l}}}^{L} \sqrt{E_{\mathrm{c}}} \cdot R_{\mathrm{d,p}}^{(1)}(m, k; l, \hat{l}) \\
&\quad \cdot [b_1(k, m) \cdot h_{1,l}(m) - b_2(k, m) \cdot h_{2,l}(m)] + \eta_{\mathrm{p}}^{(1)}(m, \hat{l})
\end{aligned} \tag{6.7}$$

$$\begin{aligned}
r_{\mathrm{p}}^{(2)}(m, \hat{l}) &= \sum_{n=0}^{N-1} \hat{u}^{(2)}(m, n) \cdot S_{\mathrm{p}}^*[(2m + 1) \cdot N + n - \lfloor \tau_{\hat{l}} / T_{\mathrm{c}} \rfloor] \\
&= \sqrt{gE_{\mathrm{c}}} \cdot N \cdot [h_{1,\hat{l}}(m) - h_{2,\hat{l}}(m)] \\
&\quad + \sum_{\substack{l=1 \\ l \neq \hat{l}}}^{L} \sqrt{gE_{\mathrm{c}}} \cdot R_{\mathrm{p,p}}^{(2)}(m; l, \hat{l}) \cdot [h_{1,l}(m) - h_{2,l}(m)] \\
&\quad + \sum_{k=1}^{K} \sum_{\substack{l=1 \\ l \neq \hat{l}}}^{L} \sqrt{E_{\mathrm{c}}} \cdot R_{\mathrm{d,p}}^{(2)}(m, k; l, \hat{l}) \\
&\quad \cdot [b_2(k, m) \cdot h_{1,l}(m) + b_1(k, m) \cdot h_{2,l}(m)] + \eta_{\mathrm{p}}^{(2)}(m, \hat{l})
\end{aligned} \tag{6.8}$$

where the background AWGN component $\eta_{\rm p}^{(j)}(m,\hat{l})$ $(j=1,2)$ is defined by

$$\eta_{\rm p}^{(j)}(m,\hat{l})=\sum_{n=0}^{N-1}\eta^{(j)}(m,n)\cdot S_{\rm p}^{*}[(2m+j-1)\cdot N+n-\lfloor\tau_{\hat{l}}/T_{\rm c}\rfloor] \tag{6.9}$$

In addition, two discrete aperiodic correlation functions of two different spreading code sequences with a length of N are defined as follows, respectively:

$$R_{\rm p,p}^{(j)}(m;l,\hat{l})=\sum_{n=0}^{N-1}S_{\rm p}[(2m+j-1)\cdot N+n-\lfloor\tau_{l}/T_{\rm c}\rfloor]\cdot S_{\rm p}^{*}[(2m+j-1)\cdot N+n-\lfloor\tau_{\hat{l}}/T_{\rm c}\rfloor] \tag{6.10}$$

$$R_{\rm d,p}^{(j)}(m,k;l,\hat{l})=\sum_{n=0}^{N-1}S_{\rm d}[k,(2m+j-1)\cdot N+n-\lfloor\tau_{l}/T_{\rm c}\rfloor]\cdot S_{\rm p}^{*}[(2m+j-1)\cdot N+n-\lfloor\tau_{\hat{l}}/T_{\rm c}\rfloor] \tag{6.11}$$

Since two different spreading code sequences with the same time delay are orthogonal over one symbol interval, when $l=\hat{l}$, $R_{\rm d,p}^{(j)}(m,k;l,\hat{l})=0$. Otherwise, when $l\neq\hat{l}$, both $R_{\rm p,p}^{(j)}(m;l,\hat{l})$ and $R_{\rm d,p}^{(j)}(m,k;l,\hat{l})$ are modeled as i.i.d. random variables with zero mean and variance $\text{Var}\{R_{\rm p,p}^{(j)}(m;l,\hat{l})\}=\text{Var}\{R_{\rm d,p}^{(j)}(m,k;l,\hat{l})\}=N$.

Similarly, assuming the local despreading code sequence is locked to the $\hat{l}$th resolvable path of the reference user $k=1$, the data channel is despread during the first and second symbol periods of the mth STBC block as follows, respectively:

$$\begin{aligned}\hat{r}_{\rm d}^{(1)}(m,\hat{l})&=\sum_{n=0}^{N-1}\hat{u}^{(1)}(m,n)\cdot S_{\rm d}^{*}(1,2mN+n-\lfloor\tau_{\hat{l}}/T_{\rm c}\rfloor)\\&=r_{\rm d}^{(1)}(m,\hat{l})+\sum_{\substack{l=1\\l\neq\hat{l}}}^{L}\sqrt{gE_{\rm c}}\cdot R_{\rm p,d}^{(1)}(m;l,\hat{l})\cdot[h_{1,l}(m)+h_{2,l}(m)]\end{aligned} \tag{6.12}$$

$$\begin{aligned}\hat{r}_{\rm d}^{(2)}(m,\hat{l})&=\sum_{n=0}^{N-1}\hat{u}^{(2)}(m,n)\cdot S_{\rm d}^{*}[1,(2m+1)\cdot N+n-\lfloor\tau_{\hat{l}}/T_{\rm c}\rfloor]\\&=r_{\rm d}^{(2)}(m,\hat{l})+\sum_{\substack{l=1\\l\neq\hat{l}}}^{L}\sqrt{gE_{\rm c}}\cdot R_{\rm p,d}^{(2)}(m;l,\hat{l})\cdot[h_{1,l}(m)-h_{2,l}(m)]\end{aligned} \tag{6.13}$$

where $r_{\rm d}^{(1)}(m,\hat{l})$ and $r_{\rm d}^{(2)}(m,\hat{l})$ are given in (5.9) and (5.10), respectively. By comparing expressions (6.12), (6.13) and (6.9), (6.10), it can be seen that the despread data channel signals suffer from one more interference term due to the pilot signals. $R_{p,d}^{(j)}(m;l,\hat{l})$ is the N-length discrete aperiodic correlation function of the pilot and data channel spreading code sequences, defined by

$$R_{\rm p,d}^{(j)}(m;l,\hat{l})=\sum_{n=0}^{N-1}S_{\rm p}[(2m+j-1)\cdot N+n-\lfloor\tau_{l}/T_{\rm c}\rfloor]\cdot S_{\rm d}^{*}[1,(2m+j-1)\cdot N+n-\lfloor\tau_{\hat{l}}/T_{\rm c}\rfloor] \tag{6.14}$$

Moreover, it is assumed that $R^{(j)}_{\mathrm{p,d}}(m;l,\hat{l})$ has the same properties as those of $R^{(j)}_{\mathrm{d,p}}(m,k;l,\hat{l})$ in (6.11).

Therefore, conditioned on the channel fading coefficients at the $\hat{l}$th path in the mth STBC coding block and the involved data bits, the mean and variance of $\hat{r}^{(j)}_{\mathrm{d}}(m,\hat{l})$ ($j = 1, 2$) are given by

$$E\{\hat{r}^{(1)}_{\mathrm{d}}(m,\hat{l})\} = \sqrt{E_{\mathrm{c}}} \cdot N \cdot [b_1(1,m) \cdot h_{1,\hat{l}}(m) - b_2(1,m) \cdot h_{2,\hat{l}}(m)] \quad (6.15)$$

$$E\{\hat{r}^{(2)}_{\mathrm{d}}(m,\hat{l})\} = \sqrt{E_{\mathrm{c}}} \cdot N \cdot [b_2(1,m) \cdot h_{1,\hat{l}}(m) + b_1(1,m) \cdot h_{2,\hat{l}}(m)] \quad (6.16)$$

$$\begin{aligned}\mathrm{Var}\{\hat{r}^{(j)}_{\mathrm{d}}(m,\hat{l})\} &= E\{[\hat{r}^{(j)}_{\mathrm{d}}(m,\hat{l}) - E\{\hat{r}^{(j)}_{\mathrm{d}}(m,\hat{l})\}] \cdot [\hat{r}^{(j)}_{\mathrm{d}}(m,\hat{l}) - E\{\hat{r}^{(j)}_{\mathrm{d}}(m,\hat{l})\}]^*\} \\ &= E_{\mathrm{c}} \cdot N \cdot (g+K) \cdot (L-1) \cdot 4\sigma^2 + N \cdot \frac{\eta_0}{2} \quad (6.17)\end{aligned}$$

On the other hand, based on the despread pilot signals in (6.7) and (6.8), the initial estimates of the pair of channels between the two transmit antennas and the receive antenna at the $\hat{l}$th resolvable path in the mth STBC block can be manipulated as

$$\begin{aligned}\hat{r}^{(1)}_{\mathrm{p}}(m,\hat{l}) &= \frac{1}{2N \cdot \sqrt{gE_{\mathrm{c}}}} \cdot [r^{(1)}_{\mathrm{p}}(m,\hat{l}) + r^{(2)}_{\mathrm{p}}(m,\hat{l})] \quad (6.18) \\ &= h_{1,\hat{l}}(m) + \frac{1}{2N \cdot \sqrt{gE_{\mathrm{c}}}} \cdot [\eta^{(1)}_{\mathrm{p}}(m,\hat{l}) + \eta^{(2)}_{\mathrm{p}}(m,\hat{l})] \\ &\quad + \frac{1}{2N} \cdot \sum_{\substack{l=1 \\ l \neq \hat{l}}}^{L} \{R^{(1)}_{\mathrm{p,p}}(m;l,\hat{l}) \\ &\qquad \cdot [h_{1,l}(m) + h_{2,l}(m)] + R^{(2)}_{\mathrm{p,p}}(m;l,\hat{l}) \cdot [h_{1,l}(m) - h_{2,l}(m)]\} \\ &\quad + \frac{1}{2N \cdot \sqrt{g}} \cdot \sum_{k=1}^{K} \sum_{\substack{l=1 \\ l \neq \hat{l}}}^{L} \left\{ \begin{array}{l} R^{(1)}_{\mathrm{d,p}}(m,k;l,\hat{l}) \\ \quad \cdot [b_1(k,m) \cdot h_{1,l}(m) - b_2(k,m) \cdot h_{2,l}(m)] \\ + R^{(2)}_{\mathrm{d,p}}(m,k;l,\hat{l}) \\ \quad \cdot [b_2(k,m) \cdot h_{1,l}(m) + b_1(k,m) \cdot h_{2,l}(m)] \end{array} \right\}\end{aligned}$$

$$\begin{aligned}\hat{r}^{(2)}_{\mathrm{p}}(m,\hat{l}) &= \frac{1}{2N \cdot \sqrt{gE_{\mathrm{c}}}} \cdot [r^{(1)}_{\mathrm{p}}(m,\hat{l}) - r^{(2)}_{\mathrm{p}}(m,\hat{l})] \quad (6.19) \\ &= h_{2,\hat{l}}(m) + \frac{1}{2N \cdot \sqrt{gE_{\mathrm{c}}}} \cdot [\eta^{(1)}_{\mathrm{p}}(m,\hat{l}) - \eta^{(2)}_{\mathrm{p}}(m,\hat{l})] \\ &\quad + \frac{1}{2N} \cdot \sum_{\substack{l=1 \\ l \neq \hat{l}}}^{L} \{R^{(1)}_{\mathrm{p,p}}(m;l,\hat{l}) \\ &\qquad \cdot [h_{1,l}(m) + h_{2,l}(m)] - R^{(2)}_{\mathrm{p,p}}(m;l,\hat{l}) \cdot [h_{1,l}(m) - h_{2,l}(m)]\} \\ &\quad + \frac{1}{2N \cdot \sqrt{g}} \cdot \sum_{k=1}^{K} \sum_{\substack{l=1 \\ l \neq \hat{l}}}^{L} \left\{ \begin{array}{l} R^{(1)}_{\mathrm{d,p}}(m,k;l,\hat{l}) \\ \quad \cdot [b_1(k,m) \cdot h_{1,l}(m) - b_2(k,m) \cdot h_{2,l}(m)] \\ - R^{(2)}_{\mathrm{d,p}}(m,k;l,\hat{l}) \\ \quad \cdot [b_2(k,m) \cdot h_{1,l}(m) + b_1(k,m) \cdot h_{2,l}(m)] \end{array} \right\}\end{aligned}$$

Similarly, conditioned on the channel fading coefficients at the $\hat{l}$th path in the mth STBC coding block, the initial channel estimates $\hat{r}_{\mathrm{p}}^{(i)}(m,\hat{l})\,(i=1,2)$ have the same variances, given by

$$\begin{aligned}\mathrm{Var}\{\hat{r}_{\mathrm{p}}^{(i)}(m,\hat{l})\} &= E\{[\hat{r}_{\mathrm{p}}^{(i)}(m,\hat{l})-E\{\hat{r}_{\mathrm{p}}^{(i)}(m,\hat{l})\}]\cdot[\hat{r}_{\mathrm{p}}^{(i)}(m,\hat{l})-E\{\hat{r}_{\mathrm{p}}^{(i)}(m,\hat{l})\}]^{*}\}\\ &= \frac{1}{gN}\cdot(g+K)\cdot(L-1)\cdot 2\sigma^{2}+\frac{1}{2gN\cdot E_{\mathrm{c}}}\cdot\frac{\eta_0}{2}\end{aligned} \tag{6.20}$$

Notice that the initial channel estimates (6.20) with the assistance of pilot transmission suffer not only from the background AWGN, but also from the multiple access and multipath interference (MPI) resulting from the data channel signals of all simultaneously active users as well as the two pilot channel signals. Therefore, it is vital to exploit some techniques to estimate channels more accurately so that the performance of the DS-CDMA system with TD-STBC can be improved significantly.

For the slow fading channel, it is assumed that the channel coefficient $h_{i,\hat{l}}$ remains invariable during several consecutive STBC blocks, thus an N_{P}-tap lowpass filter (LPF) can be used to improve the accuracy of the pilot-assisted channel estimation. In other words, the improved channel estimation can be obtained through time averaging,

$$\hat{h}_{i,\hat{l}}(m)=\frac{1}{N_{\mathrm{P}}}\cdot\sum_{m\in\Theta}\hat{r}_{\mathrm{p}}^{(i)}(m,\hat{l}) \tag{6.21}$$

where any one of the example sets given below might be adopted as the timing average window:

$$\Theta\equiv\{m-N_{\mathrm{P}}+1,m\} \tag{6.22}$$

$$\Theta\equiv\left\{m-\left\lfloor\frac{N_{\mathrm{P}}-1}{2}\right\rfloor,m+\left\lfloor\frac{N_{\mathrm{P}}}{2}\right\rfloor\right\} \tag{6.23}$$

Assuming that the noise and interference terms in successive $\hat{r}_{\mathrm{p}}^{(i)}(m,\hat{l})$ with different m are independent from each other, one obtains

$$E\{\hat{h}_{i,\hat{l}}(m)\}=E\{\hat{r}_{\mathrm{p}}^{(i)}(m,\hat{l})\}=h_{i,\hat{l}}(m) \tag{6.24}$$

$$\mathrm{Var}\{\hat{h}_{i,\hat{l}}(m)\}=\frac{1}{N_{\mathrm{P}}}\cdot\mathrm{Var}\{\hat{r}_{\mathrm{p}}^{(i)}(m,\hat{l})\} \tag{6.25}$$

With respect to the $\hat{l}$th resolvable path, the branch random variables of two estimated data bits of the first user in the mth STBC block can be constructed through STBC decoding, as follows:

$$\hat{d}_1(m,\hat{l})=\hat{r}_{\mathrm{d}}^{(1)}(m,\hat{l})\cdot\hat{h}_{1,\hat{l}}^{*}(m)+\hat{r}_{\mathrm{d}}^{(2)^{*}}(m,\hat{l})\cdot\hat{h}_{2,\hat{l}}(m) \tag{6.26}$$

$$\hat{d}_2(m,\hat{l})=\hat{r}_{\mathrm{d}}^{(2)}(m,\hat{l})\cdot\hat{h}_{1,\hat{l}}^{*}(m)-\hat{r}_{\mathrm{d}}^{(1)^{*}}(m,\hat{l})\cdot\hat{h}_{2,\hat{l}}(m) \tag{6.27}$$

Since the pair of channel coefficients $h_{1,\hat{l}}(m)$ and $h_{2,\hat{l}}(m)$ are independent from each other and the discrete aperiodic correlation functions of two different spreading code sequences are independent random variables, all of $\hat{r}_{\mathrm{d}}^{(1)}(m,\hat{l})$, $\hat{r}_{\mathrm{d}}^{(2)}(m,\hat{l})$, $\hat{h}_{1,\hat{l}}(m)$ and $\hat{h}_{2,\hat{l}}(m)$ are mutually independent. Moreover, the two branch random variables of the estimated data bits $\hat{d}_1(m,\hat{l})$ and $\hat{d}_2(m,\hat{l})$ are the sum of many independent random variables and thus can be approximated as Gaussian variables, conditioned on the channel fading coefficients of the $\hat{l}$th path $h_{i,\hat{l}}(m)$ $(i = 1, 2)$. Therefore, the conditional means of $\hat{d}_j(m, \hat{l})$ $(j = 1, 2)$ are given by, respectively,

$$\begin{aligned} E\{\hat{d}_1(m,\hat{l})\} &= E\{\hat{r}_{\rm d}^{(1)}(m,\hat{l})\} \cdot E\{\hat{h}^*_{1,\hat{l}}(m)\} + E\{\hat{r}_{\rm d}^{(2)^*}(m,\hat{l})\} \cdot E\{\hat{h}_{2,\hat{l}}(m)\} \\ &= \sqrt{E_{\rm c}} \cdot N \cdot b_1(1,m) \cdot [|h_{1,\hat{l}}(m)|^2 + |h_{2,\hat{l}}(m)|^2] \end{aligned} \quad (6.28)$$

$$\begin{aligned} E\{\hat{d}_2(m,\hat{l})\} &= E\{\hat{r}_{\rm d}^{(2)}(m,\hat{l})\} \cdot E\{\hat{h}^*_{1,\hat{l}}(m)\} - E\{\hat{r}_{\rm d}^{(1)^*}(m,\hat{l})\} \cdot E\{\hat{h}_{2,\hat{l}}(m)\} \\ &= \sqrt{E_{\rm c}} \cdot N \cdot b_2(1,m) \cdot [|h_{1,\hat{l}}(m)|^2 + |h_{2,\hat{l}}(m)|^2] \end{aligned} \quad (6.29)$$

Meanwhile, the conditional variances of $\hat{d}_j(m, \hat{l})$ can be obtained as follows, respectively:

$$\begin{aligned} {\rm Var}\{\hat{d}_1(m,\hat{l})\} &= \left|E\{\hat{r}_{\rm d}^{(1)}(m,\hat{l})\}\right|^2 \cdot {\rm Var}\{\hat{h}^*_{1,\hat{l}}(m)\} + |E\{\hat{h}^*_{1,\hat{l}}(m)\}|^2 \cdot {\rm Var}\{\hat{r}_{\rm d}^{(1)}(m,\hat{l})\} \\ &\quad + {\rm Var}\{\hat{r}_{\rm d}^{(1)}(m,\hat{l})\} \cdot {\rm Var}\{\hat{h}^*_{1,\hat{l}}(m)\} + \left|E\{\hat{r}_{\rm d}^{(2)*}(m,\hat{l})\}\right|^2 \cdot {\rm Var}\{\hat{h}_{2,\hat{l}}(m)\} \\ &\quad + |E\{\hat{h}_{2,\hat{l}}(m)\}|^2 \cdot {\rm Var}\{\hat{r}_{\rm d}^{(2)*}(m,\hat{l})\} + {\rm Var}\{\hat{r}_{\rm d}^{(2)*}(m,\hat{l})\} \cdot {\rm Var}\{\hat{h}_{2,\hat{l}}(m)\} \\ &= N \cdot \left(1 + \frac{1}{gN_{\rm P}}\right) \cdot \left[E_{\rm c} \cdot (g+K) \cdot (L-1) \cdot 4\sigma^2 + \frac{\eta_0}{2}\right] \\ &\quad \cdot [|h_{1,\hat{l}}(m)|^2 + |h_{2,\hat{l}}(m)|^2] \\ &\quad + \frac{1}{gN_{\rm P} \cdot E_{\rm c}} \cdot \left[E_{\rm c} \cdot (g+K) \cdot (L-1) \cdot 4\sigma^2 + \frac{\eta_0}{2}\right]^2 \end{aligned} \quad (6.30)$$

$$\begin{aligned} {\rm Var}\{\hat{d}_2(m,\hat{l})\} &= \left|E\{\hat{r}_{\rm d}^{(2)}(m,\hat{l})\}\right|^2 \cdot {\rm Var}\{\hat{h}^*_{1,\hat{l}}(m)\} + \left|E\{\hat{h}^*_{1,\hat{l}}(m)\}\right|^2 \cdot {\rm Var}\{\hat{r}_{\rm d}^{(2)}(m,\hat{l})\} \\ &\quad + {\rm Var}\{\hat{r}_{\rm d}^{(2)}(m,\hat{l})\} \cdot {\rm Var}\{\hat{h}^*_{1,\hat{l}}(m)\} + \left|E\{\hat{r}_{\rm d}^{(1)*}(m,\hat{l})\}\right|^2 \cdot {\rm Var}\{\hat{h}_{2,\hat{l}}(m)\} \\ &\quad + |E\{\hat{h}_{2,\hat{l}}(m)\}|^2 \cdot {\rm Var}\{\hat{r}_{\rm d}^{(1)*}(m,\hat{l})\} + {\rm Var}\{\hat{r}_{\rm d}^{(1)*}(m,\hat{l})\} \cdot {\rm Var}\{\hat{h}_{2,\hat{l}}(m)\} \\ &= N \cdot \left(1 + \frac{1}{gN_{\rm P}}\right) \cdot \left[E_{\rm c} \cdot (g+K) \cdot (L-1) \cdot 4\sigma^2 + \frac{\eta_0}{2}\right] \\ &\quad \cdot [|h_{1,\hat{l}}(m)|^2 + |h_{2,\hat{l}}(m)|^2] \\ &\quad + \frac{1}{gN_{\rm P} \cdot E_{\rm c}} \cdot \left[E_{\rm c} \cdot (g+K) \cdot (L-1) \cdot 4\sigma^2 + \frac{\eta_0}{2}\right]^2 \end{aligned} \quad (6.31)$$

It is worth noting that the variance of $\hat{d}_j(m, \hat{l})$ $(j = 1, 2)$ is the same through comparing (6.30) and (6.31), i.e. ${\rm Var}\{\hat{d}_1(m, \hat{l})\} = {\rm Var}\{\hat{d}_2(m, \hat{l})\}$. After despreading and STBC decoding at each path, a 2D-Rake receiver is exploited to combine the signal energy from $L_{\rm c}$ $(L_{\rm c} \le L)$ selected paths, which collects both the spatial diversity gain and path diversity gain benefiting from TD-STBC and DS-CDMA merits, respectively.

6.3.2 EGC 2D-Rake receiver

In the EGC 2D-Rake receiver, the signals of the first $L_{\rm c}$ arriving paths among L resolvable paths are selected and combined. Under the assumption of mutually independent channel fading along multiple paths, the decision variables in different branches of the 2D-Rake receiver are independent from each other and the output of the combiner can be represented as

$$\hat{d}_j(m) = \sum_{l=1}^{L_c} \hat{d}_j(m, l) \tag{6.32}$$

When $L_c = L$, the signals from all resolvable paths are combined. Conditioned on the channel fading coefficients and the relevant data bit, the decision variable $\hat{d}_j(m)$ $(j = 1, 2)$ is a Gaussian variable with conditional mean and variance given by, respectively,

$$E\{\hat{d}_j(m)\} = \sum_{l=1}^{L_c} E\{\hat{d}_j(m, l)\} = \sqrt{E_c} \cdot N \cdot b_j(1, m) \cdot \zeta \tag{6.33}$$

$$\begin{aligned} \mathrm{Var}\{\hat{d}_j(m)\} &= \sum_{l=1}^{L_c} \mathrm{Var}\{\hat{d}_j(m, l)\} \\ &= N \cdot \left(1 + \frac{1}{gN_P}\right) \cdot \left[E_c \cdot (g + K) \cdot (L - 1) \cdot 4\sigma^2 + \frac{\eta_0}{2}\right] \cdot \zeta \\ &\quad + \frac{L_c}{gN_P \cdot E_c} \cdot \left[E_c \cdot (g + K) \cdot (L - 1) \cdot 4\sigma^2 + \frac{\eta_0}{2}\right]^2 \end{aligned} \tag{6.34}$$

where ζ is defined as in (5.28). It can be seen that the variance in (6.34) will reduce to the expression of (6.27) when the parameter $N_P \to \infty$ and $g \ll K$. In other words, when the number of taps of the LPF tends to infinity, the perfect channel estimation is roughly achieved under the assumption of static channels. Therefore, the system performance will approximate to the results presented in Section 5.3.2.

Since it can be obtained that $|E\{\hat{d}_1(m)\}|^2 = |E\{\hat{d}_2(m)\}|^2$ from (6.28), (6.29) and (6.33), while $\mathrm{Var}\{\hat{d}_1(m)\} = \mathrm{Var}\{\hat{d}_2(m)\}$ from (6.30), (6.31) and (6.34), $\hat{d}_1(m)$ and $\hat{d}_2(m)$ have the same SINR properties. Therefore, conditioned on the instantaneous fading channel amplitudes of the selected paths, the BER is the same for $\hat{d}_j(m)$ $(j = 1, 2)$ and can be represented by [2]

$$\hat{P}_{e,EGC}(\hat{\zeta}) = Q\left\{\left[\frac{|E\{\hat{d}_j(m)\}|^2}{\mathrm{Var}\{\hat{d}_j(m)\}}\right]^{1/2}\right\} = Q\left[\left(\frac{\hat{\zeta}^2}{A \cdot \hat{\zeta} + L_c \cdot B}\right)^{1/2}\right] \tag{6.35}$$

where, by the definitions,

$$\hat{\zeta} = \frac{1}{2\sigma^2} \cdot \sum_{l=1}^{L_c} [|h_{1,l}(m)|^2 + |h_{2,l}(m)|^2] \tag{6.36}$$

$$A = \left(\frac{1}{N} + \frac{1}{gN_P \cdot N}\right) \cdot \left[2 \cdot (g + K) \cdot (L - 1) + \frac{N}{2\bar{\gamma}_b}\right] \tag{6.37}$$

$$B = \frac{1}{gN_P \cdot N^2} \cdot \left[2 \cdot (g + K) \cdot (L - 1) + \frac{N}{2\bar{\gamma}_b}\right]^2 \tag{6.38}$$

where $\bar{\gamma}_b$ is the average SNR per bit defined in (5.35) and rewritten here

$$\bar{\gamma}_b = \frac{2E_c \cdot N \cdot \sigma^2}{\eta_0} \tag{6.39}$$

Since the channel fading coefficients, $h_{i,l}(m)$, are assumed to be i.i.d. complex Gaussian distributed random variables, the normalized random variable $\hat{\zeta}$ in (6.36) follows a chi-square distribution with $4L_c$ degrees of freedom, and its PDF is given by

$$p_{\hat{\zeta}}(\hat{\zeta}) = \frac{1}{\Gamma(2L_c)} \cdot \hat{\zeta}^{2L_c-1} \cdot e^{-\hat{\zeta}} \tag{6.40}$$

Therefore, the final resultant BER can be obtained by averaging the conditional BER $\hat{P}_{e,EGC}(\hat{\zeta})$ in (6.35) over the PDF of $\hat{\zeta}$ in (6.40), i.e.

$$\begin{aligned}
\hat{P}_{e,EGC} &= \int_0^\infty \hat{P}_{e,EGC}(\hat{\zeta}) \cdot p_{\hat{\zeta}}(\hat{\zeta}) \cdot d\hat{\zeta} \\
&= \int_0^\infty Q\left[\left(\frac{\hat{\zeta}^2}{A \cdot \hat{\zeta} + L_c \cdot B}\right)^{1/2}\right] \cdot p_{\hat{\zeta}}(\hat{\zeta}) \cdot d\hat{\zeta} \\
&= \frac{1}{\Gamma(2L_c)} \cdot \int_0^\infty Q\left(\sqrt{\frac{y^2}{A \cdot y + L_c \cdot B}}\right) \cdot y^{2L_c-1} \cdot e^{-y} \cdot dy
\end{aligned} \tag{6.41}$$

Furthermore, it is seen that for a given average SNR $\bar{\gamma}_b$ and number of active users, the parameter of power ratio g might impact significantly the system performance of the 2D-Rake receiver. When g is very large, the pilot channel signal might introduce severe multipath interference to all active users. On the other hand, when g is very small, the channel estimation quality becomes poor so that the BER degrades. Therefore, there exists an optimal value of power ratio g under certain system scenarios, which would maximize the SINR or minimize the BER. This is because the Q-function is a monotonously decreasing function.

By setting $\partial(A \cdot \hat{\zeta} + L_c \cdot B)/\partial g = 0$, conditioned on $\hat{\zeta}$, the optimal g can be derived as

$$g_{opt}(\hat{\zeta}) = \sqrt{\frac{a_1 \cdot \hat{\zeta} + b_1}{a_2 \cdot \hat{\zeta} + b_2}} \tag{6.42}$$

where, by the definitions,

$$a_1 = 8K \cdot (L-1) \cdot N + 2N^2 \cdot \bar{\gamma}_b^{-1} \tag{6.43}$$

$$b_1 = L_c \cdot \left[8K \cdot N \cdot (L-1) \cdot \bar{\gamma}_b^{-1} + N^2 \cdot \bar{\gamma}_b^{-2} + 16K^2 \cdot (L-1)^2\right] \tag{6.44}$$

$$a_2 = 8 \cdot (L-1) \cdot N_P \cdot N \tag{6.45}$$

$$b_2 = 16L_c \cdot (L-1)^2 \tag{6.46}$$

Similarly, the average g_{opt} for the EGC 2D-Rake receiver can be obtained by averaging the conditional $g_{opt}(\hat{\zeta})$ in (6.42) over the PDF of $\hat{\zeta}$ in (6.40) [3], i.e.

$$\begin{aligned}
g_{opt} &= \int_0^\infty g_{opt}(\hat{\zeta}) \cdot p_{\hat{\zeta}}(\hat{\zeta}) \cdot d\hat{\zeta} \\
&= \frac{1}{\Gamma(2L_c)} \cdot \int_0^\infty \hat{\zeta}^{2L_c-1} \cdot \sqrt{\frac{a_1 \cdot \hat{\zeta} + b_1}{a_2 \cdot \hat{\zeta} + b_2}} \cdot e^{-\hat{\zeta}} \cdot d\hat{\zeta}
\end{aligned} \tag{6.47}$$

6.3.3 Generalized selection combining 2D-Rake receiver

Similarly, after the despreading STBC decoding at each resolvable path, a GSC 2D-Rake receiver is exploited to adaptively select and combine the branch decision variables from the $L_c \le L$ paths with highest SINR among the L resolvable paths available.

Since it is noticed that $|E\{\hat{d}_1(m,l)\}|^2 = |E\{\hat{d}_2(m,l)\}|^2$ from (6.28) and (6.29) while $\mathrm{Var}\{\hat{d}_1(m,l)\} = \mathrm{Var}\{\hat{d}_2(m,l)\}$ from (6.30) and (6.31), $\hat{d}_1(m,l)$ and $\hat{d}_2(m,l)$ have the same SINR properties. Therefore, conditioned on the channel fading coefficients of the lth path, the instantaneous SINR of the branch decision random variable $\hat{d}_j(m,l)\,(j=1,2)$ in the lth $(l=1,2,\ldots,L)$ resolvable path is the same, and can be derived as

$$\hat{\gamma}_l = \frac{|E\{\hat{d}_j(m,l)\}|^2}{\mathrm{Var}\{\hat{d}_j(m,l)\}} = \frac{\hat{\xi}_l^2}{A\cdot\hat{\xi}_l + B} \tag{6.48}$$

where, by the definition,

$$\hat{\xi}_l = \frac{1}{2\sigma^2}\cdot[|h_{1,l}(m)|^2 + |h_{2,l}(m)|^2] \tag{6.49}$$

It can be proven that the variable $\hat{\gamma}_l$ in (6.48) is the monotonously increasing function of the variable $\hat{\xi}_l$ in (6.49). Therefore, selecting the paths with highest instantaneous value of $\hat{\xi}_l$ is equivalent to selecting the paths with highest instantaneous SINR $\hat{\gamma}_l$. Moreover, $\hat{\xi}_{1:L} \ge \hat{\xi}_{2:L} \ge \cdots \ge \hat{\xi}_{L:L}$ are defined as the order statistic of the variables, obtained by arranging $\{\hat{\xi}_l \mid l = 1, 2, \ldots, L\}$ in descending order of magnitude. Similarly, the joint PDF of the order statistic variables [4], [5] $\hat{\xi}_{l:L}$ $(l = 1, 2, \ldots, L_c)$ is given by

$$\begin{aligned} &p_{\hat{\xi}_{1:L},\hat{\xi}_{2:L},\ldots,\hat{\xi}_{L_c:L}}(\hat{\xi}_{1:L},\hat{\xi}_{2:L},\ldots,\hat{\xi}_{L_c:L}) \\ &\quad = L_c!\cdot\binom{L}{L_c}\cdot[F_{\hat{\xi}_l}(\hat{\xi}_{L_c:L})]^{L-L_c}\cdot\prod_{l=1}^{L_c} p_{\hat{\xi}_l}(\hat{\xi}_{l:L}) \end{aligned} \tag{6.50}$$

where the relevant PDF and CDF of $\hat{\xi}_l$ are given by, respectively,

$$p_{\hat{\xi}_l}(\hat{\xi}_l) = \hat{\xi}_l\cdot\exp(-\hat{\xi}_l) \tag{6.51}$$

$$F_{\hat{\xi}_l}(\hat{\xi}_l) = 1 - (1+\hat{\xi}_l)\cdot\exp(-\hat{\xi}_l) \tag{6.52}$$

It is assumed that the channel fading along multiple paths are mutually independent and that the decision random variables in distinct branches of that 2D-Rake receiver are independent from each other. Therefore, based on the branch SINR in (6.48) and conditioned on the random variables $\{\hat{\xi}_{l:L} \mid l = 1, 2, \ldots, L_c\}$ of the selected paths, the overall output SINR of the GSC 2D-Rake receiver that selects and combines L_c strongest paths can be represented as [6]

$$\hat{\gamma}_{\mathrm{GSC}} = \frac{\left(\sum_{l=1}^{L_c}\hat{\xi}_{l:L}\right)^2}{A\cdot\sum_{l=1}^{L_c}\hat{\xi}_{l:L} + L_c\cdot B} \tag{6.53}$$

Furthermore, conditioned on $\{\hat{\xi}_{l:L} \mid l = 1, 2, \ldots, L_c\}$, the BER can be obtained by using

$$\hat{P}_{e,GSC}(\hat{\gamma}_{GSC}) = Q(\sqrt{\hat{\gamma}_{GSC}}) \tag{6.54}$$

Therefore, when the L_c strongest paths are selected and combined by the GSC 2D-Rake receiver, the average BER can be obtained by averaging conditional BER (6.54) over the joint PDF of the selected fading paths in (6.50), i.e.

$$\begin{aligned}\hat{P}^{<L_c>}_{e,GSC} &= \int_0^\infty \int_{\hat{\xi}_{L_c:L}}^\infty \cdots \int_{\hat{\xi}_{2:L}}^\infty \hat{P}_{e,GSC}(\hat{\gamma}_{GSC}) \cdot p_{\hat{\xi}_{1:L}, \hat{\xi}_{2:L}, \ldots, \hat{\xi}_{L_c:L}} \\ &\quad \cdot (\hat{\xi}_{1:L}, \hat{\xi}_{2:L}, \ldots, \hat{\xi}_{L_c:L}) \cdot d\hat{\xi}_{1:L} \cdot d\hat{\xi}_{2:L} \cdots d\hat{\xi}_{L_c:L} \\ &= L_c! \cdot \binom{L}{L_c} \cdot \int_0^\infty \int_{\hat{\xi}_{L_c:L}}^\infty \cdots \int_{\hat{\xi}_{2:L}}^\infty [1 - (1 + \hat{\xi}_{L_c:L}) \cdot \exp(-\hat{\xi}_{L_c:L})]^{L-L_c} \\ &\quad \cdot \prod_{l=1}^{L_c} \hat{\xi}_{l:L} \cdot \exp\left(-\sum_{l=1}^{L_c} \hat{\xi}_{l:L}\right) \\ &\quad \cdot Q\left[\left(\sum_{l=1}^{L_c} \hat{\xi}_{l:L}\right) \Big/ \left(A \cdot \sum_{l=1}^{L_c} \hat{\xi}_{l:L} + L_c \cdot B\right)^{\frac{1}{2}}\right] \cdot d\hat{\xi}_{1:L} \cdot d\hat{\xi}_{2:L} \cdots d\hat{\xi}_{L_c:L}\end{aligned} \tag{6.55}$$

6.3.4 Conventional Rake receiver without TD-STBC

Since only one transmit antenna is deployed at the base station, the structure of the transmitter is the same as the branch of the first transmit antenna shown in the Figure 6.1, thus, the transmit signal is similar to that given by (6.1), i.e.

$$\begin{aligned}\hat{s}_{1Tx}(t) &= \sum_{n=-\infty}^{\infty} \left[\sqrt{E_{c,1Tx}} \cdot \sum_{k=1}^{K} b(k, \lfloor n/N \rfloor) \cdot S_d(k, n) + \sqrt{gE_{c,1Tx}} \cdot S_p(n)\right] \\ &\quad \cdot g\,(t - nT_c)\end{aligned} \tag{6.56}$$

where it is also assumed that there are K active CDMA users and a common pilot channel, thus, a total of $K + 1$ different spreading code sequences are needed. Similarly, the discrete tap-delay-line channel model is characterized by

$$h_{1Tx}(t) = \sum_{l=1}^{L} h_{1,l} \cdot \delta(t - \tau_l) \tag{6.57}$$

Thus, the received signal at any one mobile station can be represented as

$$\hat{r}_{1Tx}(t) = \sum_{l=1}^{L} h_{1,l} \cdot \hat{s}_{1Tx}(t - \tau_l) + \eta(t) \tag{6.58}$$

By sampling the output of the pulse-matching filter, the received signal during the mth symbol interval can be obtained as

$$\hat{u}_{1\mathrm{Tx}}(m,n)=\sum_{k=1}^{K}\sum_{l=1}^{L}\sqrt{E_{\mathrm{c},1\mathrm{Tx}}}\cdot S_{\mathrm{d}}\left(k,mN+n-\lfloor\tau_l/T_{\mathrm{c}}\rfloor\right)\cdot b(k,m)\cdot h_{1,l}(m)$$
$$+\sum_{l=1}^{L}\sqrt{gE_{\mathrm{c},1\mathrm{Tx}}}\cdot S_{\mathrm{p}}\left(mN+n-\lfloor\tau_l/T_{\mathrm{c}}\rfloor\right)\cdot h_{1,l}(m)+\eta(m,n) \quad (6.59)$$

where $n=0,1,\ldots,N-1$; and $b(k,m)$ stands for the data bit of the kth active user transmitted during the mth symbol interval; $h_{1,l}(m)$ represents the fading coefficient of the lth resolvable path of the channel between the transmit and receive antenna in the mth symbol period; and $\eta(m,n)$ is the sampled AWGN in the mth symbol period.

It is assumed that the chip timing synchronization is perfect and the local despreading code sequence is locked to the $\hat{l}$th ($\hat{l}=1,2,\ldots,L$) resolvable path. During the mth symbol period, the pilot channel at the $\hat{l}$th path can be despread from (6.59) as follows:

$$r_{\mathrm{p},1\mathrm{Tx}}(m,\hat{l})=\sum_{n=0}^{N-1}\hat{u}_{1\mathrm{Tx}}(m,n)\cdot S_{\mathrm{p}}^{*}\left(mN+n-\lfloor\tau_{\hat{l}}/T_{\mathrm{c}}\rfloor\right) \quad (6.60)$$
$$=\sqrt{gE_{c,1\mathrm{Tx}}}\cdot N\cdot h_{1,\hat{l}}(m)+\sum_{\substack{l=1\\ l\neq\hat{l}}}^{L}\sqrt{gE_{\mathrm{c},1\mathrm{Tx}}}\cdot R_{\mathrm{p,p}}(m;l,\hat{l})\cdot h_{1,l}(m)$$
$$+\sum_{k=1}^{K}\sum_{\substack{l=1\\ l\neq\hat{l}}}^{L}\sqrt{E_{\mathrm{c},1\mathrm{Tx}}}\cdot R_{\mathrm{d,p}}(m,k;l,\hat{l})\cdot b(k,m)\cdot h_{1,l}(m)+\eta_{\mathrm{p}}(m,\hat{l})$$

where the background AWGN component $\eta_{\mathrm{p}}(m,\hat{l})$ is defined as

$$\eta_{\mathrm{p}}(m,\hat{l})=\sum_{n=0}^{N-1}\eta(m,n)\cdot S_{\mathrm{p}}^{*}\left(mN+n-\lfloor\tau_{\hat{l}}/T_{\mathrm{c}}\rfloor\right) \quad (6.61)$$

In addition, two discrete aperiodic correlation functions of two different spreading code sequences with the length of N are defined as, respectively,

$$R_{\mathrm{p,p}}(m;l,\hat{l})=\sum_{n=0}^{N-1}S_{\mathrm{p}}\left[mN+n-\lfloor\tau_l/T_{\mathrm{c}}\rfloor\right]\cdot S_{p}^{*}\left[mN+n-\lfloor\tau_{\hat{l}}/T_{\mathrm{c}}\rfloor\right] \quad (6.62)$$

$$R_{\mathrm{d,p}}(m,k;l,\hat{l})=\sum_{n=0}^{N-1}S_{\mathrm{d}}\left[k,mN+n-\lfloor\tau_l/T_{\mathrm{c}}\rfloor\right]\cdot S_{\mathrm{p}}^{*}\left[mN+n-\lfloor\tau_{\hat{l}}/T_{\mathrm{c}}\rfloor\right] \quad (6.63)$$

Similarly, when $l=\hat{l}$, $R_{\mathrm{d,p}}(m,k;l,\hat{l})=0$. Otherwise, when $l\neq\hat{l}$, both $R_{\mathrm{p,p}}(m;l,\hat{l})$ and $R_{\mathrm{d,p}}(m,k;l,\hat{l})$ are modeled as i.i.d. random variables with zero mean and variance $\mathrm{Var}\{R_{\mathrm{p,p}}(m;l,\hat{l})\}=\mathrm{Var}\{R_{\mathrm{d,p}}(m,k;l,\hat{l})\}=N$.

Again, the first user ($k=1$) is assumed to be the desired user. The receiver structure is the same as shown in Figure 6.2 but without the STBC decoder. Assuming perfect chip timing synchronization and that the local despreading code sequence is locked to the $\hat{l}$th ($\hat{l}=1,2,\ldots,L$) resolvable path, the data channel of the first user at the $\hat{l}$th path is despread during the mth symbol period as follows:

$$\hat{r}_{\mathrm{d,1Tx}}(m,\hat{l}) = \sum_{n=0}^{N-1} \hat{u}_{\mathrm{1Tx}}(m,n) \cdot S_{\mathrm{d}}^{*}(1, mN+n-\lfloor \tau_{\hat{l}} / T_{\mathrm{c}} \rfloor) \tag{6.64}$$

$$= r_{\mathrm{d,1Tx}}(m,\hat{l}) + \sum_{\substack{l=1 \\ l\neq\hat{l}}}^{L} \sqrt{gE_{\mathrm{c,1Tx}}} \cdot R_{\mathrm{p,d}}(m;l,\hat{l}) \cdot h_{1,l}(m)$$

where $r_{\mathrm{d,1Tx}}(m,\hat{l})$ is given by (5.82) and $R_{\mathrm{p,d}}(m;l,\hat{l})$ is the N-length discrete aperiodic correlation function of two different spreading code sequences, given by

$$R_{\mathrm{p,d}}(m;l,\hat{l}) = \sum_{n=0}^{N-1} S_{\mathrm{p}}(mN+n-\lfloor \tau_l / T_{\mathrm{c}} \rfloor) \cdot S_{\mathrm{d}}^{*}[1, mN+n-\mathrm{left}\lfloor \tau_{\hat{l}} / T_{\mathrm{c}} \rfloor] \tag{6.65}$$

Similarly, it is assumed that $R_{\mathrm{p,d}}(m;l,\hat{l})$ has the same properties as those of $R_{\mathrm{d,p}}(m,k;l,\hat{l})$ in (6.63). Therefore, conditioned on the relevant data bit and the channel fading coefficient of the $\hat{l}$th path during the mth symbol period, the mean and variance of $\hat{r}_{d,\mathrm{1Tx}}(m,\hat{l})$ are given by, respectively,

$$E\{\hat{r}_{\mathrm{d,1Tx}}(m,\hat{l})\} = \sqrt{E_{\mathrm{c,1Tx}}} \cdot N \cdot b(1,m) \cdot h_{1,\hat{l}}(m) \tag{6.66}$$

$$\begin{aligned} \mathrm{Var}\{\hat{r}_{d,\mathrm{1Tx}}(m,\hat{l})\} &= E\{[\hat{r}_{\mathrm{d,1Tx}}(m,\hat{l}) - E\{\hat{r}_{\mathrm{d,1Tx}}(m,\hat{l})\}] \\ &\quad \cdot [\hat{r}_{\mathrm{d,1Tx}}(m,\hat{l}) - E\{\hat{r}_{\mathrm{d,1Tx}}(m,\hat{l})\}]^{*}\} \\ &= E_{\mathrm{c,1Tx}} \cdot N \cdot (g+K) \cdot (L-1) \cdot 2\sigma^2 + N \cdot \frac{\eta_0}{2} \end{aligned} \tag{6.67}$$

Based on the despread pilot signal in (6.60), the initial estimate of the channel coefficient at the $\hat{l}$th resolvable path during the mth symbol period can be manipulated as follows:

$$\begin{aligned} \hat{r}_{\mathrm{p,1Tx}}(m,\hat{l}) &= \frac{1}{N \cdot \sqrt{gE_{\mathrm{c,1Tx}}}} \cdot r_{\mathrm{p,1Tx}}(m,\hat{l}) \\ &= h_{1,\hat{l}}(m) + \frac{1}{N} \cdot \sum_{\substack{l=1 \\ l\neq\hat{l}}}^{L} R_{\mathrm{p,p}}(m;l,\hat{l}) \cdot h_{1,l}(m) + \frac{1}{N \cdot \sqrt{gE_{\mathrm{c,1Tx}}}} \cdot \eta_{\mathrm{p}}(m,\hat{l}) \\ &\quad + \frac{1}{N \cdot \sqrt{g}} \cdot \sum_{k=1}^{K} \sum_{\substack{l=1 \\ l\neq\hat{l}}}^{L} R_{\mathrm{d,p}}(m,k;l,\hat{l}) \cdot b(k,m) \cdot h_{1,l}(m) \end{aligned} \tag{6.68}$$

Similarly, conditioned on the channel fading coefficient at the $\hat{l}$th path in the mth symbol period, the variance of the initial channel estimate $\hat{r}_{\mathrm{p,1Tx}}(m,\hat{l})$ is given by

$$\begin{aligned} \mathrm{Var}\{\hat{r}_{\mathrm{p,1Tx}}(m,\hat{l})\} &= E\{[\hat{r}_{\mathrm{p,1Tx}}(m,\hat{l}) - E\{\hat{r}_{\mathrm{p,1Tx}}(m,\hat{l})\}] \\ &\quad \cdot [\hat{r}_{\mathrm{p,1Tx}}(m,\hat{l}) - E\{\hat{r}_{\mathrm{p,1Tx}}(m,\hat{l})\}]^{*}\} \\ &= \frac{1}{gN} \cdot (g+K) \cdot (L-1) \cdot 2\sigma^2 + \frac{1}{gN \cdot E_{\mathrm{c,1Tx}}} \cdot \frac{\eta_0}{2} \end{aligned} \tag{6.69}$$

It is noticed that the above initial channel estimates with the assistance of pilot transmission suffer not only from the background AWGN, but also from the multiple access and multipath interference resulting from the data channels of all simultaneously active users as well as the pilot channel. It is seen that the variance in (6.69) is equivalent to the variance in (6.20) under the assumption of constant total transmit power. Thus, from the statistical perspective the channel estimation scheme in the conventional DS-CDMA system with only one transmit antenna suffers from as much MPI as that for the channel estimation scheme adopted in the system with two transmit antennas and TD-STBC.

For the slow fading channel, it is assumed that the channel coefficient $h_{1,l}(m)$ remains invariable during several consecutive symbol intervals, thus an N_{P}-tap lowpass filter is used to improve the accuracy of the pilot-assisted channel estimation. In other words, the improved channel estimation can be obtained through time averaging,

$$\hat{h}_{1,\hat{l}}(m) = \frac{1}{N_{\mathrm{P}}} \cdot \sum_{m \in \Theta} \hat{r}_{\mathrm{p,1Tx}}(m, \hat{l}) \tag{6.70}$$

where any one of the example sets given by (6.22) or (6.23) might be adopted as the timing average window. Assuming that the noise and interference terms in successive $\hat{r}_{\mathrm{p,1Tx}}(m, \hat{l})$ with different m are independent from each other, one obtains

$$E\{\hat{h}_{1,\hat{l}}(m)\} = E\{\hat{r}_{\mathrm{p,1Tx}}(m, \hat{l})\} = h_{1,\hat{l}}(m) \tag{6.71}$$

$$\mathrm{Var}\{\hat{h}_{1,\hat{l}}(m)\} = \frac{1}{N_{\mathrm{P}}} \cdot \mathrm{Var}\{\hat{r}_{p,\mathrm{1Tx}}(m, \hat{l})\} \tag{6.72}$$

With respect to the $\hat{l}$th resolvable path, the branch random variable of the estimated data bit of the first user in the mth symbol period can be constructed through coherent reception as

$$\hat{d}_{\mathrm{1Tx}}(m, \hat{l}) = \hat{r}_{\mathrm{d,1Tx}}(m, \hat{l}) \cdot \hat{h}^*{}_{1,\hat{l}}(m) \tag{6.73}$$

Since the data bits from K active users and the discrete aperiodic correlation functions of two different spreading code sequences are independent random variables, $\hat{r}_{\mathrm{d,1Tx}}(m, \hat{l})$ and $\hat{h}_{1,\hat{l}}(m)$ are independent from each other. Moreover, the branch random variable of the estimated data bit $\hat{d}_{\mathrm{1Tx}}(m, \hat{l})$ is the sum of many independent random variables and can be approximated as a Gaussian variable conditioned on the channel fading coefficient at the $\hat{l}$th path, $h_{1,\hat{l}}(m)$. Therefore, the conditional mean and variance of $\hat{d}_{\mathrm{1Tx}}(m, \hat{l})$ are derived as follows, respectively:

$$\begin{aligned} E\{\hat{d}_{\mathrm{1Tx}}(m, \hat{l})\} &= E\{\hat{r}_{\mathrm{d,1Tx}}(m, \hat{l})\} \cdot E\{\hat{h}^*{}_{1,\hat{l}}(m)\} \\ &= \sqrt{E_{\mathrm{c,1Tx}}} \cdot N \cdot b(1, m) \cdot \left|h_{1,\hat{l}}(m)\right|^2 \end{aligned} \tag{6.74}$$

$$\begin{aligned} \mathrm{Var}\{\hat{d}_{\mathrm{1Tx}}(m, \hat{l})\} &= |E\{\hat{r}_{\mathrm{d,1Tx}}(m, \hat{l})\}|^2 \cdot \mathrm{Var}\{\hat{h}^*{}_{1,\hat{l}}(m)\} + |E\{\hat{h}^*{}_{1,\hat{l}}(m)\}|^2 \cdot \mathrm{Var}\{\hat{r}_{\mathrm{d,1Tx}}(m, \hat{l})\} \\ &\quad + \mathrm{Var}\{\hat{r}_{\mathrm{d,1Tx}}(m, \hat{l})\} \cdot \mathrm{Var}\{\hat{h}^*{}_{1,\hat{l}}(m)\} \\ &= N \cdot \left(1 + \frac{1}{gN_{\mathrm{P}}}\right) \cdot \left[E_{\mathrm{c,1Tx}} \cdot (g + K) \cdot (L-1) \cdot 2\sigma^2 + \frac{\eta_0}{2}\right] \cdot |h_{1,\hat{l}}(m)|^2 \\ &\quad + \frac{1}{gN_{\mathrm{P}} \cdot E_{\mathrm{c,1Tx}}} \cdot \left[E_{\mathrm{c,1Tx}} \cdot (g + K) \cdot (L-1) \cdot 2\sigma^2 + \frac{\eta_0}{2}\right]^2 \end{aligned} \tag{6.75}$$

After despreading at each path, a conventional Rake receiver is used to combine the signal energy from L_c ($L_c \leq L$) selected paths. This can obtain the path diversity gain benefiting from DS-CDMA merits.

EGC Rake receiver

The EGC Rake receiver selects and combines signals of the first L_c arriving paths among the L available resolvable paths. Assuming that the fading of each path is mutually independent and the decision random variables in distinct branches are independent from each other, the output of the Rake combiner can be represented as

$$\hat{d}_{1\text{Tx}}(m) = \sum_{l=1}^{L_c} \hat{d}_{1\text{Tx}}(m, l) \tag{6.76}$$

When $L_c = L$, the signals from all resolvable paths are combined. The decision variable $\hat{d}_{1\text{Tx}}(m)$ is a Gaussian variable with the conditional mean and variance given by, respectively,

$$E\{\hat{d}_{1\text{Tx}}(m)\} = \sum_{l=1}^{L_c} E\{\hat{d}_{1\text{Tx}}(m, l)\} = \sqrt{E_{c,1\text{Tx}}} \cdot N \cdot b(1, m) \cdot \zeta_{1\text{Tx}} \tag{6.77}$$

$$\begin{aligned} \text{Var}\{\hat{d}_{1\text{Tx}}(m)\} &= \sum_{l=1}^{L_c} \text{Var}\{\hat{d}_{1\text{Tx}}(m, l)\} \\ &= N \cdot \left(1 + \frac{1}{gN_P}\right) \cdot \left[E_{c,1\text{Tx}} \cdot (g + K) \cdot (L-1) \cdot 2\sigma^2 + \frac{\eta_0}{2}\right] \cdot \zeta_{1\text{Tx}} \\ &\quad + \frac{L_c}{gN_P \cdot E_{c,1\text{Tx}}} \cdot \left[E_{c,1\text{Tx}} \cdot (g + K) \cdot (L-1) \cdot 2\sigma^2 + \frac{\eta_0}{2}\right]^2 \end{aligned} \tag{6.78}$$

where $\zeta_{1\text{Tx}}$ is defined as in (5.96). It can be observed that the variance in (6.78) will reduce to the expression of (5.95) when the parameter $N_P \to \infty$ and $g \ll K$. In other words, when the number of taps of the LPF tends to infinity, the perfect channel estimation is roughly achieved under the assumption of static channels. Therefore, the system performance will approximate to the corresponding results presented in Section 5.3.4.

Therefore, conditioned on the instantaneous fading channel amplitudes of the multipaths, the BER can be obtained by [1], [2]

$$\begin{aligned} \hat{P}_{e,\text{EGC},1\text{Tx}}(\hat{\zeta}_{1\text{Tx}}) &= Q\left\{\left[\frac{\left|E\{\hat{d}_{1\text{Tx}}(m)\}\right|^2}{\text{Var}\{\hat{d}_{1\text{Tx}}(m)\}}\right]^{1/2}\right\} \\ &= Q\left[\left(\frac{\hat{\zeta}_{1\text{Tx}}^2}{A_{1\text{Tx}} \cdot \hat{\zeta}_{1\text{Tx}} + L_c \cdot B_{1\text{Tx}}}\right)^{1/2}\right] \end{aligned} \tag{6.79}$$

where, by the definitions,

$$\hat{\zeta}_{1\text{Tx}} = \frac{1}{2\sigma^2} \cdot \sum_{l=1}^{L_c} |h_{1,l}(m)|^2 \tag{6.80}$$

$$A_{1\text{Tx}} = \left(\frac{1}{N} + \frac{1}{gN_P \cdot N}\right) \cdot \left[(g + K) \cdot (L-1) + \frac{N}{2\bar{\gamma}_{b,1\text{Tx}}}\right] \tag{6.81}$$

$$B_{1\text{Tx}} = \frac{1}{gN_P \cdot N^2} \cdot \left[(g + K) \cdot (L-1) + \frac{N}{2\bar{\gamma}_{b,1\text{Tx}}}\right]^2 \tag{6.82}$$

where $\bar{\gamma}_{\mathrm{b,1Tx}}$ is the average SNR per bit defined in (5.103) and rewritten here:

$$\bar{\gamma}_{\mathrm{b,1Tx}} = \frac{2E_{\mathrm{c,1Tx}} \cdot N \cdot \sigma^2}{\eta_0} \tag{6.83}$$

Since the channel fading coefficient, $h_{1,l}(m)$, is an i.i.d. complex Gaussian distributed random variable, the normalized random variable $\hat{\zeta}_{\mathrm{1Tx}}$ in (6.80) follows a chi-square distribution with $2L_{\mathrm{c}}$ degrees of freedom, and its PDF is given by

$$p_{\hat{\zeta},\mathrm{1Tx}}(\hat{\zeta}_{\mathrm{1Tx}}) = \frac{1}{\Gamma(L_{\mathrm{c}})} \cdot \hat{\zeta}_{\mathrm{1Tx}}^{L_{\mathrm{c}}-1} \cdot \mathrm{e}^{-\hat{\zeta}_{\mathrm{1Tx}}} \tag{6.84}$$

Therefore, the resultant BER can be obtained by averaging the conditional BER $\hat{P}_{\mathrm{e,EGC,1Tx}}(\hat{\zeta}_{\mathrm{1Tx}})$ in (6.79) over the PDF of $\hat{\zeta}_{\mathrm{1Tx}}$ in (6.84), i.e.

$$\begin{aligned}
\hat{P}_{\mathrm{e,1Tx}} &= \int_0^{\infty} \hat{P}_{\mathrm{e,EGC,1Tx}}(\hat{\zeta}_{\mathrm{1Tx}}) \cdot p_{\hat{\zeta},\mathrm{1Tx}}(\hat{\zeta}_{\mathrm{1Tx}}) \cdot \mathrm{d}\hat{\zeta}_{\mathrm{1Tx}} \\
&= \int_0^{\infty} Q\left[\left(\frac{\hat{\zeta}_{\mathrm{1Tx}}^2}{A_{\mathrm{1Tx}} \cdot \hat{\zeta}_{\mathrm{1Tx}} + L_{\mathrm{c}} \cdot B_{\mathrm{1Tx}}}\right)^{1/2}\right] \cdot p_{\hat{\zeta},\mathrm{1Tx}}(\hat{\zeta}_{\mathrm{1Tx}}) \cdot \mathrm{d}\hat{\zeta}_{\mathrm{1Tx}} \\
&= \frac{1}{\Gamma(L_{\mathrm{c}})} \cdot \int_0^{\infty} Q\left(\sqrt{\frac{y^2}{A_{\mathrm{1Tx}} \cdot y + L_{\mathrm{c}} \cdot B_{\mathrm{1Tx}}}}\right) \cdot y^{L_{\mathrm{c}}-1} \cdot \mathrm{e}^{-y} \cdot \mathrm{d}y
\end{aligned} \tag{6.85}$$

Again, let us review and compare this with the EGC 2D-Rake receiver discussed in Section 6.3.2. When $h_{1,l}(m) = h_{2,l}(m)$ and $E_{\mathrm{c,1Tx}} = 2E_{\mathrm{c}}$, it is noticed that the conditional BER performance of the EGC 2D-Rake receiver given by (6.35) degrades to the BER of the conventional EGC Rake receiver in the CDMA system without TD-STBC given by (6.79).

Generalized selection combining Rake receiver

The GSC Rake receiver is exploited to adaptively select and combine the signals from the L_{c} paths with highest SINR among the L available resolvable paths. Conditioned on the channel fading coefficient of the lth path, the instantaneous SINR of the branch decision random variable $\hat{d}_{\mathrm{1Tx}}(m, l)$ $(l = 1, 2, \ldots, L)$ in the lth resolvable path can be induced from (6.74) and (6.75) as follows:

$$\hat{\gamma}_{l,\mathrm{1Tx}} = \frac{\left|E\{\hat{d}_{\mathrm{1Tx}}(m,l)\}\right|^2}{\mathrm{Var}\{\hat{d}_{\mathrm{1Tx}}(m,l)\}} = \frac{\hat{\zeta}_{l,1Tx}^2}{A_{\mathrm{1Tx}} \cdot \hat{\zeta}_{l,\mathrm{1Tx}} + L_{\mathrm{c}} \cdot B_{\mathrm{1Tx}}} \tag{6.86}$$

where, by the definition,

$$\hat{\zeta}_{l,\mathrm{1Tx}} = \frac{1}{2\sigma^2} \cdot |h_{1,l}(m)|^2 \tag{6.87}$$

It can be proven that the variable $\hat{\gamma}_{l,\mathrm{1Tx}}$ in (6.86) is the monotonously increasing function of the variable $\hat{\zeta}_{l,\mathrm{1Tx}}$ in (6.87). Therefore, selecting the paths with highest instantaneous value of $\hat{\zeta}_{l,\mathrm{1Tx}}$ is equivalent to selecting the paths with highest instantaneous SINR $\hat{\gamma}_{l,\mathrm{1Tx}}$. Moreover, $\hat{\zeta}_{1:L,\mathrm{1Tx}} \geq \hat{\zeta}_{2:L,\mathrm{1Tx}} \geq \cdots \geq \hat{\zeta}_{L:L,\mathrm{1Tx}}$ are defined as the order statistic of the variables, obtained by arranging $\{\hat{\zeta}_{l,\mathrm{1Tx}} \mid l = 1, 2, \ldots, L\}$ in descending order of magnitude. Similarly, the joint PDF of the order statistic variables [4], [5] $\hat{\zeta}_{l:L,1\text{Tx}}$ $(l = 1, 2, \ldots, L_c)$ is given by

$$\begin{aligned} & p_{\hat{\zeta}_{1:L,1\text{Tx}}, \hat{\zeta}_{2:L,1\text{Tx}}, \ldots, \hat{\zeta}_{L_c:L,1\text{Tx}}}(\hat{\zeta}_{1:L,1\text{Tx}}, \hat{\zeta}_{2:L,1\text{Tx}}, \ldots, \hat{\zeta}_{L_c:L,1\text{Tx}}) \\ & \quad = L_c! \cdot \binom{L}{L_c} \cdot \left[F_{\hat{\zeta}_{l,1\text{Tx}}}(\hat{\zeta}_{L_c:L,1\text{Tx}})\right]^{L-L_c} \cdot \prod_{l=1}^{L_c} p_{\hat{\zeta}_{l,1\text{Tx}}}(\hat{\zeta}_{l:L,1\text{Tx}}) \end{aligned} \tag{6.88}$$

where the relevant PDF and CDF of the variable $\hat{\zeta}_{l,1\text{Tx}}$ in (6.87) are given by, respectively,

$$p_{\hat{\zeta}_{l,1\text{Tx}}}(\hat{\zeta}_{l,1\text{Tx}}) = \exp(-\hat{\zeta}_{l,1\text{Tx}}) \tag{6.89}$$

$$F_{\hat{\zeta}_{l,1\text{Tx}}}(\hat{\zeta}_{l,1\text{Tx}}) = 1 - \exp(-\hat{\zeta}_{l,1\text{Tx}}) \tag{6.90}$$

It is assumed that the channel fading along multiple paths are mutually independent and that the decision random variables in distinct branches of the Rake receiver are independent from each other. Similarly, based on the branch SINR in (6.86) and conditioned on the random variables $\{\hat{\zeta}_{l:L,1\text{Tx}} \mid l = 1, 2, \ldots, L_c\}$ of the selected paths, the total output SINR of the GSC Rake receiver that selects and combines L_c strongest paths can be represented as [4]

$$\hat{\gamma}_{\text{GSC},1\text{Tx}} = \frac{\left(\sum_{l=1}^{L_c} \hat{\zeta}_{l:L,1\text{Tx}}\right)^2}{A_{1\text{Tx}} \cdot \sum_{l=1}^{L_c} \hat{\zeta}_{l:L,1\text{Tx}} + L_c \cdot B_{1\text{Tx}}} \tag{6.91}$$

Furthermore, conditioned on $\{\hat{\zeta}_{l:L,1\text{Tx}} \mid l = 1, 2, \ldots, L_c\}$, the BER can be obtained by

$$\hat{P}_{\text{e,GSC}}(\hat{\gamma}_{\text{GSC},1\text{Tx}}) = Q(\sqrt{\hat{\gamma}_{\text{GSC},1\text{Tx}}}) \tag{6.92}$$

Thus, after the L_c strongest paths are selected and combined by the GSC Rake receiver, the average BER can be obtained by averaging the conditional BER in (6.92) over the joint PDF of the selected fading paths in (6.88):

$$\begin{aligned} \hat{P}_{\text{e},1\text{Tx}}^{<L_c>} &= \int_0^\infty \int_{\hat{\zeta}_{L_c:L,1\text{Tx}}}^\infty \cdots \int_{\hat{\zeta}_{2:L,1\text{Tx}}}^\infty \hat{P}_{\text{e,GSC}}(\hat{\gamma}_{\text{GSC},1\text{Tx}}) \cdot p_{\hat{\zeta}_{1:L,1\text{Tx}}, \hat{\zeta}_{2:L,1\text{Tx}}, \ldots, \hat{\zeta}_{L_c:L,1\text{Tx}}} \\ & \quad \cdot (\hat{\zeta}_{1:L,1\text{Tx}}, \hat{\zeta}_{2:L,1\text{Tx}}, \ldots, \hat{\zeta}_{L_c:L,1\text{Tx}}) \cdot d\hat{\zeta}_{1:L,1\text{Tx}} \cdot d\hat{\zeta}_{2:L,1\text{Tx}} \cdots d\hat{\zeta}_{L_c:L,1\text{Tx}} \\ &= L_c! \cdot \binom{L}{L_c} \cdot \int_0^\infty \int_{\hat{\zeta}_{L_c:L,1\text{Tx}}}^\infty \cdots \int_{\hat{\zeta}_{2:L,1\text{Tx}}}^\infty [1 - \exp(-\hat{\zeta}_{L_c:L,1\text{Tx}})]^{L-L_c} \\ & \quad \cdot \exp\left(-\sum_{l=1}^{L_c} \hat{\zeta}_{l:L,1\text{Tx}}\right) \\ & \quad \cdot Q\left[\left(\sum_{l=1}^{L_c} \hat{\zeta}_{l:L,1\text{Tx}}\right) \Big/ \left(A_{1\text{Tx}} \cdot \sum_{l=1}^{L_c} \hat{\zeta}_{l:L,1\text{Tx}} + L_c \cdot B_{1\text{Tx}}\right)^{\frac{1}{2}}\right] \\ & \quad \cdot d\hat{\zeta}_{1:L,1\text{Tx}} \cdot d\hat{\zeta}_{2:L,1\text{Tx}} \cdots d\hat{\zeta}_{L_c:L,1\text{Tx}} \end{aligned} \tag{6.93}$$

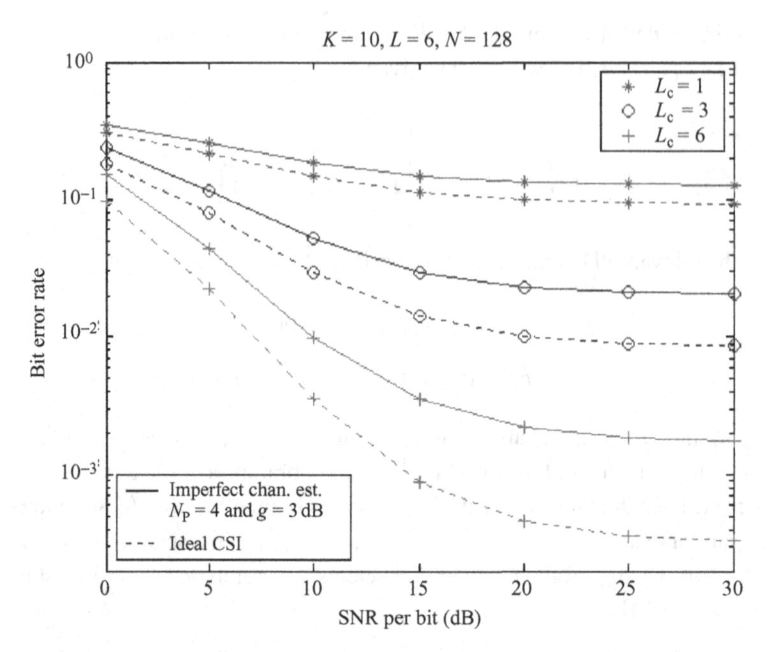

Figure 6.3 BER comparison of EGC 2D-Rake receiver with imperfect channel estimation and ideal CSI available.

6.4 Numerical results and discussion

In this section, the BER performance of the DS-CDMA system with TD-STBC and the effect of parameters on the system performance are numerically evaluated for different system configurations. In order to demonstrate the spatial and path diversity gain provided by the 2D-Rake receiver, the total transmit power is restricted as constant, irrespective of the number of transmit antennas and the number of resolvable multipaths. Therefore, the parameter of SNR per bit per antenna per path $\bar{\gamma}_{\mathrm{p}}$ in (6.37) is used to calculate BER while the SNR per bit $\bar{\gamma}_{\mathrm{b}}$ in (6.39) is used to plot the performance figures. Unless noted otherwise, the number of resolvable multipaths $L = 6$, the number of active users $K = 10$, the spreading factor $N = 128$, the power ratio of pilot to data channel $g = 3$ dB and the number of taps of LPF adopted $N_{\mathrm{P}} = 4$.

6.4.1 BER performance of the 2D-Rake receiver

In Figure 6.3, the BER of the EGC 2D-Rake receiver is illustrated versus the average SNR per bit $\bar{\gamma}_{\mathrm{b}}$ for different number of Rake fingers, i.e. $L_{\mathrm{c}} = 1,\ 3,\ 6$, respectively. For comparison, the BER results of both imperfect channel estimation and ideal CSI are shown. The latter is discussed in Chapter 5. It is clearly seen that the Rake receiver improves the performance for both imperfect channel estimation and ideal CSI when the number of fingers increases. However, the improvement in BER for imperfect channel estimation is not as much as that for ideal CSI. A similar trend in BER can be seen and

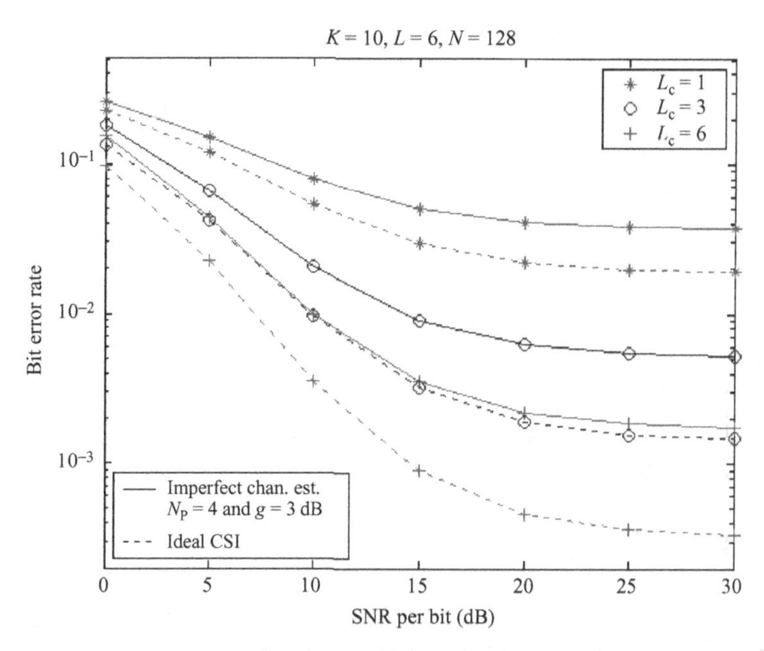

Figure 6.4 BER comparison of GSC 2D-Rake receivers with imperfect channel estimation and ideal CSI available.

the same conclusion can be obtained from Figure 6.4 when GSC is used instead of EGC. In order to clearly show the difference of BER performance between EGC and GSC, the two idle curves ($L_c = 3$) from both Figures 6.3 and 6.4 are placed into Figure 6.5. It can be seen that the GSC outperforms the EGC. However, the cost of this out-performance is more implementation complexity.

Figure 6.6 shows directly the BER performance of the full (i.e. $L_c = L = 6$) GSC or EGC 2D-Rake receiver under perfect and imperfect channel estimation for a different number of active users. It can be seen that the gap in BER between the imperfect channel estimation and ideal CSI decreases when the number of users increases. However, the required SNR to achieve a desired BER actually increases with K, therefore the accurate channel estimation becomes increasingly more important as more users are involved. From Figures 6.3–6.6, it is seen that the BER performance is significantly degraded due to the imperfect channel estimation.

In Figure 6.7, the BER of both the EGC 2D-Rake receiver and the conventional EGC Rake receiver are illustrated versus the average SNR per bit $\bar{\gamma}_b$ for different numbers of Rake fingers, i.e. $L_c = 1, 2, 4, 6$. By comparison of the dotted and solid-line curves, it is observed that even for the imperfect channel estimation, the TD-STBC provides significant spatial diversity gain with respect to the BER.

In Figure 6.8, the BER of the full 2D-Rake receiver is plotted versus the number of resolvable multipaths L for various numbers of users. The concerned average SNR per bit $\bar{\gamma}_b = 10\,\text{dB}$ is in the middle of the noise-limited range and no LPF is adopted. When the total bandwidth (and L) increases, the SF increases as $64L$ since the data

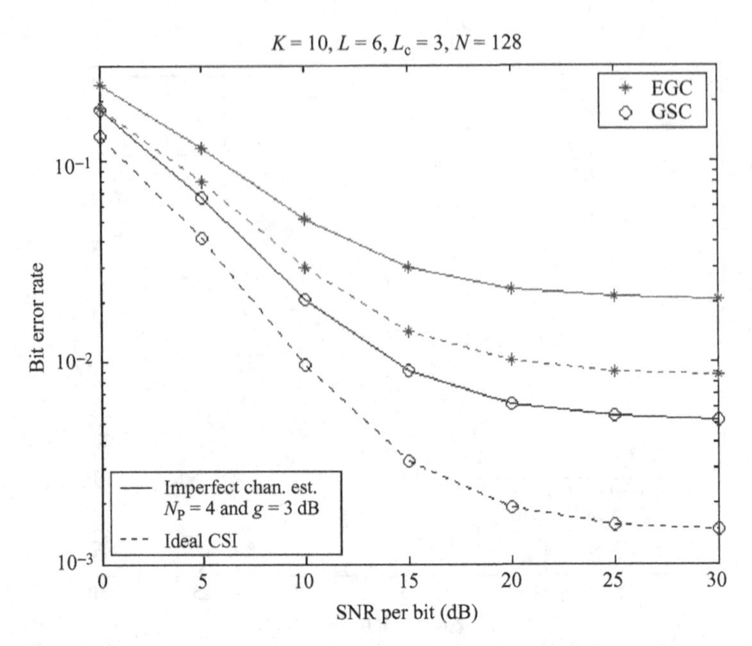

Figure 6.5 BER comparison of EGC and GSC 2D-Rake receivers with $L_c = 3$ for both imperfect channel estimation and ideal CSI available.

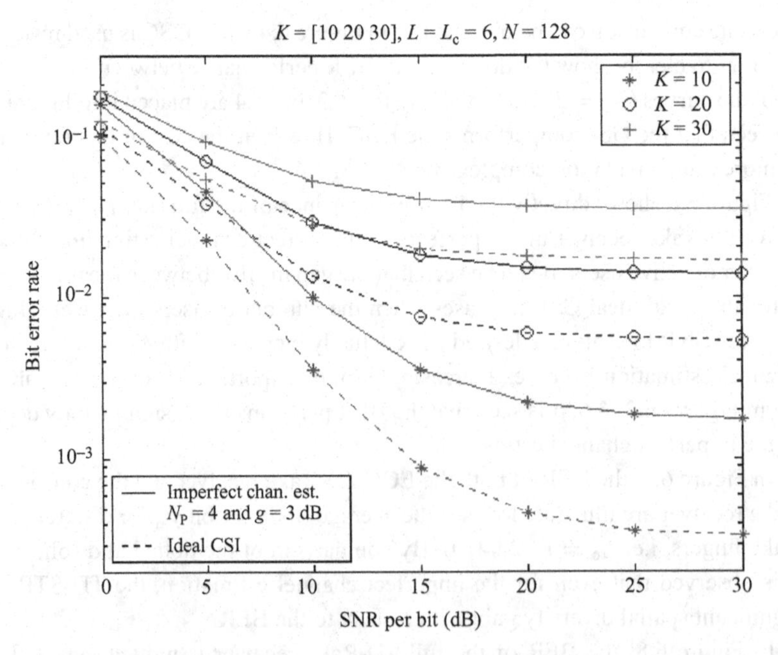

Figure 6.6 BER of full GSC/EGC 2D-Rake receiver with different number of users for both imperfect channel estimation and ideal CSI available.

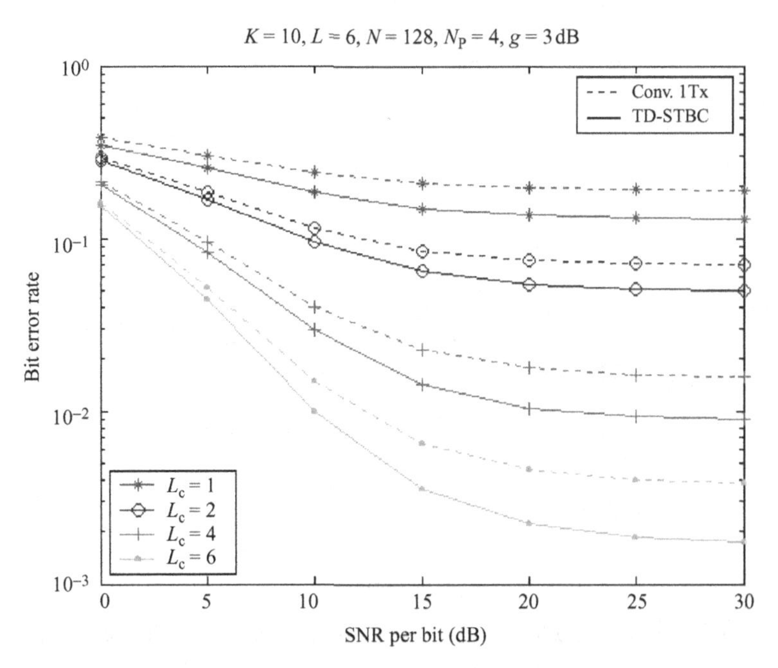

Figure 6.7 BER comparison of EGC 2D-Rake receiver with TD-STBC and conventional EGC RAKE receiver with only one transmit antenna for imperfect channel estimation.

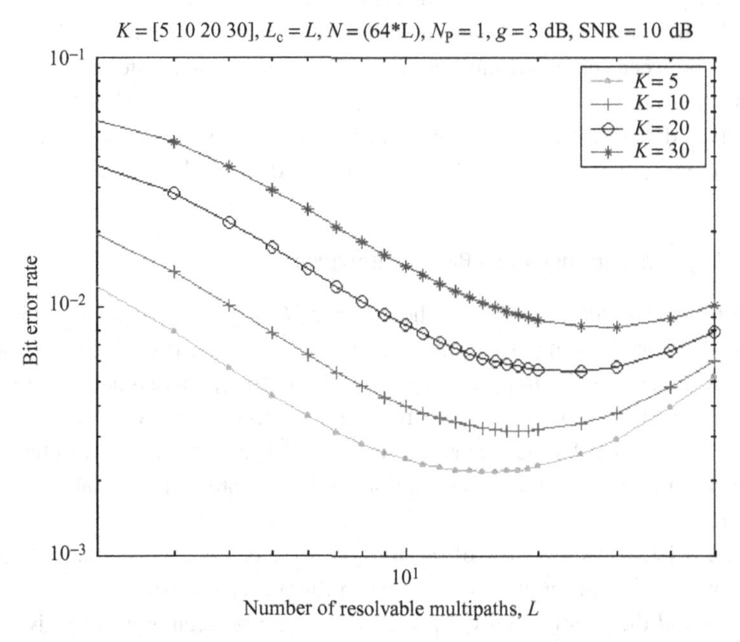

Figure 6.8 BER of full 2D-Rake receiver versus the number of resolvable multipaths with fixed data rate for imperfect channel estimation.

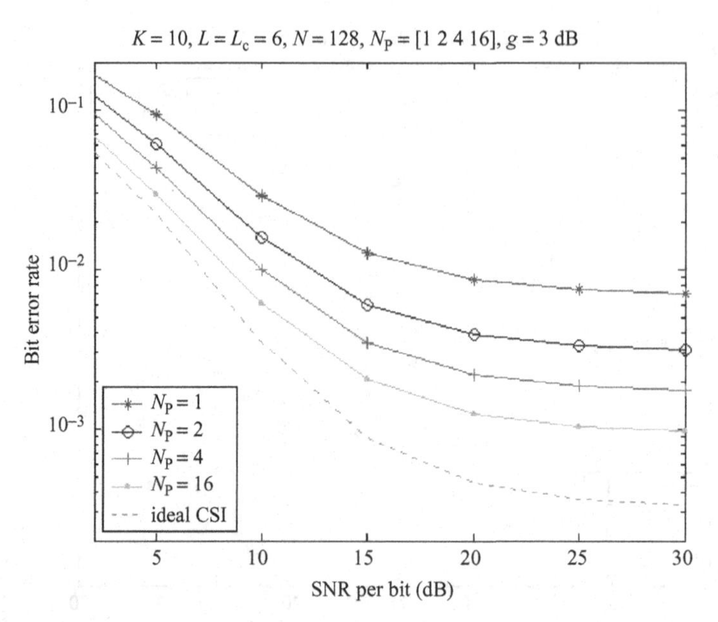

Figure 6.9 BER of full 2D-Rake receiver with a different quality of channel estimation.

rate is fixed. On the other hand, since the total transmit power remains constant, the power in each resolvable path decreases. Therefore, as L increases, the system benefits from increased processing gain and increased multipath diversity, but suffers from larger channel estimation error and more MPI. From the convex-cup shape of the curves, the trade-off between increased multipath diversity and less accurate channel estimates is shown and a varying optimal L for different system scenarios can be observed. Because of the fixed data rate, transmitting with a bandwidth greater than the optimal bandwidth results in not only increased BER, but also reduced spectral efficiency.

6.4.2 Effect of system parameters on BER performance

Figure 6.9 illustrates the BER of the full (i.e. $L_c = L = 6$) GSC or EGC 2D-Rake receiver for various numbers of taps, i.e. $N_P = 1, 2, 4, 16$. It can be seen that the system performance can be improved significantly when N_P increases in a static (or slow) fading channel. This is because increasing N_P means improving the quality of the channel estimation. As the number of taps of LPF increases, the resultant BER performance for imperfect channel estimation tends asymptotically to that for the ideal CSI scenario.

Figure 6.10 shows the BER of the full 2D-Rake receiver versus the number of taps of LPF with different number of users when the power ratio of pilot to data channel is $g = 0$ dB and the average SNR, $\overline{\gamma}_b = 20$ dB. It can be seen that for a given K, when N_P increases from a small number, the performance of imperfect channel estimation improves significantly. However, further increasing N_P only improves the performance slightly, e.g., increasing N_P beyond 20 improves BER performance insignificantly.

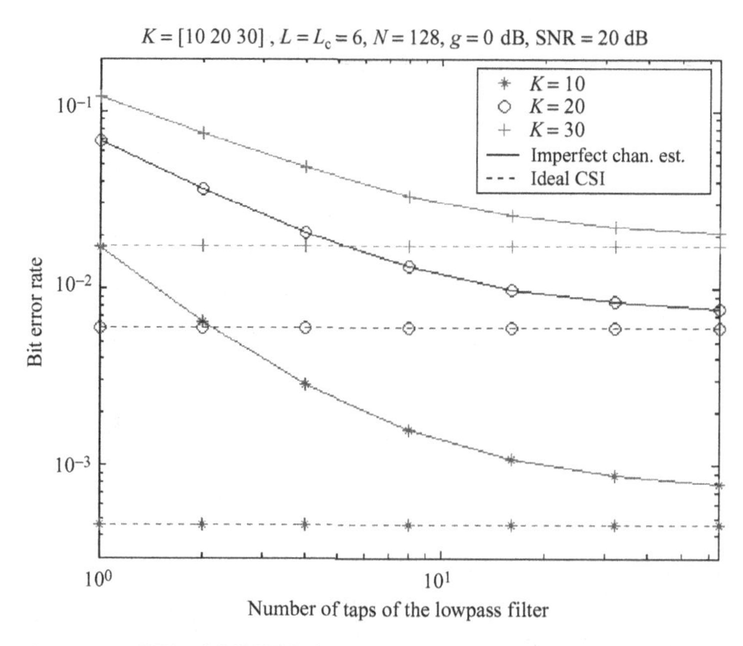

Figure 6.10 BER of full 2D-Rake receiver versus the number of taps of LPF with a different number of users.

Figure 6.11 illustrates the BER of the full 2D-Rake receiver versus the power ratio g with different numbers of active users. The number of taps of LPF is $N_P = 4$ and $\bar{\gamma}_b = 20$ dB. It can be observed that from the BER perspective an optimal power ratio g exists for a given K and increases as the number of active users increases. This is because when the number of users is large, the multiple access and multipath interference increases with respect to the pilot channel. In order to maintain the quality of channel estimates, more pilot power is needed. For $K = 10$, the optimal power ratio g is around 3 dB. Therefore, this implies that the BS should adjust dynamically the power ratio of the pilot to data channels according to the number of total active system users.

In Figure 6.12, the system capacity (number of active users, K) is demonstrated versus the power ratio g for the full 2D-Rake receiver. The average SNR per bit is $\bar{\gamma}_b = 20$ dB. For a given BER, it is also observed that an optimal value of g exists but it is rather diverse for different BER levels. From the system capacity perspective, the optimal value of g slightly decreases when the value of the given BER decreases.

Furthermore, in Figure 6.13, for the full 2D-Rake receiver the average optimal power ratio g_{opt} defined in (6.47) is plotted versus the number of taps of the LPF with different numbers of active users. Again, the average SNR per bit is large, i.e. $\bar{\gamma}_b = 20$ dB. It can be seen that the optimal value, g_{opt}, decreases monotonously as the number of taps of LPF increases. This is because increasing N_P can improve the channel estimation so that the transmit power of the pilot signal can be proportionally reduced to maintain a given channel estimation quality. Moreover, the optimal ratio g_{opt} increases as the number of active users increases. This is consistent with Figure 6.11.

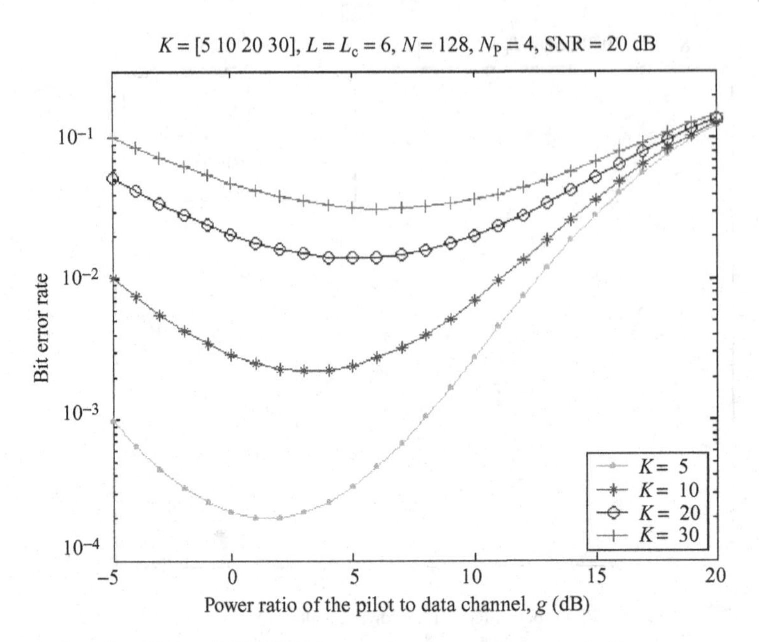

Figure 6.11 BER of full 2D-Rake receiver versus the power ratio g.

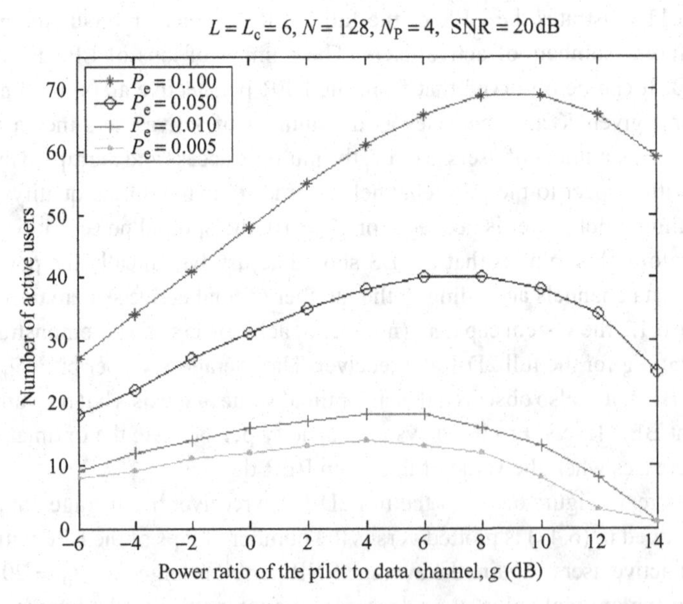

Figure 6.12 Number of active users versus the power ratio g with different BER for full 2D-Rake receiver.

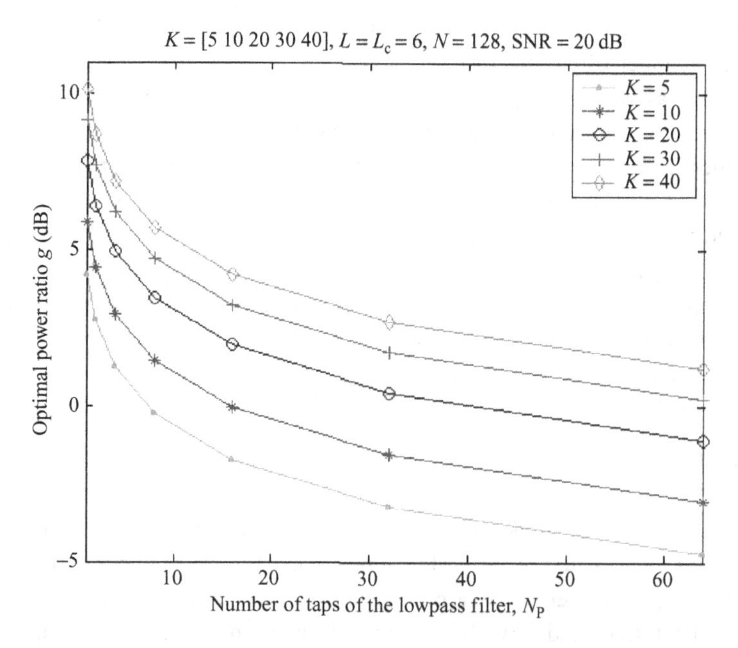

Figure 6.13 Optimal power ratio g_{opt} versus the number of taps of LPF.

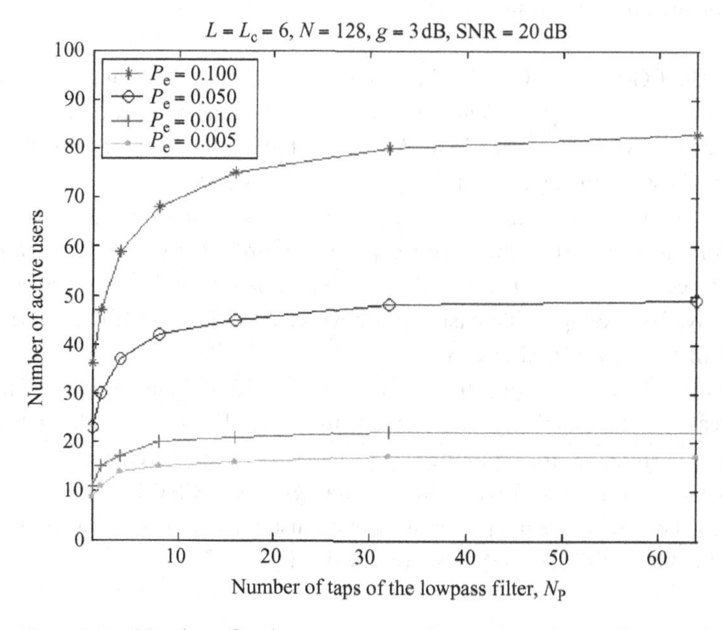

Figure 6.14 Number of active users versus the number of taps of LPF with different BER for full 2D-Rake receiver.

Finally, in Figure 6.14, the system capacity (number of active users, K) is demonstrated versus the number of taps of LPF with different values of BER for the full 2D-Rake receiver. The average SNR per bit is also $\bar{\gamma}_b = 20$ dB. It can be seen that the system capacity is significantly increased when N_P increases from a small number. However, further increasing N_P only improves performance slightly, e.g., increasing N_P beyond 20 improves system capacity insignificantly. This is consistent with Figure 6.10. On the other hand, the relatively large BERs can be actually reduced to acceptable levels by using channel coding (e.g. convolutional or Turbo coding). That is, the system performance can be significantly improved by taking advantage of LPFs with a large value of N_P in conjunction with proper channel coding.

6.5 Summary

In this chapter, the downlink BER performance of CDMA systems with and without TD-STBC is investigated for imperfect channel estimation. Without demand for additional spreading code sequence, a pilot-signal-assisted channel estimation approach is exploited. Both EGC and GSC 2D-Rake receivers are investigated. It is shown that the imperfect channel estimation degrades significantly the performance of the TD-STBC system. Furthermore, the LPF can be employed to increase the accuracy of channel estimates, thus improving the BER performance under the slow fading channel scenario. The following conclusions are drawn:

(1) Both the EGC and GSC 2D-Rake receivers with imperfect channel estimation also improve the system performance significantly when the number of fingers increases. However, GSC outperforms EGC. Moreover, the improvement in BER for imperfect channel estimation is not as much as that for ideal CSI.
(2) When the number of taps of LPF, N_P, increases from a small number, the BER performance, as well as the system capacity, for imperfect channel estimation can be improved significantly. However, further increasing N_P only improves performance slightly. For example, increasing N_P beyond 20 improves BER performance and system capacity insignificantly.
(3) From the BER perspective, the optimal power ratio of the pilot to one data channel increases as the number of active users increases. For $K = 10$, the optimal power ratio g is around 3 dB. From the system capacity perspective, an optimal value of g also exists but is rather diverse for different given BER levels.
(4) When the total transmit power remains constant, there exists an optimal spread bandwidth for the DS-CDMA system with TD-STBC.

References

[1] Third generation partnership project, www.3gpp.org.
[2] J. G. Proakis, *Digital Communications*, 4th edition. New York: McGraw Hill, 2001.

[3] J. Wang and J. Chen, "Performance of wideband CDMA systems with complex spreading and imperfect channel estimation," *IEEE J. Select. Areas Comm.*, vol. 19, no. 1, pp. 152–163, Jan. 2001.

[4] H. A. David, *Order Statistics*, 2nd edition. New York: John Wiley & Sons, 1981.

[5] B. C. Arnold, N. Balakrishnan and H. N. Nagaraja, *A First Course in Order Statistics*. New York: John Wiley & Sons, 1992

[6] X. Y. Wang, "Transmit diversity in CDMA for wireless communications," Ph.D. Thesis, University of Hong Kong, 2003.

7 QAM with antenna diversity

The coherent MRC reception of PSAM MQAM systems with antenna diversity is studied in this chapter. A general fast time varying fading channel model is assumed. Pilot symbols are periodically inserted during the transmission of data symbols, which are used to track the time varying fading, and to provide channel estimation for data decisions at the receiver. Based on a digital implementation, a coherent demodulation scheme is presented. Channel estimation error due to fast fading and additive noise is studied. System performance is evaluated in terms of BER. The analysis shows that in perfect channel estimation cases, with the antenna diversity technique, the BER performance improves significantly, and higher-order QAM can be employed for higher throughput. It is also found that inaccurate channel estimation limits the benefit of diversity when the modulation order is large. By increasing the length of the channel estimator and the amplitude of the pilot symbol, more accurate channel estimation can be achieved, so that the BER performance is improved. Moreover, when the Doppler frequency is less than $1/2ST$, where S is the number of symbols per time slot and T is the symbol duration, the performance is flat since the channel estimator is robust to fading rate.

7.1 Introduction

During the past several decades, MQAM has been considered for high rate data transmission over wireless links due to its high spectral efficiency [1–4]. To achieve best system throughput, higher-order MQAM can be employed in good channel conditions, e.g. when the received signal power or SNR is high. Moreover, due to the poor performance of higher-order modulations in fading channels, especially in low SNR cases, antenna diversity is employed in this chapter to improve the BER performance. In this chapter, PSAM (pilot symbol assisted modulation) is adopted, where pilot symbols are periodically inserted during the transmission of data symbols. The received pilot information can then be used to provide channel estimation for data decision at the receiver.

In this chapter, a coherent MRC receiver for PSAM MQAM systems with antenna diversity is studied. In particular, the channel estimation error is analyzed in fast fading channels, which leads to a better understanding of the nature of the channel estimation error. The BER performance has also been evaluated with different system parameters, such as the parameters of the channel estimator (the length of the channel estimator and

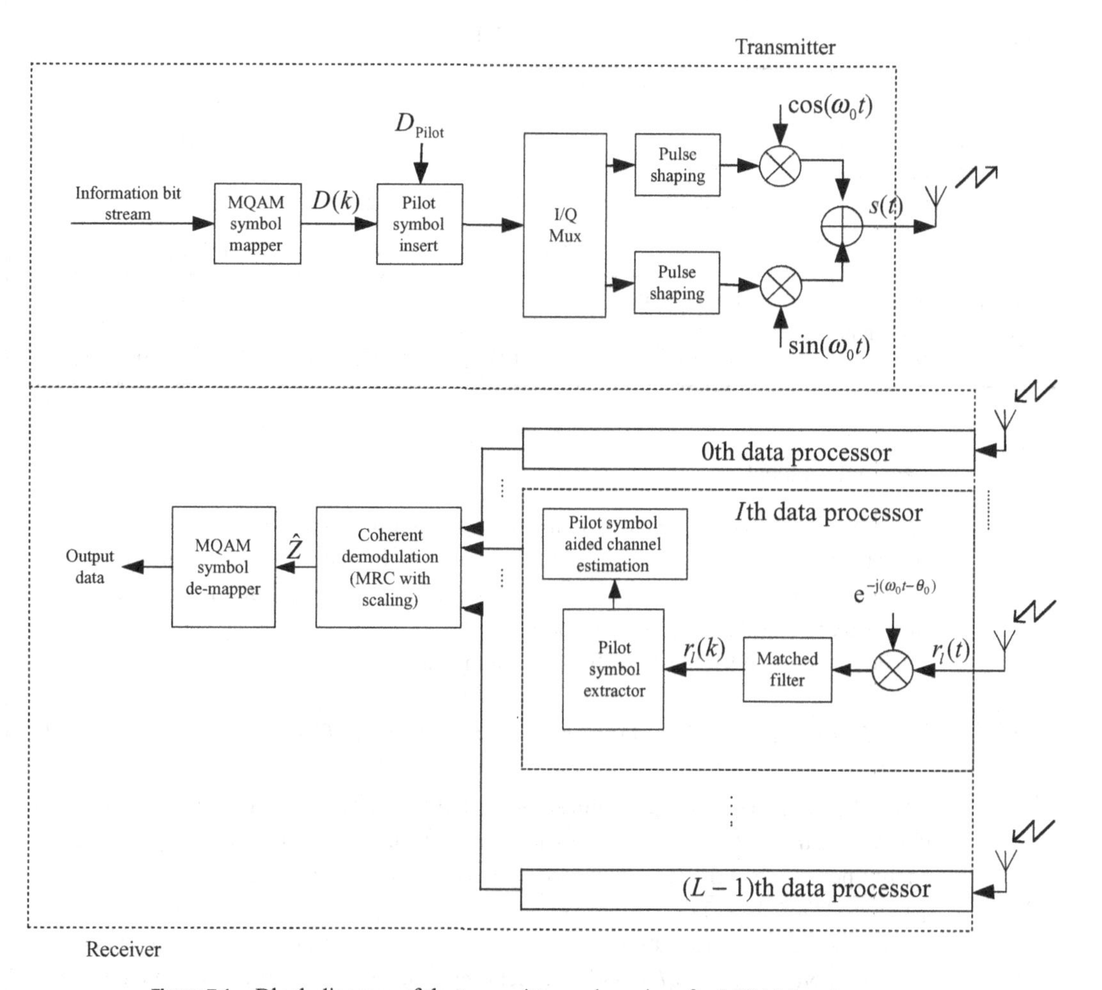

Figure 7.1 Block diagram of the transmitter and receiver for MQAM systems.

the amplitude of the pilot symbol), the Doppler frequency, the modulation order and the antenna diversity order, etc., for insightful study of the effect of imperfect channel estimation on the system performance.

This chapter is organized as follows. In Section 7.2, the transmitter model, channel model and receiver structure are presented. An analytical BER expression in the presence of channel estimation error is given in Section 7.3. Numerical results are discussed in Section 7.4. Finally, conclusions are drawn in Section 7.5.

7.2 System models

A block diagram of the transmitter and receiver of MQAM systems is shown in Figure 7.1.

Table 7.1 The input/output relationship of the gray coder for MQAM

	2-bit gray coder for 16-QAM ($d=\sqrt{2/5}$)			
Input bits (code word g_0g_1)	00	01	11	10
Output $\sqrt{M}$-AM symbol	$-3d$	$-d$	d	$3d$
Symbol label (b_0b_1)	0 (00)	1 (01)	2 (10)	3 (11)

	3-bit gray coder for 64-QAM ($d=\sqrt{1/7}$)							
Input bits (code word $g_0g_1g_2$)	000	001	011	010	110	111	101	100
Output $\sqrt{M}$-AM symbol	$-7d$	$-5d$	$-3d$	$-d$	d	$3d$	$5d$	$7d$
Symbol label ($b_0b_1b_2$)	0 (000)	1 (001)	2 (010)	3 (011)	4 (100)	5 (101)	6 (110)	7 (111)

7.2.1 Transmitter model

As shown in Figure 7.1, the data stream in the transmitter is first fed into an MQAM mapper. At the MQAM mapper, the data bit stream is split into the in-phase (I) and quadrature (Q) bit streams, which are separately Gray coded as $\sqrt{M}$-amplitude modulation (AM) signals and mapped to complex symbols.

The transmitted MQAM symbols are defined as

$$D(k_1,k_2)=\pm(2k_1+1)d\pm j(2k_2+1)d,\ k_1,k_2\in\{0,1,\ldots,\sqrt{M}/2-1\} \tag{7.1}$$

where the in-phase and quadrature parts of the complex MQAM symbol are from the set $\{\pm d,\pm 3d,\ldots,\pm(\sqrt{M}-1)d\}$, and in order to guarantee equal average energy per bit for fair performance comparison with different modulation order, $d=\sqrt{(3\log_2 M)/2(M-1)}$.

The Gray coding procedure for each $\sqrt{M}$-AM signal is as in [4]: sort elements in the set $\{\pm d,\pm 3d,\ldots,\pm(\sqrt{M}-1)d\}$ in ascending order; then label them with integers from 0 to $\sqrt{M}-1$; and finally convert the integer labels to their binary form. For the kth symbol corresponding to the element $(2k+1-\sqrt{M})d$, where $k=0,1,\ldots,\sqrt{M}-1$, letting its $(log_2 M)/2$-digit binary equivalently be $b_{0,k}b_{1,k}\ldots b_{\frac{\log_2 M}{2}-1,k}$, one obtains $b_{i,k}=\lfloor k/2^{(\log_2 M)/2-i-1}\rfloor \bmod 2$ for $i=0,1,\ldots,(\log_2 M)/2-1$. Then the corresponding Gray code $g_{0,k}g_{1,k}\ldots g_{(\log_2 M)/2-1,k}$ is given by [4]

$$\begin{aligned} g_{0,k}&=b_{0,k}\\ g_{i,k}&=b_{i,k}\oplus b_{i-1,k} \qquad i=1,2,\ldots,\frac{\log_2 M}{2}-1 \end{aligned} \tag{7.2}$$

where $\oplus$ represents modulo-2 addition. The relationship of the input and output of the encoder is listed in Table 7.1 for $M=16$ and 64 with constellations shown in Figure 7.2.

In PSAM systems, pilot symbols are periodically inserted. The time-multiplexed pilot symbol D_{Pilot} is known to the receiver. As shown in [2], a pilot symbol is inserted between every $S-1$ data symbols, by which an S-symbol slot is formatted.

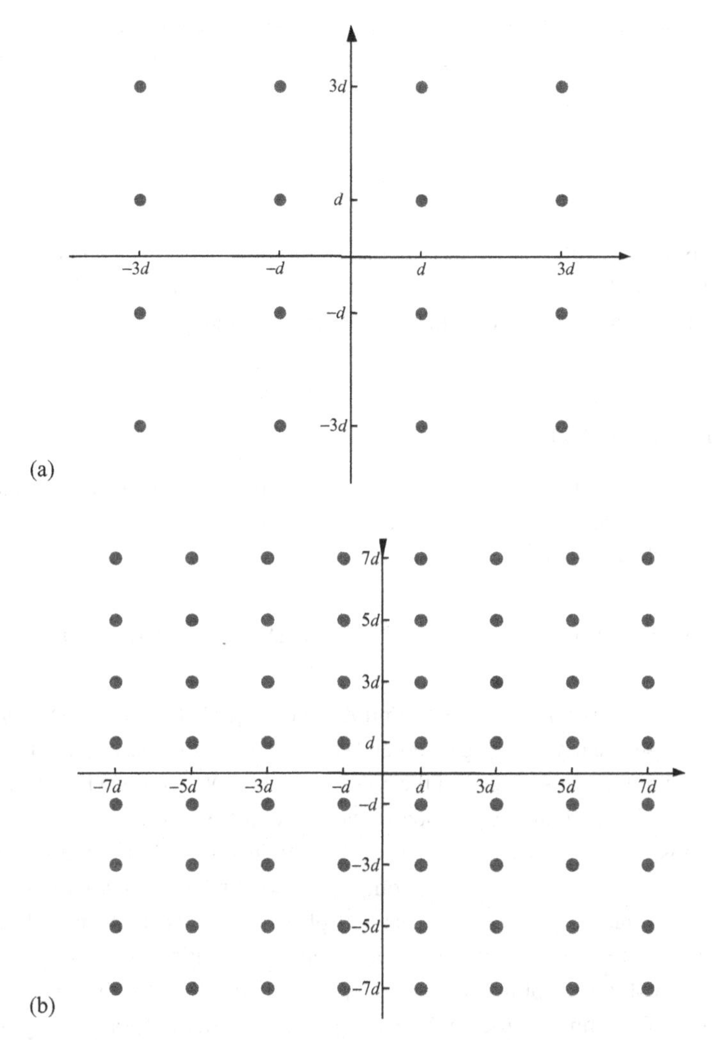

Figure 7.2 The MQAM constellation. (a) $M = 16$ ($d = \sqrt{2/5}$), (b) $M = 64$ ($d = \sqrt{1/7}$).

7.2.2 Channel model

In an L-order antenna diversity system, it is assumed that the channel for each receiver antenna is an independent flat-fading channel at a rate much slower than the symbol rate, so the channel remains constant over one symbol duration. For the lth ($l = 0, 1, \ldots, L - 1$) antenna, the phase $\theta_l(t)$ is a random variable uniformly distributed in $[0, 2\pi)$, and the fading amplitude $\alpha_l(t)$ is a Rayleigh random variable with probability density function (PDF) [5]

$$p(\alpha_l(t)) = \frac{2\alpha_l(t)}{\Omega_l} \mathrm{e}^{-(\alpha_l^2(t)/\Omega_l)}, \quad \alpha_l(t) > 0 \tag{7.3}$$

where Ω_l is the average fading power. Assuming that an equal power fading channel, Ω_l is the same for different l, i.e. $\Omega_l = \Omega$, then the auto-correlation of the channel is represented by [5]

$$E\left\{\alpha_l(t)e^{j\theta_l(t)}\left(\alpha_{\hat{l}}(\tau)e^{j\theta_{\hat{l}}(\tau)}\right)^*\right\} = \Omega_l J_0(2\pi f_D(t-\tau))\delta(l-\hat{l}) \tag{7.4}$$

where f_D is the Doppler frequency of the communication environment, $J_0(\cdot)$ is the zeroth-order Bessel function, and $\delta(l-\hat{l})$ is equal to 1 when $l = \hat{l}$, and zero otherwise. $\theta_l(t)$ and $\alpha_l(t)$ are assumed to be independent for different l.

7.2.3 Coherent receiver structure

Assuming perfect carrier and symbol timing synchronization, the baseband signal at the output of the matched filter for the lth antenna and the mth symbol ($m = 0, 1, \ldots, S-1$) transmitted in the nth slot ($n = 0, 1, \ldots$) may be written as

$$r_l(nS+m) = \alpha_l(nS+m)e^{j\xi_l(nS+m)}D(nS+m) + N_l(nS+m) \tag{7.5}$$

where $D(nS+m)$ is the transmitted data/pilot symbol, $\xi_l(nS+m) = \theta_l(nS+m) + \theta_0$ is the residue phase due to the fading channel $\theta_l(nS+m)$ and the phase difference between the oscillators of the transmitter and the receiver θ_0, and $N_l(nS+m)$ is the background complex AWGN with double-sided power spectral density $N_0/2$.

Without loss of generality, it is assumed that the nSth symbol is the received pilot symbol. From (7.5), the desired signal component (the first term on the right-hand side) is related to the fading gain $\alpha_l(nS+m)$ and phase $\xi_l(nS+m)$. If the fading gain or phase is estimated incorrectly, the improper coherent operating and scaling can lead to incorrect MQAM de-mapping even in the absence of AWGN from the wireless channel [2]. Thus, reliable communication in MQAM systems requires accurate fading estimation and scaling techniques at the receiver.

A coherent MRC receiver with pilot symbol assisted channel estimator is employed. The pilot symbol assisted channel estimator provides the estimated fading gain and phase to eliminate the phase rotation and scale the output signal. Channel fading parameters are extracted by dividing the output of the complex matched filter for the pilot symbol by the known complex pilot symbol. In PSAM systems, since the fading is time-variant, fading parameters in the pilot symbol duration are not the same as those in the other data symbols' duration. Moreover, the received signal suffers from AWGN. Thus, channel estimation error is caused because of the time varying fading parameters and AWGN. Several channel estimation algorithms have been proposed in [2], [3], including the optimum Wiener filter interpolation, low-order Gaussian interpolation (up to second order) or lowpass sinc interpolation. In general, all of them can be regarded as an FIR lowpass filter (LPF) with different sets of filter coefficients. In this chapter, the BER performance will be calculated with a general FIR lowpass filter of

W taps. Channel estimation for the mth data symbol transmitted in the nth slot duration is given by

$$\begin{aligned}\hat{r}(nS+m) &= \sum_{\mathrm{tap}=-\lfloor (W-1)/2\rfloor}^{\lfloor W/2\rfloor} \frac{f_{\mathrm{tap}}^{m} r((n+\mathrm{tap})S)}{D_{\mathrm{Pilot}}} \\ &= \sum_{\mathrm{tap}=-\lfloor (W-1)/2\rfloor}^{\lfloor W/2\rfloor} f_{\mathrm{tap}}^{m} \alpha((n+tap)S) e^{j\xi((n+\mathrm{tap})S)} \\ &+ \sum_{\mathrm{tap}=-\lfloor (W-1)/2\rfloor}^{\lfloor W/2\rfloor} f_{\mathrm{tap}}^{m} \frac{N((n+\mathrm{tap})S)}{D_{\mathrm{Pilot}}}\end{aligned} \tag{7.6}$$

Note that the coefficients set $\left\{f_{\mathrm{tap}}^{m}\right\}$ is different for different m in the slot. So the estimation $\hat{r}(nS+m)$ varies for different m. As stated in [2], the lowpass sinc interpolation can approach almost the same performance as the optimum Wiener interpolation but with less complexity. However, ideal lowpass sinc interpolation is impractical due to the infinite length of the filter. Considering the truncation effect of the finite length of FIR LPF for channel estimation, the normalized interpolation coefficients of FIR filter are defined as

$$f_{\mathrm{tap}}^{m} = \frac{\mathrm{sinc}\left(\frac{m}{S} - \mathrm{tap}\right)}{\sum_{\mathrm{tap1}=-\lfloor (W-1)/2\rfloor}^{\lfloor W/2\rfloor} \mathrm{sinc}\left(\frac{m}{S} - \mathrm{tap1}\right)} \tag{7.7}$$

where $\mathrm{sinc}\,(x) = \dfrac{\sin(\pi x)}{\pi x}$.

According to [3], MRC gives the maximum SNR of the combined signal. Therefore, the scaled output of the coherent MRC receiver for the mth data symbol transmitted in the nth slot duration is

$$\hat{Z}^{I}(nS+m) = \frac{\frac{1}{2}\sum_{l=0}^{L-1}\left(r_l(nS+m)\hat{r}_l^{*}(nS+m) + r_l^{*}(nS+m)\hat{r}_l(nS+m)\right)}{\sum_{l=0}^{L-1}|\hat{r}_l(nS+m)|^2} \tag{7.8}$$

and

$$\hat{Z}^{Q}(nS+m) = \frac{\frac{1}{2j}\sum_{l=0}^{L-1}\left(r_l(nS+m)\hat{r}_l^{*}(nS+m) - r_l^{*}(nS+m)\hat{r}_l(nS+m)\right)}{\sum_{l=0}^{L-1}|\hat{r}_l(nS+m)|^2} \tag{7.9}$$

which are used to demodulate the in-phase and quadrature channel symbols corresponding to the noise-free MQAM constellation as shown in Figure 7.2, respectively.

7.3 BER Performance analysis

7.3.1 General BER performance formula derivation

Since in MQAM systems the outputs of the in-phase and quadrature channels are symmetric, the output of the in-phase channel in the receiver, $\hat{Z}^I(nS+m)$, is studied. The BER performance is obtained by averaging the conditional BER over all possible MQAM constellation points in the first quadrant. Moreover, considering the fact that the quality of channel estimation indicated by cross coefficient ρ in [2] may be different for different m ($m = 1, 2, \ldots, S-1$), the BER for a different m in the slot may be different. Thus, the overall BER for MQAM p_b is given by

$$p_\mathrm{b} = \frac{4}{M(S-1)} \sum_{m=1}^{S-1} \sum_{q_1=0}^{\sqrt{M}/2-1} \sum_{q_2=0}^{\sqrt{M}/2-1} p\,(e|D(nS+m) \\ = (2q_1+1)\,d + j\,(2q_2+1)\,d) \tag{7.10}$$

Concerning the conditional probability $p\,(e\,|D(nS+m))$ in (7.10), it can be easily evaluated by averaging over all possibilities of the decision variable $\hat{Z}^I(nS+m)$. It can be seen that for the real part of a given transmitted symbol, i.e. $D^I(nS+m)$, when $\hat{Z}^I(nS+m) \geq (\sqrt{M}-2)d$, the receiver will detect the real part of the transmitted symbol as $(\sqrt{M}-1)d$. Then there is $Hd\big((\sqrt{M}-1)d, D^I(nS+m)\big)$-bit error among the $(\log_2 M)/2$ transmitted bits being represented by the real part of the transmitted symbol. Here, $Hd(x,y)$ is the Hamming distance between the code words represented by $\sqrt{M}$ -AM symbols x and y. For the Gray coding scheme shown in (7.2), $Hd(x,y)$ can be expressed as

$$Hd(x,y) = \sum_{i=0}^{\frac{\log_2 M}{2}-1} g_{i,k_x} \oplus g_{i,k_y} \tag{7.11}$$

where $x = (2k_x+1-\sqrt{M})d$ and $y = (2k_y+1-\sqrt{M})d$.

Then, with the same method to evaluate the rest of the possibilities of $\hat{Z}^I(nS+m)$, the conditional error probability for $p_\mathrm{e}\,(e\,|D(nS+m))$ can be written as

$$\begin{aligned} &p_\mathrm{e}\,(e\,|D(nS+m)) \\ &= \frac{1}{(\log_2 M)/2} \Bigg\{ Hd((\sqrt{M}-1)d, D^I)p(\hat{Z}^I \geq (\sqrt{M}-2)d|D) \\ &\quad + \sum_{q=-\sqrt{M}/2+1}^{\sqrt{M}/2-2} Hd((2q+1)d, D^I)p(2qd \leq \hat{Z}^I < 2(q+1)d|D) \\ &\quad + Hd((-\sqrt{M}+1)d, D^I)p(\hat{Z}^I < (-\sqrt{M}+2)d|D) \Bigg\} \end{aligned} \tag{7.12}$$

where D and $\hat{Z}^I$ denote $D(nS+m)$ and $\hat{Z}^I(nS+m)$ for simplicity of notation, respectively. Therefore, for MQAM, the focus is on calculating the conditional error probability of $p(\hat{Z}^I(nS+m) < \tilde{B}|D(nS+m))$, where $\tilde{B} = \{0, \pm 2d, \ldots, \pm(\sqrt{M}-2)d\}$. Note that it is a special case of the general quadratic form in Appendix B of [5], where $A = 0$, $B = -\tilde{B}$ and $C = 1/2$.

7.3.2 Conditional BER performance with imperfect channel estimation

Because of time varying fading and AWGN, the channel estimation $\hat{r}_l(nS+m)$ obtained by (7.6) is not the real channel information $\alpha_l(nS+m)e^{j\xi_l(nS+m)}$. Hence, $\hat{r}_l(nS+m)$ obtained by (7.6) can only be regarded as an imperfect channel estimation determined by the length of the estimator, the filter coefficients, the value of the pilot symbol, the Doppler frequency (fading rate), and the additive noise. In this subsection, the BER performance with imperfect channel estimation $\hat{r}_l(nS+m)$ is derived.

Applying the characteristic function method in Appendix B of [5], with equal fading power assumption of the fading channel, one obtains

$$p(\hat{Z}^I(nS+m) < \tilde{B}\,|D(nS+m)) = \frac{\sum_{l=0}^{L-1}\binom{2L-1}{l}\left(-\frac{v_2}{v_1}\right)^l}{\left(1-\frac{v_2}{v_1}\right)^{2L-1}} \tag{7.13}$$

where

$$v_1 = v_0 - \sqrt{v_0^2 + \frac{1}{R_{r|D(nS+m)}R_{\hat{r}|D(nS+m)} - |R_{r\hat{r}|D(nS+m)}|^2}} \tag{7.14}$$

$$v_2 = v_0 + \sqrt{v_0^2 + \frac{1}{R_{r|D(nS+m)}R_{\hat{r}|D(nS+m)} - |R_{r\hat{r}|D(nS+m)}|^2}} \tag{7.15}$$

and

$$v_0 = \frac{-\tilde{B}R_{\hat{r}|D(nS+m)} + \frac{R^*_{r\hat{r}|D(nS+m)} + R_{r\hat{r}|D(nS+m)}}{2}}{R_{r|D(nS+m)}R_{\hat{r}|D(nS+m)} - |R_{r\hat{r}|D(nS+m)}|^2} \tag{7.16}$$

The second moment functions of the received signal $r_l(nS+m)$ and the estimated channel information $\hat{r}_l(nS+m)$, $R_{r|D(nS+m)}$, $R_{\hat{r}|D(nS+m)}$ and $R_{r\hat{r}|D(nS+m)}$ can be expressed as

$$\begin{aligned}
&R_{r|D(nS+m)}\\
&= \frac{1}{2}E\{r_l(nS+m)r_l^*(nS+m)\,|D(nS+m)\}\\
&= \frac{1}{2}\Big\{E\{|D(nS+m)|^2\,\alpha_l(nS+m)e^{j\xi_l(nS+m)}\big(\alpha_l(nS+m)e^{j\xi_l(nS+m)}\big)^*\}\\
&\quad + E\{N_l(nS+m)N_l^*(nS+m)\}\Big\}\\
&= \frac{\Omega\,|D(nS+m)|^2}{2} + \frac{N_0}{2}
\end{aligned} \tag{7.17}$$

where the desired signal is uncorrelated with AWGN components,

$$
\begin{aligned}
&R_{\hat{r}|D\,(nS+m)} \\
&= \frac{1}{2} E\{\hat{r}_l(nS+m)\hat{r}_l^*(nS+m)\,|\,D(nS+m)\} \\
&= \frac{\Omega}{2} \sum_{\text{tap1}=-\lfloor (W-1)/2 \rfloor}^{\lfloor W/2 \rfloor} \sum_{\text{tap2}=-\lfloor (W-1)/2 \rfloor}^{\lfloor W/2 \rfloor} f_{\text{tap1}}^m f_{\text{tap2}}^m J_0(2\pi f_D(\text{tap1} - \text{tap2})ST) \\
&+ \sum_{\text{tap}=-\lfloor (W-1)/2 \rfloor}^{\lfloor W/2 \rfloor} \left(f_{\text{tap}}^m\right)^2 \frac{N_0}{2\,|D_{\text{Pilot}}|^2}
\end{aligned}
\tag{7.18}
$$

where T is the symbol period, f_D denotes the Doppler frequency and

$$
\begin{aligned}
R_{r\hat{r}|D(nS+m)} &= \frac{1}{2} E\{r_l(nS+m)\hat{r}_l^*(nS+m)\,|\,D(nS+m)\} \\
&= \frac{\Omega D(nS+m)}{2} \sum_{\text{tap}=-\lfloor (W-1)/2 \rfloor}^{\lfloor W/2 \rfloor} f_{\text{tap}}^m J_0(2\pi f_D(m - \text{tap} \cdot S)T)
\end{aligned}
\tag{7.19}
$$

7.3.3 Conditional BER performance with perfect channel estimation

In this section, to investigate the performance degradation due to imperfect channel estimation, BER performance with perfect channel estimation is derived, although it cannot be achieved in real communication systems due to the fact that no channel estimator can perfectly estimate the fading parameters.

In perfect channel estimation, the channel estimation $\hat{r}_l(nS+m)$ is assumed to be the real channel information $\alpha_l(nS+m)e^{j\xi_l(nS+m)}$. Assigning $\hat{r}_l(nS+m)$ with $\alpha_l(nS+m)e^{j\xi_l(nS+m)}$, $R_{\hat{r}|D(nS+m)}$ and $R_{r\hat{r}|D(nS+m)}$ can be replaced with

$$
R_{\hat{r}|D(nS+m)} = \frac{1}{2} E\{\hat{r}_l(nS+m)\hat{r}_l^*(nS+m)\,|\,D(nS+m)\} = \frac{\Omega}{2} \tag{7.20}
$$

and

$$
R_{r\hat{r}|D(nS+m)} = \frac{1}{2} E\{r_l(nS+m)\hat{r}_l^*(nS+m)\,|\,D(nS+m)\} = \frac{\Omega D(nS+m)}{2} \tag{7.21}
$$

whereas the other second moment function $R_{r|D(nS+m)}$ is the same as (7.17).

7.4 Numerical results

In this section, the effect of different system parameters (such as modulation, diversity orders, pilot symbol, length of FIR filter for channel estimation, Doppler frequency, etc.) on the BER performance of MQAM systems is investigated. The performance for imperfect channel estimation cases is for systems employing the channel estimator with (7.6), whereas the performance for perfect channel estimation cases is obtained by assuming the estimated channel information $\hat{r}_l(nS+m)$ is the true channel information $\alpha_l(nS+m)e^{j\xi_l(nS+m)}$. Unless noted otherwise, during the performance evaluation by simulation or analytical method, the Rayleigh fading power is assumed to be unitary, i.e. $\Omega = 1$, pilot symbol $D_{\text{Pilot}} = 1 + j$, Doppler frequency f_D is 240 Hz at the symbol transmission rate of 16k symbol/s, i.e. the normalized Doppler spread $f_D T = 0.015$, the length of the slot S is 16, the channel estimator length W is 15, and the diversity order L is 2.

First, the BER performance of MQAM systems without antenna diversity is presented. Figure 7.3 shows the BER performance of MQAM (M = 4, 16, 64, 256) versus SNR with imperfect channel estimation (Figure 7.3(a)) and perfect channel estimation (Figure 7.3(b)). To illustrate the accuracy of the analytical results, simulation results are plotted in Figure 7.3 as well. From the figures, it can be seen that analytical results can accurately predict the system performance. Moreover, the BER performance improves as SNR increases. But the relative performance degradation due to imperfect channel estimation becomes greater for higher SNR cases. This is because in low SNR cases, the AWGN component seriously corrupts the received signals, which results in poor BER performance. Moreover, the channel estimation is very poor due to the effect of both time varying fading channel and serious AWGN. Note that the channel estimation error resulting from the time varying fading channel, which is contained in the second line of (7.6), doesn't change with different SNR. It only dominates system performance in high SNR cases, where the AWGN effect is slight. Then the residual crosstalk between the in-phase and quadrature components due to inaccurate channel estimation results in performance degradation. Since the channel estimation error remains almost constant when the SNR is sufficiently high, performance curves reach error floors. From the figure, it can also be seen that the performance of the 256-QAM system is the worst even with perfect channel estimation. This is because the constellation map of higher-order QAM is more sensitive than that of lower order modulation, i.e. the minimum Euclidean distance of constellation points ($2d$) is smaller for higher-order modulation. Therefore, for a given SNR, the BER performance degrades as M increases.

Meanwhile, it can be seen from Figure 7.3 that to achieve a BER performance of about 10^{-2}, which is acceptable with the help of the forward error correcting code (FEC), the average SNR per bit should be greater than 17 dB, 20 dB and 25 dB for 16-QAM, 64-QAM and 256-QAM with perfect channel estimation, respectively. In imperfect channel estimation cases, however, extra SNR per bit should be invested, e.g. the required SNR are at about 23 dB and 29 dB for 16-QAM and 64-QAM, respectively, while for 256-QAM, the SNR should be greater than 30 dB. This implies that more SNR per bit should be invested as M increases.

As discussed above, in imperfect channel estimation cases, higher SNR is required for higher order modulation, e.g. 16-QAM, 64-QAM and 256-QAM. To improve the system throughput, and to employ higher-order QAM in lower SNR cases, antenna diversity is employed. Theoretically, the performance improves with the increase of the diversity order in perfect channel estimation cases. Here, to investigate the effect of the imperfect channel estimation on the performance for MQAM systems with antenna diversity,

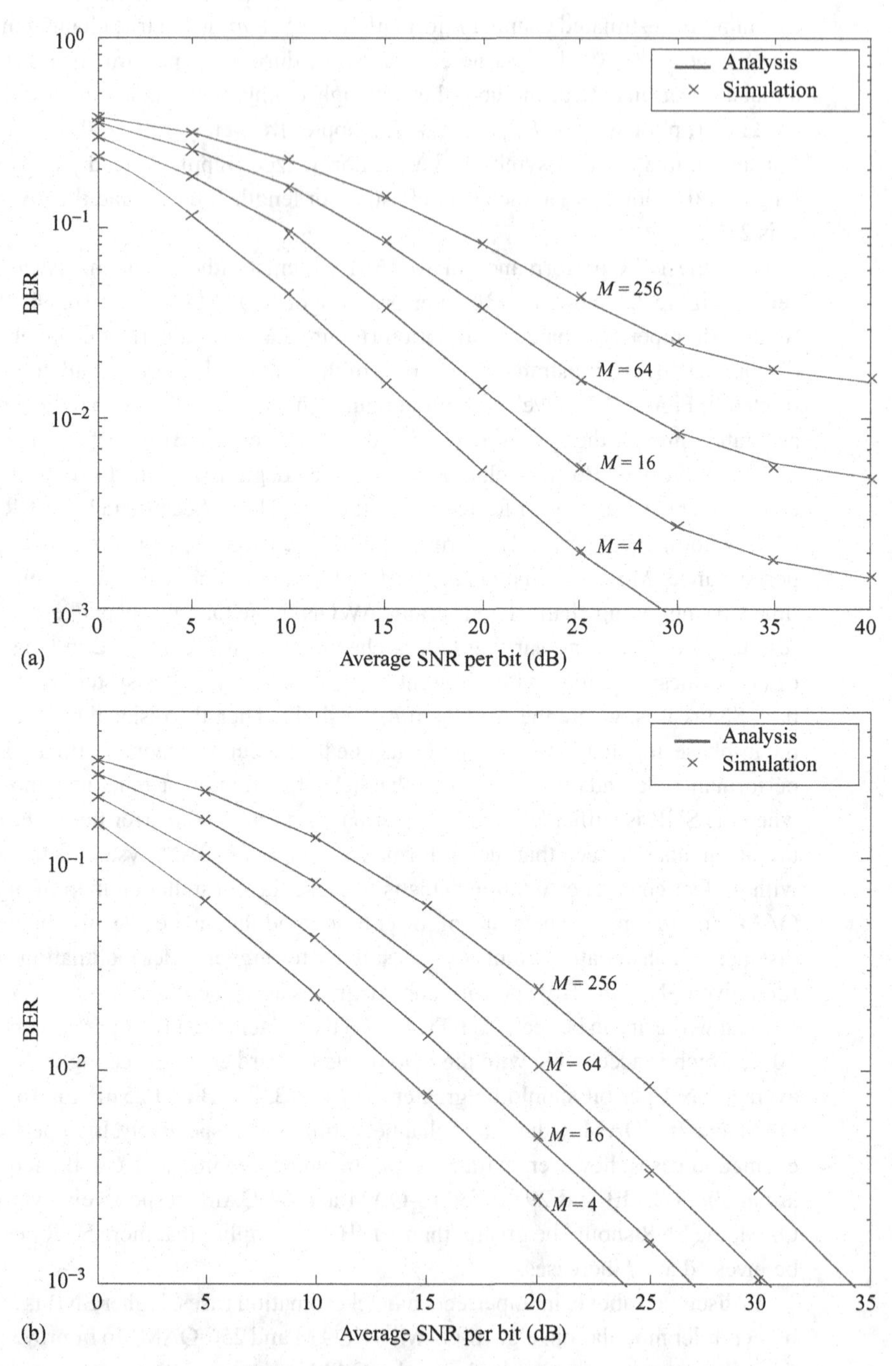

Figure 7.3 Analytical and simulation BER performance of MQAM without diversity. (a) Imperfect channel estimation with FIR channel estimator (7.6). (b) Perfect channel estimation.

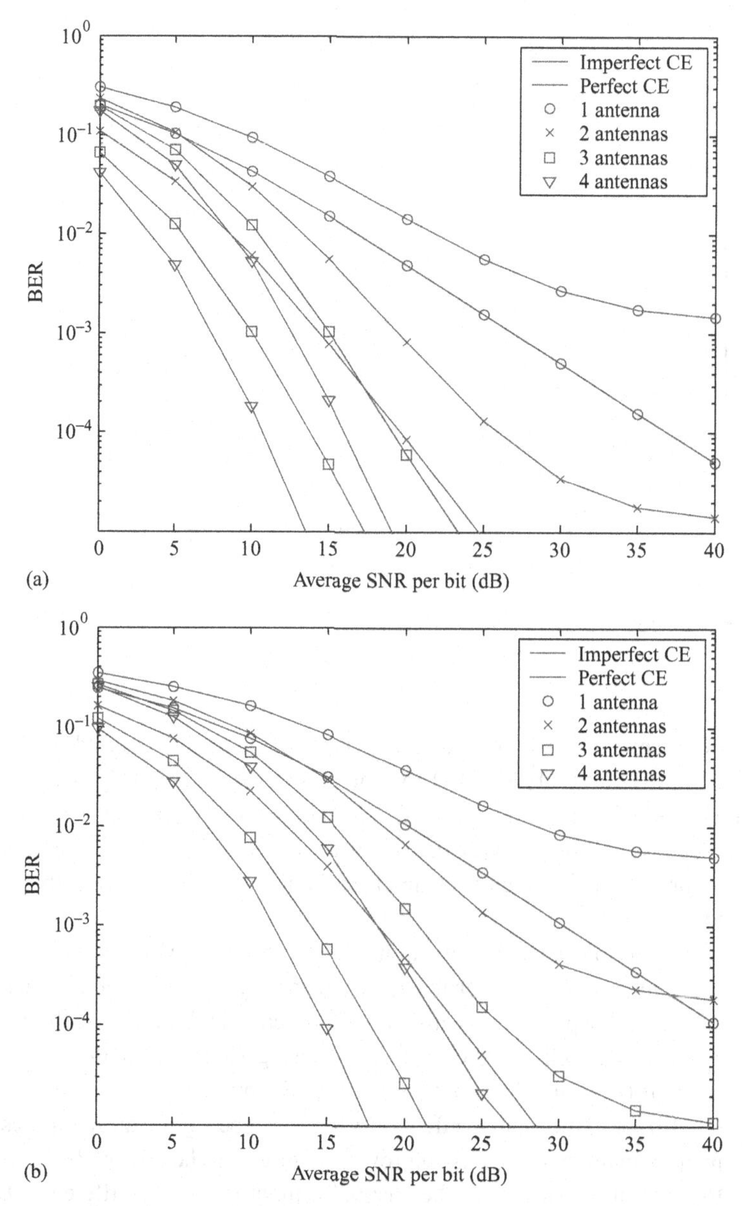

Figure 7.4 BER performance of MQAM with different antenna diversity order. (a) $M = 16$, (b) $M = 64$, (c) $M = 256$.

Figure 7.4 illustrates the performance of 16-QAM, 64-QAM and 256-QAM systems with different diversity orders. It can be seen that the performance improves rapidly with the increase of the diversity order from 1 to 3, but the improvement becomes slighter when the diversity order is larger. For a given SNR, the relative performance degradation due to imperfect channel estimation becomes larger with the increase of diversity order and modulation order, which indicates that the inaccurate channel estimation limits the

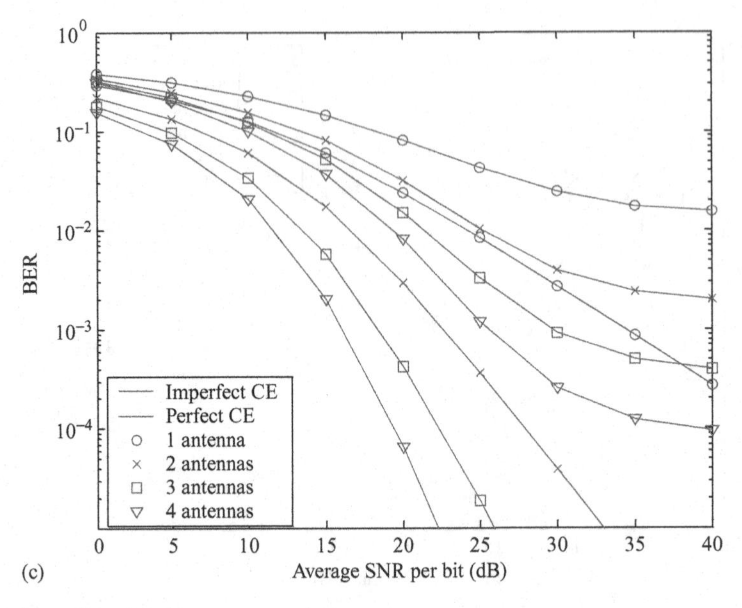

Figure 7.4 *(cont.)*

benefit of antenna diversity. In the case of two antennas, the required SNR are 14 dB, 19 dB and 25 dB for 16-QAM, 64-QAM and 256-QAM, respectively. In perfect channel estimation cases, however, the required SNR significantly decreases to 9 dB, 12 dB and 16 dB respectively. This implies that if an advanced channel estimator is employed, which can provide more accurate channel estimation, the transmission efficiency can be improved.

Next the effect of channel estimation quality on 64-QAM systems is investigated. Basically, by increasing the filter length for channel estimation, the channel estimator approaches ideal lowpass filter with cutoff frequency at $1/2ST$, which can track the time varying fading well. In other words, increasing the filter length means improving the accuracy of the channel estimation. Meanwhile, by increasing the amplitude of the pilot symbol $|D_{\text{Pilot}}|$, the relative effect of AWGN can be significantly suppressed. Then the channel estimation quality is improved. Figure 7.5 shows the BER performance of 64-QAM versus the length of the channel estimator W with different $|D_{\text{Pilot}}|$ when the average SNR per bit is 20 dB. It can be seen that performance improves with an increase in W . However, the performance reaches an error floor after W increases to a balanced point, because after that the AWGN component becomes dominant to the channel estimation quality. Moreover, by increasing $|D_{\text{Pilot}}|$, performance improves due to significant suppression of AWGN component with respect to pilot components. From the figure, it can be found that optimum W is associated with the pilot symbol amplitude. However, too large a $|D_{\text{Pilot}}|$ will decrease the transmission efficiency due to power loss for pilot symbols. Therefore, tradeoff should be made between the transmission efficiency and BER performance.

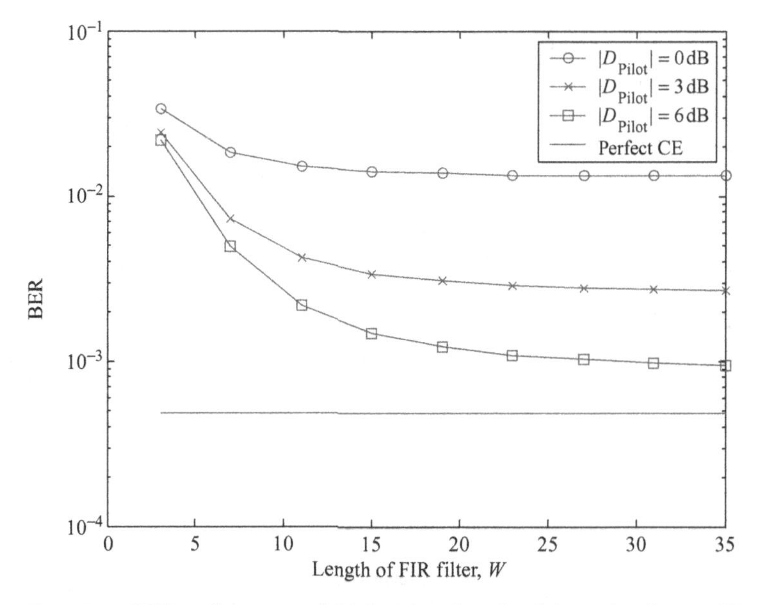

Figure 7.5 BER performance of 64-QAM vs. length of channel estimator W.

Finally, the effect of fast channel fading on the system performance is evaluated. Figure 7.6 shows the BER performance of 64-QAM versus Doppler frequency with different $|D_{\text{Pilot}}|$ when the average SNR per bit is 20 dB. It can be seen that the performance curves are almost flat when the Doppler frequency is less than 400 Hz, except for the case of $|D_{\text{Pilot}}| = 6$ dB, where the performance degrades slightly with an increase in Doppler frequency. This is because, in this case, the AWGN component in channel estimation error is minor, but a slight channel estimation error due to fast fading still degrades the system performance. As the performance degradation is acceptable, the lowpass sinc channel estimator is preferred due to its low complexity. But it can also be seen that the performance degrades abruptly when Doppler frequency is greater than 400 Hz, because the sinc interpolation method is derived from the Nyquist filter [3], which requires the sampling rate $1/ST$ to be greater than double the Doppler frequency to track the time-variant fading parameters. While, when $f_{\text{D}} > 500$ Hz, $2f_{\text{D}}$ is beyond the sampling rate, which is the bandwidth of the ideal lowpass filter. The time varying channel fading cannot be tracked well and consequently the performance degrades abruptly. In addition, the requirement for the proposed algorithm will be stricter than that for the ideal sinc interpolation method, because of the truncation effect of the finite length LPF implementation for channel estimation. Therefore, as shown in Figure 7.6, when the Doppler frequency is greater than 400 Hz, performance degradation becomes significant, and the benefit from larger $|D_{\text{Pilot}}|$ is cancelled out by the failure in tracking the fast fading. Thus, it can be concluded that when the Doppler frequency is less than $1/2ST$, the performance is flat due to the robustness of the channel estimator to fading rate.

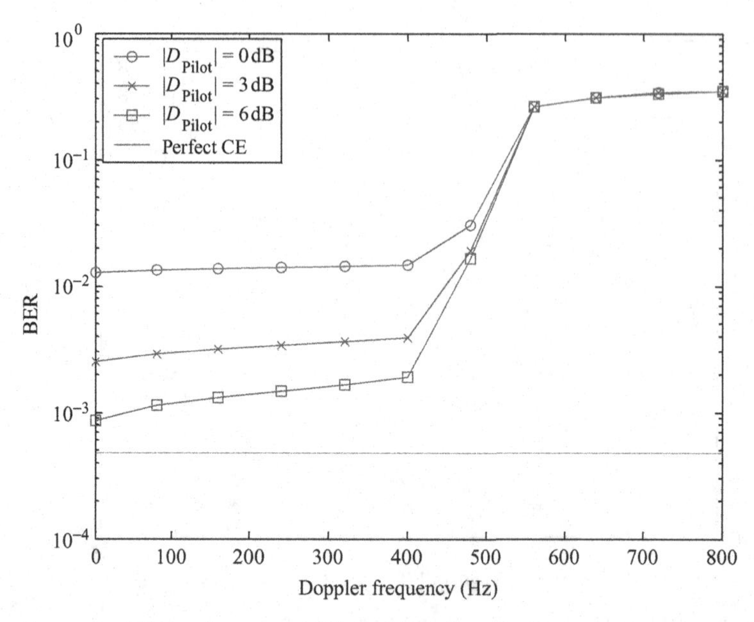

Figure 7.6 BER performance of 64-QAM vs. Doppler frequency.

7.5 Conclusions

In this chapter, the effect of antenna diversity on the BER performance of MQAM systems in Rayleigh fading channels with imperfect channel estimation has been studied. Simulation results show that the BER performance of MQAM systems can be accurately evaluated by the analytical method in this chapter. The main conclusions of this chapter can be summarized as follows.

(1) QAM is very sensitive to the quality of channel estimation. The performance degradation of higher-order QAM is more serious than that of lower-order QAM. In the case of perfect channel estimation, to obtain the acceptable BER performance (10^{-2}), the required SNR per bit for 16-QAM, 64-QAM and 256-QAM are 17 dB, 20 dB and 25 dB, respectively. In the case of imperfect channel estimation, however, extra SNR per bit should be invested.
(2) Antenna diversity is an efficient technique to improve the BER performance of MQAM systems. With the help of antenna diversity, the required SNR for higher-order QAM dramatically decreases, especially in perfect channel estimation cases.
(3) The channel estimation error limits the benefit of antenna diversity. If the channel estimator can provide more accurate channel estimation, the system throughput can be improved significantly by using higher-order modulation, and the required SNR decreases. Channel estimation can be improved by increasing the length of the channel estimator and the amplitude of the pilot symbol. Both should be determined by balancing the transmission efficiency and BER performance.

(4) The lowpass sinc channel estimator can track fading well when the Doppler frequency is less than $1/2ST$. The BER performance is flat in slow fading channels, but it degrades sharply in fast fading channels due to failure in tracking the varying fading.

References

[1] L. Yang and L. Hanzo, "A recursive algorithm for the error probability evaluation of M-QAM," *IEEE Commun. Lett.*, vol. 4, pp. 304–306, Oct. 2000.

[2] X. Tang, M.-S. Alouini and A. J. Goldsmith, "Effect of channel estimation error on M-QAM BER performance in Rayleigh fading," *IEEE Trans. Commun.*, vol. 47, pp. 1856–1864, Dec. 1999.

[3] B. Xia, "Enhanced techniques for broadband wireless communications," Ph.D. Thesis, University of Hong Kong, 2004.

[4] L. Hanzo, R. Steel and P. M. Fortune, "A subband coding, BCH coding, and 16-level QAM system for mobile radio communications," *IEEE Trans. Veh. Technol.*, vol. 39, pp. 327–339, Nov. 1990.

[5] J. G. Proakis, *Digital Communications,* 4th edition. New York: McGraw Hill, 2001.

8 QAM for multicode CDMA with interference cancellation

This chapter studies MQAM for downlink multicode CDMA systems with interference cancellation to support high data rate services. In the current 3G WCDMA systems, in addition to multicode transmission, MQAM is employed for HSDPA due to its high spectral efficiency. In frequency selective fading channels, multipath interference seriously degrades the system performance. In this chapter, theoretical analysis is presented to show that with the help of interference cancellation technique, MQAM may be employed in high SNR cases to increase system throughput. Moreover, it is found that when using the interference cancellation technique, extra pilot power should be invested for more accurate channel estimation, and consequently better BER performance can be achieved.

8.1 Introduction

MQAM modulated multicode CDMA is proposed for HSDPA in the 3G standards, by which the throughput can be increased without extra bandwidth investment. As mentioned in Chapter 1, the introduction of multicode transmission causes multipath interference in frequency selective fading channels due to multipath propagation delays. In this chapter, a coherent Rake receiver with interference cancellation is studied.

Moreover, in WCDMA systems, a common pilot channel is used for channel estimation at the receiver. However, channel estimation error occurs since the received pilot channel signal suffers from the multipath interference and AWGN noise, which affects the coherent data decision and the regeneration of multipath interference, and thus degrades the system performance. The effects of imperfect channel estimation and additive multipath interference on system performance are investigated. The power ratio of pilot channel to the total transmitted power has been evaluated for better performance.

The chapter is organized as follows. In Section 8.2, system models are presented, including transmitter model, channel model and the coherent Rake receiver with interference cancellation technique. In Section 8.3, an analytical BER performance is derived. In Section 8.4, comparison and discussions on numerical BER results are presented. Finally, some conclusions are drawn in Section 8.5.

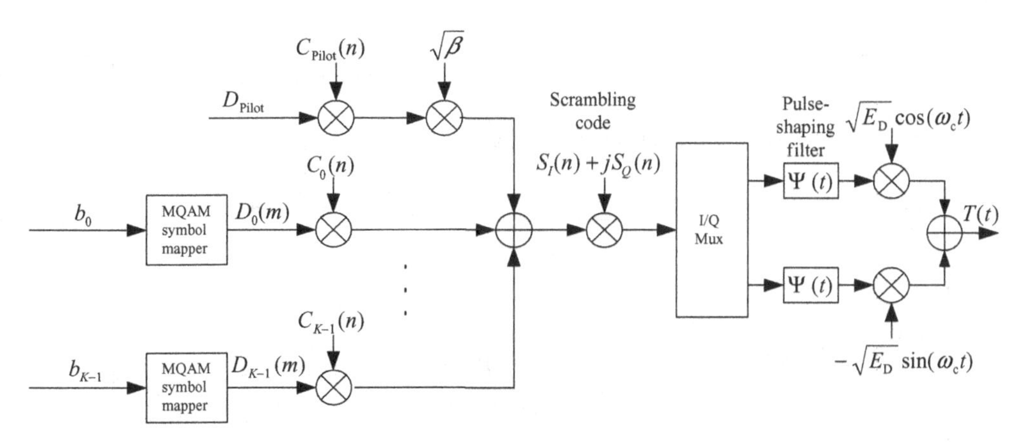

Figure 8.1 Transmitter diagram of a QAM modulated multicode CDMA system.

8.2 System model

As proposed in [1], the transmitter diagram of a downlink MQAM modulated multicode WCDMA system is shown in Figure 8.1. The source data is divided into K streams and fed into K MQAM symbol mappers before spreading. The complex data symbols, $D_k(m)$, are multiplied with real-valued OVSF codes, $C_k(n)$, for the channelization operation. Note that spreading factors for data channels and the pilot channel are N and N_{Pilot}, respectively. The resultant signals, including the signal of a code-multiplexed common pilot channel, are multiplied by a complex-valued code, $S(n)$, for scrambling. Then, the scrambled signal is shaped by a chip pulse-shaping filter, $\Psi(t)$, and up-converted to form a transmitted signal.

An L-path fading channel is assumed, whose equivalent impulse response is expressed as

$$h(t) = \sum_{l=0}^{L-1} \alpha_l(t)\delta(t-\tau_l(t))e^{j\theta_l(t)} \tag{8.1}$$

where $\tau_l(t)$ is the delay of the lth path, $\theta_l(t)$ is the phase, uniformly distributed in $[0, 2\pi)$, and $\alpha_l(t)$ is the Rayleigh distributed fading amplitude with fading power of Ω_l, i.e. $E\{\alpha_l^2(t)\} = \Omega_l$. Since a mobile terminal does not move fast for high data rate transmission, the channel is assumed to be slow fading and remain constant over a few symbols' duration. So t is omitted for simplicity of notation.

Although the transmitted signal experiences an L-path fading channel, only the $L_R (L_R \leq L)$ strongest paths are used at the receiver for Rake combining. As shown in Figure 8.2, the down-converted signal is fed into L_R despreading branches corresponding to the L_R strongest paths. In each branch, the sampled signal after chip matched filter is descrambled and then passes through $K+1$ correlators (for K data channels and one pilot channel) for despreading. Finally, the despreading outputs are fed into a QAM Rake receiver with an interference cancellation block, which makes use of the regeneration property of multipath interference and mitigates the multipath effect. The

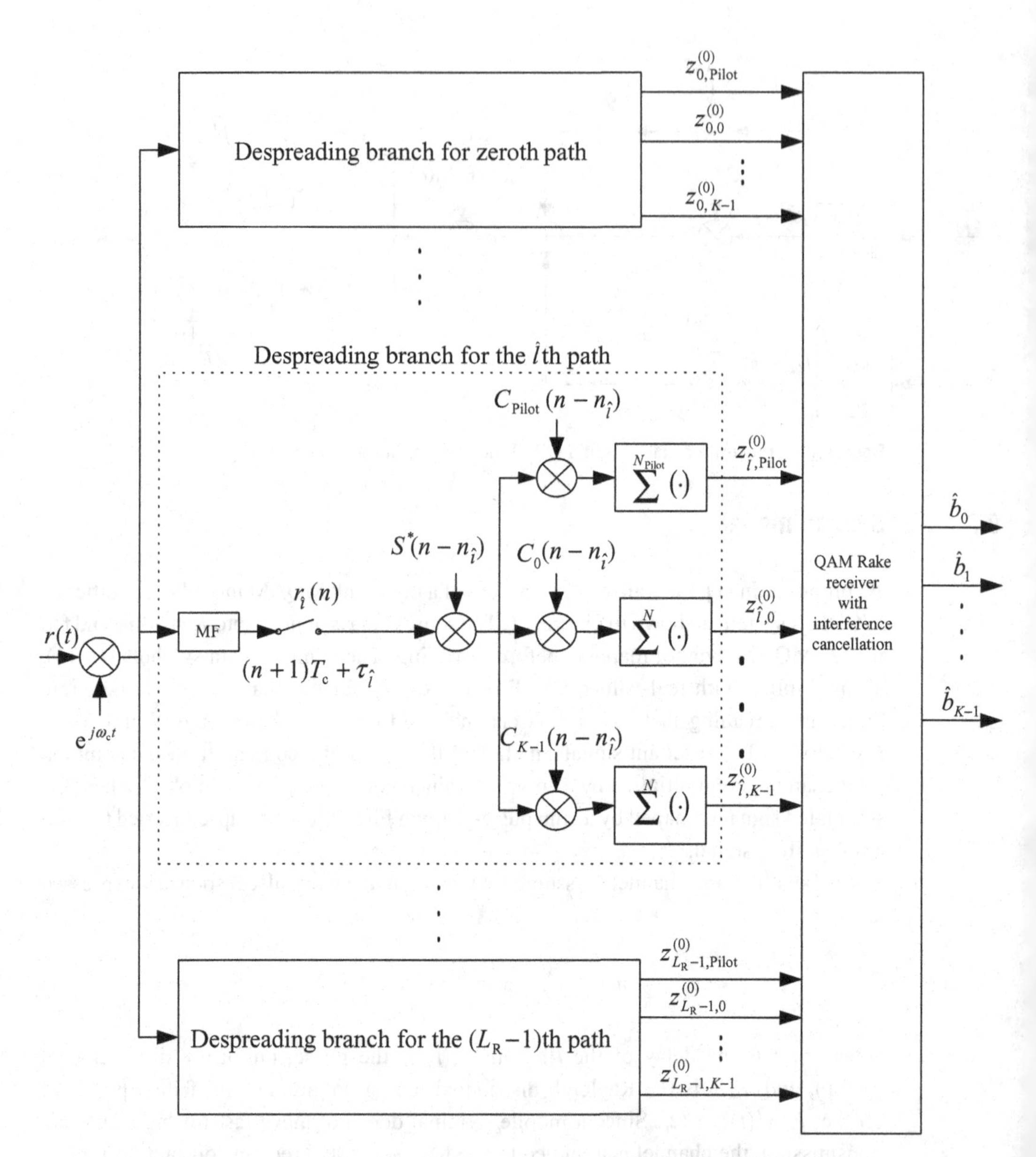

Figure 8.2 Receiver diagram of QAM modulated multicode CDMA systems.

detailed diagram of the QAM Rake receiver is shown in Figure 8.3, where the estimated channel fading and phase of the sth stage for the $\hat{l}$th path, $\hat{z}^{(s)}_{\hat{l},\text{Pilot}}$, are used to eliminate the phase error and achieve MRC with scaling for MQAM symbol demodulation. The despreading outputs $z^{(0)}_{\hat{l},\hat{k}}$ are original inputs to all stages for the $\hat{k}$th desired code. In the sth interference cancellation stage ($s > 0$), $z^{(0)}_{\hat{l},\hat{k}}$ pass through the interference cancellation block, where the multipath interference is regenerated by using the channel

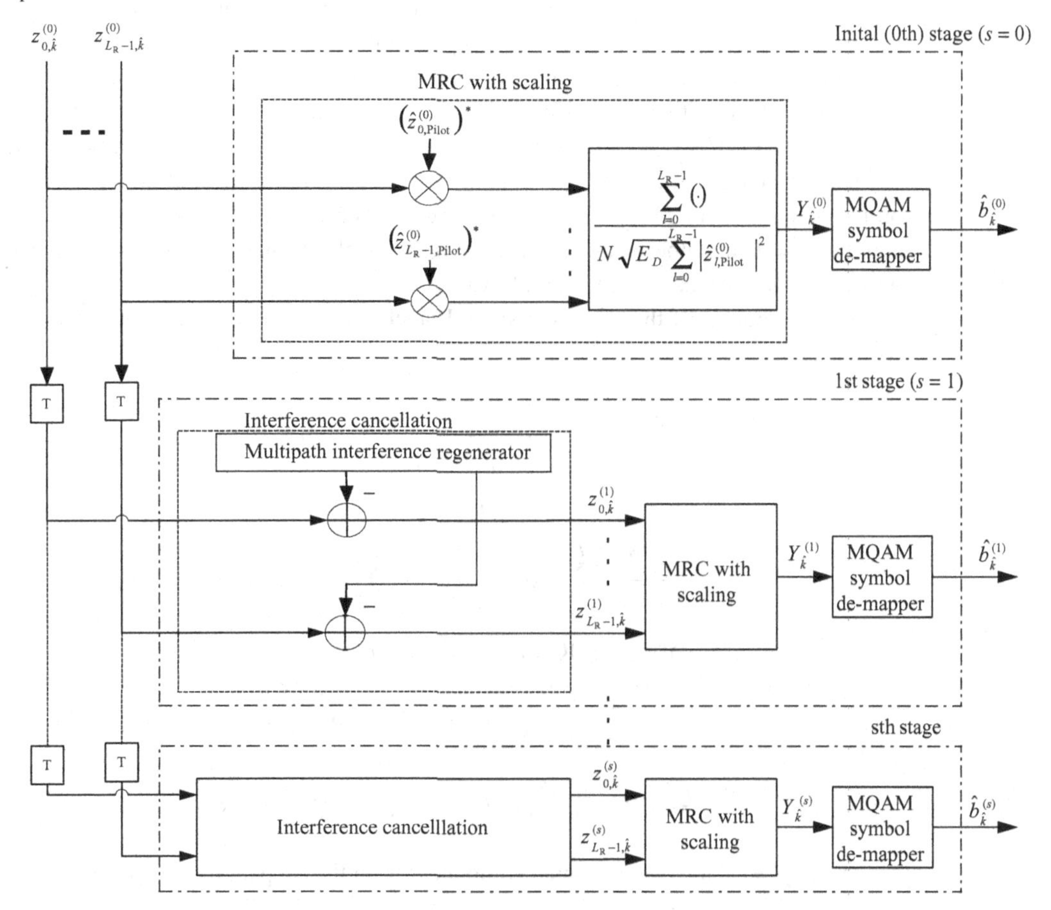

Figure 8.3 Detailed block diagram of QAM Rake receiver with interference cancellation.

estimation and data decision in the previous stage. Then despreading outputs after interference cancellation, $z_{\hat{l},\hat{k}}^{(s)}$, are obtained.

Neglecting the high-frequency components, the output of the chip-matched filter for the $\hat{l}$th path is

$$r_{\hat{l}}(n) = \frac{\sqrt{E_{\mathrm{D}}}}{2} \sum_{l=0}^{L-1} \alpha_l e^{j\xi_l} \sum_{i=-\infty}^{\infty} R(\bar{\tau}_{l,\hat{l}} - iT_{\mathrm{c}}) \times \left(\sqrt{\beta} D_{\mathrm{Pilot}} + \sum_{k=0}^{K-1} D_k(m_{l,i}) C_k(n - n_{l,i}) \right) S(n - n_{l,i}) + \eta_{\hat{l}}(n) \quad (8.2)$$

where $\xi_l = \theta_l - \theta$ is the residue phase due to the fading channel θ_l and the phase difference between the oscillators of the transmitter and the receiver θ; E_{D} is the average chip energy of each data channel; β is the power ratio of the pilot channel to one data channel; $D_k(m)$ is the mth complex-valued MQAM symbol in the kth code channel, whose real and imaginary parts are from the set $\{\pm d, \pm 3d, \ldots, \pm(\sqrt{M}-1)d\}$ with $d = \sqrt{3\log_2 M/2(M-1)}$; $D_{\text{Pilot}} = 1 + j$ is the symbol in the pilot channel for channel estimation and pre-processing at the receiver; the channelization code for the pilot channel, C_{Pilot}, as shown in Figure 8.1, is an all-1 sequence; $\eta_{\hat{l}}$ is the complex Gaussian random noise with double-sided power spectral density of $N_0/2$; $n_{l,i} = \lfloor \tau_l/T_c \rfloor - i$; $m_{l,i} = \lfloor (n - n_{l,i})/N \rfloor$; $\bar{\tau}_{l,\hat{l}} = \tau_l - \tau_{\hat{l}} - \lfloor (\tau_l - \tau_{\hat{l}})/T_c \rfloor$; T_c is the chip timing error, where $\lfloor x \rfloor$ takes the largest integer no greater than x, and $R\left(\bar{\tau}_{l,\hat{l}}\right)$ is the effect of timing error $\bar{\tau}_{l,\hat{l}}$ on the output, where $R(\bar{\tau}_{l,\hat{l}}) = \int_{-\infty}^{\infty} |\Psi(f)|^2 \cos(2\pi f \bar{\tau}_{l,\hat{l}}) \mathrm{d}f$ [2]. Note that $R(iT_c) = 1$ and 0 for $i = 0$ and other non-zero integers, respectively. Actually, $R(t)$ is very small as long as t is greater than a few chip intervals.

Then, the output of the $\hat{l}$th despreading branch for the $\hat{k}$th desired code $z_{\hat{l},\hat{k}}^{(1)}$ can be written as

$$\begin{aligned}
z_{\hat{l},\hat{k}}^{(0)}(m) = {} & \frac{\sqrt{E_D}}{2} \sum_{k=0}^{K-1} \sum_{l=0}^{L-1} \alpha_l e^{j\xi_l} \sum_{i=-\infty}^{\infty} \sum_{n=mN+n_{\hat{l}}}^{(m+1)N+n_{\hat{l}}-1} R(\bar{\tau}_{l,\hat{l}} - iT_c) D_k(m_{l,i}) \\
& \cdot C_k(n - n_{l,i}) S(n - n_{l,i}) C_{\hat{k}}(n - n_{\hat{l}}) S^*(n - n_{\hat{l}}) \\
& + \frac{\sqrt{\beta E_D}}{2} \sum_{l=0}^{L-1} \alpha_l e^{j\xi_l} \sum_{i=-\infty}^{\infty} \sum_{n=mN+n_{\hat{l}}}^{(m+1)N+n_{\hat{l}}-1} R(\bar{\tau}_{l,\hat{l}} - iT_c) D_{\text{Pilot}} \\
& \cdot S(n - n_{l,i}) C_{\hat{k}}(n - n_{\hat{l}}) S^*(n - n_{\hat{l}}) \\
& + \sum_{n=mN+n_{\hat{l}}}^{(m+1)N+n_{\hat{l}}-1} \eta_{\hat{l}}(n) C_{\hat{k}}(n - n_{\hat{l}}) S^*(n - n_{\hat{l}}) \\
= {} & U_{s,\hat{l},\hat{k}}(m) + U_{p,\hat{l},\hat{k}}^{(0)}(m) + U_{\text{Pilot},\hat{l},\hat{k}}^{(0)}(m) + U_{\text{AWGN},\hat{l},\hat{k}}(m)
\end{aligned} \tag{8.3}$$

where $U_{s,\hat{l},\hat{k}}(m)$ is the desired signal component from the $\hat{l}$th path of the $\hat{k}$th code,

$$U_{s,\hat{l},\hat{k}}(m) = N\alpha_{\hat{l}} \sqrt{E_D} D_{\hat{k}}(m) e^{j\xi_{\hat{l}}} \tag{8.4}$$

Obviously, in (8.4), the transmitted QAM symbol is multiplied with the channel fading. Thus, the effect of channel fading should be compensated before the MQAM symbol de-mapping. The MQAM symbol de-mapper is accomplished by choosing the symbol whose constellation point is closest to the input of the de-mapper. So, if the channel fading gain or phase is estimated incorrectly, the improper phase and/or amplitude compensation will lead to incorrect MQAM demodulation [4].

In (8.4), $U_{p,\hat{l},\hat{k}}^{(0)}(m)$ is the multipath interference from data channels, given by

$$\begin{aligned}
U_{p,\hat{l},\hat{k}}^{(0)}(m) = {} & \frac{\sqrt{E_D}}{2} \sum_{k=0}^{K-1} \sum_{\substack{l=0 \\ l \neq \hat{l}}}^{L-1} \alpha_l e^{j\xi_l} \sum_{i=-\infty}^{\infty} \sum_{n=mN+n_{\hat{l}}}^{(m+1)N+n_{\hat{l}}-1} \\
& \cdot R(\bar{\tau}_{l,\hat{l}} - iT_c) D_k(m_{l,i}) C_k(n - n_{l,i}) \\
& \cdot S(n - n_{l,i}) C_{\hat{k}}(n - n_{\hat{l}}) S^*(n - n_{\hat{l}})
\end{aligned} \tag{8.5}$$

$U^{(0)}_{\text{Pilot},\hat{l},\hat{k}}(m)$ is the multipath interference from the pilot channel, given by

$$U^{(0)}_{\text{Pilot},\hat{l},\hat{k}}(m) = \frac{\sqrt{\beta E_{\text{D}}}}{2} \sum_{\substack{l=0 \\ l \neq \hat{l}}}^{L-1} \alpha_l e^{j\xi_l} \sum_{i=-\infty}^{\infty} \sum_{n=mN+n_{\hat{l}}}^{(m+1)N+n_{\hat{l}}-1} R(\bar{\tau}_{l,\hat{l}} - iT_{\text{c}}) D_{\text{Pilot}} \cdot S(n - n_{l,i}) C_{\hat{k}}(n - n_{\hat{l}}) S^*(n - n_{\hat{l}}) \tag{8.6}$$

$U_{\text{AWGN},\hat{l},\hat{k}}(m)$ is the noise component, given by

$$U_{\text{AWGN},\hat{l},\hat{k}}(m) = \sum_{n=mN+n_{\hat{l}}}^{(m+1)N+n_{\hat{l}}-1} \eta_{\hat{l}}(n) C_{\hat{k}}(n - n_{\hat{l}}) S^*(n - n_{\hat{l}}) \tag{8.7}$$

In the sth stage ($s > 0$), the mulitpath interference from the data channels and the pilot channel are regenerated with tentative data decision and the estimated fading obtained from the previous stage. The regenerated interference terms are

$$\hat{U}^{(s)}_{\text{Pilot},\hat{l},\hat{k}}(m) = \frac{\sqrt{\beta E_{\text{D}}}}{2} \sum_{\substack{l=0 \\ l \neq \hat{l}}}^{L_{\text{R}}-1} \hat{z}^{(s-1)}_{l,\text{Pilot}} \sum_{i=-\infty}^{\infty} \sum_{n=mN+n_{\hat{l}}}^{(m+1)N+n_{\hat{l}}-1} \cdot R(\bar{\tau}_{l,\hat{l}} - iT_c) D_{\text{Pilot}} S(n - n_{l,i}) C_{\hat{k}}(n - n_{\hat{l}}) S^*(n - n_{\hat{l}}) \tag{8.8}$$

and

$$\hat{U}^{(s)}_{p,\hat{l},\hat{k}}(m) = \frac{\sqrt{E_{\text{D}}}}{2} \sum_{k=0}^{K-1} \sum_{\substack{l=0 \\ l \neq \hat{l}}}^{L_R-1} \hat{z}^{(s-1)}_{l,\text{Pilot}} \sum_{i=-\infty}^{\infty} \sum_{n=mN+n_{\hat{l}}}^{(m+1)N+n_{\hat{l}}-1} R(\bar{\tau}_{l,\hat{l}} - iT_c) \cdot \hat{D}^{(s-1)}_k(m_{l,i}) C_k(n - n_{l,i}) S(n - n_{l,i}) C_{\hat{k}}(n - n_{\hat{l}}) S^*(n - n_{\hat{l}}) \tag{8.9}$$

respectively, where $\hat{D}^{(s-1)}_k$ is the data decision from the $(s-1)$th stage for the kth code channel, and $\hat{z}^{(s-1)}_{\hat{l},\text{Pilot}}$ is the estimated channel fading in the $(s-1)$th stage. Note that different from [5], where the interference from all paths is regenerated, only the interference from the L_{R} strongest paths can be regenerated in this chapter since the number of resolvable paths L_{R} is pre-determined by the system.

Then, the output of the interference cancellation block of the sth stage is

$$\begin{aligned} z^{(s)}_{\hat{l},\hat{k}}(m) &= U_{s,\hat{l},\hat{k}}(m) + U^{(0)}_{p,\hat{l},\hat{k}}(m) - \hat{U}^{(s)}_{p,\hat{l},\hat{k}}(m) + U^{(0)}_{\text{Pilot},\hat{l},\hat{k}}(m) \\ &\quad - \hat{U}^{(s)}_{\text{Pilot},\hat{l},\hat{k}}(m) + U_{\text{AWGN},\hat{l},\hat{k}}(m) \\ &= U_{s,\hat{l},\hat{k}}(m) + U^{(s)}_{p,\hat{l},\hat{k}}(m) + U^{(s)}_{\text{Pilot},\hat{l},\hat{k}}(m) + U_{\text{AWGN},\hat{l},\hat{k}}(m) \end{aligned} \tag{8.10}$$

where $U^{(s)}_{p,\hat{l},\hat{k}}(m)$ and $U^{(s)}_{\text{Pilot},\hat{l},\hat{k}}(m)$ are residual multipath interference from data channels and the pilot channel, respectively.

Similarly, in the sth stage, the output of the pilot channel for the $\hat{l}$th path during the transmission of the n_pth pilot symbol is given by

$$\begin{aligned} z^{(s)}_{\hat{l},\text{Pilot}}(n_p) &= N_{\text{Pilot}} \alpha_{\hat{l}} \sqrt{\beta E_{\text{D}}} D_{\text{pilot}} e^{j\xi_{\hat{l}}} + U^{(s)}_{p,\hat{l},\text{Pilot}}(n_p) \\ &\quad + U^{(s)}_{\text{Pilot},\hat{l},\text{Pilot}}(n_p) + U_{\text{AWGN},\hat{l},\text{Pilot}}(n_p) \end{aligned} \tag{8.11}$$

Note that, different from the data channels, for the pilot channel, the correlator length is N_{Pilot}.

In (8.11), $U_{\text{AWGN},\hat{l},\text{Pilot}}(n_p)$ is the noise component, given by

$$U_{\text{AWGN},\hat{l},\text{Pilot}}(n_p) = \sum_{n=n_p N_{\text{Pilot}}+n_{\hat{l}}}^{(n_p+1)N_{\text{Pilot}}+n_{\hat{l}}-1} \eta_{\hat{l}}(n) S^*(n - n_{\hat{l}}) \tag{8.12}$$

$U^{(s)}_{p,\hat{l},\text{Pilot}}(n_p)$ and $U^{(s)}_{\text{Pilot},\hat{l},\text{Pilot}}(n_p)$ are residual multipath interference from the data channels and the pilot channel, respectively. For the initial stage, these two types of interference cannot be canceled, and are written as

$$U^{(0)}_{p,\hat{l},\text{Pilot}}(n_p) = \frac{\sqrt{E_{\text{D}}}}{2} \sum_{k=0}^{K-1} \sum_{\substack{l=0 \\ l \neq \hat{l}}}^{L-1} \alpha_l e^{j\xi_l} \sum_{i=-\infty}^{\infty} \sum_{n=n_p N_{\text{Pilot}}+n_{\hat{l}}}^{(n_p+1)N_{\text{Pilot}}+n_{\hat{l}}-1} \cdot R(\bar{\tau}_{l,\hat{l}} - iT_{\text{c}}) D_k(m_{l,i}) C_k(n - n_{l,i}) S(n - n_{l,i}) S^*(n - n_{\hat{l}}) \tag{8.13}$$

and

$$U^{(0)}_{\text{Pilot},\hat{l},\text{Pilot}}(n_p) = \frac{\sqrt{\beta E_{\text{D}}}}{2} \sum_{\substack{l=0 \\ l \neq \hat{l}}}^{L-1} \alpha_l e^{j\xi_l} \sum_{i=-\infty}^{\infty} \sum_{n=n_p N_{\text{Pilot}}+n_{\hat{l}}}^{(n_p+1)N_{\text{Pilot}}+n_{\hat{l}}-1} \cdot R(\bar{\tau}_{l,\hat{l}} - iT_{\text{c}}) D_{\text{Pilot}} S(n - n_{l,i}) S^*(n - n_{\hat{l}}) \tag{8.14}$$

In the sth stage, the residual multipath interference from data channels and the pilot channel are

$$\begin{aligned} U^{(s)}_{p,\hat{l},\text{Pilot}}(n_p) = {} & \frac{\sqrt{E_{\text{D}}}}{2} \sum_{k=0}^{K-1} \sum_{\substack{l=0 \\ l \neq \hat{l}}}^{L-1} \alpha_l e^{j\xi_l} \sum_{i=-\infty}^{\infty} \sum_{n=n_p N_{\text{Pilot}}+n_{\hat{l}}}^{(n_p+1)N_{\text{Pilot}}+n_{\hat{l}}-1} \\ & \cdot R(\bar{\tau}_{l,\hat{l}} - iT_{\text{c}}) D_k(m_{l,i}) C_k(n - n_{l,i}) S(n - n_{l,i}) S^*(n - n_{\hat{l}}) \\ & - \frac{\sqrt{E_D}}{2} \sum_{k=0}^{K-1} \sum_{\substack{l=0 \\ l \neq \hat{l}}}^{L_{\text{R}}-1} \hat{z}^{(s-1)}_{l,\text{Pilot}} \sum_{i=-\infty}^{\infty} \sum_{n=n_p N_{\text{Pilot}}+n_{\hat{l}}}^{(n_p+1)N_{\text{Pilot}}+n_{\hat{l}}-1} \\ & \cdot R(\bar{\tau}_{l,\hat{l}} - iT_{\text{c}}) \hat{D}^{(s-1)}_k(m_{l,i}) C_k(n - n_{l,i}) S(n - n_{l,i}) S^*(n - n_{\hat{l}}) \end{aligned} \tag{8.15}$$

and

$$\begin{aligned} U^{(s)}_{\text{Pilot},\hat{l},\text{Pilot}}(n_p) = {} & \frac{\sqrt{\beta E_{\text{D}}}}{2} \sum_{\substack{l=0 \\ l \neq \hat{l}}}^{L-1} \alpha_l e^{j\xi_l} \sum_{i=-\infty}^{\infty} \sum_{n=n_p N_{\text{Pilot}}+n_{\hat{l}}}^{(n_p+1)N_{\text{Pilot}}+n_{\hat{l}}-1} \\ & \cdot R(\bar{\tau}_{l,\hat{l}} - iT_{\text{c}}) D_{\text{Pilot}} S(n - n_{l,i}) S^*(n - n_{\hat{l}}) \\ & - \frac{\sqrt{\beta E_{\text{D}}}}{2} \sum_{\substack{l=0 \\ l \neq \hat{l}}}^{L_{\text{R}}-1} \hat{z}^{(s-1)}_{l,\text{Pilot}} \sum_{i=-\infty}^{\infty} \sum_{n=n_p N_{\text{Pilot}}+n_{\hat{l}}}^{(n_p+1)N_{\text{Pilot}}+n_{\hat{l}}-1} \\ & \cdot R(\bar{\tau}_{l,\hat{l}} - iT_{\text{c}}) D_{\text{Pilot}} S(n - n_{l,i}) S^*(n - n_{\hat{l}}) \end{aligned} \tag{8.16}$$

respectively.

Then, the fading during the transmission of the n_pth pilot symbol is estimated by $\frac{z^{(s)}_{\hat{l},\mathrm{Pilot}}(n_p)}{N_{\mathrm{Pilot}}\sqrt{\beta E_{\mathrm{D}}}D_{\mathrm{Pilot}}}$. This can be used in the sth stage to detect the data symbols transmitted within the n_pth pilot symbol duration. So the estimated channel fading for the transmission of the mth data symbol is $\hat{z}^{(s)}_{\hat{l},\mathrm{Pilot}}(m) = \frac{z^{(s)}_{\hat{l},\mathrm{Pilot}}(n_p)}{N_{\mathrm{Pilot}}\sqrt{\beta E_{\mathrm{D}}}D_{\mathrm{Pilot}}}$, where $\left\lfloor \frac{n_p N_{\mathrm{Pilot}}}{N} \right\rfloor \le m < \left\lfloor \frac{(n_p+1)N_{\mathrm{Pilot}}}{N} \right\rfloor$. For example, when $N_{\mathrm{Pilot}} = 256$ and $N = 16$, the channel estimation for coherent detection of the mth data symbol is: $\hat{z}^{(s)}_{\hat{l},\mathrm{Pilot}}(m) = \frac{z^{(s)}_{\hat{l},\mathrm{Pilot}}(0)}{N_{\mathrm{Pilot}}\sqrt{\beta E_{\mathrm{D}}}D_{\mathrm{Pilot}}}$ for $0 \le m < 16$, and $\hat{z}^{(s)}_{\hat{l},\mathrm{Pilot}}(m) = \frac{z^{(s)}_{\hat{l},\mathrm{Pilot}}(1)}{N_{\mathrm{Pilot}}\sqrt{\beta E_{\mathrm{D}}}D_{\mathrm{Pilot}}}$ for $16 \le m < 32$, etc.

In order to improve the channel estimation, a W-tap FIR filter is used, given by

$$\hat{z}^{(s)}_{\hat{l},\mathrm{Pilot}}(m) = \sum_{\mathrm{tap}=-\lfloor (W-1)/2 \rfloor}^{\lfloor W/2 \rfloor} \frac{z^{(s)}_{\hat{l},\mathrm{Pilot}}\left(\left\lfloor \frac{mN}{N_{\mathrm{Pilot}}} \right\rfloor - \mathrm{tap}\right)}{WN_{\mathrm{Pilot}}\sqrt{\beta E_{\mathrm{D}}}D_{\mathrm{Pilot}}} \tag{8.17}$$

Therefore, the channel fading compensation can be achieved by scaling the coherent MRC output. For the $\hat{k}$th code channel in the sth stage, the decision variable for QAM symbol detection is $Y^{(s)}_{\hat{k}}(m)$, given by

$$Y^{(s)}_{\hat{k}}(m) = \frac{\sum_{l=0}^{L_{\mathrm{R}}-1} z^{(s)}_{l,\hat{k}}(m)\left(\hat{z}^{(s)}_{l,\mathrm{Pilot}}(m)\right)^*}{N\sqrt{E_{\mathrm{D}}}\sum_{l=0}^{L_R-1}\left|\hat{z}^{(s)}_{l,\mathrm{Pilot}}(m)\right|^2} \tag{8.18}$$

8.3 Performance analysis

Without loss of generality, it is assumed that the zeroth code channel is the desired code channel. Let D^I_0 and D^Q_0 be the real and imaginary parts of D_0, respectively. Referring to [6], the BER of the sth stage $p^{(s)}_b$ for MQAM is given by

$$p^{(s)}_b = \frac{4}{M}\sum_{q_1=0}^{\sqrt{M}/2-1}\sum_{q_2=0}^{\sqrt{M}/2-1} p^{(s)}\left(e\,\middle|\,D_0 = D^I_0 + j\,D^Q_0 = (2q_1+1) + j\,(2q_2+1)\,d\right) \tag{8.19}$$

where the conditional BER $p^{(s)}(e\,|\,D_0)$ is given by

$$\begin{aligned} p^{(s)}(e\,|\,D_0) &= \frac{1}{(\log_2 M)/2}\Big\{Hd\big((\sqrt{M}-1)d,\, D^I_0\big)p\big(\Re\{Y^{(s)}_0\} \ge (\sqrt{M}-2)d|D_0\big) \\ &+ \sum_{q=-\sqrt{M}/2+1}^{\sqrt{M}/2-2} Hd\left((2q+1)d,\, D^I_0\right)p\big(2qd \le \Re\{Y^{(s)}_0\} < 2(q+1)d|D_0\big) \\ &+ Hd\big((-\sqrt{M}+1)d,\, D^I_0\big)p\big(\Re\{Y^{(s)}_0\} < (-\sqrt{M}+2)d|D_0\big)\Big\} \end{aligned} \tag{8.20}$$

where $\Re\{x\}$ takes the real part of x, and $Hd\,(\cdot)$ is the Hamming distance defined in (7.11).

From (8.20), to obtain the BER performance for MQAM, the conditional probability $p\big(\Re\{Y^{(s)}_0\} < \tilde{B}|D_0\big)$ should be calculated, where $\tilde{B} = \{0, \pm 2d, \ldots, \pm(\sqrt{M}-2)d\}$. By directly applying the characteristic function method from [7], the characteristic function in Rayleigh fading channels is

$$\phi(jv) = \prod_{l=0}^{L_R-1} \frac{v_{1l}v_{2l}}{(v + jv_{1l})(v - jv_{2l})} \tag{8.21}$$

where

$$v_{1l}, v_{2l} = mv_{3l} + \sqrt{v_{3l}^2 + \frac{1}{R^{(s)}_{z_l z_l | D_0} R^{(s)}_{\hat{z}_l \hat{z}_l | D_0} - \left| R^{(s)}_{z_l \hat{z}_l | D_0} \right|^2}} \tag{8.22}$$

and

$$v_{3l} = \frac{-\tilde{B}N\sqrt{E_D} R^{(s)}_{\hat{z}_l \hat{z}_l | D_0} + \Re\left\{ R^{(s)}_{z_l \hat{z}_l | D_0} \right\}}{R^{(s)}_{z_l z_l | D_0} R^{(s)}_{\hat{z}_l \hat{z}_l | D_0} - \left| R^{(s)}_{z_l \hat{z}_l | D_0} \right|^2} \tag{8.23}$$

$R^{(s)}_{z_l z_l | D_0}$, $R^{(s)}_{\hat{z}_l \hat{z}_l | D_0}$ and $R^{(s)}_{z_l \hat{z}_l | D_0}$ are the second moment statistic functions of $z^{(s)}_{l,0}$ and $\hat{z}^{(s)}_{l,\mathrm{Pilot}}$. With Gaussian approximation of residual multipath interference, following [3], one obtains

$$\begin{aligned} R^{(0)}_{z_l z_l | D_0} &= \frac{E\left\{ z^{(0)}_{l,0} \left(z^{(0)}_{l,0} \right)^* \right\}}{2} = \frac{N^2 E_D |D_0|^2 \Omega_l}{2} \\ &+ \frac{NE_D(K \log_2 M + 2\beta)}{3} \sum_{\substack{l1=0 \\ l1 \neq l}}^{L-1} \Omega_{l1} + NN_0 \end{aligned} \tag{8.24}$$

$$\begin{aligned} R^{(0)}_{\hat{z}_l \hat{z}_l | D_0} &= \frac{E\left\{ \hat{z}^{(0)}_{l,\mathrm{Pilot}} \left(\hat{z}^{(0)}_{l,\mathrm{Pilot}} \right)^* \right\}}{2} = \frac{\Omega_l}{2} \\ &+ \frac{(K \log_2 M + 2\beta)}{6WN_{\mathrm{Pilot}}\beta} \sum_{\substack{l1=0 \\ l1 \neq l}}^{L-1} \Omega_{l1} + \frac{N_0}{2WN_{\mathrm{Pilot}}\beta E_D} \end{aligned} \tag{8.25}$$

and

$$R^{(0)}_{z_l \hat{z}_l | D_0} = \frac{E\left\{ z^{(0)}_{l,0} \left(\hat{z}^{(0)}_{l,\mathrm{Pilot}} \right)^* \right\}}{2} = \frac{N\sqrt{E_D} D_0 \Omega_l}{2} \tag{8.26}$$

where the interference from the other code channels in the same path is zero due to the orthogonality of the channel codes.

In the following stages, the second moment statistic function $R^{(s)}_{z_l \hat{z}_l | D_0}$ is the same as (8.26). Assuming that the channel estimation and the tentative data decisions are uncorrelated as in [3], one obtains

$$\begin{aligned} R^{(s)}_{z_l z_l | D_0} &= \frac{N^2 E_D |D_0|^2 \Omega_l}{2} + \frac{NE_D}{3} \sum_{k=0}^{K-1} \sum_{\substack{l1=0 \\ l1 \neq l}}^{L_R-1} \left(E\left\{ \left| D_k - \hat{D}_k^{(s-1)} \right|^2 \right\} \Omega_{l1} \right. \\ &\left. + 8\left(\sigma^{(s-1)}_{e,l1}\right)^2 \log_2 M \right) + \frac{16N\beta E_D}{3} \sum_{\substack{l1=0 \\ l1 \neq l}}^{L_R-1} \left(\sigma^{(s-1)}_{e,l1}\right)^2 \\ &+ \frac{NE_D(K \log_2 M + 2\beta)}{3} \sum_{l2=L_R}^{L-1} \Omega_{l,2} + NN_0 \end{aligned} \tag{8.27}$$

where the variance of the tentative data decision is

$$E\{|D_k - \hat{D}_k^{(s-1)}|^2\} = \frac{1}{M^2}\sum_{D_k}\sum_{\hat{D}_k^{(s-1)}} |D_k - \hat{D}_k^{(s-1)}|^2 p(\hat{D}_k^{I(s-1)}|D_k)p(\hat{D}_k^{Q(s-1)}|D_k) \tag{8.28}$$

and the channel estimation error in the previous stage is a zero mean Gaussian variable with variance of $(\sigma_{e,l}^{(s-1)})^2 = R_{\hat{z}_l\hat{z}_l|D_0}^{(s-1)} - (\Omega_l/2)$. The second statistical function of the estimated fading is given by

$$\begin{aligned} R_{\hat{z}_l\hat{z}_l|D_0}^{(s)} = \frac{\Omega_l}{2} &+ \frac{1}{6WN_{\text{Pilot}}\beta}\sum_{k=0}^{K-1}\sum_{\substack{l1=0\\ l1\neq l}}^{L_R-1}\left(E\{|D_k - \hat{D}_k^{(s-1)}|^2\}\Omega_{l1}\right. \\ &\left.+ 8\log_2 M(\sigma_{e,l1}^{(s-1)})^2\right) + \frac{8}{3WN_{\text{Pilot}}}\sum_{\substack{l1=0\\ l1\neq l}}^{L_R-1}(\sigma_{e,l1}^{(s-1)})^2 \\ &+ \frac{K\log_2 M + 2\beta}{6WN_{\text{Pilot}}\beta}\sum_{l2=L_R}^{L-1}\Omega_{l2} + \frac{N_0}{2WN_{\text{Pilot}}\beta E_D} \end{aligned} \tag{8.29}$$

Then the conditional probability can be obtained by

$$p(\Re\{Y_0^{(s)}\} < \tilde{B}|D_0) = -\frac{1}{2\pi j}\int_{-\infty+j\sigma}^{\infty+j\sigma} \frac{\phi(jv)}{v}\mathrm{d}v \tag{8.30}$$

After calculating (8.30) by using the Gauss–Chebychev quadrature method [8], the conditional BER, given by (8.20), is obtained, hence the overall BER, given by (8.19), is achieved.

8.4 Numerical results and discussions

In this section, a six-path channel (indoor channel B model in [9]) is used. Unless noted otherwise, for analytical results, the spreading factors for the data channels and the pilot channel are 16 and 256, respectively, the number of code channels K is 6, the number of Rake fingers L_R is 3, the channel estimator length W is 3, and the power ratio of the pilot channel to one data channel β is 2 dB. During the simulation, the scrambling code and channelization codes are truncated gold sequence and OVSF codes, respectively, and the Doppler frequency is assumed to be 80 Hz for a slow fading channel.

Figure 8.4 shows the BER performance versus SNR per bit for different modulations with perfect channel estimation. In this figure, the channel estimation is assumed to be the true channel fading information, so the variance of channel estimation error $\sigma_{e,l}^{(s)}$ is zero for any s and $0 \le l < L_R$. Here, the inaccurate multipath regeneration only results from the incorrect symbol decision in the previous stage. From the figure, it can be seen that the BER performance in the initial stage is seriously limited by the multipath interference. The interference cancellation method dramatically improves the performance in the

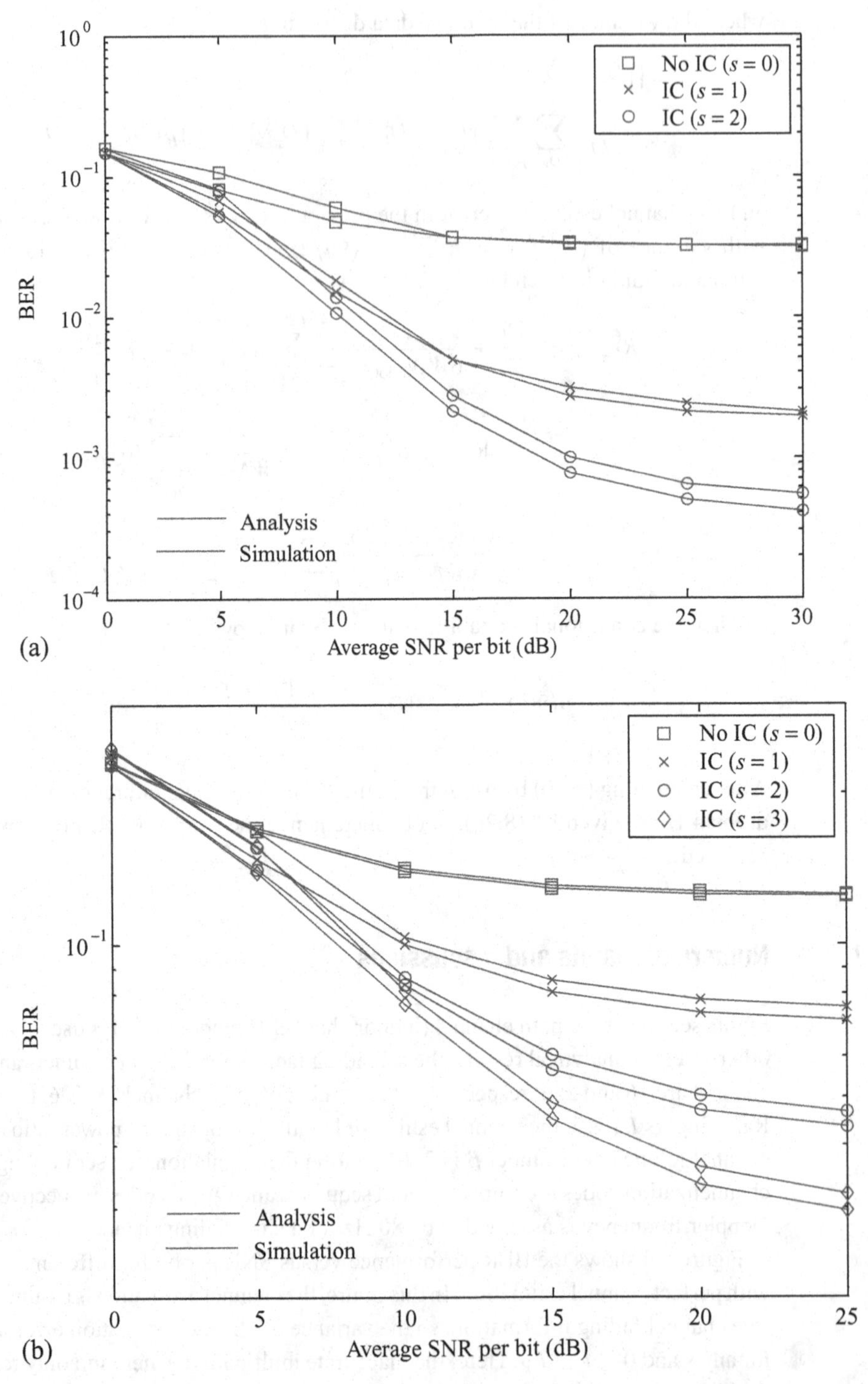

Figure 8.4 BER performance for MQAM multicode CDMA systems with perfect channel estimation. (a) 4-QAM, (b) 16-QAM.

following stages. Comparing the performance curves of 4-QAM with those of 16-QAM, it can be seen that 16-QAM requires a much greater number of interference cancellation stages. Because the minimal distance, $2d$, between constellation points is smaller, the higher-order QAM system is more sensitive to the additive multipath interference. It is found that in higher SNR cases, higher-order modulation can be employed to increase the system throughput.

Meanwhile, to illustrate the accuracy of the analytical method, simulation results are also plotted in Figure 8.4. It can be observed that simulation results are close to the analytical ones for the first few stages. The gap between analytical and simulation curves becomes larger as the number of interference cancellation stages increases. This is because of the inaccurate Gaussian approximation of the residual multipath interference when the number of stages is greater.

In the following, the effect of imperfect channel estimation obtained by the channel estimator (8.17) is investigated. Figure 8.5 shows the BER performance for 4-QAM (Figure 8.5(a)) and 16-QAM (Figure 8.5(b)) CDMA systems with imperfect channel estimation, where the length of the FIR filter for channel estimation W is 3, and the power ratio of the pilot channel to one data channel β is 2 dB. Comparing the performance curves with those in Figure 8.4 for perfect channel estimation cases, it can be seen that the inaccurate channel estimation degrades the system performance, since the existence of channel estimation error not only affects the fading compensation for coherent demodulation, but also degrades the accuracy of the regenerated multipath interference. Moreover, since effects of both inaccurate data decision and channel estimation have been considered during the analysis on the residual multipath interference, it can be observed that even in the presence of imperfect channel estimation, the analytical method is still a good estimate of system performance over the first few stages.

Basically, by increasing the pilot power, the multipath interference from data channels to the pilot channel is decreased, which results in more accurate channel estimation. However, too much pilot power will cause much residual multipath interference from the pilot channel in the despreading outputs for the data channels, which degrades the system performance. Therefore, a tradeoff should be made in the allocation of the pilot channel power. Figure 8.6 shows the BER performance of systems without (Figure 8.6(a)) and with (Figure 8.6(b)) interference cancellation (IC) for 16-QAM when the average SNR per bit is 20 dB. It can be seen that the performance first improves as β increases from -10 dB. After the BER reaches its minimum value, the performance then degrades with a further increase in β. Moreover, channel estimation not only affects the fading compensation but also determines the accuracy of regenerated interference in the following stages. It can be observed that without interference cancellation ($s = 0$), the optimum β are about -1 dB and 0 dB for $K = 4$ and 6, respectively, and when $s = 1$, the optimum β are about 3 dB and 4 dB for $K = 4$ and 6, respectively. It implies for 16-QAM, the power ratio of pilot channel to the total transmitted power $\beta/(\beta + 2K)$ are about 8% and 18% for $s = 0$ and 1, respectively. So to achieve more accurate channel estimation, the power ratio for systems with interference cancellation should be much larger than that for systems without interference cancellation.

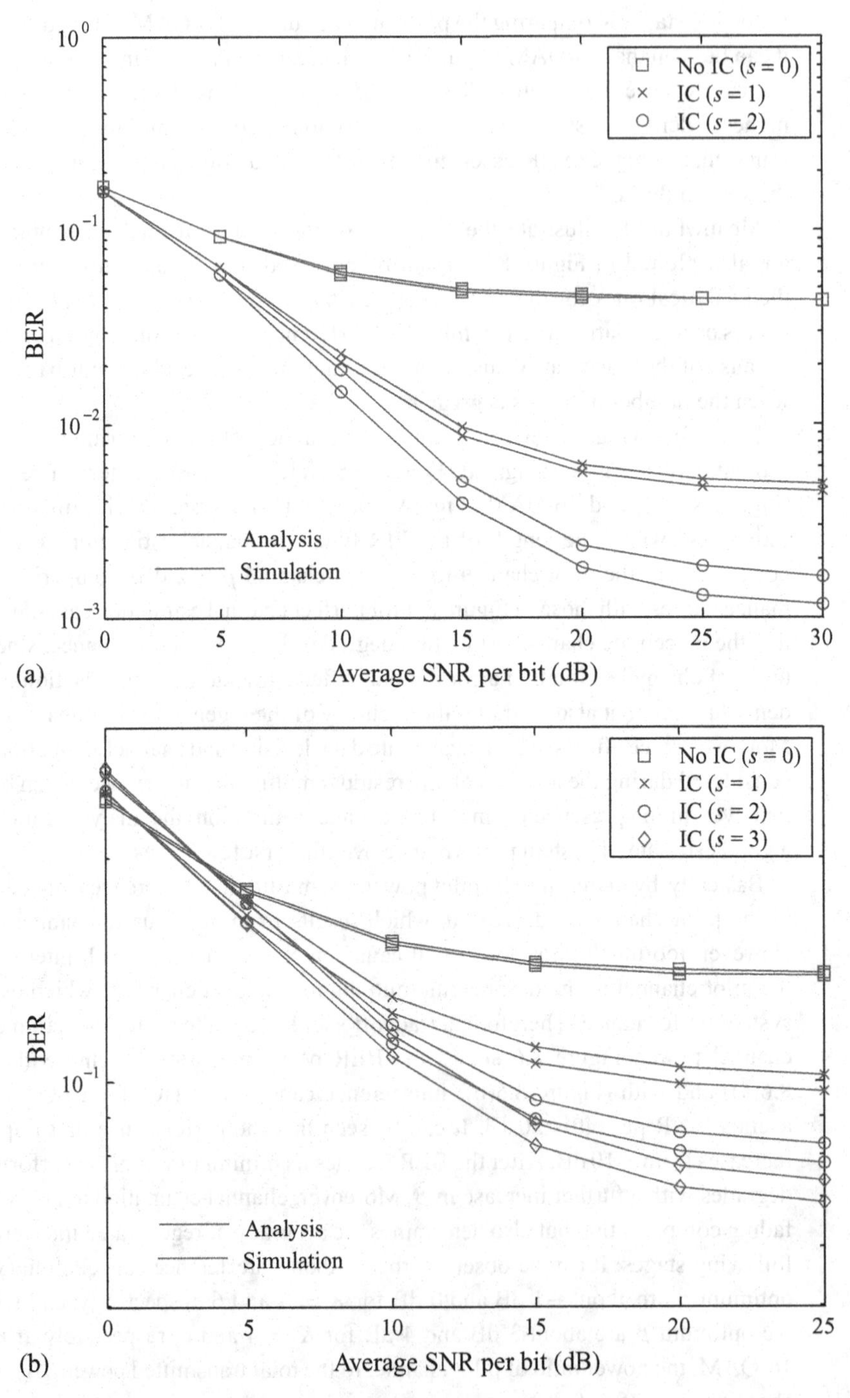

Figure 8.5 BER performance for MQAM multicode CDMA systems with imperfect channel estimation by (8.17): (a) 4-QAM, (b) 16-QAM.

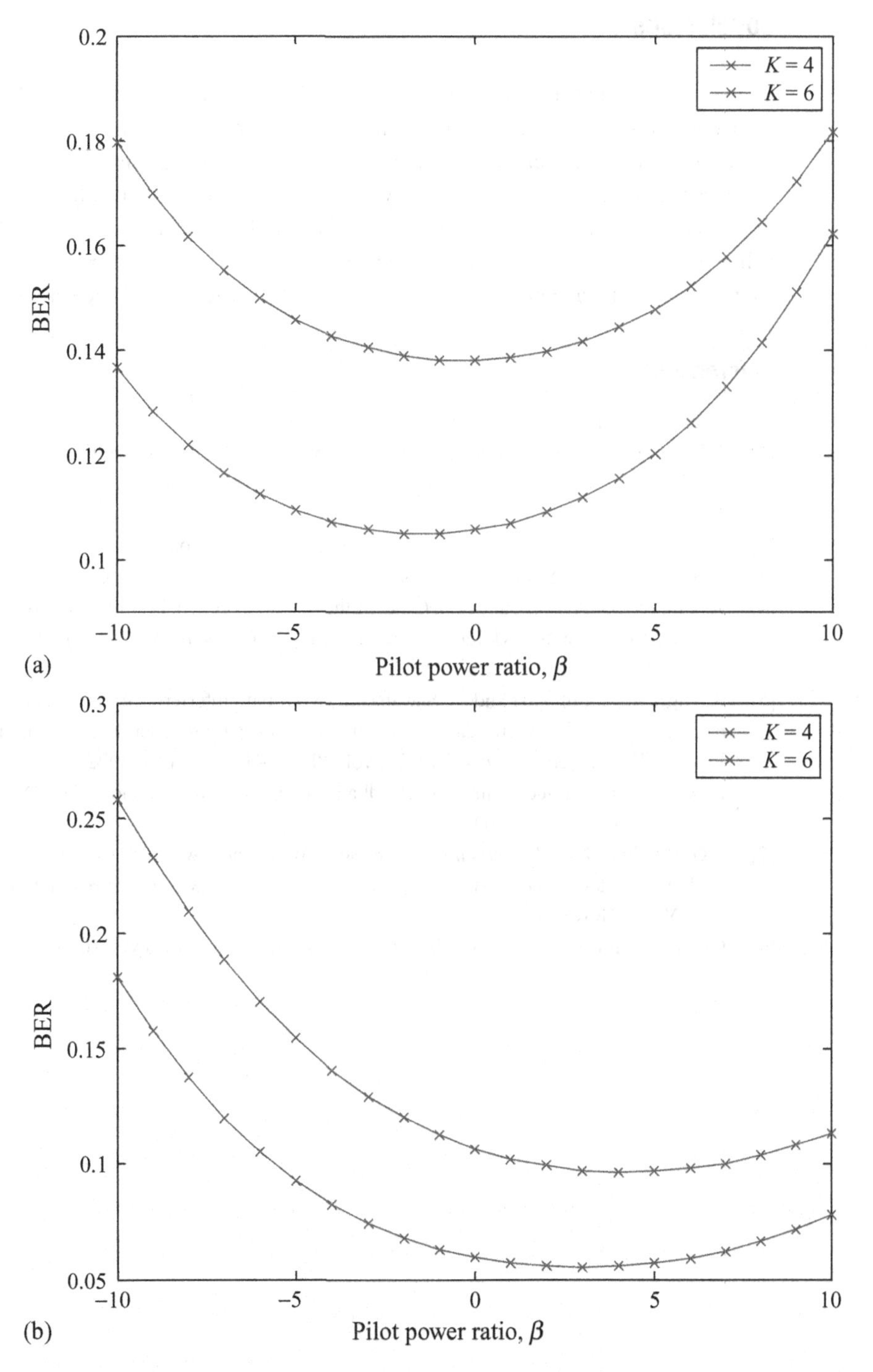

Figure 8.6 Effect of pilot channel power on 16-QAM multicode CDMA systems. (a) Without interference cancellation ($s = 0$). (b) With interference cancellation ($s = 1$).

8.5 Conclusions

This chapter presents the performance of MQAM modulated multicode CDMA systems with interference cancellation for high data rate services. An analytical BER in the presence of imperfect channel estimation is derived. Simulation results are presented to illustrate the accuracy of the analytical results. Numerical results also show that interference cancellation is an efficient technique for MQAM multicode CDMA systems. When using the interference cancellation technique, extra pilot power should be invested for more accurate channel estimation, so that better BER performance can be achieved.

References

[1] Third generation partnership project, www.3gpp.org.

[2] J. Wang and J. Chen, "Performance of wideband CDMA systems with complex spreading and imperfect channel estimation," *IEEE J. Select. Areas Commun.*, vol. 19, pp. 152–163, Jan. 2001.

[3] J. Chen, J. Wang and M. Sawahashi, "MCI cancellation for multicode wideband CDMA systems," *IEEE J. Select. Areas Commun.*, vol. 20, pp. 450–462, Feb. 2002.

[4] X. Tang, M.-S. Alouini, and A. J. Goldsmith, "Effect of channel estimation error on M-QAM BER performance in Rayleigh fading," *IEEE Trans. Commun.*, vol. 47, pp. 1856–1864, Dec. 1999.

[5] K. Higuchi, A. Fujiwara and M. Sawahashi, "Multipath interference canceller for high-speed packet transmission with adaptive modulation and coding scheme in W-CDMA forward link," *IEEE J. Select. Areas Commun.*, vol. 20, pp. 419–432, Feb. 2002.

[6] B. Xia, "Enhanced techniques for broadband wireless communications," Ph.D. Thesis, University of Hong Kong, 2004.

[7] J. G. Proakis, *Digital Communications*. New York: McGraw-Hill, 1995.

[8] S. Benedetto and E. Biglieri, *Principles of Digital Transmission with Wireless Applications*. New York: Kluwer, 1999.

[9] Guidelines for evaluation of radio transmission technology for IMT-2000. ITU-R M.1225, 1997.

Part IV

4G mobile communications

9 Optimal and MMSE detection for downlink OFCDM

The performances of optimum and per sub-carrier MMSE (or pcMMSE) detectors for OFCDM systems are compared in this chapter. In OFCDM systems, the existence of MCI in the frequency domain due to a frequency selective channel on different sub-carriers dramatically degrades the system performance. To suppress MCI, an optimum or MMSE detector is employed in this chapter. A quasi-analytical BER expression is derived in the presence of imperfect channel estimation. Numerical results show that with a linear computation complexity, the MMSE detector can improve the system performance by suppressing MCI, although it cannot perform as well as the optimum detector. Thus, in systems with a small number of code channels, the optimum detector can be employed to achieve better performance, whereas the MMSE detector is more suitable for systems with a large number of code channels. The MMSE detector is also more robust to different configurations of system parameters than the optimum detector. Moreover, it is found that pilot channel power should be carefully determined by making trade-off between the channel estimation quality and received SNR for each data channel.

9.1 Introduction

OFCDM is proposed for future broadband wireless communications. Inherited from OFDM, in OFCDM systems a high-speed data stream is divided into several parallel low-speed substreams, whose bandwidth is far smaller than the channel bandwidth, so each substream can be regarded as passing through a frequency nonselective (flat) fading. As a result, the multipath interference in frequency selective fading channels is effectively avoided. But as mentioned in Chapter 1, due to a frequency selective channel on different sub-carriers, MCI in the frequency domain occurs. In this chapter, optimum and MMSE detectors are analytically studied to combat MCI.

Moreover, since the signal is distorted due to the amplitude and phase fluctuations of the fading channel, the fading effect should be compensated at the receiver. In [1], a pilot assisted channel estimator is presented. The effect of imperfect channel estimation by the channel estimator is investigated by means of simulation. In this chapter, the pilot assisted channel estimator is analytically studied, and the noise power estimator presented in [1] is modified for unbiased estimation. Furthermore, the BER performance of two detectors in the presence of imperfect channel estimation is derived.

The chapter is organized as follows. In Section 9.2, system models are presented. Besides the description of transmit model and the scheme for the assignment of spreading codes, two detector structures (optimum and MMSE detectors) are described in detail, including the proposed pilot assisted channel and noise power estimation algorithms. In Section 9.3, analytical expressions for optimum and MMSE detectors are derived. Comparison and discussion on numerical BER results with various parameters for different detectors are presented in Section 9.4. Finally, some conclusions are drawn in Section 9.5.

9.2 System description

9.2.1 Transmitter model

In OFCDM, for each code channel with a spreading factor of $N = N_{\mathrm{T}} \times N_{\mathrm{F}}$, the transmitter performs two-dimensional spreading by using the time domain spreading code with length (or spreading factor) N_{T}, and the frequency domain spreading code with length (or spreading factor) N_{F}. Both time and frequency domain spreading codes are generated from OVSF codes. Spreading factors for time and frequency domain spreading can be varied according to channel conditions to achieve high system capacity. It is desired that channels with different time domain spreading codes are orthogonal to each other, i.e. the MCI from code channels with different time domain spreading codes is zero. Frequency diversity is achieved by frequency domain spreading. Similarly, the frequency domain spreading codes are orthogonal to each other. In Gaussian channels, there is no MCI among frequency domain spreading codes. However, because of independent fading on sub-carriers, orthogonality in the frequency domain no longer maintains among code channels in the received signal. Thus, MCI in the frequency domain occurs. Figure 9.1 shows one example of the two-dimensional spreading operation.

Similar to [2], the transmitter block diagram for the downlink of a single-cell OFCDM system is shown in Figure 9.2. Consider the kth ($k = 0, 1, \ldots, K-1$) data stream, where $K$is the number of code channels. The symbol sequence is first serial-to-parallel converted to M/N_{F} (M is the total number of sub-carriers, in VSF-OFCDM, M/N_{F} should be an integer) parallel sequences. Then each sequence $d_{b,k}$ ($b = 0, 1, \ldots, M/N_F - 1$) is spread in the time domain with time domain spreading code $c_{\mathrm{T},k}^{(\mathrm{CH})}$. After time domain spreading, each time domain spreading signal is duplicated into N_{F} parallel copies for N_{F} sub-carriers. In each sub-carrier, the time domain spreading signal is multiplied by a chip of the frequency domain spreading code, which is the combination of a channelization code $c_{\mathrm{F},k}^{(\mathrm{CH})}$ and a cell-specific scrambling code $c^{(\mathrm{SC})}$. In this chapter, for simple description they are of length N_{F}. Therefore, all M/N_{F} sequences for the kth channel are spread with the same frequency domain spreading code. For example, as shown in Figure 9.2, the zeroth copies of all M/N_{F} time domain spreading signals are multiplied by $c_{\mathrm{F},k}^{(\mathrm{CH})}(0)c^{(\mathrm{SC})}(0)$, the first copies of all M/N_F time domain spreading signals are multiplied by $c_{\mathrm{F},k}^{(\mathrm{CH})}(1)c^{(\mathrm{SC})}(1)$, and so on. Here, $c_{\mathrm{F},k}^{(\mathrm{CH})}$ and $c_{\mathrm{T},k}^{(\mathrm{CH})}$ are real-valued binary channelization codes taking the value ± 1, whereas $c^{(\mathrm{SC})}$ is the real-valued binary scrambling code being the same for all code channels. K code-multiplexed channels are assigned with

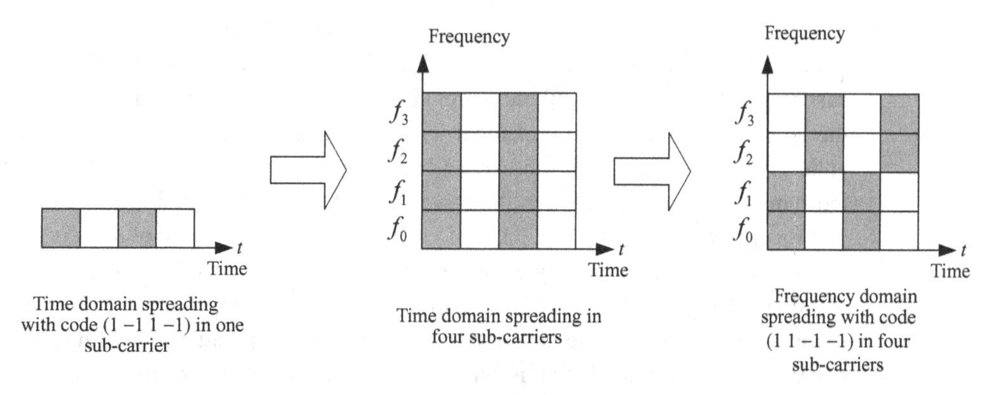

Figure 9.1 Example for two-dimensional spreading with $N = N_\mathrm{T} \times N_\mathrm{F} = 4 \times 4$ (■ stands for 1, □ stands for −1).

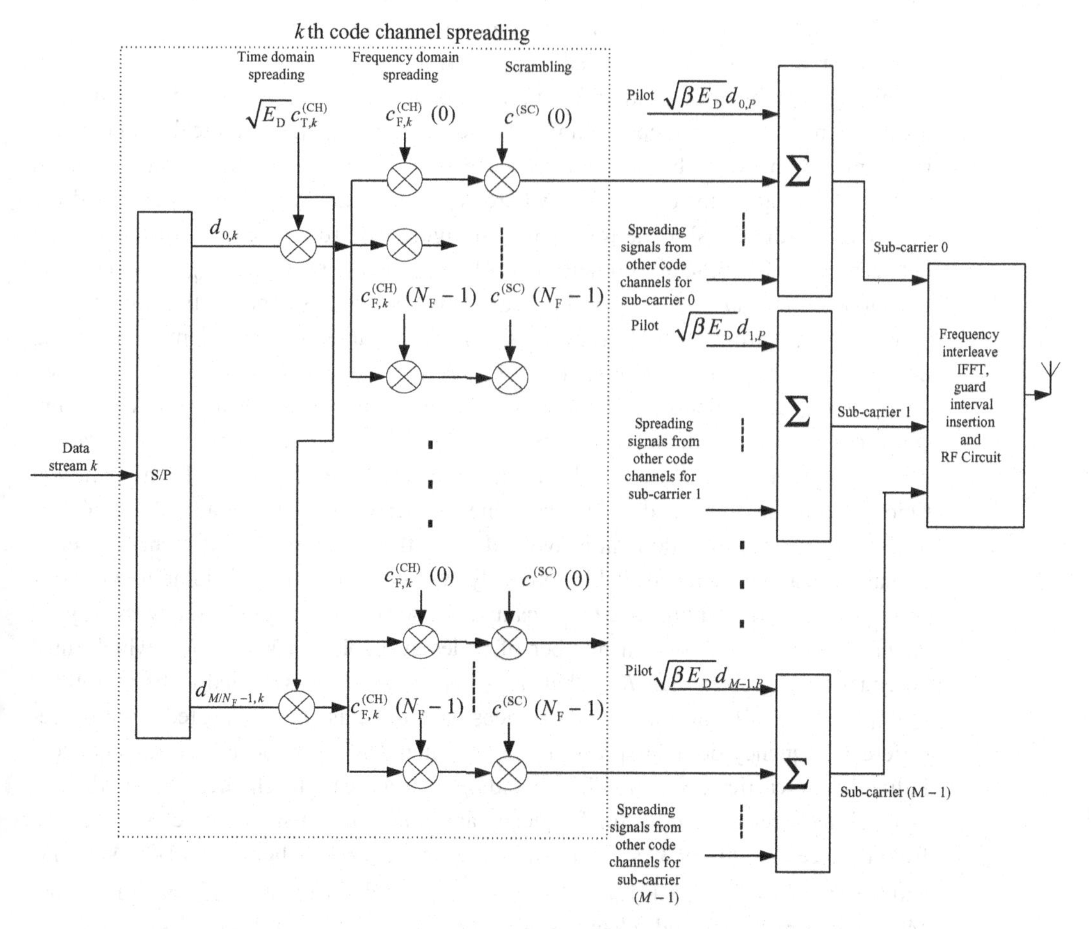

Figure 9.2 Transmitter diagram of the VSF-OFCDM system (M/N_F is an integer).

different combinations of time and frequency domain channelization codes $\{c_{\mathrm{T},k}^{(\mathrm{CH})}, c_{\mathrm{F},k}^{(\mathrm{CH})}\}$. Therefore, in the OFCDM, each data symbol is impressed over N_{F} sub-carriers by N_{T} OFCDM symbols (chips) in each sub-carrier. For channel estimation at the receiver, a common pilot channel with spreading factor of N_{Pilot} is employed and code-multiplexed over each sub-carrier. Note that E_{D} is the chip energy of the transmitted symbol on data channel, β is the power ratio of the pilot channel to one data channel, and for the mth ($m = 0, 1, \ldots, M-1$) sub-carrier, the pilot symbol $d_{m,P}$ is known at the receiver and the spreading code for pilot channels is an all-1 sequence. In order to exploit the frequency diversity, a frequency interleaver is employed before the frequency modulation. Therefore, the largest possible frequency separation between sub-carriers carrying the same information is achieved. For example, sub-carriers $\{0, 1, \ldots N_{\mathrm{F}} - 1\}$ bear the same information, they are separated as much as possible in the frequency domain to achieve the maximum frequency diversity gain. After frequency interleaving using an inverse fast Fourier transform (IFFT) the spread signals occupy all M sub-carriers. Similar to OFDM, in the transmitter, a guard interval is inserted between the OFCDM symbols to avoid ISI caused by multipath propagation.

In order to meet the orthogonality property of different codes, it is important to assign code combinations with different time domain spreading codes, but the same frequency domain spreading code. Similar to the example in [2], one typical pilot channel is assumed with a spreading factor of $N_{\mathrm{Pilot}} = 16 \times 1$ and K code channels are assumed with a spreading factor of $4 \times N_{\mathrm{F}}$, where $N_{\mathrm{T}} = 4$. The pilot channel is assigned the code $C_{16,0}$ in the OVSF code tree shown in Figure 9.3. To maintain the orthogonality between code channels, all mother codes of $C_{16,0}$, including $C_{8,0}$, $C_{4,0}$, $C_{2,0}$ and $C_{1,0}$, cannot be used. Thus, for $K \leq 3$, the K code channels are assigned with different time spreading codes from the set $\{C_{4,1}, C_{4,2}, C_{4,3}\}$ and the same frequency domain spreading code $C_{N_{\mathrm{F}},0}$. Note that the code set $\{C_{4,1}, C_{4,2}, C_{4,3}\}$ is orthogonal. When $K > 3$, however, some code channels will be assigned the same time domain spreading code from $\{C_{4,1}, C_{4,2}, C_{4,3}\}$, and distinguished by assigning different frequency domain spreading codes. In general, for $K < N_{\mathrm{T}}$, the K code channels can be assigned with different time domain spreading codes, but the same frequency domain spreading code $C_{N_{\mathrm{F}},0}$, so that MCI in the time domain is avoided. Note that there are N_{F} different frequency domain spreading codes available, but only $N_{\mathrm{T}} - 1$ different time domain spreading codes, since the remaining code $C_{N_{\mathrm{T}},0}$ cannot be used due to its connection with the pilot channel. Thus, the maximum number of codes available is $(N_{\mathrm{T}} - 1)N_{\mathrm{F}}$, which must be equal to or greater than K. When $K \geq N_{\mathrm{T}}$, where K is assumed to be an integer multiple of $N_{\mathrm{T}} - 1$, the same $N_{\mathrm{T}} - 1$ codes have to be assigned repeatedly with other different frequency domain spreading codes. Then, MCI may occur over independent fading in sub-carriers. So, with this spreading code allocation scheme, for the kth code channel, its spreading codes for frequency and time domain spreading can be derived from the tree in Figure 9.3 as C_{N_{F},k_f} and $C_{N_{\mathrm{T}},\hat{k}}$, respectively, where $k_f = \lfloor k/(N_{\mathrm{T}} - 1) \rfloor$ and $\hat{k} = k - k_f(N_{\mathrm{T}} - 1) + 1$. For the kth code channel, the received signal suffers from MCI only from $K_{\mathrm{c}} - 1$ code channels, where $K_{\mathrm{c}} = \lceil K/(N_T - 1) \rceil$ is the number of code channels employing the same time domain spreading code $C_{N_{\mathrm{T}},\hat{k}}$, and $\lceil x \rceil$ denotes the smallest integer not less than x. Referring to the original OFCDM system in [1], where

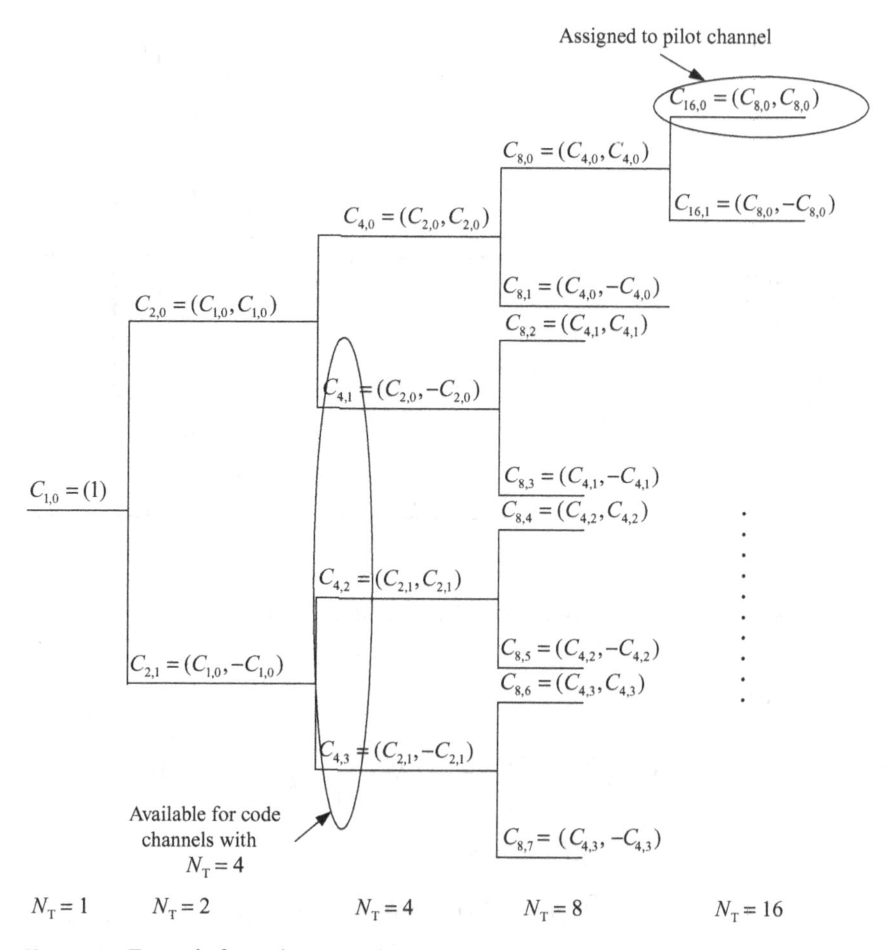

Figure 9.3 Example for assignment of time domain spreading codes for VSF-OFCDM.

no time domain spreading is employed, the MCI is from all other $K - 1$ code channels. In the VSF-OFCDM system, however, the MCI is only from a small number of code channels ($K_c - 1$). Thus, the MCI is reduced, but at the cost of small frequency diversity for a given spreading factor. Therefore, time and frequency domain spreading factors should be optimized by considering both effects of MCI and frequency diversity.

9.2.2 Receiver model

At the receiver, shown in Figure 9.4, after guard interval deletion, FFT and frequency deinterleaving, the received baseband signals on all sub-carriers $\{r_m\}$ are obtained. Since information of the bth data sequence for the kth code channel ($\{d_{b,k}\}$) is contained in N_F received signals, $\{r_m,\ bN_F \le m < (b+1)\,N_F\}$, the received signals on the mth sub-carrier ($bN_F \le m < (b+1)\,N_F$) for the λth ($\lambda = 0, 1, \ldots$) OFCDM symbol is

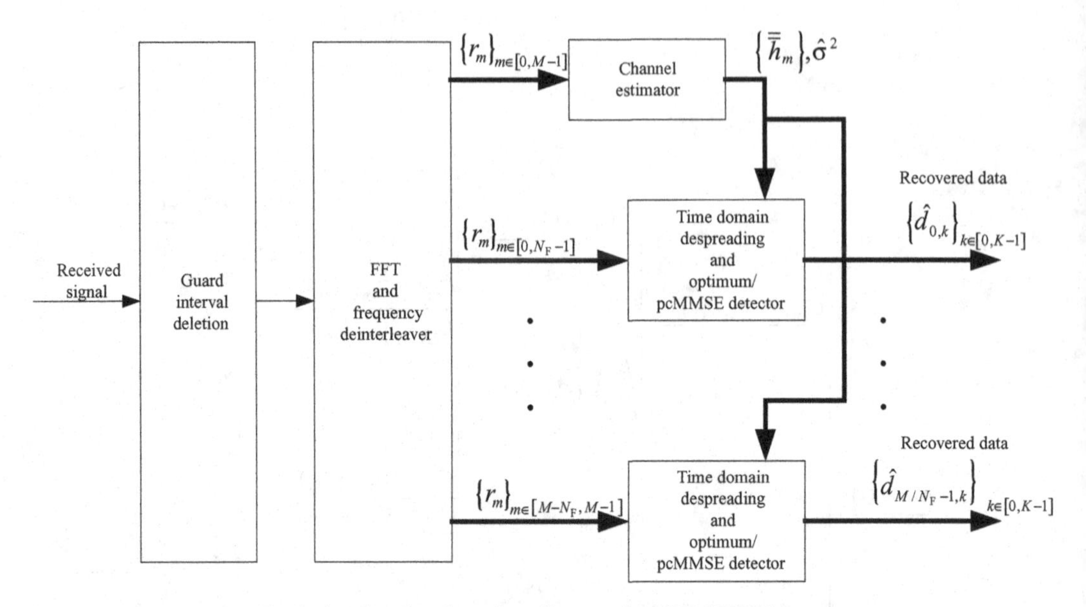

Figure 9.4 Equivalent baseband receiver diagram for VSF-OFCDM.

given by

$$
\begin{aligned}
r_m(\lambda) &= \sqrt{E_D}h_m c^{(SC)}(m-bN_F)\sum_{k=0}^{K-1} c_{F,k}^{(CH)}(m-bN_F)c_{T,k}^{(CH)}(\lambda)d_{b,k}(\lambda_D) \\
&\quad + \sqrt{\beta E_D}h_m d_{m,P} + n_m(\lambda) \\
&= \sqrt{E_D}h_m c^{(SC)}(m-bN_F)\sum_{k=0}^{K-1} C_{N_F,k_f}(m-bN_F)C_{N_T,\hat{k}} \\
&\quad \times(\lambda-(\lambda_D-1)N_T)d_{b,k}(\lambda_D) + \sqrt{\beta E_D}h_m d_{m,P} + n_m(\lambda)
\end{aligned} \tag{9.1}
$$

where h_m is the channel impulse response of the mth sub-carrier. Since frequency interleaving is employed at the transmitter, h_m is assumed to be a co-independent complex Gaussian random variable with zero mean and unit variance the OVSF codes, $C_{N_T,\hat{k}}$ and C_{N_F,k_f}, are the time and frequency domain spreading codes for the kth code channel, respectively, $k_f = \lfloor k/(N_T-1)\rfloor$, $\hat{k} = k - k_f(N_T-1)+1$, $d_{b,k}(\lambda_D)$ is the λ_D^{th} transmitted symbol in the kth data channel with $\lambda_D = \lfloor \lambda/N_T \rfloor$, $d_{m,P}$ is the pilot symbol for the mth sub-carrier and $n_m(\lambda)$ is the zero mean AWGN noise with variance

$$
\sigma_n^2 = \frac{1}{2}E\{n_m(\lambda)n_{m'}^*(\lambda')\} = \sigma^2\delta(m-m')\delta(\lambda-\lambda') \tag{9.2}
$$

where $\delta(m-m') = 1$ and 0 for $m = m'$ and $m \neq m'$, respectively.

Without loss of generality, the data symbols ($d_{0,k}$, $k = 0, 1, \ldots, K-1$) transmitted from the 0th sub-carrier to the (N_F-1)th sub-carrier are assumed to be the desired symbols. By despreading the received signal (9.1) with different time domain spreading codes $C_{N_T,\hat{k}}$ for $\hat{k} = 1, 2, \ldots, N_T-1$, the information transmitted on all code channels can be recovered by processing in the frequency domain. The time domain despreading output of the mth ($m = 0, 1, \ldots, N_F - 1$) sub-carrier is

$$y_{m,\hat{k}}(\lambda_D) = \sum_{\lambda=\lambda_D N_T}^{(\lambda_D+1)N_T-1} r_m(\lambda) C_{N_T,\hat{k}}(\lambda - (\lambda_D - 1)N_T)$$
$$= \sqrt{E_D} N_T h_m c^{(SC)}(m) \sum_{\hat{k}_f=0}^{K_c-1} C_{N_F,\hat{k}_f}(m) d_{0,\hat{k}-1+\hat{k}_f(N_T-1)}(\lambda_D)$$
$$+ \eta_{m,\hat{k}}(\lambda_D) \tag{9.3}$$

where

$$\eta_{m,\hat{k}}(\lambda_D) = \sum_{\lambda=\lambda_D N_T}^{(\lambda_D+1)N_T-1} n_m(\lambda) C_{N_T,\hat{k}}(\lambda - \lambda_D N_T) \tag{9.4}$$

In (9.3), as code channels with different time domain spreading codes are orthogonal to each other, the MCI is only from code channels with the same time domain spreading code and different frequency domain spreading codes. Then the data transmitted on code channels with time domain spreading code $C_{N_T,\hat{k}}$ can be recovered from (9.3). For convenience, the time domain despreading output is rewritten in vector form as

$$\mathbf{Y}_{\hat{k}}(\lambda_D) = [y_{0,\hat{k}}(\lambda_D),\ \ y_{1,\hat{k}}(\lambda_D), \ldots,\ \ y_{N_F-1,\hat{k}}(\lambda_D)]^T \tag{9.5}$$

In Figure 9.4, the channel estimator provides channel estimation $\{\bar{\bar{h}}_m\}$ and noise power estimation $\hat{\sigma}^2$. Using these, the optimum detector or MMSE detector is employed for symbol detection.

9.2.3 Pilot assisted channel estimator

As the channel and noise power estimations are required for both optimum and MMSE detectors, a pilot assisted channel estimator will be described first.

Similar to the time domain despreading for data channels, the time domain despreading output for the pilot channel on the mth sub-carrier can be expressed by

$$y_{m,0}(\lambda_D) = \sum_{\lambda=\lambda_D N_T}^{(\lambda_D+1)N_T-1} r_m(\lambda)$$
$$= \sqrt{\beta E_D} N_T h_m d_{m,P} + \sum_{\lambda=\lambda_D N_T}^{(\lambda_D+1)N_T-1} n_m(\lambda) \tag{9.6}$$

where the second term is the AWGN component, which can be regarded as a zero mean Gaussian variable with variance of $N_T\sigma^2$. Then one obtains $E\{y_{m,0}(\lambda_D)\} = \sqrt{\beta E_D} N_T h_m d_{m,P}$. Thus, the channel estimation for the mth sub-carrier can be obtained by

$$\bar{h}_m(\lambda_D) = \frac{y_{m,0}(\lambda_D)}{\sqrt{\beta E_D} N_T d_{m,P}} = h_m + \sum_{\lambda=\lambda_D N_T}^{(\lambda_D+1)N_T-1} \frac{n_m(\lambda)}{\sqrt{\beta E_D} N_T d_{m,P}}$$
$$= h_m + \Delta \bar{h}_m(\lambda_D) \tag{9.7}$$

where $\Delta\bar{h}_m(\lambda_D)$ is given by

$$\Delta\bar{h}_m(\lambda_D) = \sum_{\lambda=\lambda_D N_T}^{(\lambda_D+1)N_T-1} \frac{n_m(\lambda)}{\sqrt{\beta E_D} N_T d_{m,P}} \tag{9.8}$$

It is obvious that the channel estimation suffers from AWGN, which results in imperfect channel estimation. Because the variance of this channel estimation error has a direct contribution to the decision noise (as will be shown in the following analysis), it must be suppressed as much as possible. Considering that the channel estimation error can be modeled as Gaussian noise with no correlation between successive pilot symbols, an LPF with a cut-off frequency much greater than the maximum Doppler frequency is able to reduce much of the noise in the channel estimation [3]. With the assumption of a slow fading channel, so that the fading remains constant in recent N_D symbols' duration, channel estimation is obtained by

$$\begin{aligned}\bar{\bar{h}}_m(\lambda_D) &= \frac{1}{N_D} \sum_{i=-\lfloor (N_D-1)/2 \rfloor}^{\lfloor N_D/2 \rfloor} \bar{h}_m(\lambda_D - i) \\ &= h_m + \frac{1}{N_D} \sum_{i=-\lfloor (N_D-1)/2 \rfloor}^{\lfloor N_D/2 \rfloor} \Delta\bar{h}_m(\lambda_D - i) = h_m + \Delta\bar{\bar{h}}_m\end{aligned} \tag{9.9}$$

where the channel estimation error is given by

$$\Delta\bar{\bar{h}}_m = \frac{1}{N_D} \sum_{i=-\lfloor (N_D-1)/2 \rfloor}^{\lfloor N_D/2 \rfloor} \Delta\bar{h}_m(\lambda_D - i) \tag{9.10}$$

Since h_m and $\Delta\bar{h}_m(\lambda_D)$ are uncorrelated Gaussian variables with zero mean, the channel estimation error is uncorrelated with h_m, and $\bar{\bar{h}}_m(\lambda_D)$ can also be regarded as a Gaussian variable. So, the conditional mean and variance of $\bar{\bar{h}}_m$ are

$$\mu_{\bar{\bar{h}}_m|h_m} = E\{\bar{\bar{h}}_m | h_m\} = h_m \tag{9.11}$$

and

$$R_{\bar{\bar{h}}_m|h_m} = \frac{1}{2} E\{(\bar{\bar{h}}_m - E\{\bar{\bar{h}}_m\})(\bar{\bar{h}}_m - E\{\bar{\bar{h}}_m\})^* | h_m\} = \frac{\sigma^2}{\beta E_D N_T N_D} \tag{9.12}$$

respectively.

The noise power (σ^2) estimation can be calculated from the time domain despreading outputs (9.6) for the pilot channel by

$$\hat{\sigma}^2 = \frac{1}{2N_T M} \sum_{m=0}^{M-1} \left| y_{m,0}(\lambda_D) - \sqrt{\beta E_D} N_T h_m d_{m,P} \right|^2 \tag{9.13}$$

Then the noise power estimation is unbiased, i.e. $E\{\hat{\sigma}^2\} = \sigma^2$. Because only estimated channel information is available at the receiver, (9.13) can be approximated as [1]

$$\hat{\sigma}^2 \approx \frac{1}{2N_{\mathrm{T}}M}\sum_{m=0}^{M-1}\left|y_{m,0}(\lambda_{\mathrm{D}}) - \sqrt{\beta E_{\mathrm{D}}}N_{\mathrm{T}}\bar{\bar{h}}_m(\lambda_{\mathrm{D}})d_{m,\mathrm{P}}\right|^2$$
$$= \frac{1}{2N_{\mathrm{T}}M}\sum_{m=0}^{M-1}\left|-\frac{1}{N_{\mathrm{D}}}\sum_{\lambda'_{\mathrm{D}}=\lambda_{\mathrm{D}}-\lfloor(N_{\mathrm{D}}-1)/2\rfloor}^{\lambda_{\mathrm{D}}+\lfloor N_{\mathrm{D}}/2\rfloor}\;\sum_{\lambda=\lambda'_{\mathrm{D}}N_{\mathrm{T}}}^{(\lambda'_{\mathrm{D}}+1)N_{\mathrm{T}}-1} n_m(\lambda)\right.$$
$$\left. + \sum_{\lambda=\lambda_{\mathrm{D}}N_{\mathrm{T}}}^{(\lambda_{\mathrm{D}}+1)N_{\mathrm{T}}-1} n_m(\lambda)\right|^2 \tag{9.14}$$

It can be seen from (9.14) that $E\{\hat{\sigma}^2\} = (N_{\mathrm{D}} - 1)\sigma^2/N_{\mathrm{D}}$, i.e. in case of imperfect channel estimation, the approximated noise power estimation is biased. To provide unbiased estimation, the channel estimator [3] should be adjusted by

$$\hat{\sigma}^2 = \frac{N_{\mathrm{D}}}{2N_{\mathrm{T}}M(N_{\mathrm{D}}-1)}\sum_{m=0}^{M-1}\left|y_{m,0}(\lambda_{\mathrm{D}}) - \sqrt{\beta E_{\mathrm{D}}}N_{\mathrm{T}}\bar{\bar{h}}_m(\lambda_{\mathrm{D}})d_{m,\mathrm{P}}\right|^2$$
$$= \sum_{m=0}^{M-1}|\hat{n}_m(\lambda_{\mathrm{D}})|^2 \tag{9.15}$$

where

$$\hat{n}_m(\lambda_{\mathrm{D}}) = \frac{-\frac{1}{N_{\mathrm{D}}}\sum_{\lambda'_{\mathrm{D}}=\lambda_{\mathrm{D}}-\lfloor(N_{\mathrm{D}}-1)/2\rfloor}^{\lambda_{\mathrm{D}}+\lfloor N_{\mathrm{D}}/2\rfloor}\;\sum_{\lambda=\lambda'_{\mathrm{D}}N_{\mathrm{T}}}^{(\lambda'_{\mathrm{D}}+1)N_{\mathrm{T}}-1} n_m(\lambda) + \sum_{\lambda=\lambda_{\mathrm{D}}N_{\mathrm{T}}}^{(\lambda_{\mathrm{D}}+1)N_{\mathrm{T}}-1} n_m(\lambda)}{\sqrt{2N_{\mathrm{T}}M(N_{\mathrm{D}}-1)/N_{\mathrm{D}}}} \tag{9.16}$$

Since $\hat{n}_m(\lambda_{\mathrm{D}})$ is the sum of independent Gaussian variables, it can be approximated as a statistically independent and identically distributed complex Gaussian variable with zero mean and a variance of σ_1^2, where

$$\sigma_1^2 = \frac{1}{2}E\{|\hat{n}_m(\lambda_D)|^2\} = \frac{\sigma^2}{2M} \tag{9.17}$$

Therefore, the noise power estimation is unbiased and can be regarded as a central chi-square distributed random variable with $2M$ degrees of freedom. Its PDF is given by

$$p\left(\hat{\sigma}^2\right) = \frac{1}{\sigma_1^{2M}2^M\Gamma(M)}\left(\hat{\sigma}^2\right)^{M-1}\exp\left(-\frac{\hat{\sigma}^2}{2\sigma_1^2}\right) \tag{9.18}$$

where $\Gamma(\cdot)$ is the gamma function [4]. For simplification, the symbol index λ_{D} will be omitted in the following.

9.2.4 Optimum detector

The optimum detector is to select the vector $\hat{\mathbf{d}}$ from all possible combinations of the transmitted vectors $\{\mathbf{d}\}$ that maximizes the a-posteriori probability $p(\mathbf{d}|\mathbf{Y}_{\hat{k}})$, where $\mathbf{d} = [d_{0,\hat{k}-1}, d_{0,\hat{k}-1+(N_{\mathrm{T}}-1)}, \cdot d_{0,\hat{k}-1+(K_c-1)(N_{\mathrm{T}}-1)}]^{\mathrm{T}}$ denotes the transmitted symbol vector. It is well known that maximizing the a-posteriori probability is equivalent to maximizing the likelihood $p(\mathbf{Y}_{\hat{k}}|\mathbf{d})$ when equiprobable symbols are transmitted.

From (9.3) and (9.5), conditioned on $\mathbf{d}$, $\mathbf{Y}_{\hat{k}}$ is Gaussian with mean and variance of

$$\mu_{\mathbf{Y}_{\hat{k}}|\mathbf{d}} = E\{\mathbf{Y}_{\hat{k}}|\mathbf{d}\}$$

$$= \begin{bmatrix} \sqrt{E_{\mathrm{D}}} N_{\mathrm{T}} h_0 c^{(\mathrm{SC})}(0) \sum_{\hat{k}_f=0}^{K_c-1} c_{N_{\mathrm{F}},\hat{k}_f}(0) d_{0,\hat{k}-1+\hat{k}_f(N_{\mathrm{T}}-1)} \\ \sqrt{E_{\mathrm{D}}} N_{\mathrm{T}} h_1 c^{(\mathrm{SC})}(1) \sum_{\hat{k}_f=0}^{K_c-1} c_{N_{\mathrm{F}},\hat{k}_f}(1) d_{0,\hat{k}-1+\hat{k}_f(N_{\mathrm{T}}-1)} \\ \vdots \\ \sqrt{E_{\mathrm{D}}} N_{\mathrm{T}} h_{N_{\mathrm{F}}-1} c^{(\mathrm{SC})}(N_{\mathrm{F}}-1) \sum_{\hat{k}_f=0}^{K_c-1} c_{N_{\mathrm{F}},\hat{k}_f}(N_{\mathrm{F}}-1) d_{0,\hat{k}-1+\hat{k}_f(N_{\mathrm{T}}-1)} \end{bmatrix} \tag{9.19}$$

and

$$\begin{aligned} R_{\mathbf{Y}_{\hat{k}}|\mathbf{d}} &= \frac{1}{2} E\{(\mathbf{Y}_{\hat{k}} - \mu_{\mathbf{Y}_{\hat{k}}|\mathbf{d}})(\mathbf{Y}_{\hat{k}} - \mu_{\mathbf{Y}_{\hat{k}}|\mathbf{d}})^H \,|\mathbf{d}\} \\ &= \frac{1}{2} E\left\{ \begin{bmatrix} \eta_{0,\hat{k}} \\ \eta_{1,\hat{k}} \\ \vdots \\ \eta_{N_F-1,\hat{k}} \end{bmatrix} \begin{bmatrix} \eta_{0,\hat{k}} \\ \eta_{1,\hat{k}} \\ \vdots \\ \eta_{N_F-1,\hat{k}} \end{bmatrix}^H \right\} \end{aligned} \tag{9.20}$$

respectively. Since the AWGN components in the outputs for different sub-carriers are uncorrelated, $R_{\mathbf{Y}_{\hat{k}}|\mathbf{d}} = N_{\mathrm{T}}\sigma^2\mathbf{I}$, where $\mathbf{I}$ is the identity matrix.

Then, the conditional Gaussian variable $\mathbf{Y}_{\hat{k}}$ has the PDF $p(\mathbf{Y}_{\hat{k}}\,|\mathbf{d})$ as

$$p(\mathbf{Y}_{\hat{k}}\,|\mathbf{d}) = \frac{1}{(2\pi N_{\mathrm{T}}\sigma^2)^{N_{\mathrm{F}}}} \times \exp\left(- \frac{\sum_{m=0}^{N_{\mathrm{F}}-1} \left| y_{m,\hat{k}} - \sqrt{E_{\mathrm{D}}} N_{\mathrm{T}} h_m c^{(\mathrm{SC})}(m) \sum_{\hat{k}_f=0}^{K_c-1} c_{N_{\mathrm{F}},\hat{k}_f}(m) d_{0,\hat{k}-1+\hat{k}_f(N_{\mathrm{T}}-1)} \right|^2}{2N_{\mathrm{T}}\sigma^2} \right) \tag{9.21}$$

Maximizing $p(\mathbf{Y}_{\hat{k}}\,|\mathbf{d})$ is equivalent to minimizing its negative exponent, i.e.

$$\Lambda(\mathbf{d}) \equiv \sum_{m=0}^{N_{\mathrm{F}}-1} \left| y_{m,\hat{k}} - \sqrt{E_{\mathrm{D}}} N_{\mathrm{T}} \bar{\bar{h}}_m c^{(\mathrm{SC})}(m) \sum_{\hat{k}_f=0}^{K_c-1} c_{N_{\mathrm{F}},\hat{k}_f}(m) d_{0,\hat{k}-1+\hat{k}_f(N_{\mathrm{T}}-1)} \right|^2 \tag{9.22}$$

where h_m is replaced by $\bar{\bar{h}}_m$ because, at the receiver, only the channel estimation is available. The metric $\Lambda(\mathbf{d})$ can be written as

$$\begin{aligned}\Lambda(\mathbf{d}) &= \sum_{m=0}^{N_F-1}\left(|y_{m,\hat{k}}|^2 + \left|\sqrt{E_D}N_T\bar{\bar{h}}_m c^{(SC)}(m)\sum_{\hat{k}_f=0}^{K_c-1} c_{N_F,\hat{k}_f}(m)d_{0,\hat{k}-1+\hat{k}_f(N_T-1)}\right|^2\right)\\ &\quad - \sum_{m=0}^{N_F-1} 2\Re\left\{y^*_{m,\hat{k}}\sqrt{E_D}N_T\bar{\bar{h}}_m c^{(SC)}(m)\sum_{\hat{k}_f=0}^{K_c-1} c_{N_F,\hat{k}_f}(m)d_{0,\hat{k}-1+\hat{k}_f(N_T-1)}\right\}\end{aligned} \tag{9.23}$$

Since $|y_{m,\hat{k}}|^2$ is constant for any data $\mathbf{d}$, one obtains

$$\begin{aligned}\bar{\Lambda}(\mathbf{d}) &= \sum_{m=0}^{N_F-1}\left|\sqrt{E_D}N_T\bar{\bar{h}}_m c^{(SC)}(m)\sum_{\hat{k}_f=0}^{K_c-1} c_{N_F,\hat{k}_f}(m)d_{0,\hat{k}-1+\hat{k}_f(N_T-1)}\right|^2\\ &\quad -2\sqrt{E_D}N_T\sum_{m=0}^{N_F-1}\Re\left\{y_{m,\hat{k}}\left(\bar{\bar{h}}_m c^{(SC)}(m)\right)^*\left(\sum_{\hat{k}_f=0}^{K_c-1} c_{N_F,\hat{k}_f}(m)d_{0,\hat{k}-1+\hat{k}_f(N_T-1)}\right)^*\right\}\end{aligned} \tag{9.24}$$

Therefore, the received signal is first coherently descrambled by multiplying with $(\bar{\bar{h}}_m c^{(SC)}(m))^*$, and then sent to a selector to find the vector $\hat{\mathbf{d}}$ with minimum metric $\bar{\Lambda}$.

9.2.5 MMSE detector

It can be seen from (9.24) that when K_c is large, the computational complexity of the optimum detector increases exponentially. In order to reduce the complexity, when K_c is large, the MMSE is studied. The MMSE detector is used to minimize the mean squared error $J_{\hat{k},\hat{k}_f} = E\{|d_{0,\hat{k}-1+\hat{k}_f(N_T-1)} - \mathbf{w}^H_{\hat{k},\hat{k}_f}\mathbf{Y}_{\hat{k}}|^2|\mathbf{d}\}$ with the assumption that signals received on different sub-carriers are uncorrelated, where $\mathbf{w}_{\hat{k},\hat{k}_f}$ is the weight vector for the kth code channel. Then data decisions can be made by

$$\hat{d}_{0,\hat{k}-1+\hat{k}_f(N_T-1)} = \mathrm{sgn}\left[\Re\left(\mathbf{w}^H_{\hat{k},\hat{k}_f}\mathbf{Y}_{\hat{k}}\right)\right] \tag{9.25}$$

where $\mathrm{sgn}(x) = -1$ and 1 for $x < 0$ and $x \geq 0$, respectively.

Solving the zero point of the differential of the mean squared error $J_{\hat{k},\hat{k}_f}$ with respect to $\mathbf{w}_{\hat{k},\hat{k}_f}$, one obtains

$$\begin{aligned}\mathbf{w}_{\hat{k},\hat{k}_f} &= E\left\{\mathbf{Y}_{\hat{k}}\mathbf{Y}^H_{\hat{k}}\right\}^{-1}d_{0,\hat{k}-1+\hat{k}_f(N_T-1)}E\left\{\mathbf{Y}_{\hat{k}}\right\}\\ &= \left(E\left\{\mu_{\mathbf{Y}_{\hat{k}}|\mathbf{d}}\mu^H_{\mathbf{Y}_{\hat{k}}|\mathbf{d}}\right\} + E\left\{\begin{bmatrix}\eta_{0,\hat{k}}\\ \eta_{1,\hat{k}}\\ \vdots\\ \eta_{N_F-1,\hat{k}}\end{bmatrix}\begin{bmatrix}\eta_{0,\hat{k}}\\ \eta_{1,\hat{k}}\\ \vdots\\ \eta_{N_F-1,\hat{k}}\end{bmatrix}^H\right\}\right)^{-1}\end{aligned}$$

$$
\times \begin{bmatrix} \sqrt{E_{\rm D}} N_{\rm T} h_0 c^{(\rm SC)}(0) C_{N_{\rm F},\hat{k}_f}(0) \\ \sqrt{E_{\rm D}} N_{\rm T} h_1 c^{(\rm SC)}(1) C_{N_{\rm F},\hat{k}_f}(1) \\ \vdots \\ \sqrt{E_{\rm D}} N_{\rm T} h_{N_{\rm F}-1} c^{(\rm SC)}(N_{\rm F}-1) C_{N_{\rm F},\hat{k}_f}(N_{\rm F}-1) \end{bmatrix} \tag{9.26}
$$

where the desired signal and AWGN components are uncorrelated with each other.

Defining $R_\mu \equiv E\{\mu_{\mathbf{Y}_{\hat{k}}|\mathbf{d}} \mu^H_{\mathbf{Y}_{\hat{k}}|\mathbf{d}}\}$, one obtains from (9.19) that

$$
R_\mu = E_{\rm D} N_{\rm T}^2 E \left\{ \begin{bmatrix} h_0 c^{(\rm SC)}(0) \sum\limits_{\hat{k}'_f=0}^{K_c-1} C_{N_{\rm F},\hat{k}'_f}(0) d_{0,\hat{k}-1+\hat{k}'_f(N_{\rm T}-1)} \\ h_1 c^{(\rm SC)}(1) \sum\limits_{\hat{k}'_f=0}^{K_c-1} C_{N_{\rm F},\hat{k}'_f}(1) d_{0,\hat{k}-1+\hat{k}'_f(N_{\rm T}-1)} \\ \vdots \\ h_{N_{\rm F}-1} c^{(\rm SC)}(N_{\rm F}-1) \sum\limits_{\hat{k}'_f=0}^{K_c-1} C_{N_{\rm F},\hat{k}'_f}(N_{\rm F}-1) d_{0,\hat{k}-1+\hat{k}'_f(N_{\rm T}-1)} \end{bmatrix} \right.
$$

$$
\left. \times \begin{bmatrix} h_0 c^{(\rm SC)}(0) \sum\limits_{\hat{k}'_f=0}^{K_c-1} C_{N_{\rm F},\hat{k}'_f}(0) d_{0,\hat{k}-1+\hat{k}'_f(N_{\rm T}-1)} \\ h_1 c^{(\rm SC)}(1) \sum\limits_{\hat{k}'_f=0}^{K_c-1} C_{N_{\rm F},\hat{k}'_f}(1) d_{0,\hat{k}-1+\hat{k}'_f(N_{\rm T}-1)} \\ \vdots \\ h_{N_{\rm F}-1} c^{(\rm SC)}(N_{\rm F}-1) \sum\limits_{\hat{k}'_f=0}^{K_c-1} C_{N_{\rm F},\hat{k}'_f}(N_{\rm F}-1) d_{0,\hat{k}-1+\hat{k}'_f(N_{\rm T}-1)} \end{bmatrix}^H \right\} \tag{9.27}
$$

Assuming signals on different sub-carriers are uncorrelated and $E|\sum_{\hat{k}'_f=0}^{K_c-1} C_{N_{\rm F},\hat{k}'_f} \times d_{0,\hat{k}-1+\hat{k}'_f(N_{\rm T}-1)}|^2 = K_c$, (9.27) is written as

$$
R_\mu = E_{\rm D} K_c N_{\rm T}^2 \begin{bmatrix} |h_0|^2 & 0 & \cdots & 0 \\ 0 & |h_1|^2 & \cdots & 0 \\ \vdots & \vdots & \ddots & \vdots \\ 0 & 0 & \cdots & |h_{N_{\rm F}-1}|^2 \end{bmatrix} \tag{9.28}
$$

Considering that at the receiver the true channel information (including noise power and channel fading) is unknown to the detector, the estimated noise power $\hat{\sigma}^2$ and estimated channel fading $\bar{\bar{h}}_m$ are used to obtain the weight vector

$$
\mathbf{w}_{\hat{k},\hat{k}_f} = \begin{bmatrix} w_{\hat{k},\hat{k}_f}(0) \\ w_{\hat{k},\hat{k}_f}(1) \\ \vdots \\ w_{\hat{k},\hat{k}_f}(N_{\rm F}-1) \end{bmatrix}
$$

$$
= \begin{bmatrix} \dfrac{\sqrt{E_D}\bar{\bar{h}}_0}{E_D N_T K_c \left|\bar{\bar{h}}_0\right|^2 + 2\hat{\sigma}^2} c^{(SC)}(0) C_{N_F,\hat{k}_f}(0) \\ \dfrac{\sqrt{E_D}\bar{\bar{h}}_1}{E_D N_T K_c \left|\bar{\bar{h}}_1\right|^2 + 2\hat{\sigma}^2} c^{(SC)}(1) C_{N_F,\hat{k}_f}(1) \\ \vdots \\ \dfrac{\sqrt{E_D}\bar{\bar{h}}_{N_F-1}}{E_D N_T K_c \left|\bar{\bar{h}}_{N_F-1}\right|^2 + 2\hat{\sigma}^2} c^{(SC)}(N_F-1) C_{N_F,\hat{k}_f}(N_F-1) \end{bmatrix} \tag{9.29}
$$

Substituting (9.29) into (9.25), the data decision of the symbol on the kth code channel can be made.

9.3 BER performance analysis

9.3.1 Performance of the optimum detector

The upper union bound on the BER of the optimum detector for the kth code channel $p_{b,k}$ is given. Letting $\mathbf{d}_l$ denote the transmitted vector, an error occurs at the receiver when the detector finds a vector $\mathbf{d}_q$ which satisfies $\bar{\Lambda}(\mathbf{d}_q) < \bar{\Lambda}(\mathbf{d}_l)$ in (9.23) with $\mathbf{d}_l \neq \mathbf{d}_q$ and with probability $p\left(\bar{\Lambda}(\mathbf{d}_q) < \bar{\Lambda}(\mathbf{d}_l)\right)$ of this pairwise. Then, the union bound can be obtained by summing $p\left(\bar{\Lambda}(\mathbf{d}_q) < \bar{\Lambda}(\mathbf{d}_l)\right)$ over all vectors (denoted as $\overline{\{\mathbf{d}_l\}}$) that differ from $\mathbf{d}_l$ in their kth position and then averaging over all possible transmitted vectors $\mathbf{d}_l$, which is written as

$$
p_{b,k} < \frac{1}{2^{K_c}} \sum_{\mathbf{d}_l} \sum_{\mathbf{d}_q \in \overline{\{\mathbf{d}_l\}}} p(\bar{\Lambda}(\mathbf{d}_q) < \bar{\Lambda}(\mathbf{d}_l) \,|\, \mathbf{d}_l) \tag{9.30}
$$

Considering the symmetric property of the transmitted vector, $p_{b,k}$ can be written as

$$
p_{b,k} < \sum_{\mathbf{d}_q \in \overline{\{\mathbf{d}_l\}}} p\left(\bar{\Lambda}(\mathbf{d}_q) < \bar{\Lambda}(\mathbf{d}_l) \,|\, \mathbf{d}_l\right) = \sum_{\mathbf{d}_q \in \overline{\{\mathbf{d}_l\}}} p\left(D_{l,q} < 0 \,|\, \mathbf{d}_l\right) \tag{9.31}
$$

where $D_{l,q} = \bar{\Lambda}(\mathbf{d}_q) - \bar{\Lambda}(\mathbf{d}_l)$. The computational complexity is significantly reduced compared to (9.30).

Defining the equivalent transmitted symbol on the mth sub-carrier $\bar{d}^{(m)}$ as

$$
\bar{d}^{(m)} = \sum_{\hat{k}'_f=0}^{K_c-1} C_{N_F,\hat{k}'_f}(m) d_{0,\hat{k}-1+\hat{k}'_f(N_T-1)} \tag{9.32}
$$

and the vector $\mathbf{z}_m = \left[y_{m,\hat{k}} \sqrt{E_D} N_T \bar{\bar{h}}_m c^{(SC)}(m)\right]^T$, one obtains $D_{l,q} = \bar{\Lambda}(\mathbf{d}_q) - \bar{\Lambda}(\mathbf{d}_l) = \sum_{m=0}^{N_F-1} \mathbf{z}_m^H \mathbf{F}_{l,q}^{(m)} \mathbf{z}_m$. Then, $p(D_{l,q} < 0|\mathbf{d}_l)$ is written as

$$
p(D_{l,q} < 0|\mathbf{d}_l) = \sum_{s \in \text{right poles of } \Phi_{D_{l,q}}(s)} \text{residue}\{\Phi_{D_{l,q}}(s)\} \tag{9.33}
$$

where $\Phi_{D_{l,q}}(s)$ is the moment generating function (MGF) of $D_{l,q}$. With the assumption of independent channels over each sub-carrier, $\Phi_{D_{l,q}}(s)$ is written as

$$\Phi_{D_{l,q}}(s) = \prod_{m=0}^{N_F-1} \frac{1}{\det\left(\mathbf{I} + 2s\mathbf{R}^{(m)}\mathbf{F}_{l,q}^{(m)}\right)} \tag{9.34}$$

where the Hermitian matrix $\mathbf{F}_{l,q}^{(m)}$ is given by

$$\begin{aligned}\mathbf{F}_{l,q}^{(m)} &= \begin{bmatrix} 1 \\ -\bar{d}_q^{(m)} \end{bmatrix}^* \begin{bmatrix} 1 & -\bar{d}_q^{(m)} \end{bmatrix} - \begin{bmatrix} 1 \\ -\bar{d}_l^{(m)} \end{bmatrix}^* \begin{bmatrix} 1 & -\bar{d}_l^{(m)} \end{bmatrix} \\ &= \begin{bmatrix} 0 & -\left(\bar{d}_q^{(m)} - \bar{d}_l^{(m)}\right)^{\mathrm{T}} \\ -\left(\bar{d}_q^{(m)} - \bar{d}_l^{(m)}\right)^* & \left(\bar{d}_q^{(m)}\right)^*\left(\bar{d}_q^{(m)}\right)^{\mathrm{T}} - \left(\bar{d}_l^{(m)}\right)^*\left(\bar{d}_l^{(m)}\right)^{\mathrm{T}} \end{bmatrix}\end{aligned} \tag{9.35}$$

and $\mathbf{R}^{(m)} = \frac{1}{2}E\left\{\mathbf{z}_m\mathbf{z}_m^H \,|\mathbf{d}_l\right\}$ is given by

$$\begin{aligned}\mathbf{R}^{(m)} &= \begin{bmatrix} \frac{1}{2}E\left\{y_{m,\hat{k}}\left(y_{m,\hat{k}}\right)^* |\mathbf{d}_l\right\} & \frac{1}{2}E\left\{y_{m,\hat{k}}\left(\sqrt{E_{\mathrm{D}}}N_{\mathrm{T}}\bar{\bar{h}}_m c^{(\mathrm{SC})}(m)\right)^* |\mathbf{d}_l\right\} \\ \frac{1}{2}E\left\{\sqrt{E_{\mathrm{D}}}N_{\mathrm{T}}\bar{\bar{h}}_m c^{(\mathrm{SC})}(m)\left(y_{m,\hat{k}}\right)^* |\mathbf{d}_l\right\} & \frac{1}{2}E\left\{\sqrt{E_{\mathrm{D}}}N_{\mathrm{T}}\bar{\bar{h}}_m c^{(\mathrm{SC})}(m)\left(\sqrt{E_{\mathrm{D}}}N_{\mathrm{T}}\bar{\bar{h}}_m c^{(\mathrm{SC})}(m)\right)^* |\mathbf{d}_l\right\} \end{bmatrix} \\ &= \begin{bmatrix} \left(E_{\mathrm{D}}N_{\mathrm{T}}^2\left|\bar{d}_l^{(m)}\right|^2 + N_{\mathrm{T}}\sigma^2\right) & E_{\mathrm{D}}N_{\mathrm{T}}^2\left(\bar{d}_l^{(m)}\right)^{\mathrm{T}} \\ E_{\mathrm{D}}N_{\mathrm{T}}^2\left(\bar{d}_l^{(m)}\right)^* & E_{\mathrm{D}}N_{\mathrm{T}}^2\left(1 + \frac{\sigma^2}{N_{\mathrm{D}}\beta E_{\mathrm{D}}N_{\mathrm{T}}}\right) \end{bmatrix}\end{aligned} \tag{9.36}$$

Now to derive $1/\det\left(\mathbf{I} + 2s\mathbf{R}^{(m)}\mathbf{F}_{l,q}^{(m)}\right)$, for all m ($m = 0, 1, \ldots, N_{\mathrm{F}} - 1$), two different cases are to be studied. Then, $\Phi_{D_{l,q}}(s)$ can be obtained. When $\bar{d}_q^{(m)} \neq \bar{d}_l^{(m)}$, it can be found that $\left|\mathbf{F}_{l,q}^{(m)}\right|$ and $\left|\mathbf{R}^{(m)}\right|$ are nonzero, and the matrix $2\mathbf{R}^{(m)}\mathbf{F}_{l,q}^{(m)}$ is full rank, i.e. $\mathrm{rank}\left\{2\mathbf{R}^{(m)}\mathbf{F}_{l,q}^{(m)}\right\} = 2$. Therefore, one obtains

$$\frac{1}{\det\left(\mathbf{I} + 2s\mathbf{R}^{(m)}\mathbf{F}_{l,q}^{(m)}\right)} = \frac{p_{l,q,1}^{(m)}p_{l,q,2}^{(m)}}{\left(s - p_{l,q,1}^{(m)}\right)\left(s - p_{l,q,2}^{(m)}\right)} \tag{9.37}$$

where $-1/p_{l,q,1}^{(m)}$ and $-1/p_{l,q,2}^{(m)}$ are the two eigenvalues of $2\mathbf{R}^{(m)}\mathbf{F}_{l,q}^{(m)}$. When $\bar{d}_q^{(m)} = \bar{d}_l^{(m)}$, $\mathbf{F}_{l,q}^{(m)} = \mathbf{0}$, and $\mathbf{I} + 2s\mathbf{R}^{(m)}\mathbf{F}_{l,q}^{(m)} = \mathbf{I}$, so that $1/\det\left(\mathbf{I} + 2s\mathbf{R}^{(m)}\mathbf{F}_{l,q}^{(m)}\right) = 1$.

Finally, using the Gauss–Chebychev quadrature method [5], (9.37) can be calculated, and the overall BER union bound (or the BER bound for one code channel) is

$$p_b = E\{p_{b,k}\} \tag{9.38}$$

9.3.2 Performance of the MMSE detector

For the MMSE detector, the BER for the kth code channel $p_{b,k}$ is $p_{b,k} = p(\hat{d}_{0,k} = -1|d_{0,k} = 1)$, where $\hat{d}_{0,k}$ is obtained by (9.25). Conditioned on channel estimation $\{\bar{\bar{h}}_m\}$, noise power estimation $\hat{\sigma}^2$ and the transmitted symbol vector $\mathbf{d}$, the weighted output on the mth sub-carrier $w_{\hat{k},\hat{k}_f}^*(m)y_{m,\hat{k}}$ is a Gaussian variable with mean and variance of

$$E\{w^*_{\hat{k},\hat{k}_f}(m)y_{m,\hat{k}}|\{\bar{\bar{h}}_m\},\hat{\sigma}^2,\mathbf{d}\} = \frac{E_\mathrm{D}N_\mathrm{T}|\bar{\bar{h}}_m|^2 C_{N_\mathrm{F},\hat{k}_f}(m)\bar{d}^{(m)}}{E_\mathrm{D}N_\mathrm{T}K_c|\bar{\bar{h}}_m|^2 + 2\hat{\sigma}^2} \tag{9.39}$$

and

$$\begin{aligned}&\mathrm{Var}\{w^*_{\hat{k},\hat{k}_f}(m)y_{m,\hat{k}}|\{\bar{\bar{h}}_m\},\hat{\sigma}^2,\mathbf{d}\}\\ &= \frac{E_\mathrm{D}|\bar{\bar{h}}_m|^2 E\{|-\sqrt{E_\mathrm{D}}N_\mathrm{T}\Delta\bar{\bar{h}}_m C_{N_\mathrm{F},\hat{k}_f}(m)\bar{d}^{(m)} + (c^{(\mathrm{SC})}(m)C_{N_\mathrm{F},\hat{k}_f}(m))^*\eta_{m,\hat{k}}|^2\}}{2(E_\mathrm{D}N_\mathrm{T}K_c|\bar{\bar{h}}_m|^2 + 2\hat{\sigma}^2)^2}\end{aligned} \tag{9.40}$$

respectively.

With the assumption of independent AWGN noise on different sub-carriers, the conditional variance is

$$\begin{aligned}&\mathrm{Var}\{w^*_{\hat{k},\hat{k}_f}(m)y_{m,\hat{k}}|\{\bar{\bar{h}}_m\},\hat{\sigma}^2,\mathbf{d}\}\\ &= \frac{E_\mathrm{D}N_\mathrm{T}|\bar{\bar{h}}_m|^2(E_\mathrm{D}N_\mathrm{T}|\bar{d}^{(m)}|^2 R_{\bar{\bar{h}}|h} + \sigma^2)}{(E_\mathrm{D}N_\mathrm{T}K_c|\bar{\bar{h}}_m|^2 + 2\hat{\sigma}^2)^2}\end{aligned} \tag{9.41}$$

Then, the signal-to-interference-plus-noise ratio (SINR) of the sum of the N_F weighted outputs is

$$\gamma_\mathrm{o} = \frac{\frac{1}{2}\left(\mathrm{Re}\left\{\sum_{m=0}^{N_\mathrm{F}-1} E\{w^*_{\hat{k},\hat{k}_f}(m)y_{m,\hat{k}}|\{\bar{\bar{h}}_m\},\hat{\sigma}^2,\mathbf{d}\}\right\}\right)^2}{\sum_{m=0}^{N_\mathrm{F}-1}\mathrm{Var}\{w^*_{\hat{k},\hat{k}_f}(m)y_{m,\hat{k}}|\{\bar{\bar{h}}_m\},\hat{\sigma}^2,\mathbf{d}\}} \tag{9.42}$$

Therefore, the conditional BER is given by

$$p(\hat{d}_{0,k} = -1|d_{0,k} = 1, \{\bar{\bar{h}}_m\},\hat{\sigma}^2,\mathbf{d}) = Q(\sqrt{2\gamma_o}) \tag{9.43}$$

where $Q(x)$ is the Q-function defined as $Q(x) = \frac{1}{\sqrt{2\pi}}\int_x^\infty \mathrm{e}^{-t^2/2}\mathrm{d}t$.

Calculating the expectation over the joint ensemble of channel, noise power estimates and all possible $\mathbf{d}$s with $d_{0,k} = 1$, the average BER for the kth code channel can be obtained by

$$\begin{aligned}p_{b,k} &= p(\hat{d}_{0,k} = -1|d_{0,k} = 1)\\ &= E_{\{\bar{\bar{h}}_m\},\hat{\sigma}^2,\mathbf{d}}\{p(\hat{d}_{0,k} = -1|d_{0,k} = 1, \{\bar{\bar{h}}_m\},\hat{\sigma}^2,\mathbf{d})\}\end{aligned} \tag{9.44}$$

After calculating (9.44) using a numerical method (Monte Carlo integration) [6], the analytical BER for the overall system is obtained from (9.38).

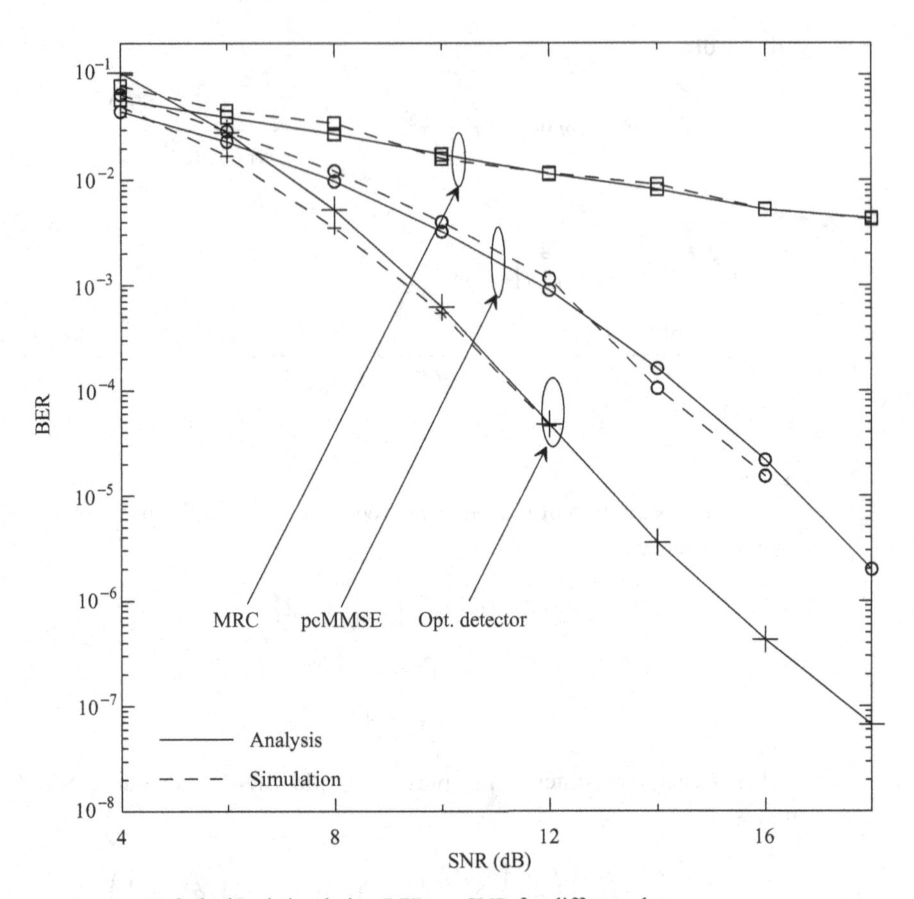

Figure 9.5 Analytical and simulation BER vs. SNR for different detectors.

9.4 Numerical results

In this section, some representative numerical results are presented to illustrate the performance of optimum and MMSE detectors for VSF-OFCDM systems. The effect of different system parameters such as spreading factors, noise and channel parameter estimations on the BER performance is investigated by numerical evaluation. The number of sub-carriers M is 768, and the OFDM symbol period (plus guard interval) is 8 μs. The ISI is efficiently suppressed by proper setting of the guard interval. Unless noted otherwise, the received signals of the most recent 48 OFDM symbols are collected for channel estimation, the power ratio of the pilot channel to all code channels $\beta/(\beta + K) = 20\%$. The average SNR is 14 dB, defined as $\text{SNR} = (1 + \beta/K)NE_{\text{D}}/\sigma_n^2$.

First, the BER performance with different detectors is investigated by means of simulation and analytical methods. Figure 9.5 shows the BER performance vs. SNR when the spreading factor is $N = 16 \times 16$ and the number of code channels is 60. It can be seen that the optimum detector significantly outperforms the MMSE (or pcMMSE) detector, especially in high SNR cases. This is mainly because the optimum detector can make a

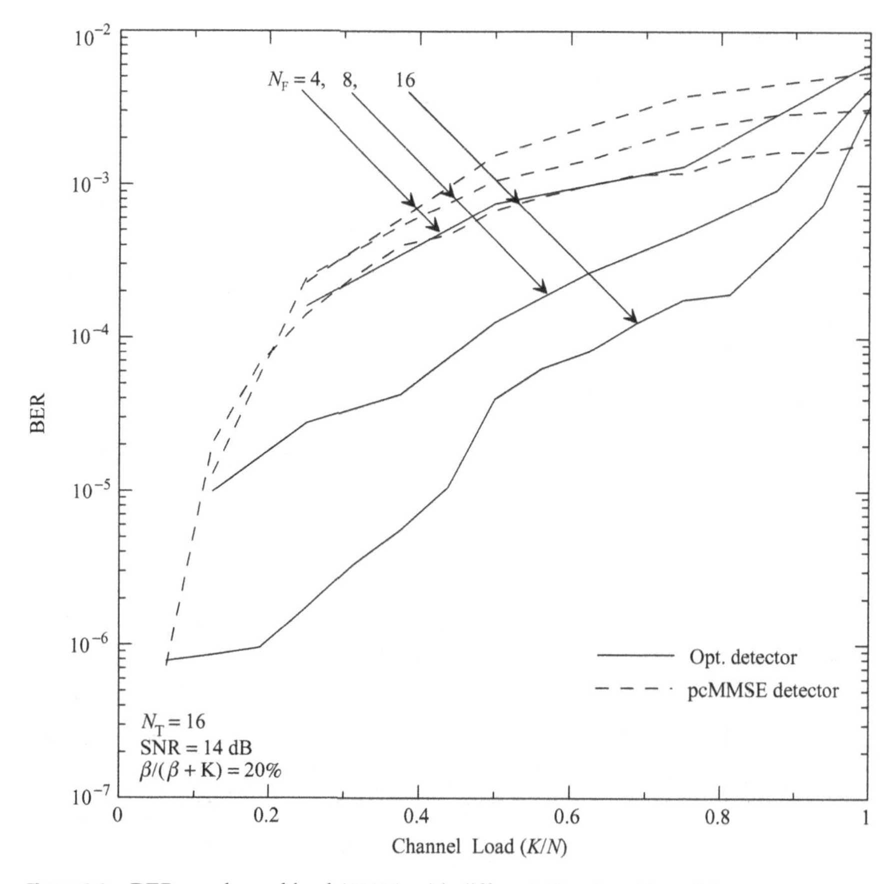

Figure 9.6 BER vs. channel load (K/N) with different N_F when $N_T = 16$.

simultaneous decision on data transmitted over all K_c code channels, which counts the effect of all transmitted code channels. Although with the additive Gaussian assumption of the MCI, the MMSE detector can suppress the interference from other code channels, it cannot make full use of the information from other code channels. Compared to the optimum detector, the performance of the MMSE detector dramatically degrades, but still outperforms the MRC detector, since, in addition to the AWGN component, MCI is another important additive interference. However, the MRC detector cannot combat MCI, whereas the MMSE detector takes into account the effect of both MCI and AWGN. Thus, the MMSE detector can provide a significant performance improvement compared to the MRC detector.

Simulation results are plotted to verify the accuracy of analytical results. During the simulations, the guard interval is set at 25% of an OFCDM symbol period as configured in [2], i.e. 2 μs, Doppler frequency is 20 Hz, a 24-path Rayleigh fading channel is assumed, and the other system parameters are the same as those for the analytical results. It can be seen that the analytical results are very close to the simulation ones for all detectors, except for the optimum detector in the case of small SNR. From the above discussion,

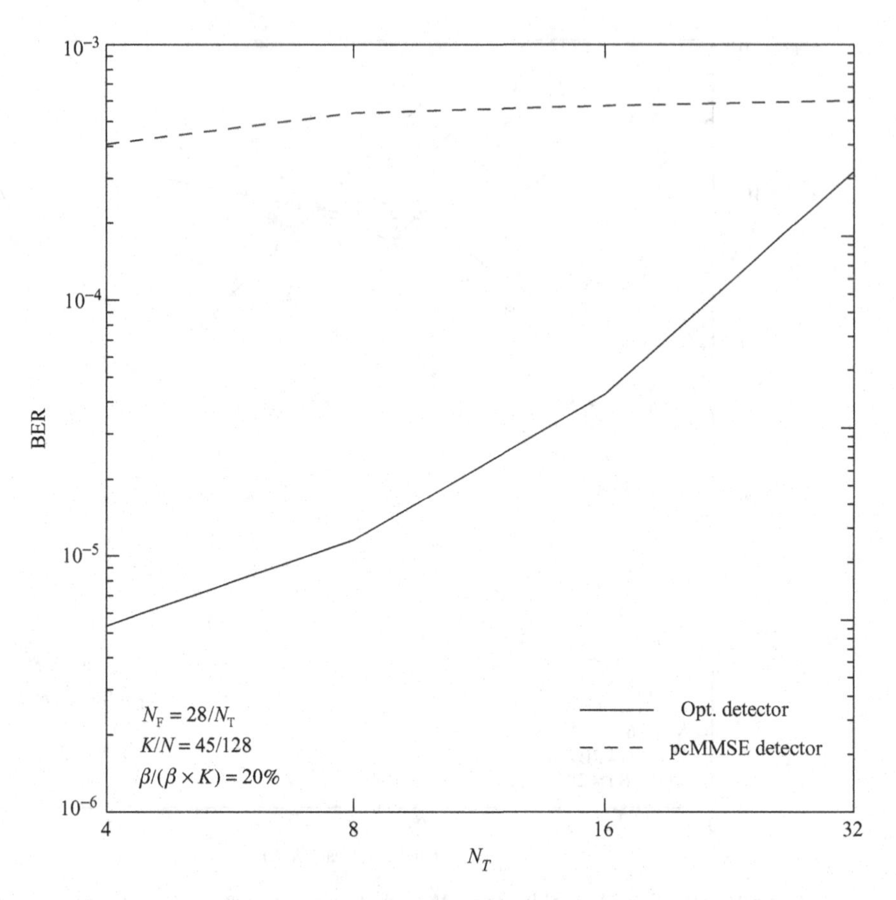

Figure 9.7 BER vs. N_T when channel load $K/N = 45/128$ and $N = 128$.

it can be found that when K_c is large, the MMSE detector is employed due to its low computational complexity.

Figure 9.6 shows the BER performance for a given channel load K/N and for various values (4, 8 and 16) of the frequency domain spreading factor N_F, when the time domain spreading factor is 16. Note that for a given K/N, K increases linearly as N increases, so that more MCI is caused. But since a larger N_F provides a higher frequency diversity gain, the performance of both detectors improves with the increase of N_F, especially in higher channel load cases, where MCI is dominant (compared to AWGN). Moreover, because the optimum detector demodulates the transmitted symbols simultaneously, the interference from other code channels can be fully exploited. Therefore, the performance gap for the optimum detector with different values of N_F is much larger that that for the MMSE detector. Moreover, it can be seen that a larger frequency domain spreading factor is preferred for both detectors. But, a larger N_F results in a larger number of code channels, which will make the optimum detection more complicated. Thus, to make the optimum detector practical, N_F should be less than or equal to 16.

Figure 9.7 shows the BER performance with variable time and frequency domain spreading factors ($N_F = N/N_T$) for a fixed channel load ($K/N = 0.352$) and a fixed

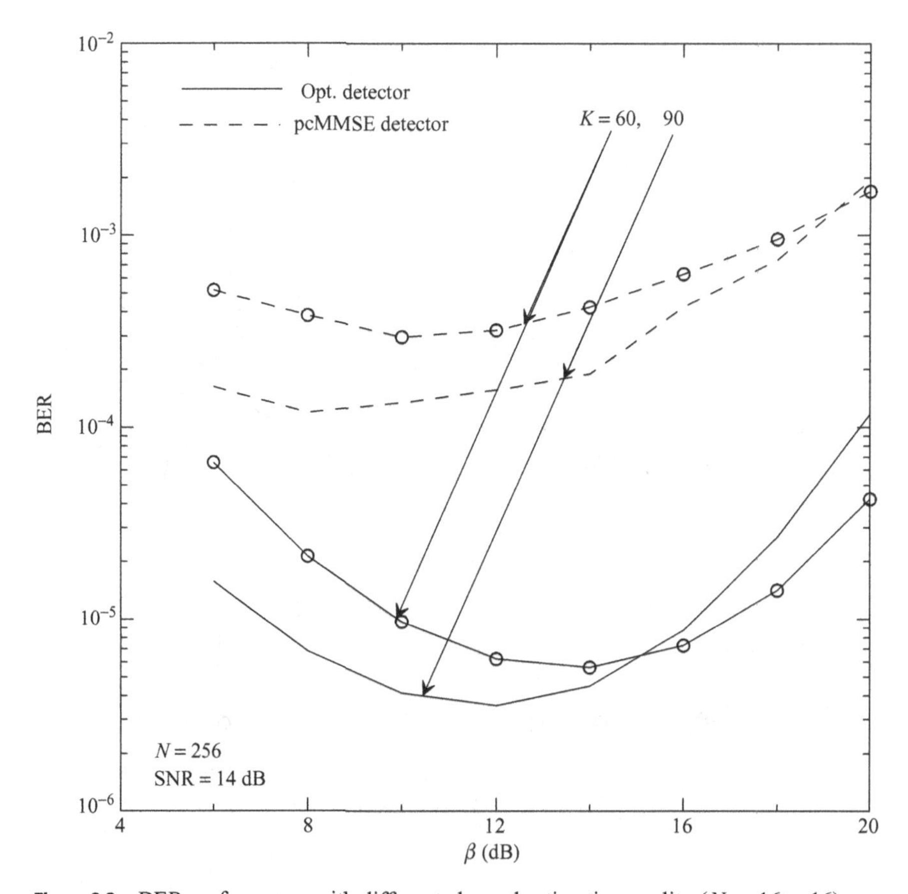

Figure 9.8 BER performance with different channel estimation quality ($N = 16 \times 16$).

overall $N = 128$. It can be seen that the BER performance of the MMSE detector is flat for different values of N_{T}. When N_{T} increases, N_{F} decreases so that fewer code channels result in the frequency domain. Note that the number of cochannel signals is $K_c = 45/(N_{\mathrm{T}} - 1)$. Thus less MCI is caused. That is, a smaller N_{F} causes less MCI and a smaller frequency diversity gain. Therefore, the BER performance of the pcMMSE is almost independent of N_{F} for a given N and K/N. With the optimum detector, performance degrades as N_{T} increases. This is because the performance of the optimum detector is mainly affected by the frequency diversity gain. In general, for a fixed overall spreading factor, a higher frequency domain spreading factor achieves a higher frequency diversity gain by employing the optimum detector.

In Figure 9.8, the effect of channel estimation quality on the system performance is investigated. In general, the channel estimation can be improved by increasing the power of the pilot channel where AWGN can be suppressed effectively. It can be seen that first when the power ratio of the pilot channel to one code channel increases, the system performance can be improved dramatically due to more accurate channel estimation. However, since the power ratio of the pilot channel to the overall transmitted power is fixed, a larger β results in a smaller received SNR for each data channel due to the decrease

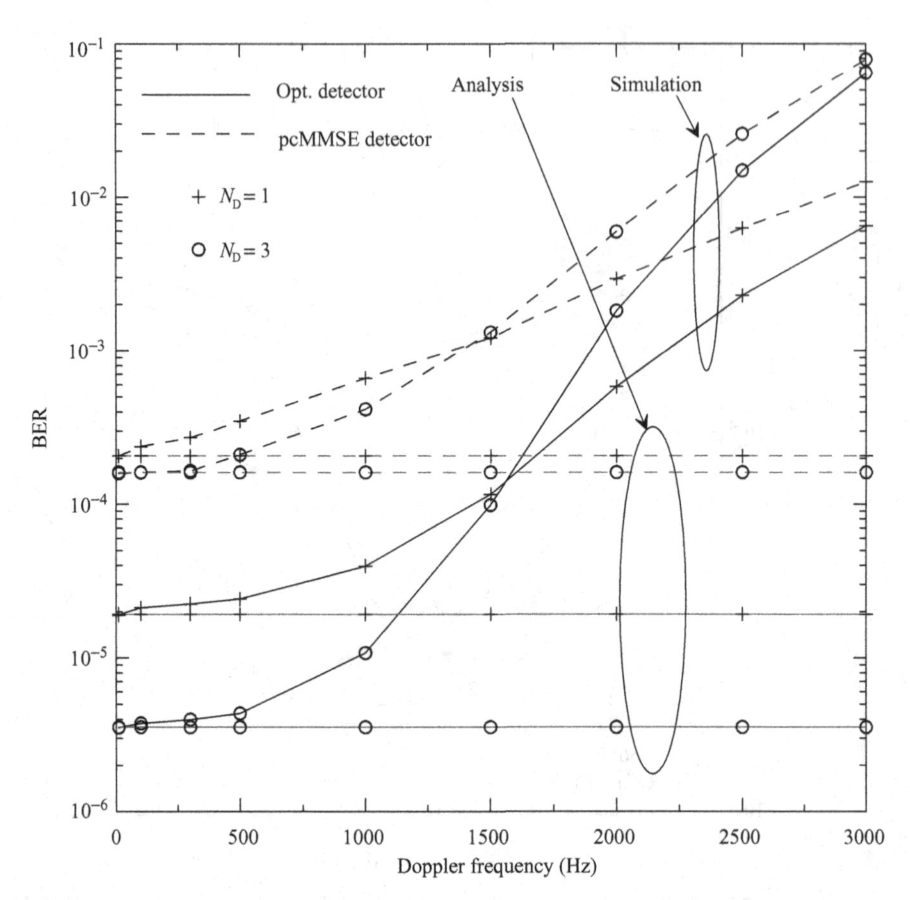

Figure 9.9 Simulation BER performance vs. Doppler frequency ($N = 16 \times 16$).

of E_D, so when β increases beyond a certain value, the benefit from more accurate channel estimation becomes slight, whereas the received SNR for each data channel is dominant to system performance. Consequently, as shown in the figure, performance degrades with the increase of β after it reaches an optimum value (approximately 10–14 dB for the MMSE detector with different K_c), i.e. the pilot channel occupies about 15 percent of the overall transmitted power. With the optimum detector, a similar scenario can be seen. Although the optimum detector jointly detects symbols transmitted on all code channels, calculation of the metric for different symbol combinations is sensitive to channel fading estimation. Thus it requires a more accurate channel estimation. Therefore, the optimum value for the optimum detector is greater than that for the MMSE detector under the same system conditions. In summary, to achieve better system performance, trade-off should be made between the channel estimation quality and received SNR for code channels.

Finally, the effect of Doppler frequency on system performance is simulated. In Figure 9.9, simulation and analytical results are shown with $N_D = 1$ and $N_D = 3$, respectively. Although by (9.15), noise power estimation cannot be achieved for $N_D = 1$, to simplify the investigation of the Doppler frequency effect on MCI and channel fading estimation, it is assumed that noise power is perfectly estimated by some other advanced algorithms for both cases in this figure. It can be seen that in slow fading cases, i.e. when the Doppler frequency is less than 500 Hz, analytical results are close to simulation results because the assumption of slow fading is satisfied so that the provided analytical method can predict system performance well. However, when the Doppler frequency increases beyond 500 Hz, code channels assigned with different time domain spreading codes are no longer orthogonal to each other due to the time varying fading for different chips, which introduces more MCI in time domain. As expected, simulation results show that BER performance degrades dramatically with an increase in Doppler frequency. Moreover, detectors with $N_{\mathrm{D}} = 3$ outperform those with $N_{\mathrm{D}} = 1$ when the Doppler frequency is less than 50 Hz, but as the Doppler frequency increases, the BER performance degrades more for $N_{\mathrm{D}} = 3$ than that for $N_{\mathrm{D}} = 1$. This is because although channel estimation in (9.9) can suppress the AWGN noise more with larger N_{D}, the desired component is distorted more in fast fading. Therefore, trade-off should be made between both effects of the noise and Doppler frequency. An optimum N_{D} should exist, which is associated with Doppler frequency.

9.5 Conclusions

In this chapter, the performance of VSF-OFCDM systems with different detectors (optimum and MMSE detectors) has been studied. As OFCDM systems seriously suffer from MCI in frequency selective fading channels, two typical detectors are compared in this chapter. Analytical BER performance is derived for both detectors considering the presence of imperfect channel estimation over frequency selective fading channels. The main conclusions of this chapter can be summarized as follows.

(1) The performance of the optimum detector varies dramatically with different configurations of system parameters, whereas the MMSE detector is more robust to different system parameters.
(2) Although the optimum detector outperforms the MMSE detector in most cases, it is too complicated, especially when there are a large number of code channels.
(3) Pilot channel power should be determined by making trade-off between the channel estimation quality and received SNR for each data channel.
(4) The system performance of both detectors is much affected by Doppler frequency, especially when the Doppler frequency is larger than 500 Hz.

References

[1] N. Maeda, H. Atarashi, S. Abeta and M. Sawahashi, "Pilot channel assisted MMSE combining in forward link for broadband OFCDM packet wireless access," *IEICE Trans. Fundamentals*, vol. E85-A, pp. 1635–1646, July 2002.
[2] N. Meada, Y. Kishiyama, H. Atarashi and M. Sawahashi, "Variable spreading factor-OFCDM with two-dimensional spreading that prioritizes time domain spreading for forward link broadband wireless access," in *Proc. IEEE Veh. Technol Conf. (VTC)*, pp. 127–132, 2003.

[3] J. Wang and J. Chen, "Performance of wideband CDMA systems with complex spreading and imperfect channel estimation,"*IEEE J. Select. Areas Commun.*, vol. 19, pp. 152–163, January 2001.

[4] J. G. Proakis, *Digital Communications*. New York: McGraw-Hill, 1995.

[5] S. Benedetto and E. Biglieri, *Principles of Digital Transmission with Wireless Applications*. New York: Kluwer, 1999.

[6] Q. Shi and M. Latva-aho, "Spreading sequences for asynchronous MC-CDMA revisited: accurate bit error rate analysis," *IEEE Trans. Commun.*, vol. 51, pp. 8–11, Jan. 2003.

10 Hybrid detection for OFCDM systems

As discussed in the last chapter, in a Gaussian or flat fading channel, multicode channels are orthogonal. In a realistic wireless channel, however, the orthogonality no longer maintains. Thus, MCI is caused. In this chapter, a novel detection method, called hybrid MCI cancellation and MMSE detection, is proposed and compared with pure MMSE detection. The system performance is analytically studied with imperfect channel estimation to show how it is affected by parameters such as the window size in the channel estimation, Doppler shift, the number of stages of the hybrid detection, the power ratio of pilot to data channels, spreading factor, and so on.

10.1 Introduction

Although MMSE detection can suppress MCI as discussed in the last chapter, so far MMSE detection has not been investigated for the OFCDM system with interference cancellation. Figure 10.1 illustrates the situation when MMSE detection is employed with MCI cancellation. It can be seen that the input to the MCI canceller is a combination of the useful signal, MCI and background noise. After MCI cancellation, the useful signal and background noise remain unchanged, but MCI is reduced. The combined signal is input to the MMSE detector with estimated signal plus interference power. Then the weighted output is generated after MMSE detection. Since the weight of MMSE is related to the input signal plus interference power, it should be updated stage by stage because the input power of MCI changes due to MCI cancellation. Therefore, when MMSE detection is employed with MCI cancellation, it is not a simple concatenation of the MCI canceller and pure MMSE detector, but a hybrid scheme. The objective of this chapter is to analytically investigate the performance of VSF-OFCDM with this hybrid MCI cancellation and MMSE detection.

For the broadband OFCDM system with high carrier frequency, fast fading may exist. For example, for a carrier frequency of 5 GHz, when a bullet train runs at the speed of 300 km/hour, the resultant maximum Doppler frequency (f_{D}) can be as high as 1500 Hz. In this case, even with a short packet length such as 0.5 ms, the channel variation in one packet duration is not negligible. Hence, in a fast fading channel, spreading in the time domain cannot preserve the orthogonality between code channels and MCI will occur in the time domain. Furthermore, the orthogonality between code channels in the frequency domain will be distorted by the frequency selectivity as well. Therefore, in a fast fading

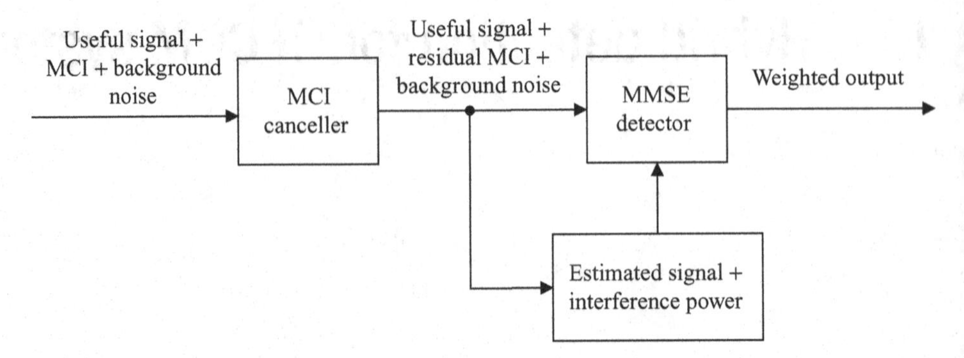

Figure 10.1 Hybrid MCI cancellation and MMSE detection.

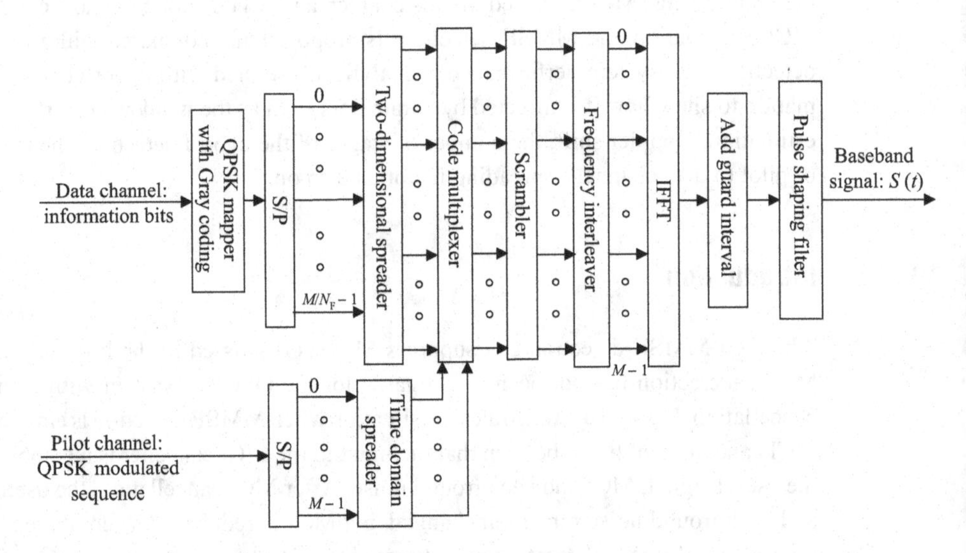

Figure 10.2 Transmitter block diagram of the OFCDM system.

broadband channel, all two-dimensional multicode channels interfere with each other and the resultant two-dimensional MCI is severe. In order to improve performance, MCI should be mitigated as much as possible. On the other hand, channel estimation is important for the system in a fast fading channel. Therefore, this chapter is devoted to investigating the performance of OFCDM in the presence of Doppler shift, when hybrid MCI cancellation and MMSE detection is employed with pilot-aided channel estimation.

10.2 System model

Note that with the introduction of the pilot channel and channel estimation, the system model is shown in Figures 10.2 and 10.3 for transmitter and receiver, respectively. At the transmitter, information bits on each data channel are first translated into Gray codes

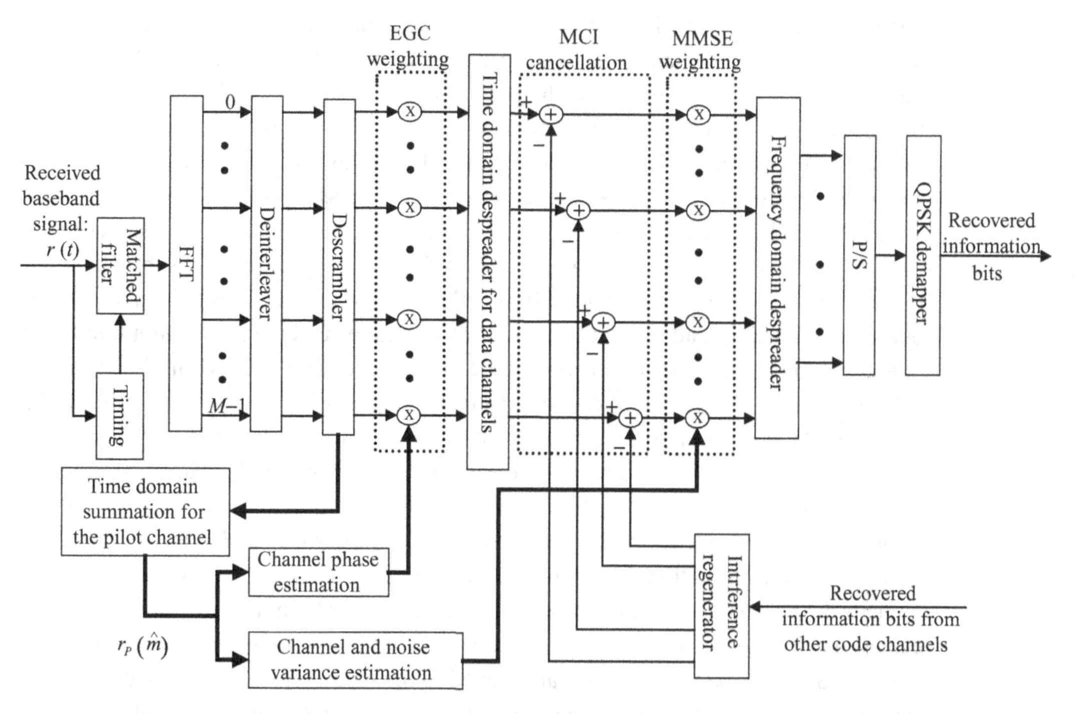

Figure 10.3 Receiver block diagram of the VSF-OFCDM system with hybrid MCI cancellation and MMSE detection.

by a QPSK mapper, and then serial-to-parallel (S/P) converted into M/N_F streams of mapped symbols, where M is the total number of sub-carriers. Every symbol of the streams will be two-dimensionally spread with N_T chips in the time domain and N_F chips in the frequency domain. At the same time, known QPSK-modulated pilot symbols are S/P converted into M streams, each one of which is spread into N_T chips in only the time domain. Then all data and pilot code channels are combined by a code multiplexer. Scrambling will be carried out on the combined signals. After scrambling, a frequency interleaver is employed to separate the N_F sub-carriers carrying the same data symbol as far as possible, so that the system can benefit from frequency diversity due to independent fading in interleaved sub-carriers. After interleaving, a total of M chips should be up-converted to M sub-carriers and transmitted in parallel. An M-point IFFT realizes this operation. At the output of IFFT, an effective OFCDM symbol with duration T_s is obtained with M samples. A guard interval called cyclic prefix with L_g samples (or duration T_g) is inserted between OFCDM symbols to prevent inter-symbol interference (ISI). Hence, a complete OFCDM symbol is composed of $M + L_g$ samples with duration $T = T_s + T_g$. Finally, OFCDM symbols pass through a pulse shaping filter, which gives rise to a transmitted baseband signal.

In the receiver, with perfect symbol timing, the received equivalent baseband signals are first processed by a matched filter where the guard interval is removed. The resultant signals are further input to the FFT block, which realizes the M-sub-carrier down-conversion. After FFT, the M chips involved with the M sub-carriers are obtained, deinterleaved and descrambled. On one hand, the output of the descrambler passes through the time domain summation for pilot channel estimation. On the other hand, using the phase information of the estimated channels, the output of the descrambler is weighted by equal gain combining (EGC) and accumulated by the time domain despreader for data channels. The resultant data signals will be further processed with the hybrid detection. When the MCI canceller is employed, the output from the time domain despreader for data channels is subtracted by a regenerated MCI. The outputs of MCI subtraction are then multiplied by weights obtained from MMSE algorithms and combined at the frequency domain despreader to get the signals of the desired code channel. After frequency domain despreading, a hard decision will be made to recover QPSK symbols. Then signals are parallel-to-serial (P/S) converted and demapped. Finally, information bits at the desired code channel are obtained. The recovered information bits from all other code channels will be used to regenerate the MCI for the desired code channel. Basically, the interference regenerator performs the operation as the transmitter except channel information. Although MCI cancellation cannot be performed at the zeroth stage due to unavailable information data, the signals of the first stage at the output of the time domain despreader for data channels will be subtracted by non-zero MCI. This cancellation process will continue in an iterative way until a specified number of stages is reached. Since the MMSE weights are related to the input power, they must be updated stage by stage due to the reduction of the MCI in each stage. Generally speaking, as the recovered information bits become more reliable stage by stage, MCI can be regenerated with higher accuracy stage by stage. After subtraction, MCI can be cancelled out much from the output of the despreader. Thus, the BER performance can be improved stage by stage.

10.3 Performance analysis

10.3.1 Transmitted signal

Using M sub-carriers, one data symbol can be modulated by N_{T} chips in the time domain over N_{F} sub-carriers in the frequency domain. Therefore, $N_{\mathrm{B}} = M/N_{\mathrm{F}}$ QPSK symbols can be modulated, where N_{B} is an integer. Using a block frequency interleaver, the frequency separation between adjacent sub-carriers impressing the same symbol is $N_{\mathrm{B}} \cdot \Delta f$ where $\Delta f = 1/T_{\mathrm{s}}$ is the sub-carrier spacing. The first of the N_{B} symbols uses the 0th, N_{B}th, $\ldots((N_{\mathrm{F}} - 1) \cdot N_{\mathrm{B}})$th sub-carriers, the second uses the 1st, $(N_{\mathrm{B}} + 1)$th, $\ldots$, $((N_{\mathrm{F}} - 1) \cdot N_{\mathrm{B}} + 1)$th sub-carriers, $\ldots$, and the last of the N_{B} symbols uses $(N_{\mathrm{B}} - 1)$st, $(2N_{\mathrm{B}} - 1)$th, $\ldots$, $(N_{\mathrm{F}} \cdot N_{\mathrm{B}} - 1)$th sub-carriers. When a high data rate is required, more than N_{B} symbols should be supported at a time. Thus, multiple codes are needed with one code for N_{B} symbols. When K codes are used for each sub-carrier, $K \cdot N_{\mathrm{B}}$ symbols can be modulated one time by the system. Thus, in forward link, the baseband transmitted data signal of the mth sub-carrier on the ith chip of the time domain can be expressed as

$$S_{m,i}(t) = \sqrt{P}\sum_{k=0}^{K-1} a_{k,m,i}e^{j2\pi f_m(t-iT)}p(t-iT) \tag{10.1}$$

where P is the signal power on one sub-carrier and $f_m = m \cdot \Delta f$ represents the baseband equivalent frequency of the mth sub-carrier. $a_{k,m,i}$ is given by

$$a_{k,m,i} = d_{k,m\%N_\mathrm{B},\lfloor i/N_\mathrm{T}\rfloor} \cdot c^{(k_\mathrm{F})}_{N_\mathrm{F},\lfloor m/N_\mathrm{B}\rfloor} \cdot c^{(k_\mathrm{T})}_{N_\mathrm{T},i\%N_\mathrm{T}} \cdot s_{m,i} \tag{10.2}$$

where $x\%y$ and $\lfloor x/y\rfloor$ represent the remainder and integer parts of x/y, respectively, $d_{k,m\%N_\mathrm{B},\lfloor i/N_\mathrm{T}\rfloor}$ is the data symbol of the kth code channel, and $s_{m,i}$ is the random binary scrambling code chip of the ith chip (or OFCDM symbol) in the mth sub-carrier. In (10.1), $p(t)$ is assumed to be a rectangular pulse with unity amplitude and duration of T.

In order to carry out effective channel estimation, a code-multiplexed pilot channel is employed on each sub-carrier. Pilot symbols are spread only in time domain with $C^{(0)}_{N_\mathrm{T}}$. The baseband pilot signal of the mth sub-carrier on the ith chip can be expressed as

$$S_{\mathrm{P},m,i}(t) = \sqrt{\beta P} \cdot a_{\mathrm{P},m,i} \cdot e^{j2\pi f_m(t-iT)}p(t-iT) \tag{10.3}$$

where β is the power ratio of pilot to one data channel and

$$a_{\mathrm{P},m,i} = d_\mathrm{P} \cdot c^{(0)}_{N_\mathrm{T},i\%N_T} \cdot s_{m,i} = d_\mathrm{P} \cdot s_{m,i} \tag{10.4}$$

where d_P is the known pilot symbol and $c^{(0)}_{N_\mathrm{T},i\%N_\mathrm{T}} = 1$ for all i of the pilot code $C^{(0)}_{N_\mathrm{T}}$. Therefore, the total baseband equivalent complex transmitted signal in N_T-chip duration is given by

$$S(t) = \sum_{m=0}^{M-1}\sum_{i=0}^{N_T-1}[S_{m,i}(t) + S_{\mathrm{P},m,i}(t)] \tag{10.5}$$

10.3.2 Channel model and received signal

In forward link, all code signals are synchronized and experience the same multipath fading channel with the lowpass equivalent impulse response

$$h(t) = \sum_{l=0}^{L-1} h_l(t)\delta(\tau - \tau_l) \tag{10.6}$$

where $h_l(t)$ is the complex time-varying channel fading for the lth path and τ_l is the delay of the lth path, uniformly distributed in $[0, T_\mathrm{g})$. The amplitude and phase of $h_l(t)$ are Rayleigh distributed with the variance $\sigma_l^2 = E\{|h_l(t)|^2\}$ and uniformly distributed in $[0, 2\pi)$, respectively. When Doppler frequency exists, the autocorrelation function of the lth path is defined as

$$\phi_l(\Delta t) = E\{h_l(t+\Delta t)\cdot h_l^*(t)\} = \sigma_l^2 J_0(2\pi f_\mathrm{D}\Delta t) \tag{10.7}$$

where $(\cdot)^*$ stands for the conjugate operation, f_D is the maximum Doppler frequency and $J_0(\cdot)$ is the correlation coefficient (zeroth-order Bessel function of the first kind [1]).

After experiencing the multipath channel, the baseband received signal inputs to the matched filter. By setting T_g larger than the maximum channel delay, there is no contribution (or interference) from adjacent symbols. With ideal symbol timing, the output of the matched filter on the $\hat{m}$th sub-carrier in the $\hat{i}$th time chip duration can be expressed as

$$\begin{aligned} r_{\hat{m}}(\hat{i}) &= \frac{1}{T_s}\int_{\hat{i}\cdot T}^{\hat{i}\cdot T+T_s} r(t)\cdot \mathrm{e}^{-j2\pi f_{\hat{m}}(t-\hat{i}\cdot T)} p(t-\hat{i}\cdot T)\mathrm{d}t \\ &= \frac{1}{T_s}\int_{\hat{i}\cdot T}^{\hat{i}\cdot T+T_s}\left[\sum_{l=0}^{L-1} h_l(t)\cdot S(t-\tau_l)+\eta(t)\right]\cdot \mathrm{e}^{-j2\pi f_{\hat{m}}(t-\hat{i}\cdot T)} p(t-\hat{i}\cdot T)\mathrm{d}t \\ &= \sum_{m=0}^{M-1}\left(\sum_{k=0}^{K-1}\sqrt{P}a_{k,m,\hat{i}}+\sqrt{\beta P}a_{P,m,\hat{i}}\right)\cdot \lambda_{\hat{m}}^{(m)}(\hat{i})+N(\hat{m},\hat{i}) \end{aligned} \tag{10.8}$$

where $r(t)$ is the received baseband signal, $\eta(t)$ is a complex additive white Gaussian noise (AWGN) with power spectral density of N_0 and

$$\lambda_{\hat{m}}^{(m)}(\hat{i}) = \frac{1}{T_s}\sum_{l=0}^{L-1}\mathrm{e}^{-j2\pi f_m \tau_l}\cdot \int_0^{T_s} h_l(t+\hat{i}\cdot T)\mathrm{e}^{j2\pi(f_m-f_{\hat{m}})t}\mathrm{d}t \tag{10.9}$$

is the factor of the contribution from the mth sub-carrier. When $m \neq \hat{m}$, $\lambda_{\hat{m}}^{(m)}(\hat{i})$ is the factor of interference whereas when $m=\hat{m}$, $\lambda_{\hat{m}}(\hat{i})=\lambda_{\hat{m}}^{(\hat{m})}(\hat{i})$ tends to be the factor of the useful component from the $\hat{m}$th sub-carrier. $N(\hat{m},\hat{i})$ is the noise component with zero mean and a variance of $\sigma_n^2 = N_0/T_s$. The output $r_{\hat{m}}(\hat{i})$ involved with the M sub-carriers will be frequency deinterleaved and then descrambled.

10.3.3 Channel estimation

In order to demodulate the data signal, the channel estimation is discussed first. The output of the descrambler inputs to the time domain summation for the pilot channel. Actually, the summation is a simple average of the output of the descrambler over γN_T chips duration, where γ is an odd integer. Therefore, the resultant output of the summation for pilot is given by

$$\begin{aligned} r_P(\hat{m}) &= \frac{1}{\gamma N_T}\sum_{\hat{i}=-(\gamma-1)N_T2}^{(\gamma+1)N_T/2-1} r_{\hat{m}}(\hat{i})\cdot s_{\hat{m},\hat{i}} \\ &= \sqrt{\beta P}d_P\cdot\left[\frac{1}{\gamma N_T}\sum_{\hat{i}=-(\gamma-1)N_T/2}^{(\gamma+1)N_T/2-1}\lambda_{\hat{m}}(\hat{i})\right]+I_1+I_2+N_P \end{aligned} \tag{10.10}$$

where the first term is the useful component and I_1 is the interference caused by all the K data channels from the $\hat{m}$th sub-carrier. In a flat fading channel ($f_D = 0$), $I_1 = 0$

due to the orthogonality of time domain spreading codes. When $f_D \neq 0$, however, the orthogonality no longer maintains, so $I_1 \neq 0$. The variance of I_1 is given by

$$\sigma_{I_1}^2 = \frac{P}{(\gamma N_T)^2} \sum_{k=0}^{K-1} \left[\sum_{\hat{i}_1=-(\gamma-1)N_T/2}^{(\gamma+1)N_T/2-1} \sum_{\hat{i}_2=-(\gamma-1)N_T/2}^{(\gamma+1)N_T/2-1} R_\lambda(\hat{i}_1, \hat{i}_2) \cdot \left(c_{N_T,\hat{i}_1\%N_T}^{(k_T)} c_{N_T,\hat{i}_2\%N_T}^{(k_T)} \right) \right] \tag{10.11}$$

where $R_\lambda(\hat{i}_1, \hat{i}_2)$ is the autocorrelation function of $\lambda_{\hat{m}}(\hat{i})$, defined as

$$\begin{aligned} R_\lambda(\hat{i}_1, \hat{i}_2) &= E\{\lambda_{\hat{m}}(\hat{i}_1) \cdot (\lambda_{\hat{m}}(\hat{i}_2))^*\} \\ &= \frac{1}{T_s^2} \int_{-T_s}^{T_s} \sum_{l=0}^{L-1} \phi_l(t + (\hat{i}_1 - \hat{i}_2)T)(T_s - |t|)\mathrm{d}t \end{aligned} \tag{10.12}$$

I_2 is the interference caused by all other sub-carriers, given by

$$\begin{aligned} I_2 &= \frac{1}{\gamma N_T} \sum_{\hat{i}=-(\gamma-1)N_T/2}^{(\gamma+1)N_T/2-1} \\ &\times \left[\sqrt{P} \sum_{k=0}^{K-1} \sum_{\substack{m=0 \\ m\neq\hat{m}}}^{M-1} \lambda_{\hat{m}}^{(m)}(\hat{i}) \cdot a_{k,m,\hat{i}} + \sqrt{\beta P} \sum_{\substack{m=0 \\ m\neq\hat{m}}}^{M-1} \lambda_{\hat{m}}^{(m)}(\hat{i}) \cdot a_{P,m,\hat{i}} \right] \cdot s_{\hat{m},\hat{i}} \end{aligned} \tag{10.13}$$

The variance of I_2 is derived in [2] and given by

$$\sigma_{I_2}^2 = \frac{2P(K+\beta)}{\gamma N_T M T_s^2} \int_0^{T_s} \sum_{l=0}^{L-1} \phi_l(t) \cdot \left[\frac{\sin^2(\pi M \Delta f t)}{\sin^2(\pi \Delta f t)} - M \right] \cdot (T_s - t)\mathrm{d}t \tag{10.14}$$

N_P is the noise term with variance

$$\sigma_{N_P}^2 = \frac{\sigma_n^2}{\gamma N_T} = \frac{N_0}{\gamma N_T T_s} \tag{10.15}$$

Therefore, the channel estimation for N_T chips ($i = 0, 1, \ldots, N_T - 1$) is given by the simple averaging estimator

$$\overline{\lambda}_{\hat{m}} = \frac{r_P(\hat{m})}{\sqrt{\beta P} d_P} = \frac{1}{\gamma N_T} \sum_{\hat{i}=-(\gamma-1)N_T/2}^{(\gamma+1)N_T/2-1} \lambda_{\hat{m}}(\hat{i}) + I_P \tag{10.16}$$

where I_P is the total interference plus noise with variance

$$\sigma_{I_P}^2 = \frac{\sigma_{I_1}^2 + \sigma_{I_2}^2 + \sigma_{N_P}^2}{\beta P} \tag{10.17}$$

Defining the channel estimation error:

$$e_i = \overline{\lambda}_{\hat{m}} - \lambda_{\hat{m}}(i), \qquad i = 0, \ldots, N_T - 1 \tag{10.18}$$

the autocorrelation function of e_i is given by

$$\begin{aligned}
R_{\mathrm{err}}(i_1, i_2) &= E\{e_{i_1} \cdot e^*_{i_2}\} \\
&= E\left\{\begin{array}{l}\left[\dfrac{1}{\gamma N_{\mathrm{T}}} \displaystyle\sum_{\hat{i}_1=-(\gamma-1)N_{\mathrm{T}}/2}^{(\gamma+1)N_{\mathrm{T}}/2-1} \lambda_{\hat{m}}(\hat{i}_1) + I_{\mathrm{P}} - \lambda_{\hat{m}}(i_1)\right] \\ \cdot\left[\dfrac{1}{\gamma N_{\mathrm{T}}} \displaystyle\sum_{\hat{i}_2=-(\gamma-1)N_{\mathrm{T}}/2}^{(\gamma+1)N_{\mathrm{T}}/2-1} \lambda_{\hat{m}}(\hat{i}_2) + I_{\mathrm{P}} - \lambda_{\hat{m}}(i_2)\right]^*\end{array}\right\} \\
&= \frac{1}{(\gamma N_{\mathrm{T}})^2} \sum_{\hat{i}_1=-(\gamma-1)N_{\mathrm{T}}/2}^{(\gamma+1)N_{\mathrm{T}}/2-1} \sum_{\hat{i}_2=-(\gamma-1)N_{\mathrm{T}}/2}^{(\gamma+1)N_{\mathrm{T}}/2-1} R_\lambda(\hat{i}_1, \hat{i}_2) \\
&\quad - \frac{1}{\gamma N_{\mathrm{T}}} \sum_{\hat{i}=-(\gamma-1)N_{\mathrm{T}}/2}^{(\gamma+1)N_{\mathrm{T}}/2-1} [R_\lambda(i_1, \hat{i}) + R_\lambda(i_2, \hat{i})] + R_\lambda(i_1, i_2) + \sigma^2_{I_{\mathrm{P}}}
\end{aligned} \tag{10.19}$$

Therefore, the variance of the estimation error, e_i, is given by

$$\sigma^2_{\mathrm{err}}(i) = E\{|e_i|^2\} = R_{\mathrm{err}}(i, i) \tag{10.20}$$

10.3.4 Time domain despreading for data channels

Suppose the $\hat{k}$th code channel is the desired data channel. A simple EGC is employed in time domain despreading to collect any useful signal from different chips. The phase of the channel estimation $\overline{\lambda}_{\hat{m}}$ is denoted as $\overline{\varphi}_{\hat{m}}$, given by

$$\overline{\varphi}_{\hat{m}} = \tan^{-1}\{\mathrm{Im}(\overline{\lambda}_{\hat{m}})/\mathrm{Re}(\overline{\lambda}_{\hat{m}})\} \tag{10.21}$$

where $\tan^{-1}(\cdot)$ stands for the reverse tan function, $\mathrm{Im}(\cdot)$ and $\mathrm{Re}(\cdot)$ represent the imaginary and real parts, respectively. Assuming that the $\hat{k}$th code channel employs two-dimensional spreading code $\{C^{(\hat{k}_{\mathrm{T}})}_{N_{\mathrm{T}}}, C^{(\hat{k}_{\mathrm{F}})}_{N_{\mathrm{F}}}\}$, the signal at the output of the time domain despreader is given by

$$\begin{aligned}
r_{\hat{k},\hat{m}} &= \frac{1}{N_{\mathrm{T}}} \sum_{\hat{i}=0}^{N_{\mathrm{T}}-1} r_{\hat{m}}(\hat{i}) \cdot \mathrm{e}^{-\mathrm{j}\overline{\varphi}_{\hat{m}}} \cdot c^{(\hat{k}_{\mathrm{T}})}_{N_{\mathrm{T}},\hat{i}} \cdot s_{\hat{m},\hat{i}} \\
&= S_{\hat{k},\hat{m}} + \mathrm{MCI}_{\hat{m},\mathrm{X}} + \mathrm{MCI}_{\hat{m},\mathrm{Y}} + \mathrm{ICI}_{\hat{m}} + N_{\hat{m}}
\end{aligned} \tag{10.22}$$

where $S_{\hat{k},\hat{m}}$ is the useful signal from the $\hat{k}$th code channel. When there are a large number of sub-carriers $M \gg N_{\mathrm{F}}$, the minimum spacing $N_{\mathrm{B}} \cdot \Delta f$ of two adjacent sub-carriers carrying the same data can be quite large and the crosstalk (or contribution) from the other $N_{\mathrm{F}} - 1$ sub-carriers (rather than the $\hat{m}$th sub-carrier) will be very small, so it is neglected. Therefore, $S_{\hat{k},\hat{m}}$ is given by

$$S_{\hat{k},\hat{m}} = \frac{\sqrt{P}}{N_{\mathrm{T}}} \sum_{\hat{i}=0}^{N_{\mathrm{T}}-1} a_{\hat{k},\hat{m},\hat{i}} \cdot c_{N_{\mathrm{T}},\hat{i}}^{(\hat{k}_{\mathrm{T}})} \cdot s_{\hat{m},\hat{i}} \cdot \lambda_{\hat{m}}(\hat{i}) \cdot \mathrm{e}^{-\mathrm{j}\overline{\varphi}_{\hat{m}}}$$
$$= \sqrt{P} \cdot d_{\hat{k}} \cdot c_{N_{\mathrm{F}},\lfloor \hat{m}/N_{\mathrm{B}} \rfloor}^{(\hat{k}_{\mathrm{F}})} \cdot \alpha(\hat{m}) \tag{10.23}$$

where $d_{\hat{k}} = d_{\hat{k},\hat{m}\%N_{\mathrm{B}},\lfloor \hat{i}/N_{\mathrm{T}} \rfloor}$ for simple notation and $\alpha(\hat{m})$ is the channel distortion, given by

$$\alpha(\hat{m}) = \frac{1}{N_{\mathrm{T}}} \sum_{\hat{i}=0}^{N_{\mathrm{T}}-1} \lambda_{\hat{m}}(\hat{i}) \cdot \mathrm{e}^{-\mathrm{j}\overline{\varphi}_{\hat{m}}} \tag{10.24}$$

$\mathrm{MCI}_{\hat{m},\mathrm{X}}$ is the interference from the K_{X} codes in Ω_{X} given by

$$\Omega_{\mathrm{X}} = \left\{C_{N_{\mathrm{T}}}^{(k_{\mathrm{T}})}, C_{N_{\mathrm{F}}}^{(\hat{k}_{\mathrm{F}})}\right\}, \hat{k}_{\mathrm{F}} = 0, 1, \ldots, N_{\mathrm{F}} - 1, \text{ but } \hat{k}_{\mathrm{F}} \neq k_{\mathrm{F}}, k_{\mathrm{T}} \neq 0 \tag{10.25}$$

which have the same time domain spreading code as the $\hat{k}$th code channel, given by

$$\mathrm{MCI}_{\hat{m},\mathrm{X}} = \sqrt{P} \sum_{k \in \Omega_{\mathrm{X}}} d_k \cdot c_{N_{\mathrm{F}},\lfloor \hat{m}/N_{\mathrm{B}} \rfloor}^{(k_{\mathrm{F}})} \cdot \alpha(\hat{m}) \tag{10.26}$$

$\mathrm{MCI}_{\hat{m},\mathrm{Y}}$ is interference from the K_{Y} interference codes in Ω_{Y} given by

$$\Omega_{\mathrm{Y}} = \left\{C_{N_{\mathrm{T}}}^{(\hat{k}_{\mathrm{T}})}, C_{N_{\mathrm{F}}}^{(\hat{k}_{\mathrm{F}})}\right\}, \hat{k}_{\mathrm{F}} = 0, 1, \ldots, N_{\mathrm{F}} - 1, \hat{k}_{\mathrm{T}}$$
$$= 1, \ldots, N_{\mathrm{T}} - 1, \text{ but } \hat{k}_{\mathrm{T}} \neq k_{\mathrm{T}} \neq 0 \tag{10.27}$$

which have different time domain spreading codes from the $\hat{k}$th code channel and from the pilot channel, given by

$$\mathrm{MCI}_{\hat{m},\mathrm{Y}} = \sqrt{P} \sum_{k \in \Omega_{Y}} d_k \cdot c_{N_{\mathrm{F}},\lfloor \hat{m}/N_{\mathrm{B}} \rfloor}^{(k_{\mathrm{F}})} \cdot \alpha_k(\hat{m}) + \sqrt{\beta P} d_{\mathrm{P}} \cdot \alpha_{\mathrm{P}}(\hat{m}) \tag{10.28}$$

where $\alpha_k(\hat{m})$ and $\alpha_{\mathrm{P}}(\hat{m})$ are defined as

$$\alpha_k(\hat{m}) = \frac{1}{N_{\mathrm{T}}} \sum_{\hat{i}=0}^{N_{\mathrm{T}}-1} \lambda_{\hat{m}}(\hat{i}) \cdot \mathrm{e}^{-\mathrm{j}\overline{\varphi}_{\hat{m}}} \cdot c_{N_{\mathrm{T}},\hat{i}}^{(k_{\mathrm{T}})} \cdot c_{N_{\mathrm{T}},\hat{i}}^{(\hat{k}_{\mathrm{T}})} \tag{10.29}$$

$$\alpha_{\mathrm{P}}(\hat{m}) = \frac{1}{N_{\mathrm{T}}} \sum_{\hat{i}=0}^{N_{\mathrm{T}}-1} \lambda_{\hat{m}}(\hat{i}) \cdot \mathrm{e}^{-\mathrm{j}\overline{\varphi}_{\hat{m}}} \cdot c_{N_{\mathrm{T}},\hat{i}}^{(\hat{k}_{\mathrm{T}})} \tag{10.30}$$

The variances of $\mathrm{MCI}_{\hat{m},\mathrm{X}}$ and $\mathrm{MCI}_{\hat{m},\mathrm{Y}}$ are given by

$$\sigma_{\mathrm{MCI}_{\hat{m},\mathrm{X}}}^2 = P \cdot K_{\mathrm{Y}} \cdot |\alpha(\hat{m})|^2 \tag{10.31}$$

$$\sigma_{\mathrm{MCI}_{\hat{m},\mathrm{Y}}}^2 = P \sum_{k \in \Omega_{\mathrm{Y}}} |\alpha_k(\hat{m})|^2 + \beta P \cdot |\alpha_{\mathrm{P}}(\hat{m})|^2 \tag{10.32}$$

$\mathrm{ICI}_{\hat{m}}$ is the interference from all other sub-carriers. Considering the sub-carriers with $\hat{m} = 0, N_\mathrm{B}, 2N_\mathrm{B}, \ldots, (N_\mathrm{F}-1)N_\mathrm{B}$, $\mathrm{ICI}_{\hat{m}}$ is given by

$$\begin{aligned}\mathrm{ICI}_{\hat{m}} &= \frac{1}{N_\mathrm{T}}\sum_{\hat{i}=0}^{N_\mathrm{T}-1}\left(\sum_{\substack{m=0\\ m\neq\hat{m}}}^{M-1}\lambda_{\hat{m}}^{(m)}(\hat{i})\left(\sqrt{P}\sum_{k=0}^{K-1}a_{k,m,\hat{i}}+\sqrt{\beta P}\cdot a_{\mathrm{P},m,\hat{i}}\right)\right)\\ &\quad\cdot \mathrm{e}^{-\mathrm{j}\overline{\varphi}_{\hat{m}}}\cdot c_{N_\mathrm{T},\hat{i}}^{(\hat{k}_\mathrm{T})}\cdot s_{\hat{m},\hat{i}} \\ &\quad-\frac{1}{N_\mathrm{T}}\sum_{\hat{i}=0}^{N_\mathrm{T}-1}\left(\sqrt{P}\sum_{\substack{m=0\\ m\cdot N_\mathrm{B}\neq\hat{m}}}^{N_\mathrm{F}-1}\lambda_{\hat{m}}^{(m\cdot N_\mathrm{B})}(\hat{i})\cdot a_{\hat{k},m\cdot N_\mathrm{B},\hat{i}}\right)\cdot \mathrm{e}^{-\mathrm{j}\overline{\varphi}_{\hat{m}}}\cdot c_{N_\mathrm{T},\hat{i}}^{(\hat{k}_\mathrm{T})}\cdot s_{\hat{m},\hat{i}}\end{aligned} \tag{10.33}$$

The reason to subtract the second term from the first term on the right-hand side is that N_F out of M sub-carriers carry the same data as the desired channel. The variance of $\mathrm{ICI}_{\hat{m}}$ can be derived [2] and is given by

$$\begin{aligned}\sigma_{\mathrm{ICI}_{\hat{m}}}^2 &= \frac{2P}{NT_\mathrm{s}^2}\int_0^{T_\mathrm{s}}\sum_{l=0}^{L-1}\phi_l(t)\\ &\quad\cdot\left\{\begin{array}{l}\left[\begin{array}{l}\dfrac{(K+\beta)\cdot\sin^2(\pi M\Delta ft)\cos(\pi(1-N_\mathrm{B})\Delta ft)}{\sin(\pi\Delta ft)\sin(\pi N_\mathrm{B}\Delta ft)}\\ -(K+\beta)\cdot N_\mathrm{F}\end{array}\right]\\ -\left[\dfrac{\sin^2(\pi M\Delta ft)}{\sin^2(\pi N_\mathrm{B}\Delta ft)}-N_\mathrm{F}\right]\end{array}\right\}\cdot(T_\mathrm{s}-t)\mathrm{d}t\end{aligned} \tag{10.34}$$

Finally, $N_{\hat{m}}$ is the background noise with zero mean and a variance of $\sigma_{N_{\hat{m}}}^2 = \sigma_n^2/N_\mathrm{T} = N_0/N_\mathrm{T}\cdot T_\mathrm{s}$.

10.3.5 Frequency domain despreading

In frequency domain despreading, the outputs of the time domain despreader will be weighted by different factors, multiplied by a frequency domain spreading code and then summed over N_F interleaved sub-carriers. In the despreading process, however, MCI is caused by independent fading on interleaved sub-carriers, so MCI cancellation plus MMSE detection must be employed.

Pure MMSE detection

At the zeroth (or original) stage of MCI cancellation, the regenerated MCI is unavailable due to detected data being unavailable. Thus pure MMSE detection is used to combine the signals from different interleaved sub-carriers. The weights of pure MMSE are given by [3]

$$\begin{aligned}\omega(\hat{m}) &= \frac{E\{\sqrt{P}d_{\hat{k}}\cdot r_{\hat{k},\hat{m}}^*|\alpha(\hat{m})\}\cdot c_{N_\mathrm{F},\lfloor\hat{m}/N_\mathrm{B}\rfloor}^{(\hat{k}_\mathrm{F})}}{E\{|r_{\hat{k},\hat{m}}|^2|\alpha(\hat{m})\}}\\ &= \frac{P\cdot\alpha^*(\hat{m})}{P\cdot|\alpha(\hat{m})|^2+\sigma_{\mathrm{MCI}_{\hat{m},\mathrm{X}}}^2+\sigma_{\mathrm{MCI}_{\hat{m},\mathrm{Y}}}^2+\sigma_{\mathrm{ICI}_{\hat{m}}}^2+\sigma_{N_{\hat{m}}}^2}\end{aligned} \tag{10.35}$$

In the evaluation of $\omega(\hat{m})$, the estimates of several parameters are required. When $\overline{\lambda}_{\hat{m}}$ in (10.16) is used to replace $\lambda_{\hat{m}}(\hat{i})$, $\alpha(\hat{m})$ is approximated as $\hat{\alpha}(\hat{m}) = |\overline{\lambda}_{\hat{m}}|$ and both (10.31) and (10.30) are zero. Thus, (10.35) reduces to

$$\omega(\hat{m}) = \frac{\hat{\alpha}(\hat{m})}{(1 + K_{\mathrm{X}}) \cdot |\hat{\alpha}(\hat{m})|^2 + \frac{\sigma^2_{\mathrm{ICI}_{\hat{m}}}}{P} + \left(\frac{N_{\mathrm{T}} E_{\mathrm{s}}}{N_0}\right)^{-1}} \tag{10.36}$$

where $E_{\mathrm{s}}/N_0 = PT_{\mathrm{s}}/N_0$.

For the first of the N_{B} symbols on the $\hat{k}$th code channel, $\hat{m} = m \cdot N_{\mathrm{B}}$ and $m = 0, 1, \ldots, N_{\mathrm{F}} - 1$. By summing the signals of the N_{F} interleaved sub-carriers, the output of the frequency domain despreader is given by

$$\begin{aligned} y_{\hat{k}} &= \frac{1}{N_{\mathrm{F}}} \cdot \sum_{m=0}^{N_{\mathrm{F}}-1} r_{\hat{k},m\cdot N_{\mathrm{B}}} \left[c_{N_{\mathrm{F}},m}^{(\hat{k}_{\mathrm{F}})} \cdot \omega(m \cdot N_{\mathrm{B}})\right] \\ &= S_{\hat{k}} + \mathrm{MCI}_{\mathrm{X}} + \mathrm{MCI}_{\mathrm{Y}} + \mathrm{ICI}_{\hat{k}} + N_{\hat{k}} \end{aligned} \tag{10.37}$$

where $S_{\hat{k}}$ is the desired data signal, given by

$$S_{\hat{k}} = \sqrt{P} \cdot d_{\hat{k}} \cdot \left[\frac{1}{N_{\mathrm{F}}} \sum_{m=0}^{N_{\mathrm{F}}-1} \alpha(m \cdot N_{\mathrm{B}}) \cdot \omega(m \cdot N_{\mathrm{B}})\right] \tag{10.38}$$

$\mathrm{MCI}_{\mathrm{X}}$ is the MCI caused by the K_{X} code channels from Ω_{X}, given by

$$\mathrm{MCI}_{\mathrm{X}} = \frac{\sqrt{P}}{N_{\mathrm{F}}} \sum_{k \in \Omega_X} d_k \cdot \left[\sum_{m=0}^{N_{\mathrm{F}}-1} c_{N_{\mathrm{F}},m}^{(\hat{k}_{\mathrm{F}})} \cdot c_{N_{\mathrm{F}},m}^{(k_{\mathrm{F}})} \cdot \alpha(m \cdot N_{\mathrm{B}}) \cdot \omega(m \cdot N_{\mathrm{B}})\right] \tag{10.39}$$

$\mathrm{MCI}_{\mathrm{Y}}$ is the interference caused by other K_{Y} code channels from Ω_{Y} and the pilot channel due to non-zero Doppler shift, given by

$$\begin{aligned} \mathrm{MCI}_{\mathrm{Y}} &= \frac{\sqrt{P}}{N_{\mathrm{F}}} \sum_{k \in \Omega_{\mathrm{Y}}} d_k \cdot \left[\sum_{m=0}^{N_{\mathrm{F}}-1} c_{N_{\mathrm{F}},m}^{(\hat{k}_{\mathrm{F}})} \cdot c_{N_{\mathrm{F}},m}^{(k_{\mathrm{F}})} \cdot \alpha_k(m \cdot N_{\mathrm{B}}) \cdot \omega(m \cdot N_{\mathrm{B}})\right] \\ &\quad + \frac{\sqrt{\beta P}}{N_{\mathrm{F}}} \sum_{m=0}^{N_{\mathrm{F}}-1} d_P \cdot \left[c_{N_{\mathrm{F}},m}^{(\hat{k}_{\mathrm{F}})} \cdot \alpha_P(m \cdot N_{\mathrm{B}}) \cdot \omega(m \cdot N_{\mathrm{B}})\right] \end{aligned} \tag{10.40}$$

$\mathrm{ICI}_{\hat{k}}$ and $N_{\hat{k}}$ are the ICI and the background noise components, respectively, with variances given by

$$\sigma^2_{\mathrm{ICI}} = \frac{\sigma^2_{\mathrm{ICI}_{\hat{m}}}}{N_{\mathrm{F}}^2} \sum_{m=0}^{N_{\mathrm{F}}-1} |\omega(m \cdot N_{\mathrm{B}})|^2 \tag{10.41}$$

$$\sigma^2_N = \frac{N_0}{T_{\mathrm{s}} N_{\mathrm{T}} N_{\mathrm{F}}^2} \sum_{m=0}^{N_{\mathrm{F}}-1} |\omega(m \cdot N_{\mathrm{B}})|^2 \tag{10.42}$$

Finally, a hard decision will be made on the output, $y_{\hat{k}}$, and a tentative decision $d_{\hat{k}}^{(0)}$ will be obtained, where the superscript "0" stands for the zeroth stage. For QPSK modulation, the probabilities of incorrect decisions of real and imaginary parts of $y_{\hat{k}}$ are the same.

Hybrid MCI cancellation and MMSE detection

Using the tentative decision $\{d_k^{(0)}, k \neq \hat{k}\}$ from the pure MMSE, MCI interference can be regenerated for the first stage. Generally speaking, with the tentative data decisions $d_0^{(s-1)}, \ldots, d_{k,k\neq\hat{k}}^{(s-1)}, \ldots, d_{K-1}^{(s-1)}$, where $k \neq \hat{k}$, of the previous stage and channel estimation $\overline{\lambda}_{\hat{m}}$, the MCI interference in (10.26) can be regenerated for the sth stage, so MCI cancellation can be carried out. The regenerated MCI for the $(m \cdot N_{\mathrm{B}})$th sub-carrier is given by

$$Q_{m\cdot N_{\mathrm{B}}}^{(s)} = \sqrt{P} \sum_{k\in\Omega_{\mathrm{X}}} d_k^{(s-1)} \cdot c_{N_{\mathrm{F}},m}^{(k_{\mathrm{F}})} \cdot \hat{\alpha}(m \cdot N_{\mathrm{B}}) \tag{10.43}$$

Note that only the interference caused by the K_{X} code channels from Ω_{X} can be regenerated. For interference $\mathrm{MCI}_{\hat{m},\mathrm{Y}}$ from Ω_{Y} and the pilot channel, regenerated MCI is zero due to orthogonality of different codes and to the fixed channel estimation over N_{T} chips.

The output of the MCI cancellation is given by

$$\begin{aligned} r_{\hat{k},m\cdot N_{\mathrm{B}}}^{(s)} &= r_{\hat{k},m\cdot N_{\mathrm{B}}} - Q_{m\cdot N_{\mathrm{B}}}^{(s)} \\ &= S_{\hat{k},m\cdot N_{\mathrm{B}}} + \mathrm{MCI}_{m\cdot N_{\mathrm{B}},\mathrm{X}}^{(s)} + \mathrm{MCI}_{m\cdot N_{\mathrm{B}},\mathrm{Y}} + \mathrm{ICI}_{m\cdot N_{\mathrm{B}}} + N_{m\cdot N_{\mathrm{B}}} \end{aligned} \tag{10.44}$$

where $S_{\hat{k},m\cdot N_{\mathrm{B}}}$ and $\mathrm{ICI}_{m\cdot N_{\mathrm{B}}}$ are given by (10.23) and (10.33), respectively. $\mathrm{MCI}_{m\cdot N_{\mathrm{B}},\mathrm{X}}^{(s)}$ is the residual MCI, given by

$$\begin{aligned} \mathrm{MCI}_{m\cdot N_{\mathrm{B}},\mathrm{X}}^{(s)} &= \mathrm{MCI}_{m\cdot N_{\mathrm{B}},X} - Q_{m\cdot N_{\mathrm{B}}}^{(s)} \\ &= \sqrt{P} \sum_{k\in\Omega_{\mathrm{X}}} \left[d_k \cdot \alpha(m \cdot N_{\mathrm{B}}) - d_k^{(s-1)} \cdot \hat{\alpha}(m \cdot N_{\mathrm{B}})\right] \cdot c_{N_{\mathrm{F}},m}^{(k_{\mathrm{F}})} \end{aligned} \tag{10.45}$$

Assuming that the data decision error $d_k - d_k^{(s-1)}$ and the error $\alpha(m \cdot N_{\mathrm{B}}) - \hat{\alpha}(m \cdot N_{\mathrm{B}})$ are independent with zero mean, the variance of residual MCI is given by

$$\begin{aligned} \sigma_{\mathrm{MCI_X}}^2(s) &= P \sum_{k\in\Omega_{\mathrm{X}}} E\left\{\left|d_k \cdot \alpha(m \cdot N_{\mathrm{B}}) - d_k^{(s-1)} \cdot \hat{\alpha}(m \cdot N_{\mathrm{B}})\right|^2\right\} \\ &\approx P \sum_{k\in\Omega_{\mathrm{X}}} \left[\begin{array}{l} E\left\{\left|\left(d_k - d_k^{(s-1)}\right) \cdot \alpha(m \cdot N_{\mathrm{B}})\right|^2\right\} \\ +E\left\{\left|d_k^{(s-1)} \cdot [\alpha(m \cdot N_{\mathrm{B}}) - \hat{\alpha}(m \cdot N_{\mathrm{B}})]\right|^2\right\} \end{array}\right] \\ &= 4P \cdot P_b^{(s-1)} \cdot K_{\mathrm{X}} \cdot |\alpha(m \cdot N_{\mathrm{B}})|^2 + \frac{PK_{\mathrm{X}}}{N_{\mathrm{T}}^2} \sum_{\hat{i}_1=0}^{N_{\mathrm{T}}-1} \sum_{\hat{i}_2=0}^{N_{\mathrm{T}}-1} R_{\mathrm{err}}(\hat{i}_1, \hat{i}_2) \end{aligned} \tag{10.46}$$

where $P_b^{(s-1)}$ is the BER of the $(s-1)$th stage and $R_{\mathrm{err}}(\hat{i}_1, \hat{i}_2)$ is given by (10.19).

Since in $r_{\hat{k},m\cdot N_{\mathrm{B}}}^{(s)}$, the powers of the useful signal, interference and noise terms are unchanged, the new weights of MMSE with MCI cancellation at the sth stage are given by (10.35) with the substitution of $\sigma^2_{\text{MCI}_{\hat{m},\text{X}}}$ by $\sigma^2_{\text{MCI}_{m\cdot N_\text{B},\text{X}}}(s)$, i.e.

$$
\begin{aligned}
&\omega^{(s)}(m\cdot N_\text{B})\\
&= \frac{P\cdot\hat{\alpha}(m\cdot N_\text{B})}{P\cdot|\hat{\alpha}(m\cdot N_\text{B})|^2+\sigma^2_{\text{MCI}_{m\cdot N_\text{B},\text{X}}}(s)+\sigma^2_{\text{ICI}_{m\cdot N_\text{B}}}+\sigma^2_{N_{m\cdot N_\text{B}}}}\\
&= \frac{\hat{\alpha}(m\cdot N_\text{B})}{(1+4P_b^{(s-1)}K_\text{X})\cdot|\hat{\alpha}(m\cdot N_\text{B})|^2+\frac{K_\text{X}}{N_\text{T}^2}\sum\limits_{\hat{i}_1=0}^{N_\text{T}-1}\sum\limits_{\hat{i}_2=0}^{N_\text{T}-1}R_\text{err}(\hat{i}_1,\hat{i}_2)+\frac{\sigma^2_{\text{ICI}_{m\cdot N_\text{B}}}}{P}+\left(\frac{N_\text{T}E_\text{s}}{N_0}\right)^{-1}}
\end{aligned}
\tag{10.47}
$$

Finally, the output signal of the frequency domain despreader after hybrid MCI cancellation and MMSE detection is expressed as

$$
\begin{aligned}
y_{\hat{k}}^{(s)} &= \frac{1}{N_\text{F}}\sum_{m=0}^{N_\text{F}-1} r^{(s)}_{\hat{k},m\cdot N_\text{B}}\left[c^{(\hat{k}_\text{F})}_{N_\text{F},m}\cdot\omega^{(s)}(m\cdot N_\text{B})\right] \\
&= S_{\hat{k}}^{(s)}+\text{MCI}_\text{X}^{(s)}+\text{MCI}_\text{Y}^{(s)}+\text{ICI}_{\hat{k}}^{(s)}+N_{\hat{k}}^{(s)}
\end{aligned}
\tag{10.48}
$$

where $S_{\hat{k}}^{(s)}$ is the desired signal component contributed from the N_T chips in time domain over N_F sub-carriers, given by (10.38) with the substitution of $\omega(m\cdot N_\text{B})$ by $\omega^{(s)}(m\cdot N_\text{B})$. $\text{MCI}_\text{X}^{(s)}$ is the residual MCI caused by the other K_X code channels from Ω_X, given by

$$
\text{MCI}_\text{X}^{(s)} = \frac{1}{N_\text{F}}\sum_{m=0}^{N_\text{F}-1}\text{MCI}^{(s)}_{m\cdot N_\text{B},\text{X}}\cdot\left[c^{(\hat{k}_\text{F})}_{N_\text{F},m}\cdot\omega^{(s)}(m\cdot N_\text{B})\right]
\tag{10.49}
$$

$\text{MCI}_\text{Y}^{(s)}$ is the MCI caused by the other K_Y code channels from Ω_Y and the pilot channel, given by (10.38) with the substitution of $\omega(m\cdot N_\text{B})$ by $\omega^{(s)}(m\cdot N_\text{B})$. $\text{ICI}_{\hat{k}}^{(s)}$ is the ICI component and $N_{\hat{k}}^{(s)}$ is the background noise component at the sth stage, with variances given respectively by (10.41) and (10.42) with the substitution of $\omega(m\cdot N_\text{B})$ by $\omega^{(s)}(m\cdot N_\text{B})$. Note that in the hybrid detection, not only the MCI but also the power or variance of the useful signal, the ICI and the background noise change with the stages because of the updated MMSE weighting factors.

10.4 Performance evaluation

Conditioned on the estimated channel fading factors $\{\overline{\lambda}_{m\cdot N_\text{B}}\}$, the decision variable, $y_{\hat{k}}^{(s)}$, at the sth stage of MCI cancellation is given by (10.48). When the number of code channels in Ω_Y is large, conditioned on $\{\overline{\lambda}_{m\cdot N_\text{B}}\}$ which is contained in $\omega^{(s)}(m\cdot N_\text{B})$, and spreading codes $\{C^{(k_\text{T})}_{N_\text{T}}, C^{(k_\text{F})}_{N_\text{F}}\}$, where $C^{(k_\text{T})}_{N_\text{T}}$ is contained in $\alpha_k(m\cdot N_\text{B})$, $\text{MCI}_\text{X}^{(s)}$ in (10.48) can be modeled as a Gaussian random variable with variance

$$
\begin{aligned}
\sigma^2_{\text{MCI}_\text{Y}}(s) &= \frac{P}{N_\text{F}^2}\sum_{k\in\Omega_\text{Y}}\left|\sum_{m=0}^{N_\text{F}-1}\left[c^{(\hat{k}_\text{F})}_{N_\text{F},m}\cdot c^{(k_\text{F})}_{N_\text{F},m}\cdot\alpha_k(m\cdot N_\text{B})\cdot\omega^{(s)}(m\cdot N_\text{B})\right]\right|^2 \\
&+\frac{\beta P}{N_\text{F}^2}\left|\sum_{m=0}^{N_\text{F}-1}\left[c^{(\hat{k}_\text{F})}_{N_\text{F},m}\cdot\alpha_P(m\cdot N_\text{B})\cdot\omega^{(s)}(m\cdot N_\text{B})\right]\right|^2
\end{aligned}
\tag{10.50}
$$

Therefore, at the sth stage, there are three Gaussian random variables in $y_{\hat{k}}^{(s)}$, i.e., $\text{MCI}_{\text{Y}}^{(s)}$, $\text{ICI}_{\hat{k}}^{(s)}$ and $N_{\hat{k}}^{(s)}$. $\text{MCI}_{\text{Y}}^{(s)}$ is the multicode interference in the time domain from code channels in Ω_{Y} and $\text{ICI}_{\hat{k}}^{(s)}$ is the inter-carrier interference from all other sub-carriers. Since $\text{MCI}_{\text{Y}}^{(s)}$ and $\text{ICI}_{\hat{k}}^{(s)}$ are generated with different principles, the correlation between $\text{MCI}_{\text{Y}}^{(s)}$ and $\text{ICI}_{\hat{k}}^{(s)}$ is negligible. Furthermore, $N_{\hat{k}}^{(s)}$ is the background noise. Therefore, $\text{MCI}_{\text{Y}}^{(s)}$, $\text{ICI}_{\hat{k}}^{(s)}$ and $N_{\hat{k}}^{(s)}$ are independent Gaussian random variables. A new Gaussian random variable as $\eta_{\hat{k}}^{(s)} = \text{MCI}_{\text{Y}}^{(s)} + \text{ICI}_{\hat{k}}^{(s)} + N_{\hat{k}}^{(s)}$ can be defined with the variance of $\sigma_\eta^2(s) = \sigma_{\text{MCI}_\text{Y}}^2(s) + \sigma_{\text{ICI}}^2(s) + \sigma_N^2(s)$. Then the decision variable $y_{\hat{k}}^{(s)}$ is written as

$$y_{\hat{k}}^{(s)} = S_{\hat{k}}^{(s)} + \text{MCI}_{\text{X}}^{(s)} + \eta_{\hat{k}}^{(s)} \tag{10.51}$$

The original data in $\text{MCI}_{\text{X}}^{(s)}$ (see (10.45) or (10.49)) is written as $d_k = (d_{I,k} + j \cdot d_{Q,k})/\sqrt{2}$, where $d_{I,k}$ and $d_{Q,k}$ take 1 or –1 with equal probability. Similarly, the tentative decisions can be written as $d_k^{(s-1)} = (d_{I,k}^{(s-1)} + j \cdot d_{Q,k}^{(s-1)})/\sqrt{2}$. Defining the error vector $e_k^{(s-1)} = d_k - d_k^{(s-1)} = e_{I,k}^{(s-1)} + j \cdot e_{Q,k}^{(s-1)}$, the error probabilities of the real and imaginary parts are supposed to be equal and given by

$$P(e_{I,k}^{(s-1)}) = \begin{cases} 1 - P_b^{(s-1)}, & e_{I,k}^{(s-1)} = 0 \\ P_b^{(s-1)}, & e_{I,k}^{(s-1)} \neq 0 \end{cases} \tag{10.52}$$

Conditioned on $\{\overline{\lambda}_{m \cdot N_\text{B}}\}$, $\{d_k, k \in \Omega_\text{X}\}$ and $\{e_k^{(s-1)} = e_{I,k}^{(s-1)} + j \cdot e_{Q,k}^{(s-1)}, k \in \Omega_\text{X}\}$, the conditional error probability at the sth stage is given by

$$\begin{aligned} & P_b^{(s)}(\{\overline{\lambda}_{m \cdot N_\text{B}}\}, \{d_k\}, \{e_k^{(s-1)}\}|_{k \in \Omega_\text{X}}) \\ &= \frac{1}{2} \cdot \Pr\left(\text{Re}(S_{\hat{k}}^{(s)} + \text{MCI}_{\text{X}}^{(s)} + \eta_{\hat{k}}^{(s)}) \leq 0|_{d_{I,k}=d_{Q,k}=1}\right) \\ &+ \frac{1}{2} \cdot \Pr\left(\text{Re}(S_{\hat{k}}^{(s)} + \text{MCI}_{\text{X}}^{(s)} + \eta_{\hat{k}}^{(s)}) \leq 0|_{d_{I,k}=1, d_{Q,k}=-1}\right) \\ &= \frac{1}{2} \cdot Q\left(\left.\frac{\text{Re}(S_{\hat{k}}^{(s)} + \text{MCI}_{\text{X}}^{(s)})}{\sigma_\eta(s)/\sqrt{2}}\right|_{d_{I,k}=d_{Q,k}=1}\right) \\ &+ \frac{1}{2} \cdot Q\left(\left.\frac{\text{Re}(S_{\hat{k}}^{(s)} + \text{MCI}_{\text{X}}^{(s)})}{\sigma_\eta(s)/\sqrt{2}}\right|_{d_{I,\hat{k}}=1, d_{Q,\hat{k}}=-1}\right) \end{aligned} \tag{10.53}$$

where $Q(z) = \frac{1}{\sqrt{2\pi}} \int_z^{+\infty} e^{-t^2/2} dt$ for $z \geq 0$. The factor of $\sqrt{2}$ is introduced because only the real part of the Gaussian noise is considered with a variance equal to half of that of the complex Gaussian noise.

In order to obtain the average bit error rate, (10.53) must be averaged over all conditions. First consider the average over $\{e_{I,k}^{(s-1)}, k \in \Omega_\text{X}\}$ and $\{e_{Q,k}^{(s-1)}, k \in \Omega_\text{X}\}$. For n bit errors out of K_X codes, the number of orders of different error vectors (or sequences) is $\binom{K_\text{X}}{n} = \frac{K_\text{X}!}{n!(K_\text{X}-n)!}$. Further, defining $\{e_{I,k}^{(s-1)}\}_{n,i}$, $i = 1, 2, \ldots, \binom{K_\text{X}}{n}$ as the ith order error vector with n bit errors, the probability of the ith order error vector with n bit errors out of K_X codes is given by

$$P(n) = \left(P_b^{(s-1)}\right)^n \cdot \left(1 - P_b^{(s-1)}\right)^{K_X - n} \tag{10.54}$$

When n is large, (10.54) is very small. Thus, in numerical calculation, only a few small values of n need to be considered. Therefore, the average of (10.53) over $\left\{e_{I,k}^{(s-1)}, k \in \Omega_X\right\}$ is given by

$$\begin{aligned} &P_b^{(s)}\left(\{\overline{\lambda}_{m \cdot N_B}\}, \{d_k\}, \left\{e_{Q,k}^{(s-1)}\right\}\Big|_{k \in \Omega_X}\right) \\ &= \sum_{n=0}^{K_X} P(n) \sum_{i=1}^{\binom{K_X}{n}} P_b^{(s)}\left(\{\overline{\lambda}_{m \cdot N_B}\}, \{d_k\}, \left\{e_{Q,k}^{(s-1)}\right\}, \left\{e_{I,k}^{(s-1)}\right\}_{n,i}\Big|_{k \in \Omega_X}\right) \end{aligned} \tag{10.55}$$

Similarly, the average of (10.55) over $\left\{e_{Q,k}^{(s-1)}, k \in \Omega_X\right\}$ is given by

$$\begin{aligned} &P_b^{(s)}(\{\overline{\lambda}_{m \cdot N_B}\}, \{d_k\}|_{k \in \Omega_X}) \\ &= \sum_{n=0}^{K_X} P(n) \sum_{i=1}^{\binom{K_X}{n}} P_b^{(s)}\left(\{\overline{\lambda}_{m \cdot N_B}\}, \{d_k\}\left\{e_{Q,k}^{(s-1)}\right\}_{n,i}\Big|_{k \in \Omega_X}\right) \end{aligned} \tag{10.56}$$

Then (10.56) should be averaged over all possible K_X interference data

$$P_b^{(s)}(\{\overline{\lambda}_{m \cdot N_B}\}) = \sum_{j=0}^{4^{K_X}-1} P_b^{(s)}(\{\overline{\lambda}_{m \cdot N_B}\}, \{d_k\}_j|_{k \in \Omega_X}) \tag{10.57}$$

where 4 stands for the number of different values of d_k and 4^{K_X} is the total number of combinations of d_k over K_X data. In the special case (i.e. the zeroth stage or pure MMSE detection), the regenerated MCI or the error vector do not exist. Thus, the error vectors in (10.53) should be removed.

Finally, $P_b^{(s)}(\{\overline{\lambda}_{m \cdot N_B}\})$ is averaged over all $\{\overline{\lambda}_{m \cdot N_B}, m = 0, 1, \ldots, N_F - 1\}$, which can be numerically evaluated using a Monte Carlo approach [4].

10.5 Numerical results

Using the Monte Carlo approach, some representative numerical results are obtained from the analytical expressions and presented in this section. A broadband system with 100 MHz is employed. A total of $M = 1024$ sub-carriers are used and the resultant sub-carrier spacing is about 97.7 kHz. The effective OFCDM symbol duration is 10.24 μs. When the delay spread of the multipath channel is around 1.0 μs [5], a guard interval T_g of 1.76 μs is adequate to combat the multipath interference. Therefore, the complete OFCDM symbol duration is 12.0 μs. Furthermore, the average received signal-to-noise ratio (SNR) per bit is defined as $\mathrm{SNR_b} = \frac{E_s}{N_0}\frac{R_\lambda(0,0)}{2}N(1 + \beta/K) = 14$ dB and the system load is defined as K/N. The times γ of N_T for channel estimation is set to be $\gamma = 3$. All the numerical results are based on these assumptions unless noted otherwise.

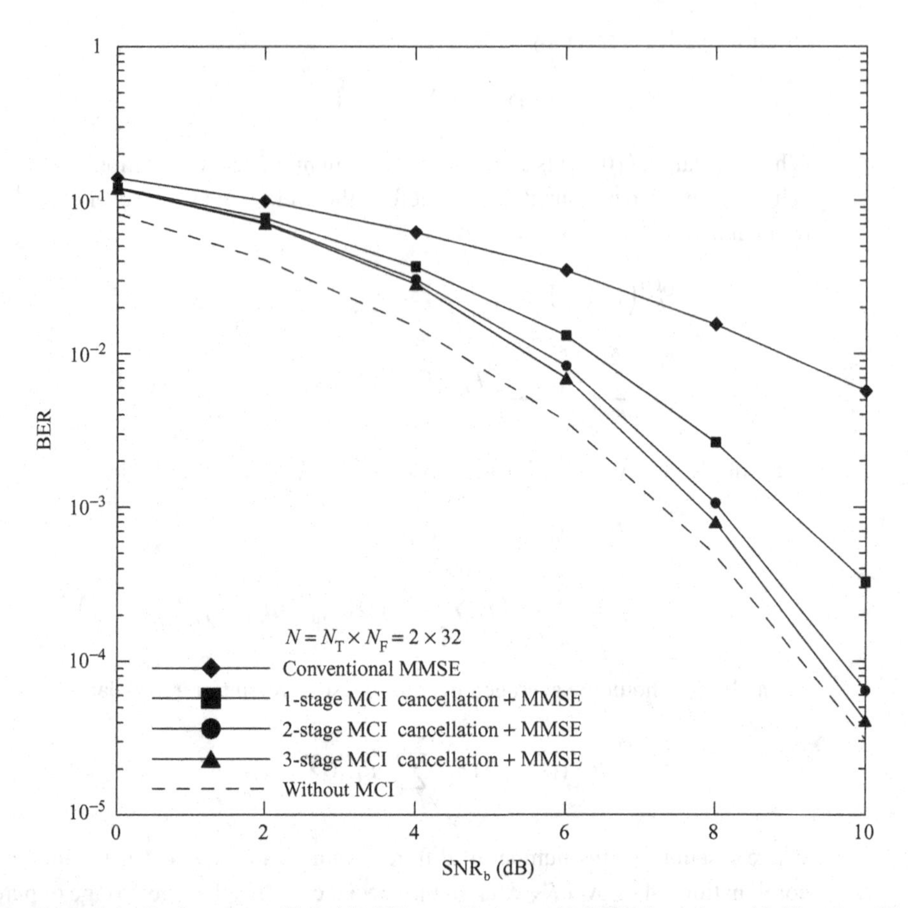

Figure 10.4 Performance of hybrid MCI cancellation and MMSE detection with $N = 2 \times 32$ and full load.

First, performance of the OFCDM system is evaluated with the assumption of ideal channel estimation. Thus, no pilot is needed. The BER performance of the hybrid MCI cancellation and MMSE detection is shown in Figure 10.4 for $N = N_T \times N_F = 2 \times 32$ and full system load, i.e. $K/N = 1.0$ when the Doppler shift is zero. A three-stage MCI canceller, including the zeroth to third stages, is considered. The system performance without MCI is also shown for comparison. It can be seen from the figure that the hybrid detection performs much better than pure MMSE when SNR_b is large. The BER decreases as the number of stages increases and gets close to the performance without MCI. The performance improvement between the pure MMSE and the hybrid detection with the first stage is significant. The gap in BER between the first stage and second stage is also large. However, the improvement beyond the second stage is insignificant. In conclusion, the BER performance improvement for the hybrid detection decreases as the number of stages increases. Considering two stages is sufficient.

Figure 10.5 shows the BER performance as a function of system load K/N for a given $SNR_b = 8$ dB and a different number of stages. It can be seen that with various

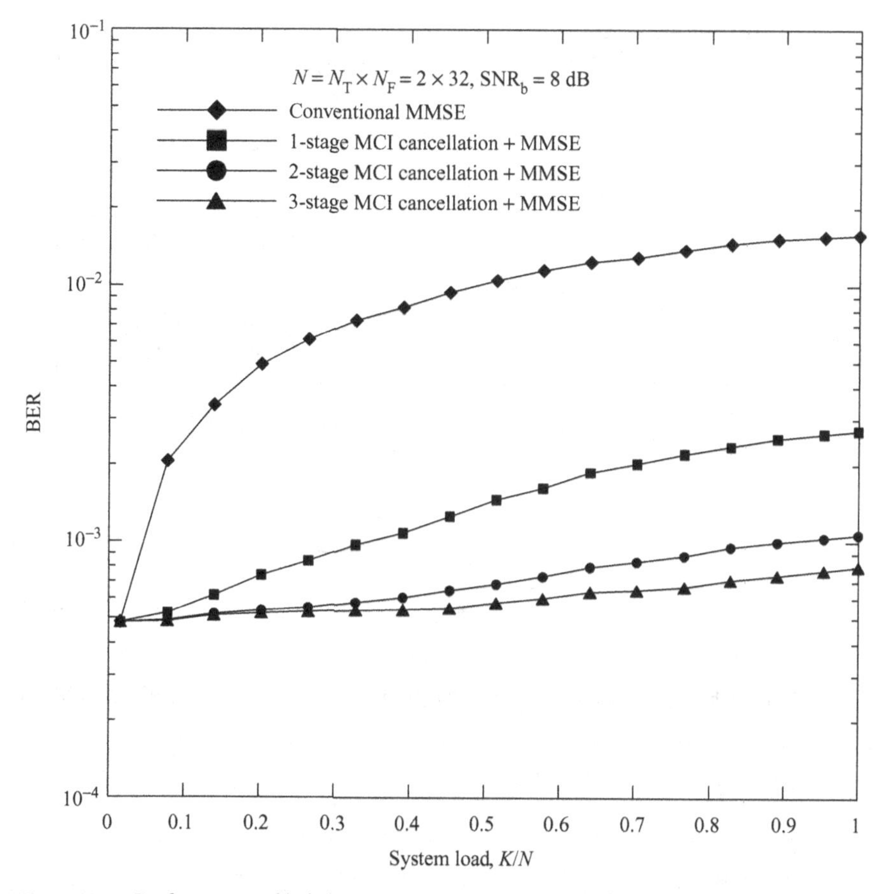

Figure 10.5 Performance of hybrid MCI cancellation and MMSE detection as a function of system load.

system loads, the hybrid detection exhibits much better performance than the conventional MMSE. When system load is light, the first stage is needed. When the system load is heavy, however, it is necessary to consider uing two stages. This is consistent with Figure 10.4.

Figure 10.6 illustrates the BER performance with the hybrid two-stage MCI cancellation and MMSE detection, and with various values of N_F, the frequency domain spreading factor, for a given total spreading factor. It can be seen that the BER of the hybrid detection decreases as N_F increases, especially for large SNR_b. This is because a greater frequency diversity gain is obtained when N_F increases. Although more MCI is caused when N_F gets larger, MCI cancellation can reject most of the MCI. In this case, frequency diversity can still provide more gain when N_F becomes larger. In summary, with hybrid detection, a large value of the frequency domain spreading factor is desired.

The BER of the hybrid detection is plotted in Figure 10.7 as a function of the frequency domain spreading factor for a given total spreading factor and for various values of the

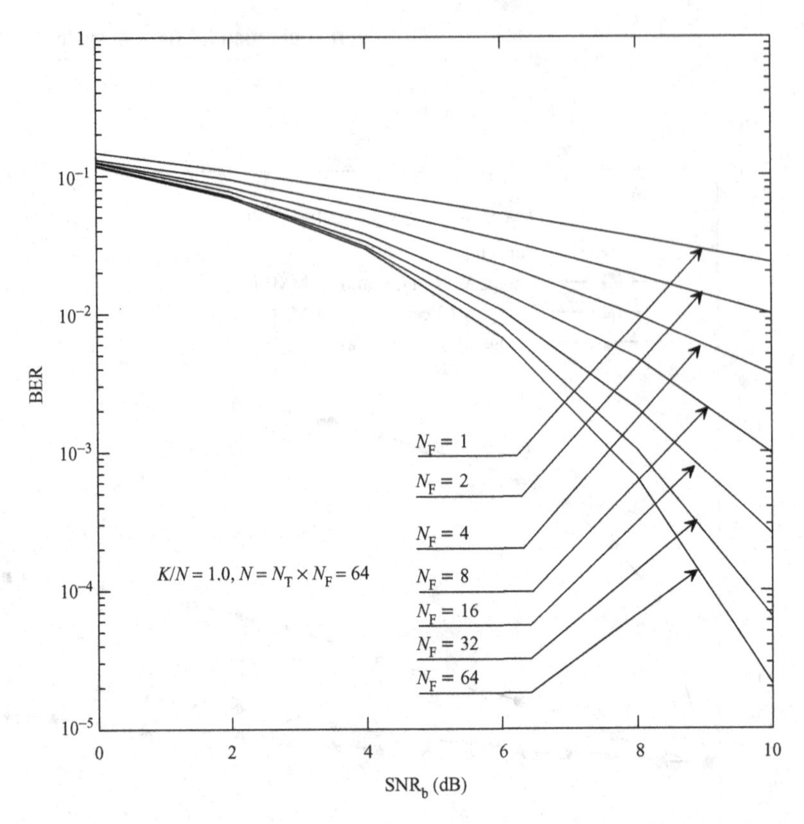

Figure 10.6 System performance with hybrid two-stage MCI cancellation and MMSE detection.

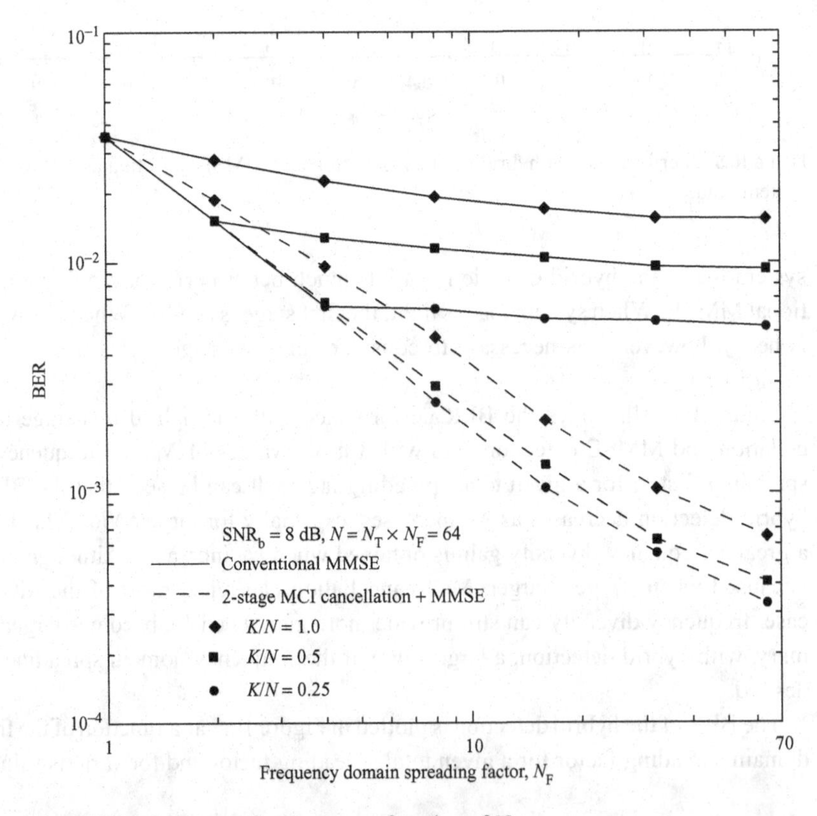

Figure 10.7 System performance as a function of N_F.

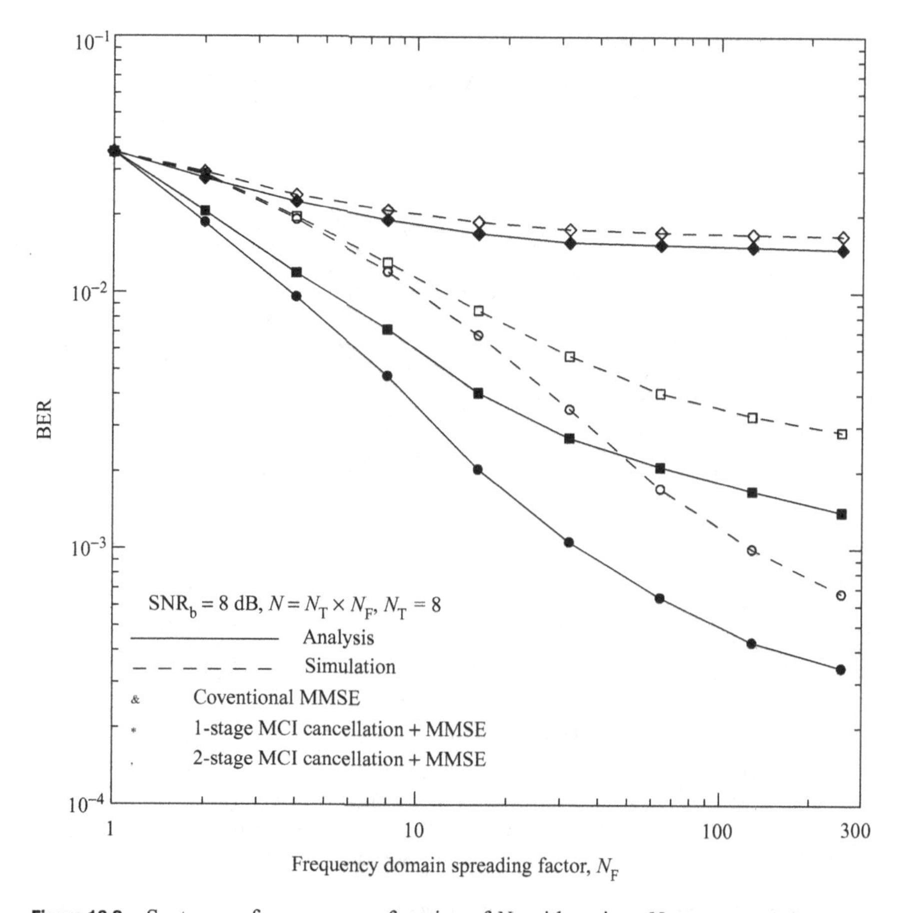

Figure 10.8 System performance as a function of N_F with a given N_T.

system load. The BER of conventional MMSE is also plotted for comparison. It can be seen that for pure MMSE, the BER decreases when N_F increases from a small value. This is because the gain of frequency diversity is more than the loss caused by the MCI. For different values of N_F, however, the pure MMSE provides very stable performance. In other words, large N_F does not provide any frequency diversity gain for pure MMSE. This is because more MCI is caused when N_F increases. The frequency diversity gain is canceled out by the MCI. However, the hybrid detection performs differently. Its BER performance improves when N_F increases for different system loads. This is similar to Figure 10.6.

The BER is plotted in Figure 10.8 for both detection schemes as a function of N_F for a given N_T, but variable $N = N_T \times N_F$. Simulation results are also shown for comparison. It can be seen that the analytical results provide correct and useful indications on the system performance as simulation results, although the analytical results are optimistic in the hybrid detection scheme. It can also be seen that for a given number of stages (first stage or second stage) in the hybrid scheme, the BER decreases as N_F increases,

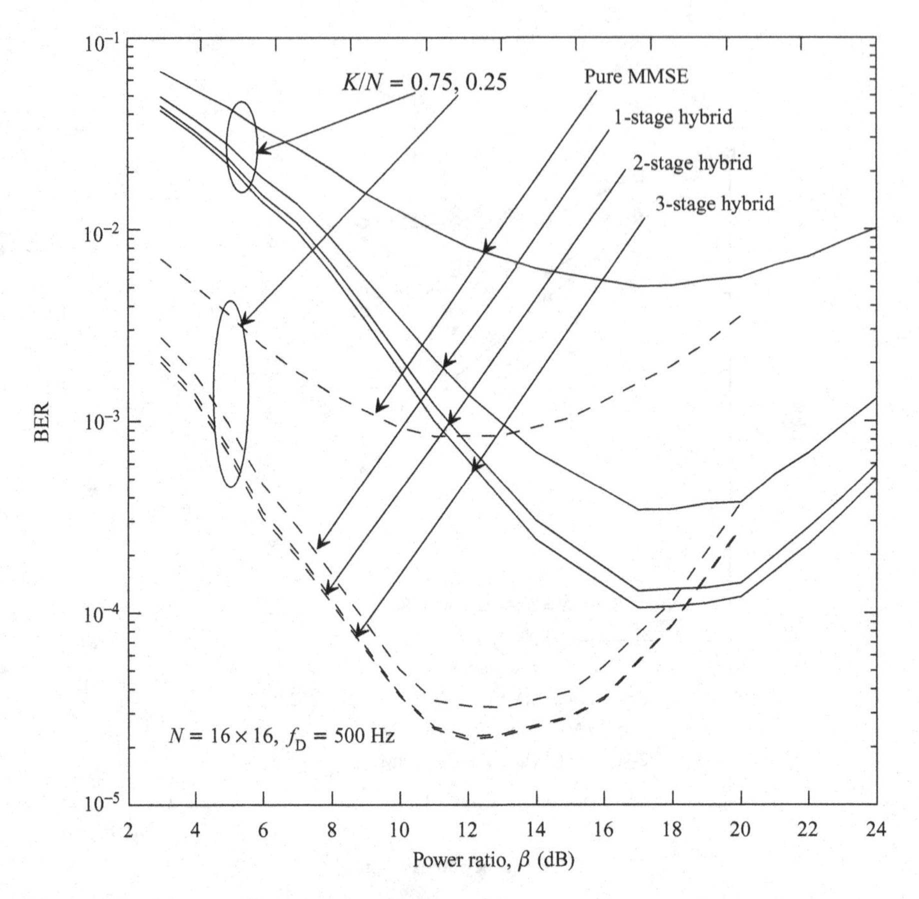

Figure 10.9 System performance as a function of β.

whereas the BER of conventional MMSE is flat for various N_F. This is consistent with Figure 10.7.

In the following, practical channel estimation is considered. Thus the code-multiplexed pilot channel is needed. Figure 10.9 illustrates the effect of the power ratio β on the performance of a two-dimensionally spreading system with the system load $K/N = 0.25$ and 0.75, respectively. The Doppler shift f_D is assumed to be 500 Hz. For comparison, the performances of pure MMSE, and one-stage, two-stage and three-stage hybrids are shown. It can be seen from the figure that when β is small, the system performance degrades by the poor channel estimation. When β increases, BER performance improves as the quality of channel estimation improves, and BER reaches a minimum for a particular value of β. Further increasing β beyond that value increases BER due to the low transmit power efficiency and more interference caused to data channels by the pilot channel. When the system load is light, i.e. $K/N = 0.25$, the optimum β is around 12 dB. However, when system load is heavy, i.e. $K/N = 0.75$, the optimum β is around 18 dB. Generally speaking, the power on pilot channel is set to be about 20 percent of the total transmit power, i.e. $\beta/(\beta + K) = 0.2$, or equivalently $\beta = 0.25K$. Moreover,

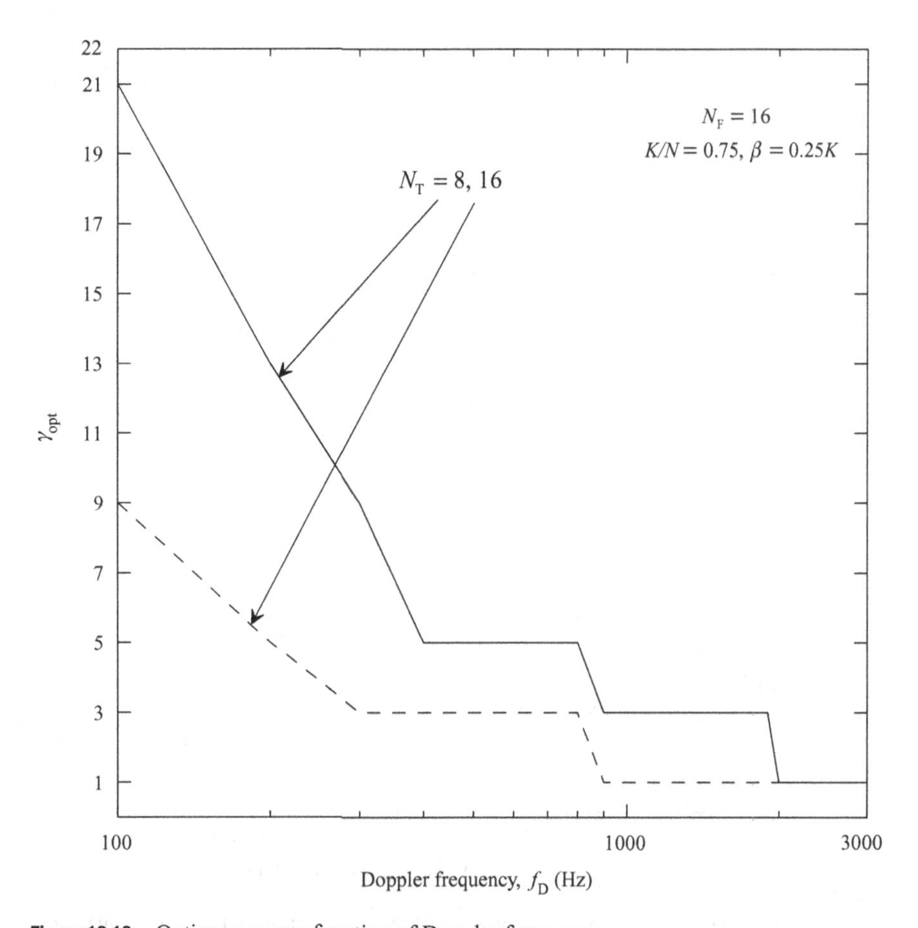

Figure 10.10 Optimum γ as a function of Doppler frequency.

the significant improvement in BER performance can be achieved by making use of the hybrid detection. For either heavy or light system loads, the most significant reduction in BER is obtained by the one-stage hybrid detection, then the improvement decreases when the number of stages increases. For the light load, a one-stage hybrid detector is sufficient, whereas for the heavy load, two-stage hybrid detection is necessary.

Given $N_F = 16$ and $\beta = 0.25K$, Figure 10.10 shows the optimal γ (γ_{opt} is the multiplier of N_T in channel estimation) as a function of f_D for various values of N_T. It can be seen from the figure that for a given N_T, γ_{opt} decreases when f_D increases. This is because when f_D is small the channel variation is negligible and a larger γ can suppress more interference and noise in channel estimation. But when f_D is large, the channel variation is obvious, so γ should not be large in order not to distort the channel estimation. Moreover, for given f_D, γ_{opt} is smaller when N_T is larger. However, γ_{opt} should be greater than one for a wide range of Doppler shift.

The BER is plotted in Figure 10.11 also as a function of f_D with $N = 16 \times 16$, $K/N = 0.75$, and $\beta = 0.25K$ for pure MMSE, and one-stage, two-stage and three-stage hybrids, respectively. It can be seen from the figure that when f_D is small ($f_D \leq 200$ Hz), the

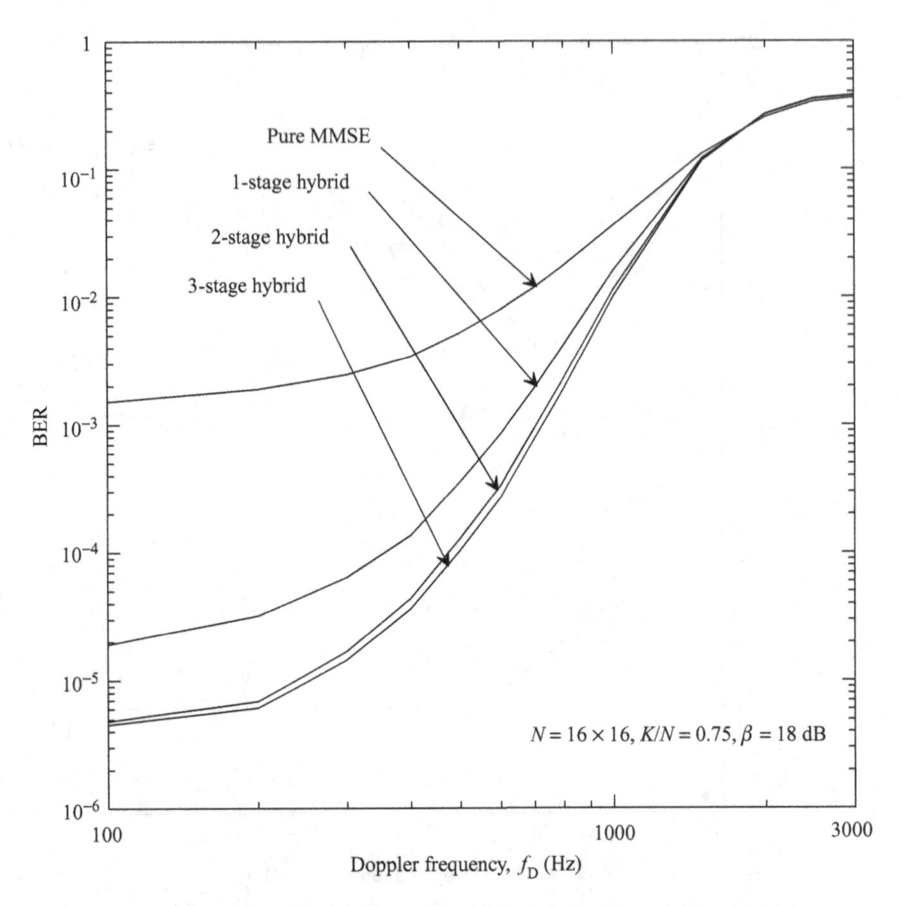

Figure 10.11 Performance of hybrid detection as a function of Doppler frequency.

performance of each detection is very stable and the performance of the hybrid is much better than that of the pure MMSE. Considering the two-stage hybrid is sufficient. When f_D increases from 200 Hz, the performance of any detection degrades dramatically due to poor channel estimation. Furthermore, the degradation in performance is much more rapid for hybrid than for pure MMSE. This is because when f_D increases, MCI in the time domain increases, which cannot be cancelled out by the hybrid detection. When f_D becomes large ($f_D \geq 1000$ Hz), the BER performances of the pure MMSE and hybrid detection are very similar. In general, as long as f_D is less than 800 Hz, the improvement in BER with hybrid detection is still significant.

The BER of the two-stage hybrid is plotted in Figure 10.12 as a function of N_F for various values of f_D. For similar channel estimation, γ is assumed to be 3 and 5 for $N_T = 16$ and $N_T = 8$, respectively. It can be seen that for a given f_D, when N_F increases, the system performance improves. This is because with hybrid detection, when N_F increases, the frequency diversity gain increases and overcomes the increased MCI in frequency domain. Note that for $N_T = 16$ and $f_D = 1000$ Hz, the system performance degrades slightly when N_F increases from a small value and remains stable. Furthermore,

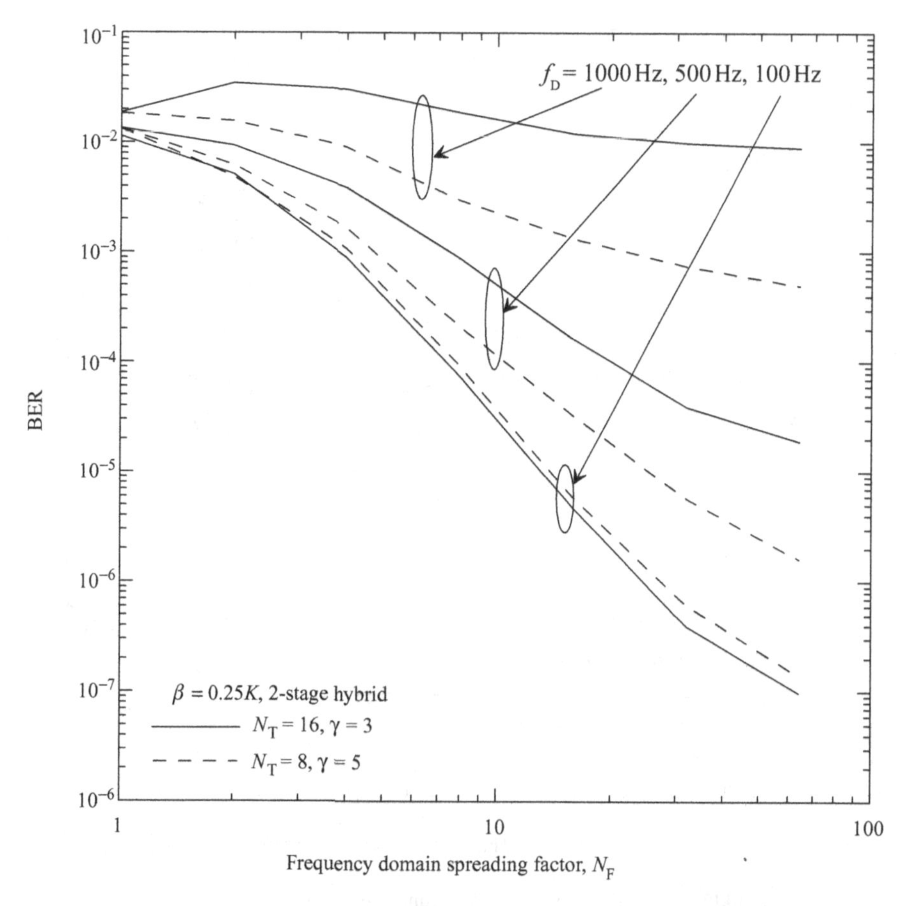

Figure 10.12 System performance as a function of N_F.

when f_D is small, the performance difference between $N_T = 16$ and $N_T = 8$ is negligible. However, the difference increases with f_D. When $f_D \geq 500$ Hz, the system with $N_T = 8$ performs much better than that with $N_T = 16$. This is because the MCI in the time domain is larger with $N_T = 16$. This MCI increases with f_D and cannot be cancelled out by the hybrid detection. In summary, with similar channel estimation quality, a larger frequency domain spreading factor is preferred, irrespective of the Doppler shift f_D. But a shorter N_T is more favorable when f_D increases.

Figure 10.13 illustrates the system performance of the two-stage hybrid as a function of N_T for various values of f_D, as well. The frequency domain spreading factor N_F is set to 8 or 16. γ is changed with N_T. Using the optimum $\gamma = 3$ when $N_T = 16$ and $f_D = 500$ Hz as a reference, to obtain similar channel estimation, γ is chosen as 11, 5, 1 and 1 for $N_T = 4$, 8, 32 and 64, respectively. It can be seen that, basically, for a given value of f_D, the BER of each detection increases when N_T increases. This is because with similar channel estimation, the MCI in the time domain increases as N_T increases, which cannot be mitigated by the hybrid detection. There are also some special points acting as

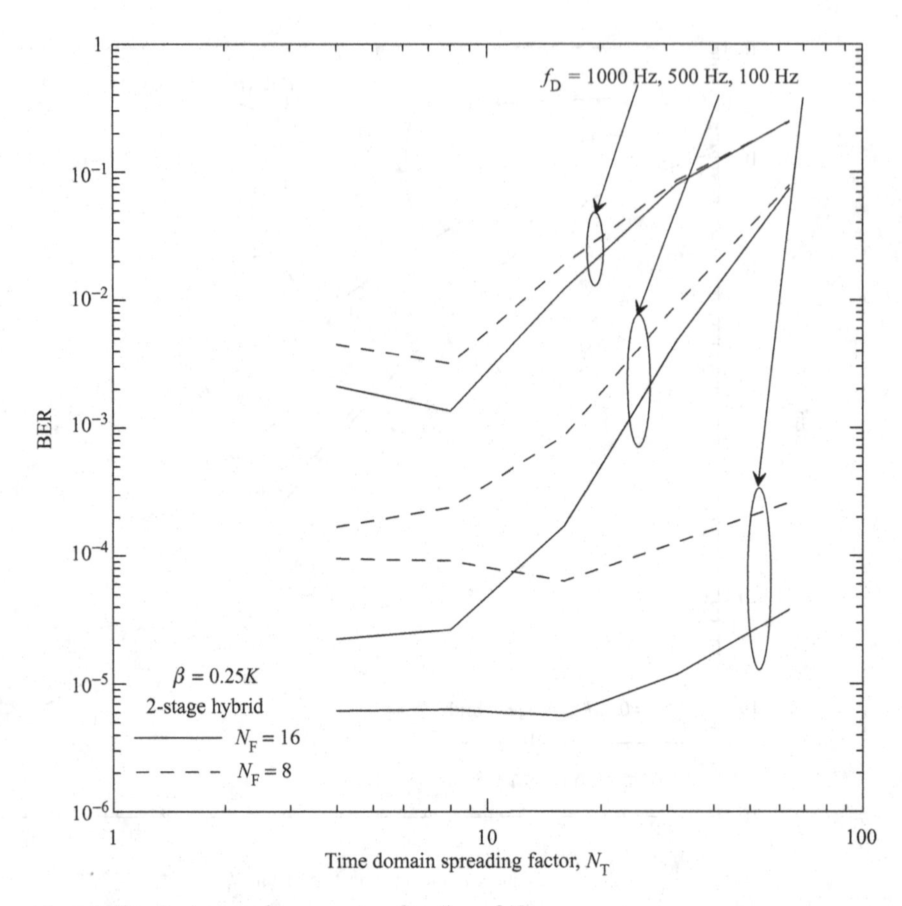

Figure 10.13 System performance as a function of N_T.

the optimum value such as $N_T = 8$ for $f_D = 1000$ Hz and $N_T = 16$ for $f_D = 100$ Hz, but the performance difference is not significant. Furthermore, when f_D is small, the system has stable performance with $N_T = 4$, 8 and 16. When f_D increases, however, the system performance with $N_T = 16$ is obviously degraded. In summary, when similar channel estimation is possible, the system with shorter N_T is always favorable, especially when the Doppler frequency is high.

10.6 Summary

The performance of OFCDM systems with hybrid MCI cancellation and MMSE detection is investigated in this chapter. The performance of conventional MMSE and the hybrid detection have been compared extensively. The following conclusions are drawn.

(1) The hybrid detection performs much better than the conventional MMSE. But the BER performance improvement for the hybrid detection decreases as the number of stages increases. Considering a two-stage MCI cancellation is sufficient.
(2) For the hybrid detection, a large value of frequency domain spreading factor, N_F, is expected. For the conventional MMSE, however, the value of N_F does not make much difference in BER performance.
(3) The quality of channel estimation is critical to the performance of hybrid detection in the presence of Doppler shift. The optimum power ratio (β) between pilot and one data channel takes a value around $0.25K$ for a wide range of system loads.
(4) The optimal times (γ) of N_T in channel estimation decreases when the Doppler shift increases, but should be greater than one for a wide range of Doppler shift.
(5) When the Doppler frequency f_D is less than 200 Hz, the effect of the Doppler frequency on system performance can be negligible. However, increasing f_D rapidly degrades the system performance. As long as $f_D < 800$ Hz, the hybrid detection still provides significant improvement in performance compared to the pure MMSE detection.
(6) In order to achieve the best performance for a wide range of Doppler shifts, a larger N_F is preferred, while a shorter N_T should be employed, especially when the Doppler frequency is high.

References

[1] W. C. Jakes, *Microwave Mobile Communications*. Piscataway, NJ: IEEE Press, 1993.
[2] Y. Q. Zhou, "Advanced techniques for high speed wireless communications," Ph.D. Thesis, University of Hong Kong, 2003.
[3] N. Yee and J. P. Linnarz, "Wiener filtering of multi-carrier CDMA in Rayleigh fading channel," in *Proc. IEEE PIMRC'94*, vol. 4, pp. 1344–1347, Sept. 1994.
[4] Z. Xing, "Parallel ensemble Monte Carlo for device simulation," *Workshop on High Performance Computing Activities in Singapore*, National Supercomputing Research Center, Sept. 1995.
[5] H. Atarashi and M. Sawahashi, "Investigation of inter-carrier interference due to Doppler spread in OFCDM broadband packet wireless accesses," *Special issue on software defined radio technologies and its applications*, 2002SRP-28.

11 Coded layered space-time-frequency architecture

For high-speed OFDM MIMO multiplexing, a new coded layered space-time-frequency (LSTF) architecture (i.e. LSTF-c) with iterative signal processing at the receiver is proposed, where multiple encoders/decoders are designed and each independent codeword is threaded in the three-dimensional (3-D) space-time-frequency (STF) transmission resource array. The iterative receiver structure is adopted consisting of a joint MMSE-SIC detector and the maximum a posteriori (MAP) convolutional decoders. Simulation results show that the proposed LSTF architecture can achieve almost the same performance as the LSTF (i.e. LSTF-a) where single coding is applied across the whole information stream. However, due to its structure of multiple parallel lower speed encoders/decoders with shorter codeword length, the proposed LSTF architecture can be more easily implemented than the LSTF-a.

11.1 Introduction

The challenge of the detection of MIMO multiplexing is to design a low complexity detector, which can efficiently suppress multi-antenna interference and approach the interference-free performance. In this chapter, an iterative processing technique for joint detection and decoding is used in the coded MIMO multiplexing. The iterative receiver contains an MMSE-SIC detector [1], the complexity of which is much lower than that of the MAP detector especially when the number of transmit antennas N_t and modulation level are large. As a constituent code, the convolutional code is used due to the computationally efficient SISO MAP decoding. The performances of the joint iterative detection/decoding schemes for both turbo code and convolutional code are quite similar for different numbers of antennas [2], [3]. However, the iterative joint scheme for convolutional code outperforms that for low density parity check (LDPC) code [4], because the former can approach the performance with zero interference while the latter is very sensitive to error propagation when the sum-product decoding algorithm is used. The objective of this chapter is to evaluate the performance of iterative signal processing for the proposed novel LSTF-c. The effects of system or channel parameters, such as the number of iterations in the joint detectors and decoders, the percentage of the pilot channel energy, the number of multipaths, the root mean square delay spread, the correlation coefficient between antennas, and the Doppler frequency shift, etc., are studied extensively.

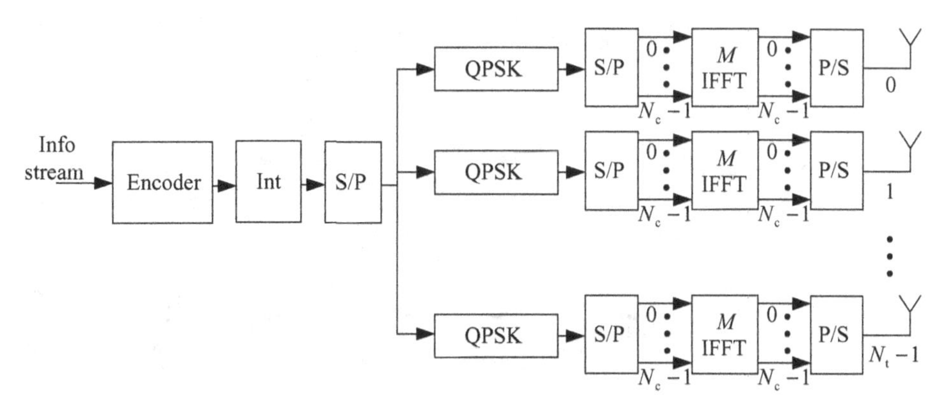

Figure 11.1 Transmitter model of LSTF-a.

The remainder of this chapter is organized as follows. The transmitter structures, channel model and iterative receiver structures of the three LSTF architectures are given in Section 11.2. The channel estimation algorithm and the MMSE-SIC detector are described in Section 11.3. Some representative simulation results are given in Section 11.4. Finally, some conclusions are drawn in Section 11.5.

11.2 System description

11.2.1 Transmitter structures

The coded LSTF architectures have N_t transmit antennas and N_r receive antennas, assuming there are N_d data symbols and N_c sub-carriers in one packet. Thus, the LSTF architectures have a 3-D space-time-frequency (i.e. $N_t \times N_d \times N_c$) transmission resource array, where N_t, N_d and N_c are the spatial, temporal and frequency spans, respectively. The convolutional-coded LSTF architectures with iterative processing are investigated in this chapter, and therefore an interleaver is required to follow each convolutional encoder in the transmitter structures.

In the LSTF-a transmitter structure shown in Figure 11.1, the information stream is first encoded by a single convolutional encoder and then interleaved. After that, it is divided into N_t sub-streams, each of which is QPSK modulated, S/P converted, up-converted to N_c sub-carriers by an M-point IFFT ($M \geq N_c$), and then transmitted through transmit antennas. In the LSTF-b transmitter structure shown in Figure 11.2, the information stream is first divided into N_t parallel sub-streams, each of which is independently encoded by a convolutional encoder, interleaved, modulated, S/P converted, up-converted to N_c sub-carriers, and transmitted through a dedicated antenna. The LSTF-c transmitter structure is shown in Figure 11.3. As in the LSTF-b, the information stream in one packet in the LSTF-c transmitter is divided into N_t sub-streams towards N_t encoders, interleavers and modulators in parallel. Then the QPSK modulated symbols in each codeword are S/P converted and then distributed according to the function F into the

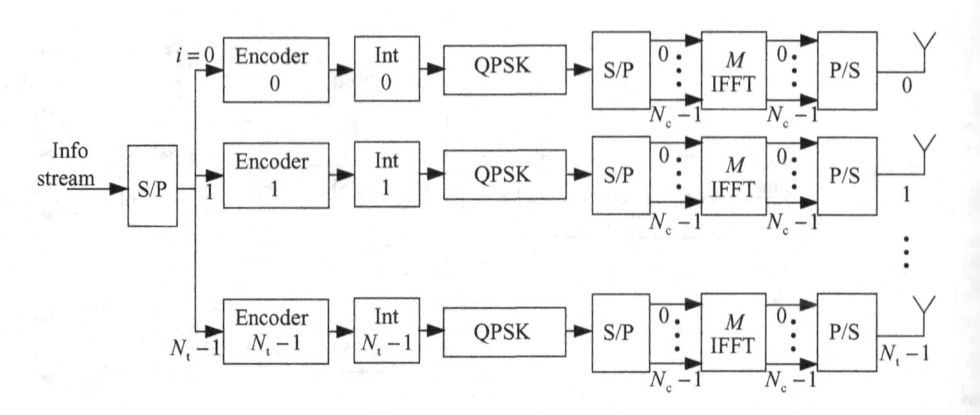

Figure 11.2 Transmitter model of LSTF-b.

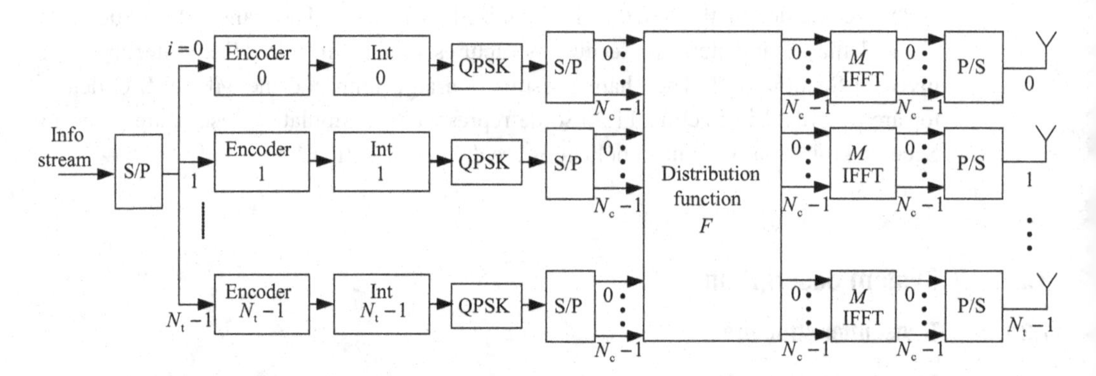

Figure 11.3 Transmitter model of LSTF-c.

$N_t \times N_d \times N_c$ 3-D transmission resource array. After that, the symbols in each transmit antenna are up-converted to N_c sub-carriers and transmitted.

In all the LSTF transmitters, after encoding and interleaving, a block of 2 bits is Gray mapped into one of the four possible signal points of a complex constellation in QPSK modulation. At the output of the M-point IFFT, an effective OFDM symbol is obtained with M samples, and a guard interval of T_g for CP is inserted between the OFDM symbols to prevent inter-symbol interference. Therefore, the duration of a complete symbol is $T = T_e + T_g$, where $T_e = MT_s$ is the effective symbol duration and T_s is the sampling time.

Our aim is to design a novel LSTF structure that can exploit spatial, temporal and frequency diversity like the LSTF-a, and also have a low processing speed and short codeword length in the channel encoder/decoder like the LSTF-b. Thus, a novel (i.e. LSTF-c) LSTF structure is proposed by introducing a distribution function F in the LSTF-b, as shown in Figure 11.1. As in the LSTF-b, the whole information stream in one packet in the LSTF-c is divided into N_t sub-streams and then encoded and modulated in parallel. After that, the modulated symbols in each codeword are S/P converted and then assigned in the $N_t \times N_d \times N_c$ 3-D transmission resource array with the distribution pattern F.

In our framework, there are N_t independent codewords, which are distributed to the 3-D transmission resource array according to the distribution function F. In the 3-D array, the symbols from the ith ($i = 0, \ldots, N_t - 1$) codeword at the input of the proposed distribution function F are expressed by an indexing set

$$F_I(n_t, n_d, n_c) = \{i, n_d, n_c\} \tag{11.1}$$

where $n_t = 0, \ldots, N_t - 1$, $n_d = 0, \ldots, N_d - 1$ and $n_c = 0, \cdots, N_c - 1$ represents the n_tth transmit antenna, the n_dth time slot and the n_cth sub-carrier, respectively. At the output of the distribution function F, the symbols from the ith codeword can be indicated by an indexing set in the 3D array as

$$F_O(n_t, n_d, n_c) = \{\lfloor n_d + n_c + i \rfloor_{N_t}, n_d, n_c\} \tag{11.2}$$

where $\lfloor \cdot \rfloor_{N_t}$ denotes the operation of modulo N_t. The cube of spatial, temporal and frequency spans is $\{N_t, N_d, N_c\}$ and it shows the capability of codewords to exploit the full spatial, temporal and frequency diversities. That is, all these codewords are distributed according to (11.2) in the 3-D array with the usage of the three resources: antennas, symbol slots and sub-carriers.

Given the parameters N_t, N_d and N_c, at the transmitter, (11.2) determines the distribution of the modulated symbols in each codeword to the 3-D array, and at the receiver the inverse distribution function F^{-1} is used to collect the symbols back in the 3-D array to form the original codeword for decoding. Figure 11.4 shows the symbol distribution with $N_t = 4$, $N_d = 12$ and $N_c = 8$, where every parallelogram denotes a symbol with a codeword number to which the symbol belongs. It can be seen from this figure that each codeword (given i) in the 3-D array shows a shape of thread, and thus this proposed novel LSTF-c is called threaded LSTF architecture. In summary, the proposed LSTF-c makes the whole 3-D array be filled by N_t disjoint codewords in order to jointly exploit the available spatial, temporal and frequency diversity for every codeword.

To carry out coherent MIMO detection, time-multiplexed pilot-aided channel estimation is considered. Figure 11.5 illustrates the packet format. One packet at each transmit antenna consists of N_d modulated data symbols, one time-multiplexed pilot symbol and $N_t - 1$ dummy symbols at every sub-carrier. The position of the pilot symbol depends on the order of transmit antennas. In order to achieve high-quality channel estimations, the dummy symbols are given to avoid interference among different transmit antennas. That is, when one transmit antenna transmits one pilot symbol, all the other transmit antennas transmit nothing. $\mathbf{x}[n_d, n_c]$ stands for the vector of N_t transmitted symbols from N_t antennas at the n_cth sub-carrier and the n_dth time slot, and $x_{n_t}[n_d, n_c]$ is the transmitted

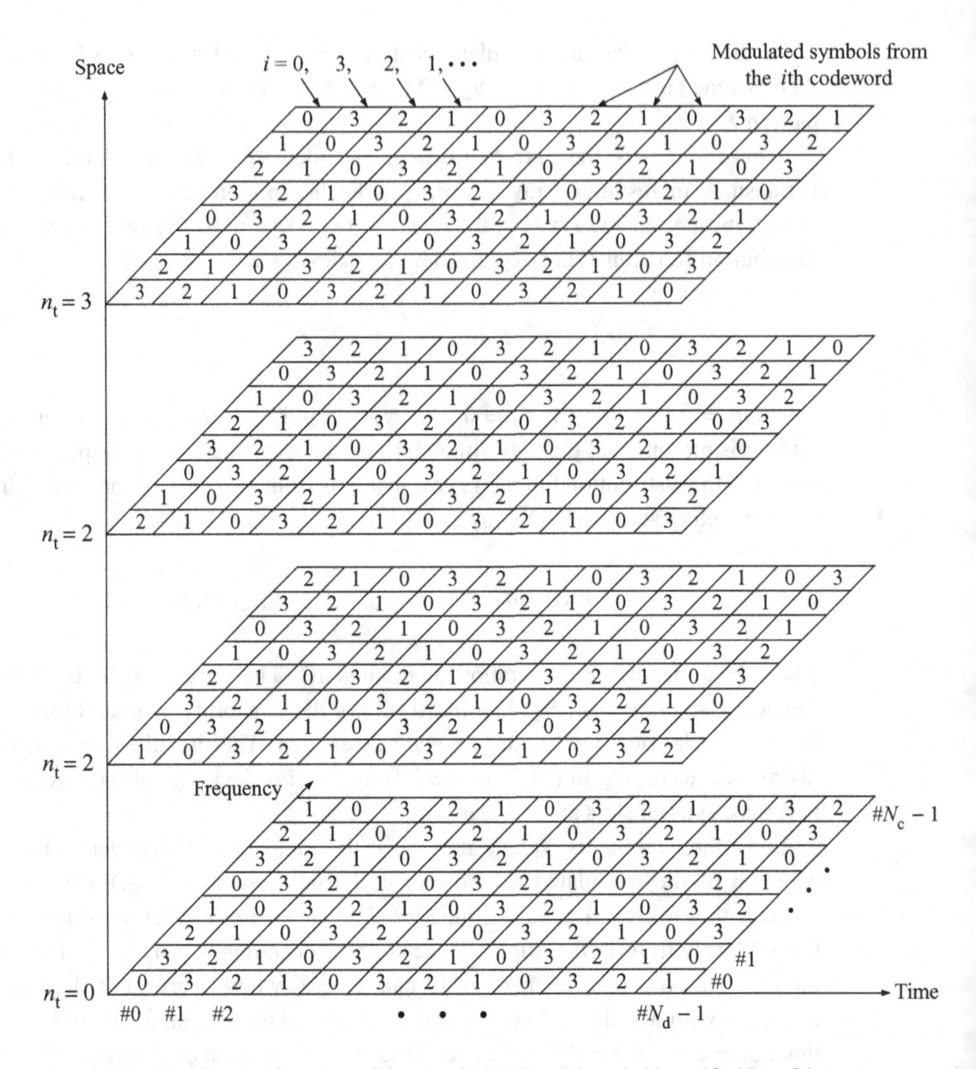

Figure 11.4 Distribution function of the novel LSTF-c. ($N_t = 4$, $N_d = 12$, $N_c = 8$.)

symbol at the n_tth antenna,

$$x_{n_t}[n_d, n_c] = \begin{cases} \sqrt{P_d} x_{n_t}^{(\text{data})}[n_d, n_c], & 0 \leq n_d \leq N_d/2 - 1 \\ \sqrt{\beta P_d} x^{(\text{pilot})}, & n_d = N_d/2 + n_t \\ \sqrt{P_d} x_{n_t}^{(\text{data})}[n_d - N_t, n_c], & N_d/2 + N_t \leq n_d \leq N_d + N_t - 1 \\ 0, & \text{otherwise} \end{cases} \tag{11.3}$$

where P_d is the signal power of data symbols at one sub-carrier, $x_{n_t}^{(\text{data})}[\cdot, \cdot]$ is the QPSK modulated data symbol with unity power at the n_tth transmit antenna, $x^{(\text{pilot})}$ is the known QPSK modulated pilot symbol with unity power, and β is the power ratio of pilot symbol to data symbol. Thus, the percentage of the pilot channel energy in one packet is $\gamma = \beta/(\beta + N_d)$.

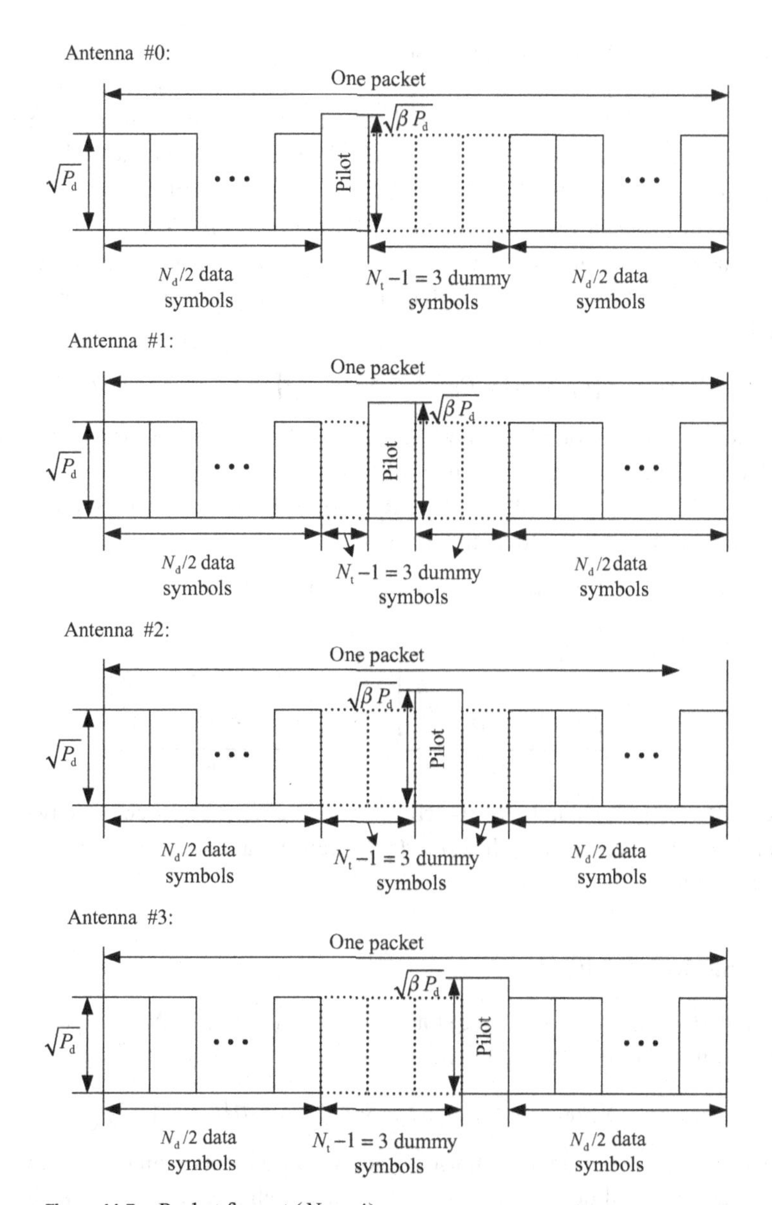

Figure 11.5 Packet format ($N_t = 4$).

11.2.2 Channel model

At the n_dth time slot and the n_cth sub-carrier, the channel response matrix with N_r rows and N_t columns, where N_r is the number of receive antennas, is given by [5]

$$\mathbf{H}[n_d, n_c] = \sum_{l=0}^{L-1} \mathbf{R}_l^{1/2} \mathbf{H}_l^{(n_d)} \mathbf{S}_l^{1/2} \exp(-j2\pi f_{n_c} \tau_l) \tag{11.4}$$

where f_{n_c} is the baseband equivalent frequency of the n_cth sub-carrier; and L is the number of resolvable paths of the frequency-selective fading channels.

The stochastic MIMO channel impulse response $\mathbf{H}_l^{(n_d)}$ of the lth path is the matrix with elements $h_l^{(n_d)}(n_r, n_t)$, where $n_r = 0, 1, \ldots, N_r - 1$ and $n_t = 0, 1, \ldots, N_t - 1$, being independently and identically distributed (i.i.d.) complex Gaussian random variables with zero mean and variance σ_l^2. The L-path channel has an exponential power decay $10\log(\sigma_l^2/\sigma_{l-1}^2) = -12/L$ dB with $\sum_{l=0}^{L-1} \sigma_l^2 = 1$, and a time delay difference of equal interval between two adjacent paths. Furthermore, $h_{l_1}^{(n_d)}(n_r, n_t)$ and $h_{l_2}^{(n_d)}(n_r, n_t)$ are assumed to be statistically independent for $l_1 \neq l_2$. The autocorrelation function of the lth path between two different symbol slots with time difference Δn_d is defined as

$$\phi_l(\Delta n_d) = E\{h_l^{(n_d)}(n_r, n_t)^* h_l^{(n_d+\Delta n_d)}(n_r, n_t)\} = \sigma_l^2 J_0(2\pi f_D \Delta n_d T) \tag{11.5}$$

where * stands for the complex conjugate, $J_0(\cdot)$ is the zeroth-order Bessel function and f_D is the maximum Doppler frequency shift.

In (11.4), $\mathbf{S}_l = \mathbf{S}_l^{1/2}\mathbf{S}_l^{1/2}$ and $\mathbf{R}_l = \mathbf{R}_l^{1/2}\mathbf{R}_l^{1/2}$ represent the transmit and receive spatial-correlation matrices for the lth path, given by

$$\mathbf{S}_l = \underbrace{\begin{bmatrix} 1 & \rho & \cdots & \rho \\ \rho & 1 & \ddots & \vdots \\ \vdots & \ddots & \ddots & \rho \\ \rho & \cdots & \rho & 1 \end{bmatrix}}_{N_t}, \mathbf{R}_l = \underbrace{\begin{bmatrix} 1 & \rho & \cdots & \rho \\ \rho & 1 & \ddots & \vdots \\ \vdots & \ddots & \ddots & \rho \\ \rho & \cdots & \rho & 1 \end{bmatrix}}_{N_r} \tag{11.6}$$

where ρ $(0 \leq \rho \leq 1)$ stands for the correlation coefficient between any two transmit antennas or receive antennas. When $\rho = 0$, the transmit antennas or the receive antennas are uncorrelated.

11.2.3 Iterative receiver structures

At the n_dth $(0 \leq n_d \leq N_d - 1)$ time slot and the n_cth $(0 \leq n_c \leq N_c - 1)$ sub-carrier, the received signal vector at the N_r antennas

$$\mathbf{y}[n_d, n_c] = \mathbf{H}[n_d, n_c]\,\mathbf{x}[n_d, n_c] + \boldsymbol{\eta}[n_d, n_c] \tag{11.7}$$

where $\boldsymbol{\eta}[n_d, n_c]$ is the additive noise vector with elements being i.i.d. AWGN with double-sided spectral density $N_0/2$. In addition, the channel estimation is first carried out on each sub-carrier by using the time-multiplexed pilot channel, and then further averaged over adjacent sub-carriers in the frequency domain. The element, $\xi_{n_r,n_t}[n_c]$, of the tentative channel estimation matrix, $\boldsymbol{\xi}[n_c]$, for the n_cth sub-carrier in one packet is given by

$$\xi_{n_r,n_t}[n_c] = \frac{1}{\sqrt{\beta P_d} x^{(\text{pilot})}} y_{n_r}[N_d/2 + n_t, n_c] \tag{11.8}$$

where $y_{n_r}[n_d, n_c]$ is the n_rth element of $\mathbf{y}[n_d, n_c]$. To improve the reliability of channel estimation, $\xi_{n_r,n_t}[n_c]$ should be averaged over adjacent sub-carriers in the frequency domain. Using a sliding window average, the elements, $\tilde{h}_{n_r,n_t}[n_c]$, of the final channel

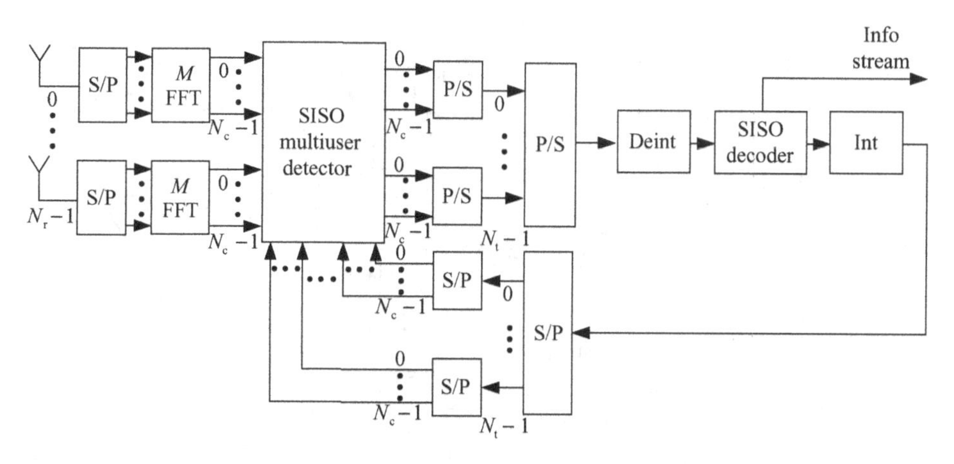

Figure 11.6 Receiver model of LSTF-a.

estimation matrix, $\tilde{\mathbf{H}}[n_c]$, for the n_cth sub-carrier is given by

$$\tilde{h}_{n_r,n_t}[n_c] = \frac{1}{2\lambda+1} \sum_{j=n_c-\lambda}^{j=n_c+\lambda} \xi_{n_r,n_t}[j] \tag{11.9}$$

where $2\lambda + 1$ is the size of the sliding window.

The receiver performance of MIMO multiplexing depends upon the efficiency of the signal processing in separating the signals from different transmit antennas. Basically, the problem of coded MIMO detection can be considered as a joint multiuser detection and decoding problem. Hence, the turbo processing can be efficiently used to develop a set of iterative multiuser detection algorithms that make a trade-off between performance and complexity.

In the receiver structure of LSTF-a as shown in Figure 11.6, the sampled OFDM symbols with M samples in every symbol interval from each receive antenna are S/P converted, down-converted from the N_c sub-carriers by an M-point FFT, and then sent to a SISO multiuser detector, which provides soft-valued estimates of the symbols from all the transmit antennas. The iterative MMSE-SIC multiuser detector consists of a feedback module which performs parallel interference cancellation and a feed-forward filter which performs interference suppression based on the MMSE criteria. In each iteration, the soft outputs from the decoders in the previous iteration are used to update the a priori likelihood ratio (LLR) values of the transmitted symbols. In the MMSE-SIC detector, these updated LLR values are then used to regenerate the soft estimates in the parallel interference cancellation, and also to calculate the MMSE filter coefficients. After the detector, the signals are P/S converted, deinterleaved, and then decoded by an SISO decoder. After that, the soft LLR outputs from the decoders are sent through interleavers and S/P converters as the a priori LLR values for the SISO multiuser detector in the next iteration. The receiver structure of the LSTF-b is shown in Figure 11.7, where at the output of the detector all soft-valued sequences for N_t codewords are independently P/S converted, deinterleaved, and then decoded by N_t parallel decoders. After that, soft

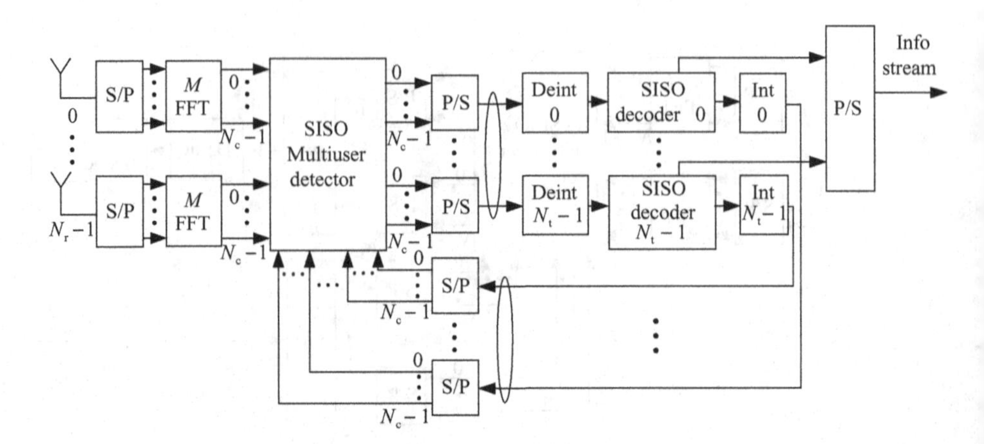

Figure 11.7 Receiver model of the LSTF-b.

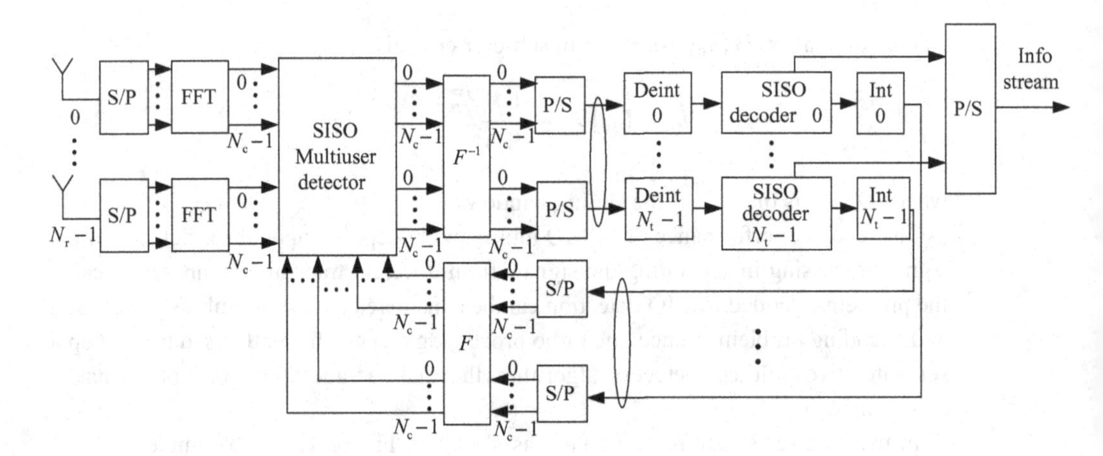

Figure 11.8 Iterative receiver of the LSTF-c.

outputs of the N_t decoders are independently interleaved, S/P converted and then sent to the detector. The receiver structure of LSTF-c is described in Figure 11.8, which is obtained from the LSTF-b receiver by inserting the distribution function F between the detector and S/P converters, and inserting the inverse distribution function F^{-1} between the detector and P/S converters. Note that, in the iterative receiver, each of the sub-streams is independently interleaved to facilitate convergence. This feature of the receiver is to ensure good convergence characteristics for the iterative processing. The SISO multiuser detector constitutes a major part of the overall complexity of the iterative receivers.

11.3 MMSE-SIC detection

For simplicity, the indices, n_d and n_c, of the symbol slots and sub-carriers in $\mathbf{x}[n_d, n_c]$, $\mathbf{y}[n_d, n_c]$, $\mathbf{H}[n_d, n_c]$ and $\eta[n_d, n_c]$, are omitted in the following. In the

MMSE-SIC detection, it is assumed that the channel response matrix is perfectly known in the receiver by using a channel estimation algorithm. The coded bit vector $\left(x_0^{(0)}, x_0^{(1)}, x_1^{(0)}, x_1^{(1)}, \ldots, x_{N_t-1}^{(0)}, x_{N_t-1}^{(1)}\right)$ from the N_t transmit antennas is mapped to the symbol vector $\mathbf{x} = (x_0, x_1, \ldots, x_{N_t-1})$, where the element x_{n_t} corresponds to one of the four possible signal points in the complex constellation of QPSK modulation. In the MMSE-SIC detector, based on the a priori LLR sequence, $\Lambda_A(2n_t + j) = \ln \Pr(x_{n_t}^{(j)} = 1)/\Pr(x_{n_t}^{(j)} = 0)$, $j = 0, 1$, of the coded bit vector given by the channel decoders at the previous iteration, the expectation and variance of the symbol x_{n_t} transmitted from the n_tth transmit antenna are given by

$$\begin{aligned} E\{x_{n_t}\} &= \sum_{\tilde{x} \in \Phi} \tilde{x} \Pr(x_{n_t} = \tilde{x}) \\ &= \sum_{\tilde{x} \in \Phi} \tilde{x} \prod_{j=0}^{1} \Pr\left(x_{n_t}^{(j)} = \tilde{x}^{(j)}\right) \\ &= \sum_{\tilde{x} \in \Phi} \tilde{x} \prod_{j=0}^{1} \frac{(1 - \tilde{x}^{(j)}) + \tilde{x}^{(j)} \exp(\Lambda_A(2n_t + j))}{1 + \exp(\Lambda_A(2n_t + j))} \\ &= \sum_{\tilde{x} \in \Phi} \tilde{x} \prod_{j=0}^{1} [1 + \exp((1 - 2\tilde{x}^{(j)}) \cdot \Lambda_A(2n_t + j))]^{-1} \end{aligned} \tag{11.10}$$

and

$$\begin{aligned} \mathrm{Var}(x_{n_t}) &= \sum_{\tilde{x} \in \Phi} |\tilde{x}|^2 \Pr(x_{n_t} = \tilde{x}) - |E\{x_{n_t}\}|^2 \\ &= \sum_{\tilde{x} \in \Phi} |\tilde{x}|^2 \prod_{j=0}^{1} [1 + \exp((1 - 2\tilde{x}^{(j)}) \cdot \Lambda_A(2n_t + j))]^{-1} - |E\{x_{n_t}\}|^2 \end{aligned} \tag{11.11}$$

where $\tilde{x}^{(j)}$ denotes the corresponding jth binary bit in symbol $\tilde{x}$. For the first iteration, assuming equally likely coded bits, i.e. no a priori information available, one obtains $\Lambda_A(2n_t + j) = 0$.

The soft estimate of the symbol vector $\mathbf{x}$ is formed as $E\{\mathbf{x}\} = (E\{x_1\}, E\{x_2\}, \ldots, E\{x_{N_t-1}\})$, and for each transmit antenna n_t, the soft interference cancellation is performed on the received signal vector $\mathbf{y}$, to obtain the interference-reduced signal

$$\tilde{\mathbf{y}}_{n_t} = \mathbf{y} - \mathbf{H}E\{\mathbf{x}\}|_{E\{x_{n_t}\}=0} = \mathbf{H}(\mathbf{x} - E\{\mathbf{x}\}|_{E\{x_{n_t}\}=0}) + \eta \tag{11.12}$$

In order to further suppress the residual interference in $\tilde{\mathbf{y}}_{n_t}$, an instantaneous linear MMSE filter is applied to $\tilde{\mathbf{y}}_{n_t}$, to obtain

$$z_{n_t} = \omega_{n_t}^* \tilde{\mathbf{y}}_{n_t} \tag{11.13}$$

where $\omega_{n_t}^*$ stands for the filter coefficient vector with N_r elements, and it minimizes the mean square error, $E\{|x_{n_t} - z_{n_t}|^2\} = E\{|x_{n_t} - \omega_{n_t}^* \tilde{\mathbf{y}}_{n_t}|^2\}$, between the transmitted symbol x_{n_t} and the filter output z_{n_t}. Solving the zero point of the differential of this mean squared error with respect to $\omega_{n_t}^*$, one obtains

$$\begin{aligned}
\omega_{n_t}^* &= E\left\{x_{n_t}\tilde{\mathbf{y}}_{n_t}^*\right\} E\left\{\tilde{\mathbf{y}}_{n_t}\tilde{\mathbf{y}}_{n_t}^*\right\}^{-1} \\
&= E\left\{x_{n_t}\left[\left(\mathbf{x}^* - E\{\mathbf{x}^*\}|_{E\{x_{n_t}\}=0}\right)\mathbf{H}^* + \boldsymbol{\eta}^*\right]\right\} \\
&\quad \times E\left\{\left[\mathbf{H}\left(\mathbf{x} - E\{\mathbf{x}\}|_{E\{x_{n_t}\}=0}\right) + \boldsymbol{\eta}\right]\left[\left(\mathbf{x}^* - E\{\mathbf{x}^*\}|_{E\{x_{n_t}\}=0}\right)\mathbf{H}^* + \boldsymbol{\eta}^*\right]\right\}^{-1} \\
&= E\left\{x_{n_t}\mathbf{x}^* - x_{n_t}E\{\mathbf{x}^*\}|_{E\{x_{n_t}\}=0}\right\} \\
&\quad \times \mathbf{H}^*\left\{E\left\{\mathbf{H}(\mathbf{x} - E\{\mathbf{x}\}|_{E\{x_{n_t}\}=0})(\mathbf{x}^* - E\{\mathbf{x}^*\}|_{E\{x_{n_t}\}=0})\mathbf{H}^*\right\} + E\left\{\boldsymbol{\eta}\boldsymbol{\eta}^*\right\}\right\}^{-1} \\
&= E\left\{|x_{n_t}|^2\right\}\mathbf{e}^*\mathbf{H}^*\left\{\mathbf{H}E\left\{\left(\mathbf{x} - E\{\mathbf{x}\}|_{E\{x_{n_t}\}=0}\right)\left(\mathbf{x}^* - E\{\mathbf{x}^*\}|_{E\{x_{n_t}\}=0}\right)\right\}\mathbf{H}^* + N_0\mathbf{I}_{N_t}\right\}^{-1} \\
&= E_s\mathbf{e}^*\mathbf{H}^*\left\{\mathbf{H}\,\mathrm{diag}\{\mathrm{Var}(x_0), \ldots, \mathrm{Var}(x_{n_t-1}),\right. \\
&\quad \left.\times\, \mathrm{Var}(x_{n_t+1}), \ldots, \mathrm{Var}(x_{N_t-1})\}\mathbf{H}^* + N_0\mathbf{I}_{N_t}\right\}^{-1}
\end{aligned} \tag{11.14}$$

where $E_s = E\{|x_{n_t}|^2\} = P_d T_e$ is the energy of each modulated data symbol, $\mathbf{e}$ denotes a vector with the n_tth element being 1 and all the other $N_t - 1$ elements being 0, $\mathrm{diag}\{\cdot\}$ stands for a diagonal matrix, and $\mathbf{I}_{n_t}$ is an identity matrix with N_t rows and columns.

Consider two special cases of the MMSE filter: in the absence of the a priori information, i.e. in the first iteration, the variances $\mathrm{Var}(x_{n_t}) = E_s$ make $\omega_{n_t}^* = E_s\mathbf{e}^*\mathbf{H}^*\{E_s\mathbf{H}\mathbf{H}^* + N_0\mathbf{I}_{N_t}\}^{-1}$, and then (11.13) is the output of the well-known MMSE filter for the n_tth transmit antenna; in the opposite case of genie a priori knowledge, the variances $\mathrm{Var}(x_{n_t}) = 0$ make $\omega_{n_t}^* = \frac{E_s}{E_s+N_0}\mathbf{e}^*$, and then (11.13) is simply a scale version of the matched filter for the n_tth transmit antenna after perfect cancellation of the interference from other transmit antennas.

The instantaneous MMSE filtering provides an efficient and accurate method of computing the extrinsic LLR, which is vital to the iterative receiver. Conditioned on x_{n_t}, it is shown in [1] that the soft instantaneous MMSE filter output z_{n_t} is well approximated as a Gaussian random variable with mean $\mu_{n_t}x_{n_t}$ and variance $\sigma_{n_t}^2$ given respectively by

$$\begin{aligned}
\mu_{n_t} &= E\left\{z_{n_t}x_{n_t}^*\right\}/E\left\{|x_{n_t}|^2\right\} \\
&= \omega_{n_t}^*\mathbf{H}\mathbf{e}/E_s \\
&= \mathbf{e}^*\mathbf{H}^*\left\{\mathbf{H}\,\mathrm{diag}\{\mathrm{Var}(x_0), \ldots, \mathrm{Var}(x_{n_t-1}), E_s, \mathrm{Var}(x_{n_t+1}), \ldots, \mathrm{Var}(x_{n_t-1})\}\mathbf{H}^*\right. \\
&\quad \left. + N_0\mathbf{I}_{n_t}\right\}\mathbf{H}\mathbf{e}
\end{aligned} \tag{11.15}$$

and

$$\begin{aligned}
\sigma_{n_t}^2 &= \mathrm{Var}(z_{n_t}) \\
&= E\{|z_{n_t}|^2\} - \mu_{n_t}^2 = \mu_{n_t} - \mu_{n_t}^2
\end{aligned} \tag{11.16}$$

Then the extrinsic LLR sequence of the coded bit vector given by the MMSE-SIC detector is

$$
\begin{aligned}
\Lambda_{\mathrm{D}}(2n_{\mathrm{t}}+j) &= \ln\frac{\Pr\left(x_{n_{\mathrm{t}}}^{(j)}=1|z_i\right)}{\Pr\left(x_{n_{\mathrm{t}}}^{(j)}=0|z_i\right)} - \ln\frac{\Pr\left(x_{n_{\mathrm{t}}}^{(j)}=1\right)}{\Pr\left(x_{n_{\mathrm{t}}}^{(j)}=0\right)} \\
&= \ln\frac{\sum\limits_{x_{n_{\mathrm{t}}}\in\Phi,x_{n_{\mathrm{t}}}^{(j)}=1}\Pr(x_{n_{\mathrm{t}}}|z_{n_{\mathrm{t}}})}{\sum\limits_{x_{n_{\mathrm{t}}}\in\Phi,x_{n_{\mathrm{t}}}^{(j)}=0}\Pr(x_{n_{\mathrm{t}}}|z_{n_{\mathrm{t}}})} - \Lambda_{\mathrm{A}}(2n_{\mathrm{t}}+j) \\
&= \ln\frac{\sum\limits_{x_{n_{\mathrm{t}}}\in\Phi,x_{n_{\mathrm{t}}}^{(j)}=1}\Pr(z_{n_{\mathrm{t}}}|x_{n_{\mathrm{t}}})\Pr(x_{n_{\mathrm{t}}})}{\sum\limits_{x_{n_{\mathrm{t}}}\in\Phi,x_{n_{\mathrm{t}}}^{(j)}=0}\Pr(z_{n_{\mathrm{t}}}|x_{n_{\mathrm{t}}})\Pr(x_{n_{\mathrm{t}}})} - \Lambda_{\mathrm{A}}(2n_{\mathrm{t}}+j) \\
&= \ln\frac{\sum\limits_{x_{n_{\mathrm{t}}}\in\Phi,x_{n_{\mathrm{t}}}^{(j)}=1}\exp\left(-\left|z_{n_{\mathrm{t}}}-\mu_{n_{\mathrm{t}}}x_{n_{\mathrm{t}}}\right|^2\Big/\sigma_{n_{\mathrm{t}}}^2+\sum\limits_{k=0}^{1}x_{n_{\mathrm{t}}}^{(k)}\Lambda_{\mathrm{A}}(2n_{\mathrm{t}}+k)\right)}{\sum\limits_{x_{n_{\mathrm{t}}}\in\Phi,x_{n_{\mathrm{t}}}^{(j)}=0}\exp\left(-\left|z_{n_{\mathrm{t}}}-\mu_{n_{\mathrm{t}}}x_{n_{\mathrm{t}}}\right|^2\Big/\sigma_{n_{\mathrm{t}}}^2+\sum\limits_{k=0}^{1}x_{n_{\mathrm{t}}}^{(k)}\Lambda_{\mathrm{A}}(2n_{\mathrm{t}}+k)\right)} \\
&\quad - \Lambda_{\mathrm{A}}(2n_{\mathrm{t}}+j)
\end{aligned}
\tag{11.17}
$$

After that, the extrinsic LLR sequence $\Lambda_{\mathrm{D}}(2n_{\mathrm{t}}+j)$ passes through the inverse distribution function F^{-1}, P/S converters, deinterleaver, and is then sent to the SISO MAP decoder as the a priori LLR input for the coded bit vector. While in the next iteration, the soft extrinsic LLR output of decoders is sent through the S/P converters, the distribution function F and interleavers are then used to regenerate the a priori LLR input of the MMSE-SIC detector for the coded bit vector.

Note that in a conventional SISO MAP decoder for convolutional code, only the information bits are concerned. Therefore, the conventional SISO MAP decoding algorithm is derived to provide LLR of each information bit. As a result, only information bits are recovered after decoding. In the SIC scheme, however, data replicas must be regenerated and then the coded bits should be recovered. Hence, only in the last iteration is the conventional SISO MAP decoder used to recover information bits, while in the iterations before the last one, the conventional SISO MAP decoder is modified to obtain LLR of the coded bits, and then the soft estimates of the coded bits can be generated in the MMSE-SIC detector.

11.4 Simulation results

Some representative simulation results are given in this section. Unless noted otherwise, the parameters given in Table 11.1 are used as follows. The number of transmit or receive antennas is $N_{\mathrm{t}}=N_{\mathrm{r}}=4$. A carrier frequency of 4.635 GHz with bandwidth of 50.75 MHz is assumed. The number of sub-carriers, N_{c}, is 384 with a sub-carrier spacing

Table 11.1 Simulation parameters for convolutional-coded LSTF architectures

Number of antennas, $N_t = N_r$	4
Carrier frequency	4.635 GHz
Bandwidth	50.75 MHz
Number of sub-carriers, N_c	384 (131.836 kHz sub-carrier spacing)
Number of IFFT/FFT points, M	512
Packet length	$N_d(= 24)$ OFDM symbol slots
Data modulation	QPSK
OFDM symbol duration, T (effective data, T_e+ guard interval, T_g)	9.259 μs (7.585 + 1.674 μs: 512 + 113 samples)
Channel coding/decoding	1/2-convolutional code/SISO MAP decoding
Number of iterations in joint detection and decoders	4
Channel model	$L(= 6)$-path exponential decay Rayleigh fading, r.m.s. delay spread $\sigma_{rms} = 0.23$ μs, Doppler frequency shift $f_D = 20$ Hz

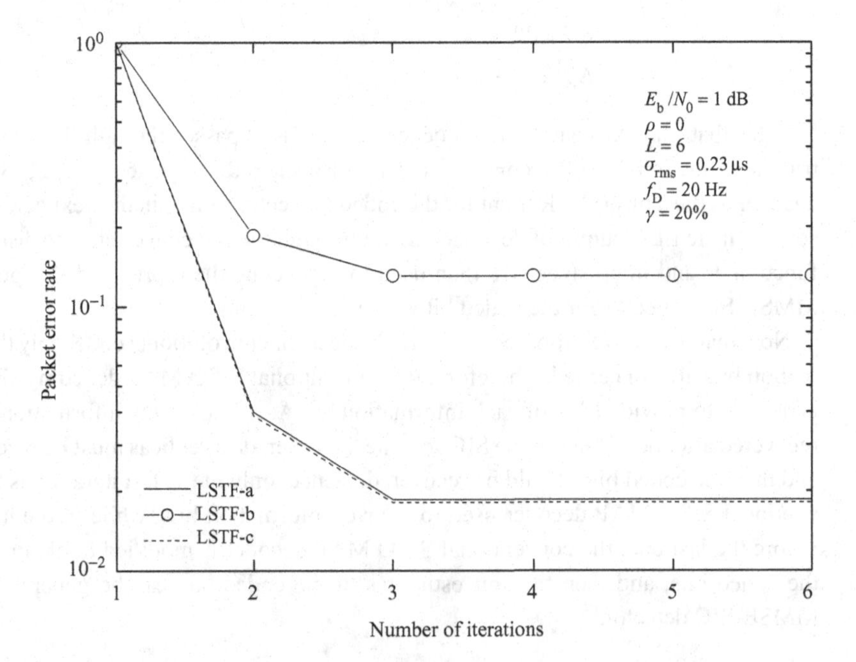

Figure 11.9 PER performance vs. the number of iterations in the joint detection and decoding.

of 131.836 kHz. The number of IFFT points, M, is set to 512, and in each packet there are $N_d = 24$ OFDM symbols. QPSK modulation is adopted. To prevent ISI, a guard interval of $T_g = 1.674$ μs for cyclic prefix is inserted between two adjacent OFDM symbols. The convolutional codes with coding rate $R = 1/2$, constraint length of 5 and generator polynomials $(23_8, 35_8)$ are adopted, and the codeword length is $2N_tN_cN_d = 55\,296$ bits in the LSTF-a but only $2N_cN_d = 18\,432$ bits in the LSTF-b and the LSTF-c. The number

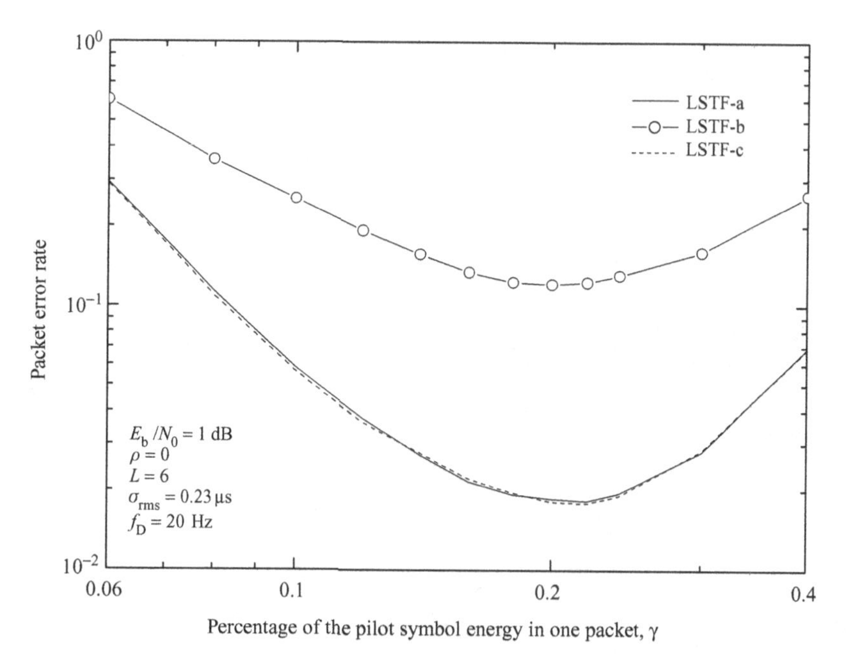

Figure 11.10 PER performance vs. the percentage of pilot channel energy in one packet.

of iterations in the joint MMSE-SIC detector and SISO MAP decoders is 4. The number of multipaths is $L = 6$, the r.m.s. delay spread $\sigma_{\text{rms}} = 0.23$ μs and the maximum Doppler frequency shift f_{D} is 20 Hz. The fading correlation coefficient ρ between antennas is 0. In channel estimation, the average window size in frequency domain is $2\lambda + 1 = 3$, and the percentage, γ, of the pilot channel energy in one packet is set to be 20 percent. Finally, because the average energy per transmitted symbol after cyclic insertion is $P_{\text{d}}T$ and the elements in each channel response matrix $\mathbf{H}[n_{\text{d}}, n_{\text{c}}]$ have unity variances, the average symbol power of every receive antenna is $N_{\text{t}}P_{\text{d}}T/(1-\gamma)$, carrying $2N_{\text{t}}R$ information bits. Then the average SNR of every receive antenna per information bit is defined as $E_{\text{b}}/N_0 = P_{\text{d}}T/2N_0R(1-\gamma)$, which is set to 1 dB.

First, the packet error rate (PER) performance is shown in Figure 11.9 as a function of the number of iterations in the joint MMSE-SIC detector and SISO MAP decoders. It can be seen from the figure that the LSTF-c outperforms the LSTF-b significantly, and has almost the same PER performance as the LSTF-a, because the LSTF-c can jointly exploit the available spatial, temporal and frequency diversity like the LSTF-a. Generally, the PER performance of all the three LSTF architectures improves dramatically as the number of iterations increases from 1 to 3. However, when the number of iterations increases further beyond 3, the PER performance of the three LSTF architectures remain flat.

Figure 11.10 investigates the effect of the percentage, γ, of the pilot channel energy in one packet on the system performance. The total transmit power and the variance of channel noise are fixed to meet E_{b}/N_0 of 1 dB. It can be seen that, for all the three LSTF architectures, when γ is small, the pilot channel has small power and this results in poor

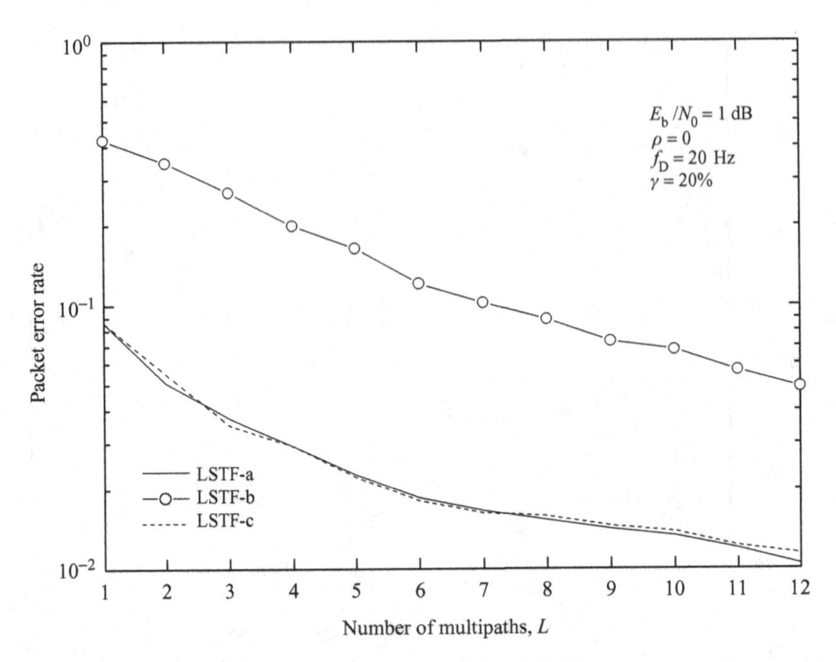

Figure 11.11 PER performance vs. the number of multipaths.

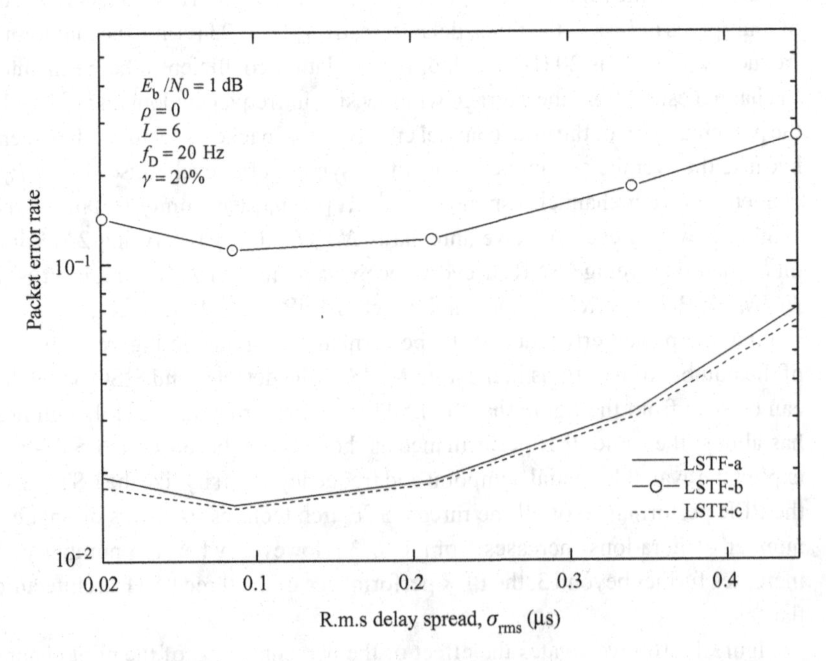

Figure 11.12 PER performance vs. the r.m.s. delay spread.

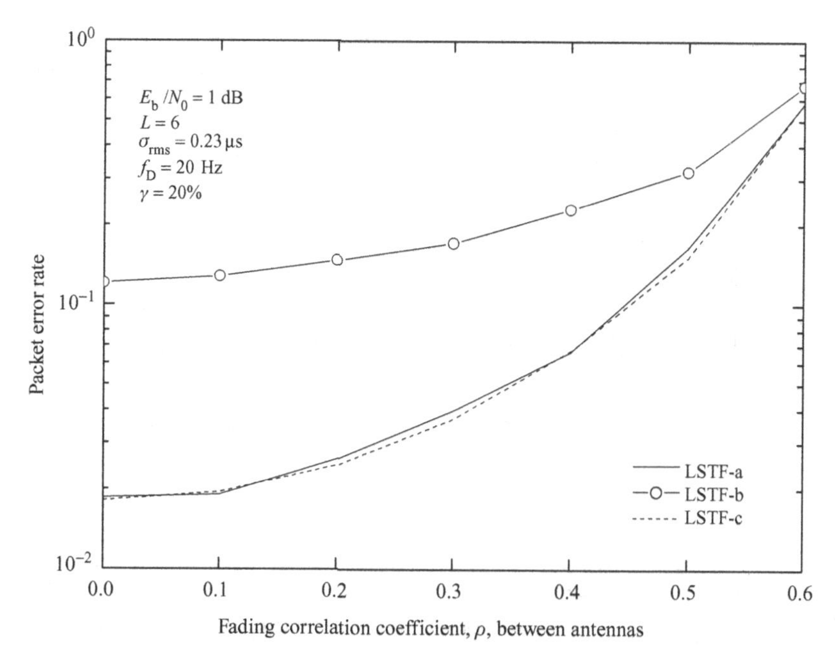

Figure 11.13 PER performance vs. the fading correlation coefficient between antennas.

channel estimation. Thus, the system performance degrades. When γ increases, PER performance improves as the quality of channel estimation improves, and PER reaches a minimum value for a particular value of γ. Further increasing γ beyond that value increases PER. This is because too much power is assigned to the pilot channel and small power to data channel so that data channel becomes vulnerable to channel noise. Thus, γ should be in the range [16%, 24%] to provide near optimum system performance. On the other hand, it can be seen that, for a given value of γ, the LSTF-c outperforms the LSTF-b significantly, and has almost the same PER performance as the LSTF-a. This is consistent with Figure 11.9.

The PER performance of the three LSTF architectures as a function of the number of multipaths, L, is shown in Figure 11.11. For different values of L from 2 to 12, the maximum time delays are set to different values to make the corresponding r.m.s. delay spreads, σ_{rms}, be approximately 0.23 µs. It can be seen that the PER performance of all the three LSTF architectures decreases almost linearly as L increases. This is because a larger L makes the channel more frequency selective, and thus achieves much more frequency diversity gain. Therefore, the instantaneous received signal power tends to vary less, i.e. the time variations in the received signal approach a constant level. Furthermore, for a given value of L, the LSTF-c achieves almost the same performance as the LSTF-a, and outperforms the LSTF-b significantly.

Figure 11.12 illustrates the PER as a function of the r.m.s. delay spread σ_{rms}. It can be seen that for all the three LSTF architectures, when σ_{rms} is very small, the high correlation between sub-carriers results in low frequency diversity gain. Thus, the system performance degrades. When σ_{rms} increases from a very small value, PER performance

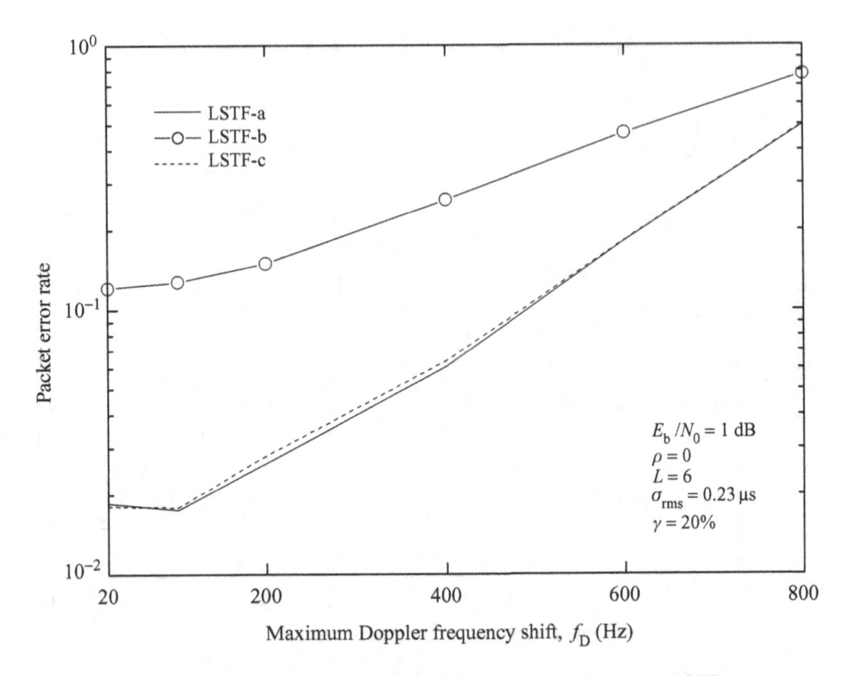

Figure 11.14 PER performance vs. the maximum Doppler frequency shift.

improves as the frequency diversity gain increases, and PER reaches a minimum value for a particular value of σ_{rms}. Further increasing σ_{rms} beyond that value increases PER. This is because the decreasing correlation between sub-carriers results in worse channel estimation error in the frequency domain.

Figure 11.13 investigates the effect of the fading correlation coefficient, ρ, between antennas on the system performance. It can be seen that, given ρ, the PER performance of the LSTF-c is almost the same as that of the LSTF-a, but much better than that of the LSTF-b. For all the three LSTF architectures, as ρ increases, the system performance degrades, and the performance difference between the LSTF-c (or LSTF-a) and the LSTF-b becomes smaller and smaller. This is because the spatial diversity gain decreases with increasing correlation between antennas.

Although in most cases the high-speed LSTF architectures experience a slow fading channel, when occasionally a large Doppler frequency shift occurs, the channel variation in one packet duration is serious. For example, for a carrier frequency of 4.635 GHz, when a mobile terminal moves at a speed of 200 km/hour, the resultant maximum Doppler frequency shift can be almost as high as 800 Hz. Therefore, it is necessary to investigate the effect of Doppler frequency shift f_D. Figure 11.14 shows the PER performance as a function of the maximum Doppler frequency shift. It can be seen that for all the three LSTF architectures, when the Doppler frequency shift is small (i.e. less than 100 Hz), the PER is almost flat. When f_D increases from 100 Hz, the system performance degrades severely. This is because the channel estimation cannot track the channel variation well. On the other hand, for a given value of f_D, the LSTF-c achieves the same performance as the LSTF-a, and outperforms the LSTF-b significantly. This is consistent with other figures. When f_D increases to 800 Hz, compared to that of the LSTF-b the performance gains of the LSTF-a and the LSTF-c become smaller due to the increasing channel estimation error.

11.5 Summary

In this chapter, the iterative receiver structure is studied for the novel LSTF architecture (i.e. LSTF-c) with each independent codeword threaded in the three-dimensional space-time-frequency transmission resource array for high-speed OFDM MIMO multiplexing over frequency-selective fading channels. This iterative receiver structure consists of the joint MMSE-SIC detector and SISO MAP convolutional decoder with time-multiplexed pilot-aided channel estimation. The following conclusions are drawn.

(1) The proposed LSTF-c architecture can get almost the same performance as the LSTF-a applying coding across the whole information stream, and considerably outperform the LSTF-b assuming each independent codeword is being transmitted through a dedicated antenna.
(2) Four iterations in joint MMSE-SIC detector and SISO MAP convolutional decoders are sufficient to achieve stable performance.
(3) To provide near optimum performance, the percentage of the pilot channel energy in one packet should be in the range [16%, 24%].
(4) As long as the Doppler frequency shift is less than 100 Hz, the performances of the LSTF architectures are flat, but they degrade dramatically when the Doppler frequency shift is large.

References

[1] X. Wang and H. V. Poor, "Iterative (Turbo) soft interference cancellation and decoding for coded CDMA," *IEEE Trans. Commun.*, vol. 47, pp. 1046–1061, July 1999.
[2] S. L. Ariyavisitakul, "Turbo space-time processing to improve wireless channel capacity," *IEEE Trans. Commun.*, vol. 48, no. 8, pp. 1347–1359, Aug. 2000.
[3] B. Vucetic and J. Yuan, *Space-Time Coding*. New York: John Wiley & Sons, 2003.
[4] V. S. Somayazulu, J. R. Foerster and S. Roy, "Design challenges for very high data rate UWB systems," in *Conference Record of the Thirty-Sixth Asilomar Conference on Signals, Systems and Computers*, vol. 1, pp. 717–721, Nov. 2002.
[5] H. Bolcskei, D. Gesbert and A. J. Paulraj, "On the capacity of OFDM-based spatial multiplexing systems," *IEEE Trans. Commun.*, vol. 50, no. 2, pp. 225–234, Feb. 2002.

12 Sub-packet transmission for hybrid ARQ systems

ARQ is a flexible and efficient technique for data transmissions. In hybrid ARQ, sub-packet schemes are more attractive for systems with burst errors than complete packet schemes. Although sub-packet schemes were proposed in ARQ systems, optimum sub-packet transmission is more effective to maximize throughput in a dynamic channel. Since convolutional codes are burst errors in decoding, the optimum sub-packet can be applied to convolutional codes. This chapter investigates the performance of sub-packet transmission for convolutionally coded systems. An efficient method is proposed to estimate the optimum number of sub-packets, and adaptive sub-packet schemes, i.e. schemes that enable a system to employ different optimum numbers of sub-packets under various conditions, are suggested to achieve maximum throughput in the system. Numerical and simulation results show that the adaptive sub-packet scheme is very effective for the convolutionally coded hybrid ARQ system, and it can provide higher throughput, smaller delay and lower dropping rate than complete packet schemes. Moreover, the adaptive sub-packet scheme can be flexibly used with packet combining techniques to further improve system throughput.

12.1 Introduction

In high-speed data communication systems, information sequences are transmitted usually in packets with fixed length. At a receiver, error correction and detection are carried out on the whole packet. If the packet is found in error, the receiver sends a request to the transmitter via a feedback channel, and then the whole packet is retransmitted. However, such a conventional complete-packet ARQ scheme is inefficient in the presence of burst errors. As shown in Figure 12.1, when the whole packet is divided into a few sub-packets, all or a majority of the bit errors may be located in one sub-packet, leaving other sub-packets error-free. But in the complete packet scheme, the whole packet will be retransmitted no matter how bit errors are distributed. This situation can be improved if sub-packet schemes are employed. In sub-packet transmissions, only those sub-packets that include errors need to be retransmitted.

Although sub-packet schemes were proposed in ARQ systems [1], there is no work that discusses the optimum number of sub-packets, and a fixed sub-packet scheme only provides improved throughput in a small range of the signal-to-noise ratio (SNR).

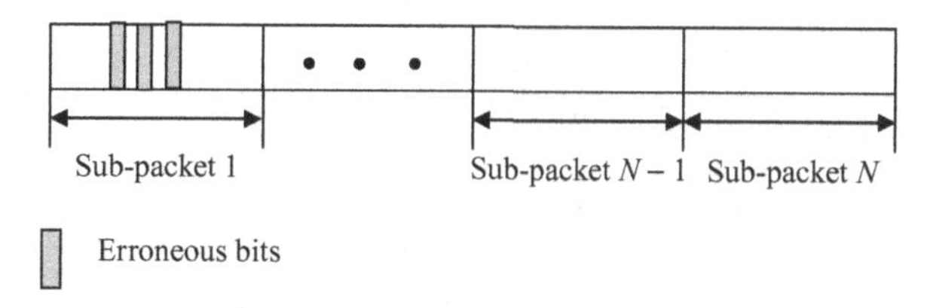

Figure 12.1 Burst packet error patterns.

Ascertaining the optimum number of sub-packets in one transmission is the objective of this chapter. Exploiting the burst error property of convolutional codes, an adaptive sub-packet scheme is proposed to provide the highest throughput in a dynamic channel. The chapter is organized as follows. In Section 12.2, the system model and sub-packet schemes are introduced. In Section 12.3, PER performance of FEC codes is analyzed. Section 12.4 is devoted to finding the optimum number of sub-packets for a system to maximize the throughput. An efficient algorithm is proposed to estimate the optimum number of sub-packets at variable SNR. Representative analytical and simulation results of throughput, delay and dropping rate are also presented. Finally, conclusions are drawn in Section 12.5.

12.2 System overview

In a hybrid ARQ system, the information data blocks are encoded with error detection and FEC codes. The coded bit stream modulates a carrier using binary phase shift keying (BPSK). After passing through an additive white Gaussian noise (AWGN) channel and being demodulated, the baseband signals are decoded by *error correction* and checked by *error detection*. Finally, acknowledgment signals, i.e. positive acknowledgement (ACK) and negative acknowledgment (NAK), are transmitted back to the transmitter over an ideal feedback channel, which is assumed to be error-free.

Suppose that an information data block of L-bit length is transmitted with ARQ techniques. The whole information data block is segmented into N sub-blocks as illustrated in Figure 12.2(a), which consists of N error detection encoders and one FEC encoder. Each detection code is an (n, k) systematic block code where n is the length of one codeword (one sub-packet) and k is the length of information bits in each sub-block ($k = L/N$ is an integer). That is, each sub-block passes through an error detection encoder that produces a sub-packet of n bits. All sub-packets are then time-multiplexed into a whole packet with length $L_{\mathrm{cb}} = n \cdot N$ for the convolutional encoder with possible punctuation. The final output is a channel-coded word of length L_{f}. The sub-packet structure is shown in Figure 12.2(b) where each sub-packet includes a systematic error detection code of $n - k$ bits. In the following analysis, two different PERs are used in the N sub-packet schemes. One is the sub-packet error rate (SPER), denoted as $P_{\mathrm{sp}}(N)$, and another one is the whole packet error rate, i.e. the codeword error rate of the FEC code, denoted as

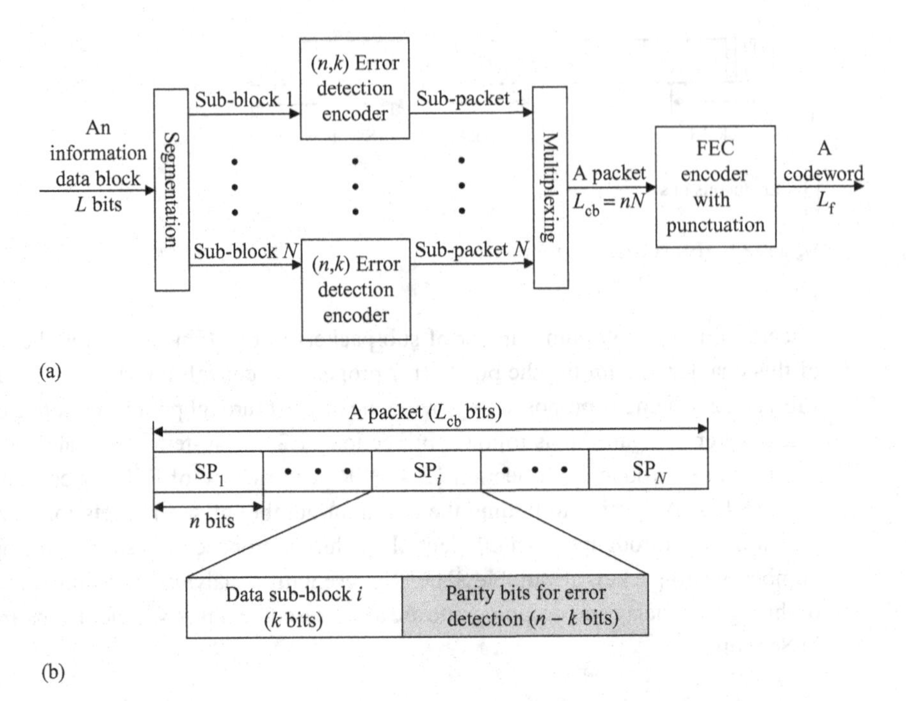

Figure 12.2 Sub-packet schemes. (a) Encoder structure of sub-packet scheme ($K = L/N$), (b) sub-packet structure.

$P_w(N)$. Note that the conventional complete-packet scheme can be treated as a special case of the sub-packet scheme with $N = 1$. That is, the PER of the complete-packet scheme is simply represented by $P_w(1)$ or $P_{sp}(1)$.

A systematic cyclic redundancy check (CRC) code with 16 parity bits ($L_{crc} = 16$) in CCITT standards is used as the error detection code [2]. Its generator polynomial is $g(D) = D^{16} + D^{12} + D^5 + 1$, where D stands for bit delay operation. The CRC code is easy to implement, but it is not easy to evaluate its performance in traditional methods such as weight distribution and minimum distance. However, there are some other meaningful measures for the performance of CRC codes, such as the "burst error detection capability" and "error detection coverage (λ)." A binary CRC code with L_{crc} parity bits can detect all burst errors of length L_{crc} or less and detect the fraction $\lambda = 1 - 2^{-L_{crc}}$ of all the error patterns. On a binary symmetric channel, when the code-dimension k and the error probability are large, the undetected error probability of CRC codes approaches $2^{-L_{crc}}$. Thus, the performance of CRC codes is mainly determined by L_{crc}. The longer L_{crc}, the better the performance. A CRC code with $L_{crc} = 16$ can provide adequate detection for most applications.

A 1/2-rate convolutional code is employed with constraint length $m_c = 4$ (or memory length of $m_c - 1 = 3$) and generator polynomial $G(D) = [1 + D + D^3, 1 + D + D^2 + D^3]$, or (64, 74) [2]. Higher-rate codes are obtainable by puncturing from the low-rate code [3]. For example, the compatible 2/3-rate and 5/6-rate codes can be obtained from the 1/2-rate code by means of the following puncturing matrices:

$$\begin{bmatrix} 1 & 1 & 0 & 1 \\ 0 & 1 & 1 & 1 \end{bmatrix} \quad \text{and} \quad \begin{bmatrix} 1\ 1\ 0\ 1, & 1\ 0\ 0\ 1, & 1\ 1\ 0\ 1, & 1\ 0\ 0\ 1, & 1\ 0\ 0\ 1 \\ 0\ 1\ 1\ 0, & 0\ 1\ 1\ 1, & 0\ 1\ 1\ 0, & 0\ 1\ 1\ 1, & 0\ 1\ 1\ 0 \end{bmatrix}$$

respectively, where "1" means that the code bit will be transmitted and "0" means that the code bit will be dropped and not transmitted.

Consider the ARQ protocol of sub-packet schemes. As shown in Figure 12.2(b), sub-packets are independent of each other in CRC coding processes. The N sub-packets named $\{\mathrm{SP}_1, \mathrm{SP}_2, \ldots, \mathrm{SP}_N\}$ are time-multiplexed into a single whole packet to be encoded by the FEC coder. At the receiver side, Viterbi decoding is carried out, and the syndrome of CRC codes is calculated for each sub-packet SP_i ($i = 1, 2, \ldots, N$). If there is no error in a sub-packet, the corresponding decoded sub-block is stored in a buffer and an ACK is sent to the transmitter. However, if there are some erroneous sub-packets after the decoding is finished, then the corresponding NAK is sent to the transmitter for retransmission of the erroneous sub-packets. All correctly detected sub-blocks are stored in a buffer, and sequentially forwarded to the higher layer. At the transmitter, since the codeword of the previous transmission has been stored, the corresponding parts in the codeword of the sub-packets to be retransmitted can be obtained and retransmitted to the receiver. At the receiver, the signals of the previous transmission are needed. The retransmitted parts will replace the erroneous parts in the previous codeword, and the whole codeword is decoded again. The maximum transmission times for a particular sub-packet are limited to M. If the sub-packet is still in error after M transmissions, then it will be dropped out.

12.3 Performance bound

The tangential sphere bound is used since it is a tight bound for binary linear codes [3–5]. Compared to the union bound [6], tangential sphere bound gets much closer to the actual performance at low to moderate SNR. The total number of codewords with Hamming weight d is defined as

$$S_d = \sum_{w=0}^{L_{\mathrm{cb}}} A_{w,d}, \quad d = 0, 1, \ldots, L_{\mathrm{f}} \tag{12.1}$$

where $A_{w,d}$ is the number of codewords with Hamming weight d, generated from information bits with Hamming weight w. Assuming that the variance of the additive channel noise is σ^2, the tangential sphere bound on the PER, P_p, is given by [5]

$$\begin{aligned} P_{\mathrm{p}} \leq \int_{-\infty}^{+\infty} \frac{\mathrm{d}z_1}{\sqrt{2\pi}\sigma} \mathrm{e}^{-z_1^2/2\sigma^2} \cdot \Bigg\{ 1 - \overline{\gamma}\left(\frac{L_{\mathrm{f}}-1}{2}, \frac{r_{z_1}^2}{2\sigma^2}\right) \\ + \sum_{\substack{d \\ \delta_d/2 < \alpha_d}} S_d \left[Q\left(\frac{\beta_d(z_1)}{\sigma}\right) - Q\left(\frac{r_{z_1}}{\sigma}\right) \right] \overline{\gamma}\left(\frac{L_{\mathrm{f}}-2}{2}, \frac{r_{z_1}^2 - \beta_d^2(z_1)}{2\sigma^2}\right) \Bigg\} \end{aligned} \tag{12.2}$$

where the sum is only taken over $\{d\}$ which satisfy $\delta_d/2 < \alpha_d$, and variables r_{z_1}, $\beta_d(z_1)$ and α_d are further defined as

$$\begin{cases} r_{z_1} = \left(1 - \frac{z_1}{\sqrt{L_f E_s}}\right) r \\ \beta_d(z_1) = \dfrac{r_{z_1}}{\sqrt{1 - \frac{\delta_d^2}{4L_f E_s}}} \cdot \dfrac{\delta_d}{2r} \\ \alpha_d = r\sqrt{1 - \dfrac{\delta_d^2}{4L_f E_s}} \end{cases} \tag{12.3}$$

where E_s is the energy per coded bit. Thus, the total energy of a codeword is $L_f E_s$. δ_d is defined as the Euclidean distance between two codewords in d different positions ($d \leq L_f$). For antipodal signals, $\delta_d = 2\sqrt{dE_s}$. Furthermore, $\overline{\gamma}(a, x)$ in (12.2) is the normalized incomplete gamma function, given by

$$\overline{\gamma}(a, x) = \frac{1}{\Gamma(a)} \int_0^x t^{a-1} e^{-t} dt, \quad a > 0, \quad x > 0 \tag{12.4}$$

where $\Gamma(a)$ is the gamma function, given by

$$\Gamma(a) = \int_0^{+\infty} t^{x-1} e^{-t} dt, \quad x \geq 0 \tag{12.5}$$

In (12.2), $Q(x)$ is the Q function, expressed as

$$Q(x) = \frac{1}{\sqrt{2\pi}} \int_x^{+\infty} e^{-t^2/2} dt, \quad x \geq 0 \tag{12.6}$$

By setting the derivative of (12.2) to zero, an optimal r can be found to achieve the tightest upper bounds [5].

To calculate the tangential sphere bound, it is necessary to evaluate (12.1) first. Although many efforts have been made by experts to find the distance spectra [7–10], it remains a problem for most codes. Fortunately, for convolutional codes with small constraint lengths, (12.1) can be quickly calculated [11] when the codeword is not too long. Since the computation complexity increases exponentially with the constraint length, the tangential sphere bound is only calculated for the 1/2-rate convolutional code with a constraint length of 4. Given $L_{cb} = 1200$, the length of the codeword is 2406, including 6 tail bits. Otherwise, an overflow will occur because the calculation involves very large values. Convolutional codes with other code rates, constraint lengths and code lengths will be studied as well.

12.4 Analysis and simulation

As explained in Section 12.1, sub-packet schemes are good schemes in the presence of burst errors. This section is devoted to finding the optimum sub-packet schemes to maximize system throughput. Consider an N-sub-packet scheme with an (n, k) error detection code. The effective code rate of the packet scheme is defined as

$$R_{\text{eff}} = \frac{L}{L_{\text{f}}} = \frac{N \cdot k}{L_{\text{f}}} = \frac{N \cdot (n - L_{\text{crc}})}{L_{\text{f}}} = \frac{L_{\text{cb}} - N \cdot L_{\text{crc}}}{L_{\text{f}}} \tag{12.7}$$

which accounts for the redundancy introduced by error detection codes. Define $\text{SNR}_{\text{b}} = E_{\text{b}}/N_0$ as the SNR per information bit, where E_{b} is the energy per information bit and N_0 is the power spectrum density of channel noise, and $\text{SNR}_{\text{c}} = R_{\text{eff}} \cdot \text{SNR}_{\text{b}}$ as the SNR per channel bit. Obviously, when all the other parameters are fixed, SNR_{c} (or R_{eff}) decreases as N increases. Given $L_{\text{crc}} = 16$ and N, one way to alleviate the effect of N on SNR_{c} is to apply sub-packet schemes in long-packet-length systems (large L_{cb}).

12.4.1 Optimum sub-packet schemes

Throughput is chosen as the criterion for optimizing the number of sub-packets, N. Consider a truncated system that employs a selective repeat-ARQ protocol with M maximum transmissions. The throughput is defined as [6]

$$\eta = R_{\text{eff}} \cdot [1 - P_{\text{sp}}(N)] = \frac{L_{\text{cb}} - N \cdot L_{\text{crc}}}{L_{\text{f}}} \cdot [1 - P_{\text{sp}}(N)] \tag{12.8}$$

When L_{cb}, L_{crc}, L_{f} (or the code rate) and SNR_{b} are given, the throughput only depends on N. When the number of sub-packets, N, is very large, R_{eff} tends to be small, so that the throughput η becomes small. When N is very small, however, the SPER $P_{\text{sp}}(N)$ tends to be large, so the throughput η becomes small as well. Therefore, there must exist a value of N at which the throughput takes a maximum value. The optimum number of sub-packets (N_{opt}) that achieves the highest throughput can be obtained by setting the derivative $\text{d}\eta/\text{d}N$ to zero. N_{opt} can then be obtained from

$$N = \frac{L_{\text{cb}} \cdot \frac{\text{d}P_{\text{sp}}(N)}{\text{d}N} + L_{\text{crc}} \cdot [1 - P_{\text{sp}}(N)]}{L_{\text{crc}} \cdot \frac{dP_{\text{sp}}(N)}{\text{d}N}} \tag{12.9}$$

If $P_{\text{sp}}(N)$ is given, then N_{opt} can be calculated from (12.9) directly. Unfortunately, it is difficult to obtain a closed form of $P_{\text{sp}}(N)$. Under the assumption of burst packet error patterns and long packet lengths, N_{opt} can be obtained from (12.9) by approximating $P_{\text{sp}}(N)$ with its average upper bound as follows.

For the hybrid ARQ system employing an N-sub-packet scheme, the error rate of the whole packet, $P_{\text{w}}(N)$, can be approximated by the tangential bound (12.2). Note that the input sequence to the FEC encoder is not pure information bits but N time-multiplexed CRC codewords. CRC encoders will change the weight distribution of the input sequence. As S_{d} in (12.1) is the number of FEC codewords with weight d, it is a function of the number of sub-packets. In addition, the effective code rate is a function of N. For long packet lengths, the effects of N on weight distribution and code rate are not significant. Moreover, the PER of convolutional codes is not sensitive to the change in SNR. Thus, $P_{\text{w}}(N) \approx P_{\text{w}}(1)$. Furthermore, since bit errors are in burst, $P_{\text{sp}}(N)$ can be approximated by $P_{\text{w}}(N)/N$:

$$P_{\text{sp}}(N) \approx \frac{P_{\text{w}}(N)}{N} \approx \frac{P_{\text{w}}(N)}{N} \approx \frac{P_{\text{p}}}{N} \tag{12.10}$$

Then, N_{opt} can be obtained from (12.9) and (12.10):

$$N_{\text{opt}} \approx \sqrt{\frac{L_{\text{cb}} \cdot P_{\text{p}}}{L_{\text{crc}}}} \tag{12.11}$$

Equation (12.11) clearly describes the relationship between N_{opt} and other system parameters. Firstly, N_{opt} is limited by the redundancy (L_{crc}) introduced by error detection codes. When the CRC is long (large L_{crc}), N_{opt} may take a value as small as one, which means that there is no benefit from sub-packet schemes due to too much redundancy introduced by the CRC. When L_{crc} is very small, N_{opt} becomes large. Thus, a small L_{crc} is required to render a sub-packet scheme advantageous. For CRC codes, however, L_{crc} should be large enough to ensure detection. Therefore, to balance both CRC detection capability and sub-packet scheme, an appropriate value of L_{crc} should be chosen. Second, when L_{crc} and $P_{\text{w}}(1)$ are given, N_{opt} increases with L_{cb}, which means more benefit from large-N sub-packet schemes in long-packet-length systems. Third, a lower bound of $P_{\text{w}}(1)$ can be obtained from (12.11) by setting $N_{\text{opt}} = 1$, and is given by

$$\underline{P_{\text{W}}} = \frac{L_{\text{crc}}}{L_{\text{cb}}} \tag{12.12}$$

Thus, a longer packet length L_{cb} can result in a smaller $\underline{P_{\text{W}}}$, which will increase the effective SNR_{b} region of sub-packet schemes (see (12.7)). In the SNR_{b} region where $P_{\text{w}}(1) < \underline{P_{\text{W}}}$, N_{opt} is set to one, which means that complete packet schemes are the best packet schemes in the region, instead of sub-packet schemes.

Given L_{crc} and L_{cb}, the optimum sub-packet scheme can be constructed as follows. First, a PER versus SNR_{b} curve is obtained from the tangential sphere bound (or simulation). Then $\underline{P_{\text{W}}}$ can be found from the curve and the corresponding SNR_{b} is taken as a threshold. For SNR_{b} where $P_{\text{w}}(1) \leq \underline{P_{\text{W}}}$, N_{opt} should be set to one, whereas for smaller SNR_{b} where $P_{\text{w}}(1) > \underline{P_{\text{W}}}$, N_{opt} is calculated by using (12.11).

The proposed method of constructing adaptive sub-packet schemes is verified with convolutional codes. Consider a convolutional code with $L_{\text{f}} = 2406$, $L_{\text{cb}} = 1200$, $m_{\text{c}} = 4$ and a code rate of 1/2, whose generator polynomial is (64, 74). The tangential sphere bound of PER is obtained and $\underline{P_{\text{W}}}$ is calculated from (12.12) with the corresponding SNR_{b} of about 5.0 dB. Therefore, when SNR_{b} is smaller than 5.0 dB, the optimum N_{opt} can be found by using (12.11), whereas when $\text{SNR}_{\text{b}} \geq 5$ dB, $N_{\text{opt}} = 1$. The adaptive sub-packet scheme (or N_{opt}) can be derived. Simulation results are also depicted in Figure 12.3 for comparison. It is seen that analytical N_{opt} derived from the tangential sphere bound is close to the simulated N_{opt} obtained from simulations at low or high SNR_{b}. At moderate SNR_{b}, however, it is larger than the simulated N_{opt}. This is because when SNR_{b} is low or high, the tangential sphere bound provides good estimation for PER, while at moderate SNR_{b}, this upper bound is much larger than the simulated PER. It is seen from Figure 12.3 that the optimum sub-packet scheme is not a constant-N scheme. Instead, N_{opt} changes with SNR_{b}. The scheme that enables a system to employ different values of N_{opt} under variable SNR_{b} is called "*an adaptive sub-packet scheme.*" Using a predefined table of N_{opt} at different SNR_{b}, the adaptive sub-packet scheme can be easily realized.

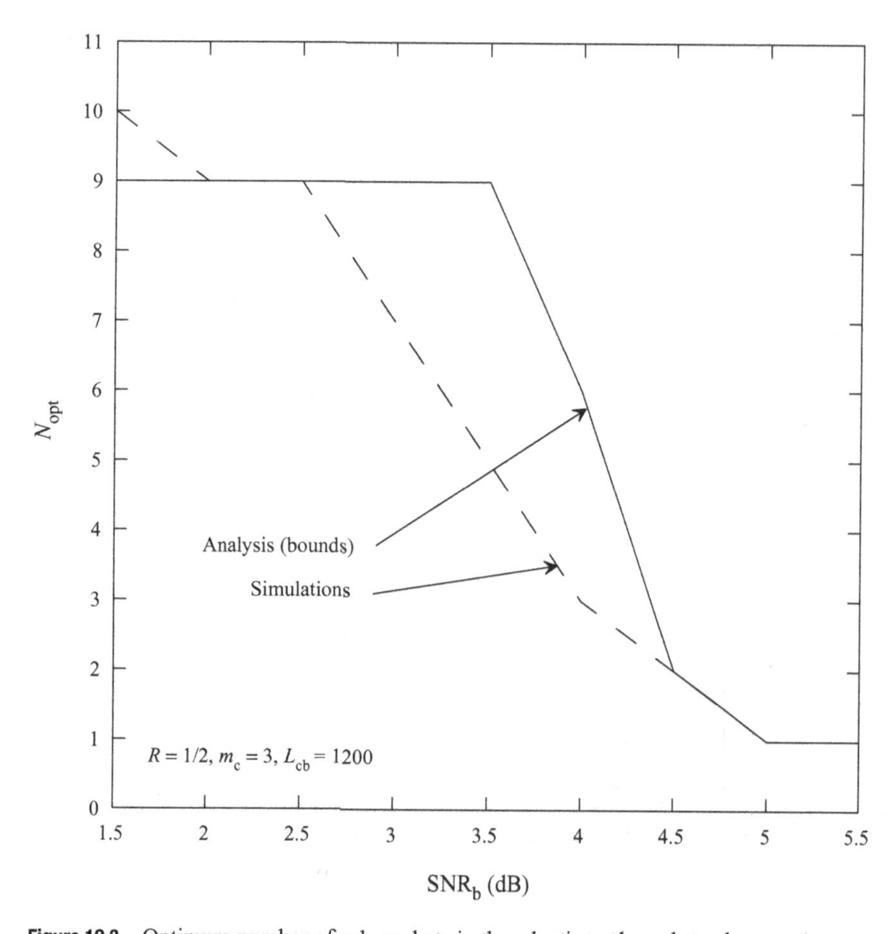

Figure 12.3 Optimum number of sub-packets in the adaptive sub-packet scheme.

Since at moderate SNR_b, N_{opt} derived from the tangential sphere bound provides a rough estimation for the real N_{opt}, it is necessary to investigate the system throughput with the adaptive optimum sub-packet scheme. The throughput of the complete packet scheme obtained from simulation is also plotted for comparison. Note that the optimum number of sub-packets is dependent on the SNR_b. It can be seen from Figure 12.4 that when SNR_b is small or moderate ($SNR_b \leq 3$ dB), the adaptive sub-packet scheme can significantly improve the system throughput compared to the complete packet scheme. Analytical and simulation results are very close even at moderate SNR_b where analytical N_{opt} is larger than simulated N_{opt} (see Figure 12.3). That is, performance degradation is small when N_{opt} is slightly over-estimated.

Simulated throughput is illustrated in Figure 12.5 with fixed-N and adaptive sub-packet schemes. It is seen that an adaptive sub-packet scheme always provides the highest throughput, whereas fixed-N sub-packet schemes provide the same throughput as adaptive only in narrow ranges of SNR_b. Specifically, when SNR_b is from 2 dB to 3 dB, or from 3 dB to 4 dB, or from 4 dB to 5 dB, the optimum numbers of sub-packets

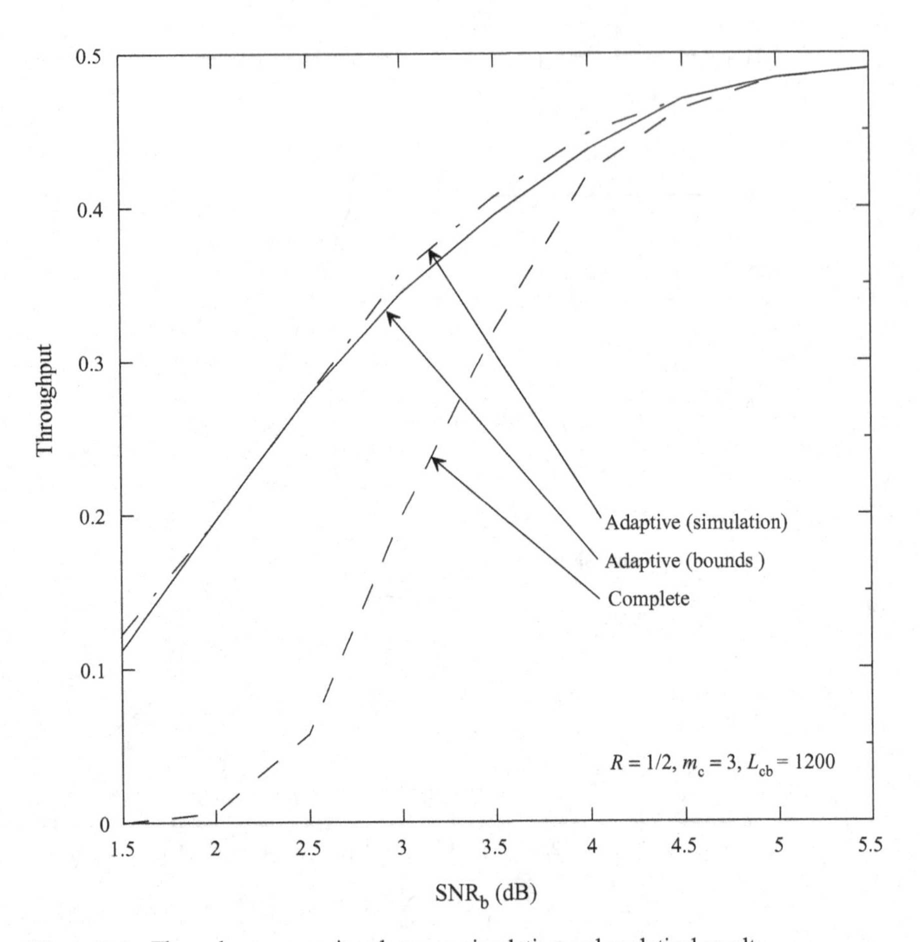

Figure 12.4 Throughput comparison between simulation and analytical results.

are about $N_{opt} = 8$, 4 and 2, respectively. When SNR$_b \geq 5$ dB, the complete packet scheme ($N_{opt} = 1$) performs the best. This is consistent with Figure 12.3. In summary, for convolutionally-coded hybrid ARQ systems, the adaptive sub-packet scheme improves the throughput significantly.

The performance of adaptive sub-packet schemes will be further investigated with different code rates and constraint lengths. As mentioned in the previous section, it is complicated to obtain the distance spectra of convolutional codes with punctuation and long constraint lengths. In order to provide the performance comparison of the adaptive sub-packet and complete packet schemes in more detail, the following results are obtained by means of simulations.

Figure 12.6 illustrates the system throughput as a function of SNR$_b$ for different code rates generated by puncturing. It can be seen that for a given scheme (adaptive or complete), the convolutional code with code rate of 1/2 provides the highest throughput at small SNR$_b$, while the code with code rate of 5/6 performs best at high SNR$_b$. For a given code rate, although adaptive and complete schemes perform the same when SNR$_b$

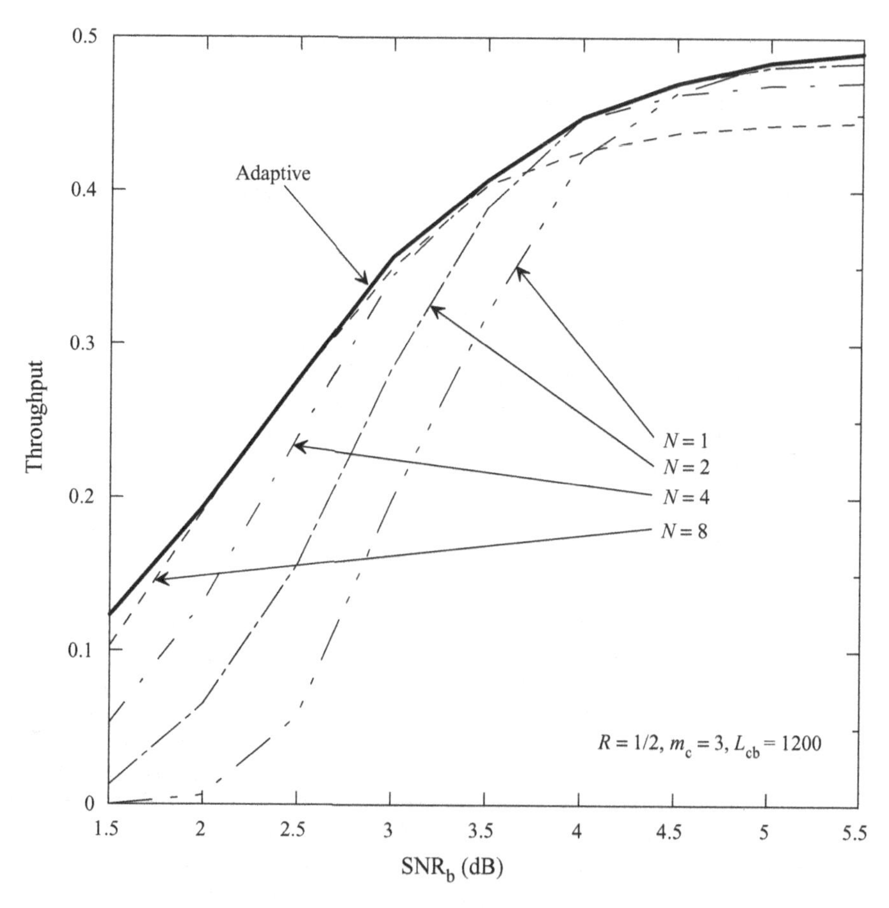

Figure 12.5 Throughput of fixed sub-packet and adaptive sub-packet schemes.

is large, the former performs much better than the latter when SNR_b is small or moderate. It is also interesting to note that in the SNR_b range of interest, the throughput of 5/6-rate code with the adaptive scheme is always higher than that of 1/2- or 2/3-rate code with the complete scheme. Although the SPER of the 5/6-rate code is higher than the complete PER of the 1/2-rate code, the former code still provides better throughput than the latter due to the higher code rate. In summary, for the adaptive scheme, the best code rates should be 1/2, 2/3 and 5/6 for SNR_b less than 3 dB, from 3 dB to 4 dB and greater than 4 dB, respectively.

In order to investigate the effect of packet length on the optimum sub-packet scheme, Figure 12.7 shows the throughput (η) as a function of the number of sub-packets (N) for $L_{\text{cb}} = 1200$ and 1800, respectively. It is seen that each curve takes a peak value at a specific value of N, which is $N_{\text{opt}} = 9$ and 12 for $L_{\text{cb}} = 1200$ and 1800, respectively. N_{opt} is roughly proportional to L_{cb}. Although the peak values of the two curves are very close, the curve of $L_{\text{cb}} = 1800$ is more robust to N. Therefore, a large value of L_{cb} is preferable.

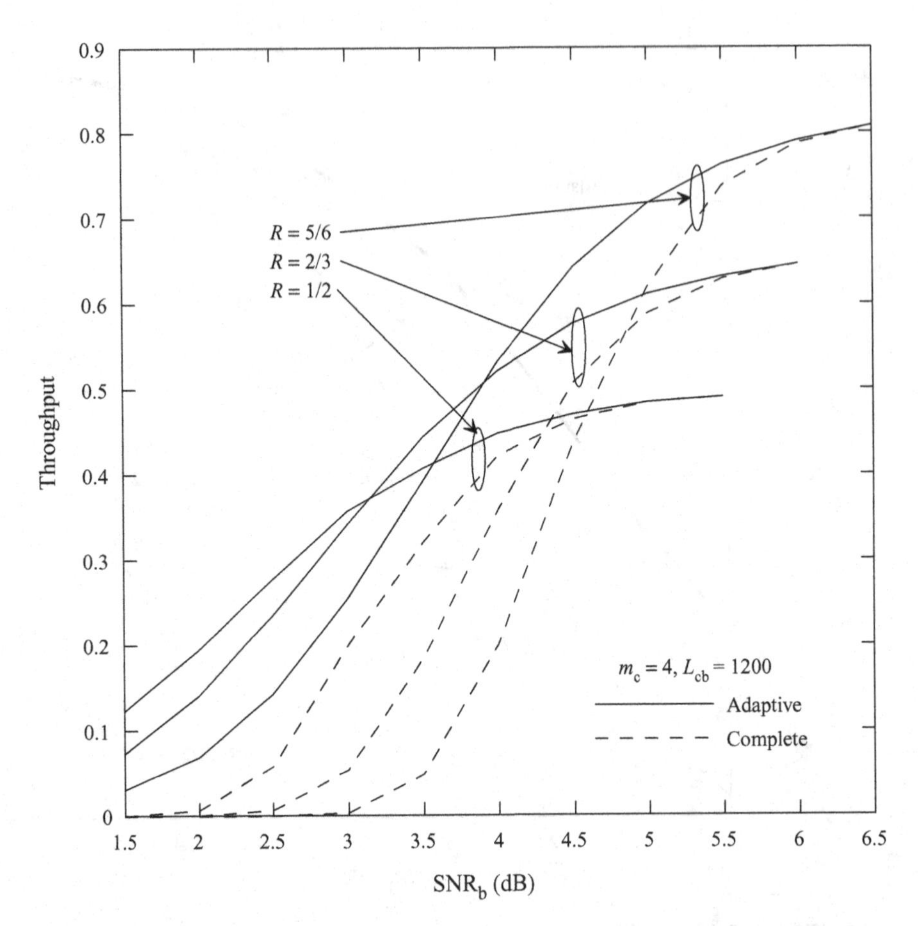

Figure 12.6 System throughput of adaptive sub-packet and complete packet schemes for different code rate.

12.4.2 Delay

Assuming that T is the time needed to encode, transmit and decode a sub-block of data, the system delay is T plus additional delay caused by retransmission. For $N = 1$, i.e. in complete packet schemes, if the packet is error free, then the system delay is T. If the packet is in error, then additional time is needed for retransmission. Assuming that feedback delay is negligible, the average additional system delay for complete packet schemes is given by

$$\text{Delay} = T \cdot P_{\text{sp}}(1) + T \cdot P_{\text{sp}}(1)^2 + \cdots + T \cdot P_{\text{sp}}(1)^{M-1} \tag{12.13}$$

When $N > 1$, the situation is quite different. For the received sub-packets SP_1 to SP_N, if the first sub-packet SP_1 is in error, then the other sub-packets must wait for it even if they are error free, because data blocks must be sent to the upper layer protocol in sequence. Hence, the system delay also depends on the position of the erroneous sub-packet. Consider the simple case of $M = 2$, where all new sub-packets are transmitted.

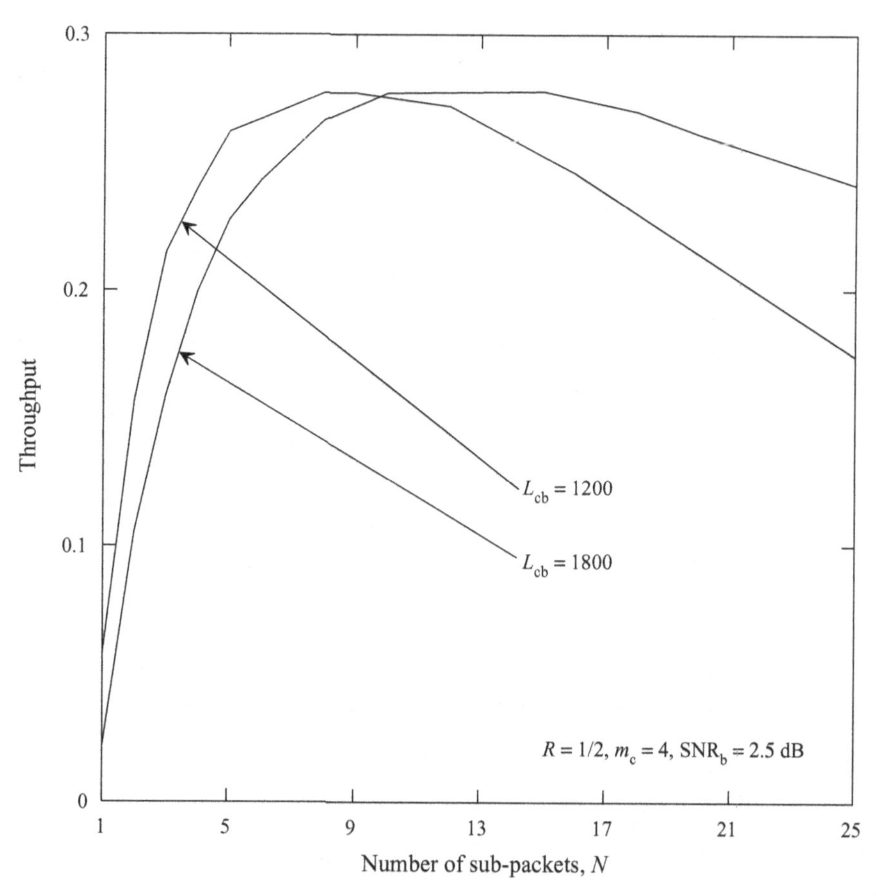

Figure 12.7 System throughput as a function of the number of sub-packets for different packet lengths.

At the receiver side, if the first sub-packet is in error, then all N sub-packets must wait for another T before being sent to the upper layer or dropped, and the average additional delay is $P_{sp}(N) \cdot T$. If the first sub-packet is error free and the second is in error, then the first is sent to the upper layer and the other $N-1$ sub-packets wait in the buffer for another T. Hence, the average additional delay is $[1-P_{sp}(N)] \cdot P_{sp}(N) \cdot (N-1)/N \cdot T$. Similarly, when the first i sub-packets are error free and the $(i+1)$th is in error, the average additional delay is $[1-P_{sp}(N)]^i \cdot P_{sp}(N) \cdot (N-1)/N \cdot T$. Then, the additional delay caused by retransmission for N-sub-packet schemes is given by

$$\begin{aligned}\text{Delay} &= \sum_{i=0}^{N-1} [1-P_{sp}(N)]^i \cdot P_{sp}(N) \cdot \frac{N-i}{N} \cdot T \\ &= T \cdot \left(P_{sp}(N) + \frac{[1-P_{sp}(N)]^{N+1} - [1-N \cdot P_{sp}(N)] \cdot [1-P_{sp}(N)]}{N \cdot P_{sp}(N)} \right)\end{aligned} \tag{12.14}$$

When $N=1$, (12.14) reduces to (12.13).

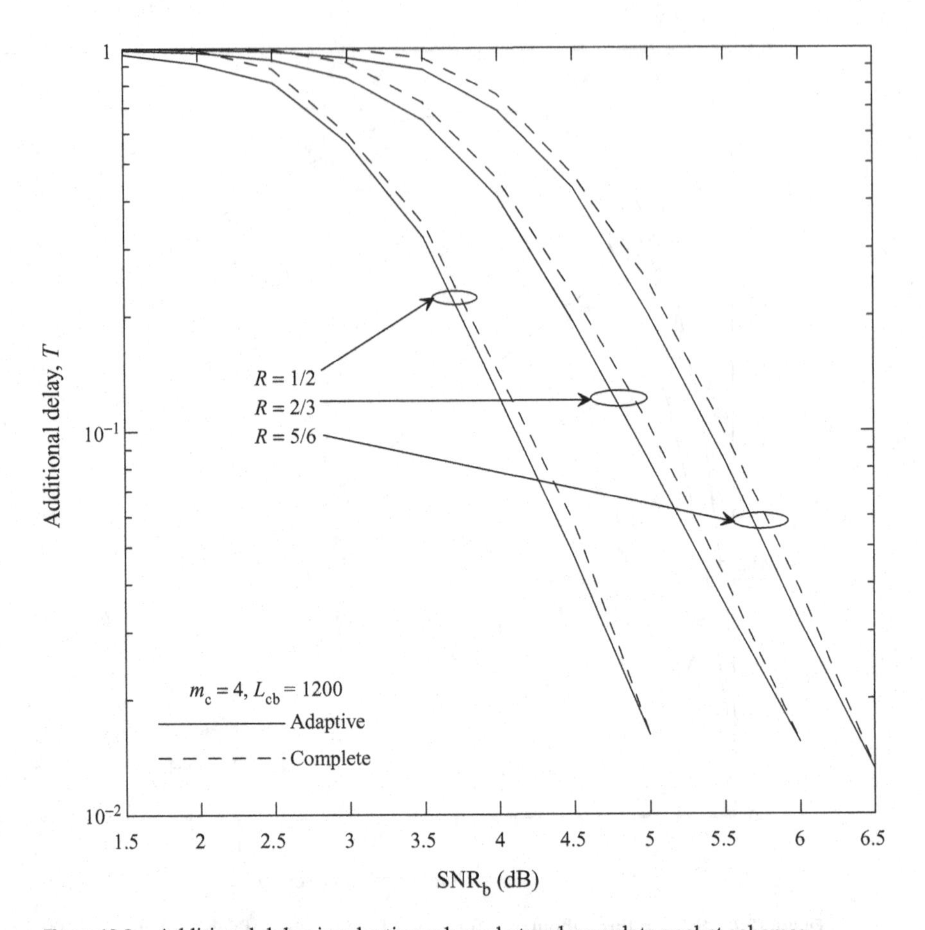

Figure 12.8 Additional delay in adaptive sub-packet and complete packet schemes.

Figure 12.8 illustrates the additional delay of complete and adaptive schemes for convolutionally coded systems. For all considered code rates, the adaptive sub-packet scheme always provides less additional delay than the complete packet scheme, although the improvement is relatively small. In this figure, L_{cb} is set to 1200, and the packet length varies with code rates. Note that in fact T should change with packet length. The same T is used in this figure, however, since the focus is on the performance difference between complete packet and adaptive sub-packet schemes for each code rate.

12.4.3 Dropping rate

When a sub-packet is unsuccessfully transmitted for M times, it will be dropped out. Therefore, in a truncated ARQ system with M maximum transmissions, the dropping rate of a sub-packet is simply a power function of $P_{\text{sp}}(N)$:

$$P_{\text{drop}} = \underbrace{P_{\text{sp}}(N) \cdot P_{\text{sp}}(N) \cdots P_{\text{sp}}(N)}_{M} = P_{\text{sp}}^{M}(N) \tag{12.15}$$

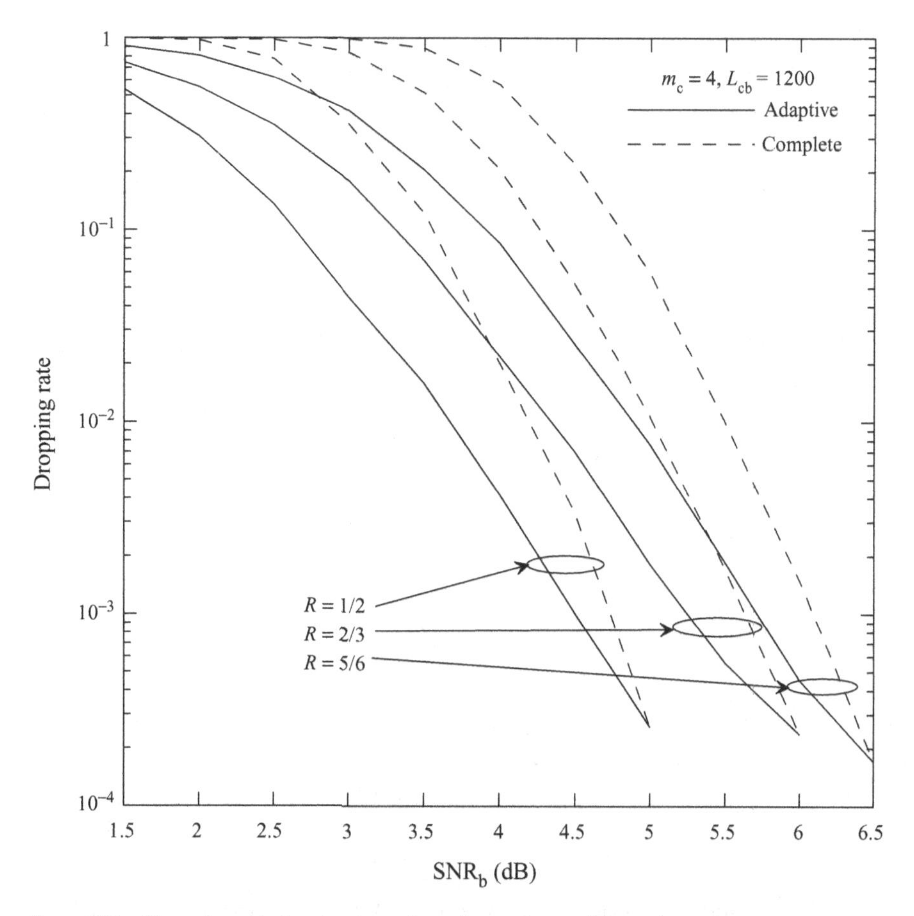

Figure 12.9 Dropping rate in adaptive sub-packet and complete packet schemes.

For $M = 2$, the dropping rates of complete packet and adaptive sub-packet schemes are shown in Figure 12.9. It can be seen that for a given code rate, the dropping rate of the adaptive sub-packet is much smaller than that of the complete packet, especially when SNR_b is moderate.

12.4.4 Sub-packet combining

By combining a previously received erroneous packet and a new packet, the system throughput can be improved significantly [12]. The adaptive sub-packet scheme can be used with packet combining to further improve the system performance. It is assumed that in total $M = 2$ transmissions are permitted. At the first transmission, the data signals are transmitted using the optimum sub-packet scheme according to the channel condition. At the receiver, all sub-packets are stored and NAK is sent to the transmitter for retransmission of the erroneous sub-packets. The received signals of the previous erroneous sub-packets and the corresponding sub-packets in the second transmission are combined and the whole codeword is decoded again. If the channel is static during the two transmissions, then the combination is simply the half sum of the two received

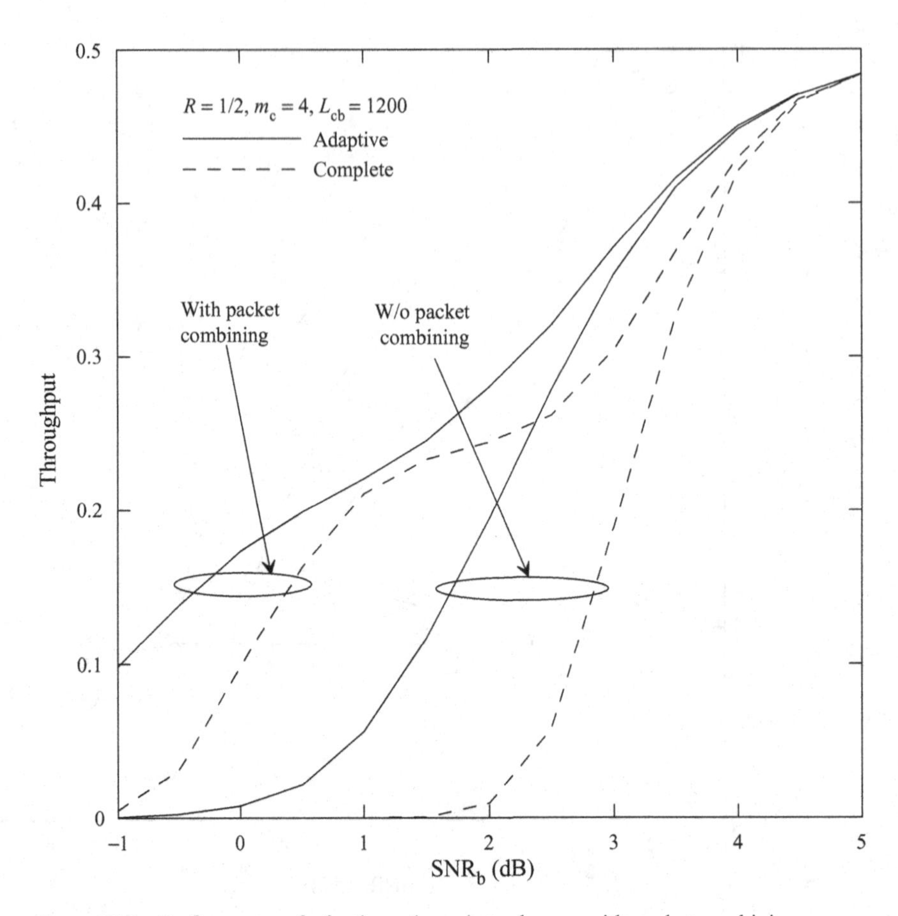

Figure 12.10 Performance of adaptive sub-packet schemes with packet combining.

signals. In the Gaussian channel, the SNR of the combined signal is improved by 3 dB and the SPER can be reduced significantly. If the sub-packet is still in error after packet combining, it will be dropped.

Figure 12.10 shows the throughput performance of the adaptive scheme with and without sub-packet combining. As a comparison, the performance of the complete scheme with and without packet combining is also shown. It can be seen that for either the adaptive or complete scheme, the system throughput is enhanced considerably by the combining algorithm, especially at low SNR. With the combining algorithm, the adaptive scheme still outperforms the complete scheme, but the performance gap is reduced when SNR_b is around 1 dB. This is because, when SNR_b is small, the SPER and PER of the first transmission is high and the system throughput is mainly determined by the SPER and PER after combining. Hence, due to two transmissions, the throughput is roughly equal to half of the throughput without combining by shifting the SNR by –3 dB. Therefore, with combining, the performance gap between the adaptive and complete schemes reduces to a small value as SNR_b increases. When SNR_b is further increased beyond 1 dB, however, the throughput in the first transmission increases for the adaptive scheme, while it is still very small for the complete scheme. Therefore, after combining, the throughput gap between the two schemes increases as $\mathrm{SNR_b}$ increases from 1 dB. Then, when $\mathrm{SNR_b}$ is moderate or high, after combining, the packet is error free in most cases. Hence, the throughput is mainly determined by the SPER and PER at the first transmission, and the throughput difference between the two schemes reduces as $\mathrm{SNR_b}$ is high.

It is interesting to note that although the adaptive scheme without combining is inferior to the complete scheme with combining at low SNR, the former outperforms the latter when $\mathrm{SNR_b} > 2$ dB. As explained before, when $\mathrm{SNR_b}$ is moderate or high, the throughput after combining is mainly determined by the PER of the complete scheme at the first transmission. Since the SPER of the adaptive scheme is much lower than the PER of the complete scheme when $\mathrm{SNR_b} > 2$ dB, it is noticeable that the adaptive scheme without combining has higher throughput than the complete scheme with combining. In summary, the adaptive sub-packet scheme can be flexibly used with the packet combining technique and still outperforms the complete scheme with combining.

12.5 Conclusions

This chapter presents an optimum sub-packet scheme for convolutionally coded ARQ systems. By means of theoretical analysis and computer simulations, the following has been shown.

(1) Sub-packet schemes are suitable for convolutionally coded ARQ systems at low to moderate $\mathrm{SNR_b}$. Significant improvement in the system throughput is obtainable with optimum sub-packet schemes.

(2) The optimum number of sub-packets (N_{opt}) can be derived under the assumption of burst packet error patterns. N_{opt} decreases with increasing $\mathrm{SNR_b}$. Adaptive sub-packet schemes provide best performance in dynamic channels (variable $\mathrm{SNR_b}$).

(3) The optimum sub-packet scheme outperforms complete packet schemes in the additional delay and dropping rate.

(4) The optimum sub-packet scheme can be flexibly used with the packet combining technique to further improve system throughput.

References

[1] N. Guo, F. Khaleghi, A. Gutierrez, J. Li and M.-H. Fong, "Transmission of high speed data in cdma2000," in *Proc. IEEE Wireless Communications and Networking Conference*, vol. 3, pp. 1442–1445, 1999.

[2] 3rd Generation Partnership Project, Technical Specification Group Radio Access Network, "Multiplexing and channel coding (FDD), (Release 1999)," ETSI TS 25 212 V3.5.0, Dec. 2000.

[3] H. Herzberg and G. Poltyrev, "The error probability of M-ary PSK block coded modulation schemes," *IEEE Trans. Commun.*, vol. 44, pp. 427–433, April 1996.

[4] G. Poltyrev, "Bounds on the decoding error probability of binary linear codes via their spectra," *IEEE Trans. Inform. Theory*, vol. 40, no. 4, pp.1282–1292, July 1994.

[5] I. Sason and S. Shamai, "Improved upper bounds on the ML decoding error probability of parallel and serial concatenated Turbo codes via their ensemble distance spectrum," *IEEE Trans. Inform. Theory*, vol. 46, pp. 24–47, Jan. 2000.

[6] S. Lin and D. J. Costello Jr., *Error Control Coding: Fundamentals and Applications*. Englewood Cliffs, NJ: Prentice-Hall, 1984.

[7] L. R. Bahl, C. D. Cullum, W. D. Frazer and F. Jelinek, "An efficient algorithm for computing the free distance," *IEEE Trans. Inform. Theory*, vol. IT-18, pp. 437–439, Nov. 1972.

[8] K. L. Larsen, "Comments on 'An efficient algorithm for computing the free distance' by Bahl et al.," *IEEE Trans. Inform. Theory*, vol. IT-19, pp. 577–579, July 1973.

[9] M. Cedervall and R. Johannesson, "A fast algorithm for computing distance spectrum of convolutional codes," *IEEE Trans. Inform. Theory*, vol. 35, pp. 1146–1159, Nov. 1989.

[10] R. J. McEliece, "How to compute weight enumerators for convolutional codes," in *Communication and Coding*, M. Darnell and B. Honary, eds. Taunton, UK: Research Studies Press, pp. 121–141, 1998.

[11] D. Chase, "Code combining – a maximum-likelihood decoding approach for combining an arbitrary number of noisy packets," *IEEE Trans. Commun.*, vol. COM-33, pp. 385–393, May 1985.

[12] B. A. Harvey and S. B. Wicker, "Packet combining systems based on the Viterbi decoder," *IEEE MILCOM'92*, pp. 757–762, Oct. 1992.

Index

Printed in the United States
By Bookmasters